MARINE BOTANY

SECOND EDITION

Clinton J. Dawes
University of South Florida
Tampa, Florida

JOHN WILEY & SONS, INC.

New York • Chichester • Weinheim • Brisbane • Singapore • Toronto

Library of Congress Cataloging-in-Publication Data
Dawes, Clinton J.
    Marine botany / Clinton J. Dawes. – 2nd ed.
       p.   cm.
    Includes bibliographical references (p.   ) and index.
    ISBN 0-471-19208-2 (cloth : alk. paper)
    1. Marine plants.   2. Plant ecology.   I. Title.
QK931.D38   1997
579′.177–dc21   97-10372

Printed in the United States of America

10 9 8 7 6 5 4

Marine plants are a diverse group that include unicellular algae, seaweeds, seagrasses, salt marshes, and mangrove forests. They carry out a variety of ecological functions and serve as the primary producers in coastal wetlands and oceanic waters. The theme that connects such a wide variety of plants is their ecology, which was also emphasized in the 1981 edition. The goal of this revision is to present taxonomic, physiological, chemical, and ecological aspects of marine plants, their adaptations, and how abiotic and biotic factors interact in their communities. The data are presented in a concise, comparative manner in order to identify similarities and differences between communities such as salt marsh and mangroves or subtidal seaweeds and seagrasses. To accomplish this, the text is organized into five chapters that introduce the marine habitats, consider abiotic and biotic factors, and anthropogenic influences on the communities followed by seven chapters that deal with microalgae, seaweeds, salt marshes, mangroves, seagrasses, and coral reefs. Two appendixes are included; one presents simple field techniques and the other is a summary of seaweed uses.

Marine Botany, therefore, covers a wide variety of plants, uses a botanical perspective, and emphasizes their adaptations and ecological functions. The text developed from an upper-level undergraduate and beginning graduate course that was first taught in 1965. The course included studies in salt marshes, seagrass beds, mangrove swamps, seaweed habitats, and coral reefs as found in south Florida. It became evident that no single text was available, which led to the publication of *Marine Botany* in 1981. To date, there is still no single text that compares the diverse marine plants, although a number of comprehensive texts dealing with individual communities are now available. These include reviews of mangroves (Tomlinson, 1986; Robertson and Alongi, 1992), salt marshes (Adam, 1990; Mitsch and Gosselink, 1993), seagrasses (Larkum et al., 1989a; Kuo et al., 1996), seaweeds (Round, 1981; Lüning, 1990; Lobban and Harrison, 1994; Little and Kitching, 1996), and coral reefs (Dubinsky, 1990; Sorokin, 1993). Another recent development is the recognition of the ecological value of marine plant communities, and thus these comprehensive reviews and specialized texts (e.g., Daiber, 1986) include considerations of restoration and management. Thus, this revision utilized original research papers, comprehensive texts, and governmental publications that demonstrated the ecological value of these marine plant communities.

While writing this revision of *Marine Botany*, a major problem was to determine what should be included and what must be left out when there are entire texts that deal with each marine plant community. Guiding principles included whether the data could be used to compare communities and their plants, personal experience working in each community, and comments from students taking the course or carrying out research in the communities. The revision retains the ecological theme of the 1981 edition, while the community chapters (Chaps. 7 to 12) are expanded in terms of plant adaptations, ecology, and resource management. Anthropogenic effects are also expanded (Chap. 5) and consider possible effects by exotic marine plants, ultraviolet light, global warming, and changing sea levels. Discussions on seaweed taxonomy are reduced and considerations of critical abiotic (Chap. 2) and biotic (Chaps. 3 and 4) factors focus the reader on the communities by using examples of all types of marine plants. The text is intended for upper-level undergraduate and graduate courses, as a supplement to marine biology, phycology, and field courses. The book can also be used as a source for high school and junior college courses as well as by researchers and managers interested in specific marine plant communities.

As with the first edition, a number of scientists were instrumental in reviewing this revision of *Marine Botany*. Dr. Arthur Mathieson (University of New Hampshire) reviewed all 12 chapters; Dr. Gabe Vargo (University of South Florida), Chapter 7 on microalgae; Dr. David Burdick (University of New Hampshire), Chapter 9 on salt marshes; Dr. Aaron Ellison (Mount Holyoke College), Chapter 10 on mangroves; Dr. David Tomasko (Southwest Florida Water Management District), Chapter 11 on seagrasses; and Dr. Pamela Hallock Muller (University of South Florida), Chapter 12 on coral reefs. Many persons furnished figures, and acknowledgments for these are given in the figure captions. Mrs. Betty Loraamm (University of South Florida) reprinted a number of the micrographs. Four artists are acknowledged for producing diagrams and habit sketches; Mrs. Linda Leatherwood (as L. Baumhart, LB), Dr. Ana-Lisa P. King (Als), Ms. Robin Holbrook (Rrh), and Mrs. Carol Torres (C. Torres). In addition, Mr. John Andorfer (ja) made a number of computerized figures.

Finally, full responsibility for the text and its contents is assumed by myself. The hope is that the comparative approach to marine plant communities will help in our understanding and stimulate further research. As with the first edition, this book is dedicated to all marine botanists who have expanded the field through their research and teaching, especially my teachers, the late Dr. George J. Hollenberg and Dr. E. Yale Dawson and friends, Dr. Arthur Mathieson and the late Dr. Michael Neushul. The text is also dedicated to my students, who have made teaching marine botany a joy, and a special acknowledgment to my wife Kathleen, who has supported me throughout my academic life.

CLINTON J. DAWES

Tampa, Florida

![] CONTENTS

CHAPTER 1

Marine Plants and Their Habitats

Earth is unique within our solar system for a number of reasons, not the least of which is the abundance of water. The ocean covers a large portion of Earth's biosphere, over 71% (360 million km^2) of the surface with an average depth of almost 4 km. Almost all of the water (98%) is either marine or frozen with only 0.4% being freshwater.

Equally impressive is the diversity of life, ranging from virus and bacteria, to fungi, plants, and animals, many of which are critical in the biogeochemical cycles of Earth. The biodiversity of marine organisms is at risk due to anthropogenic activities of fishing, pollution, habitat alteration, introduction of exotic species, and global climate change (Committee on Biological Diversity, 1995). Singly or in combination, such human activities can change energy-flow patterns, alter the structure of marine communities, and result in global extinction of marine organisms. Some examples of this include the invasion of the Chinese clam, *Potamocorbula amurensis*, in California; the deterioration of coral reefs in the Caribbean; the decline of the brown seaweed, *Fucus vesiculosus*, the Baltic Sea; and the extinction of oyster reefs in Chesapeake Bay.

The variety of marine plants that grow in the sea range from unicellular to macroalgae, to flowering plants, including seagrasses, salt marsh plants, and mangroves. It is from the primeval oceans that all life including flowering plants probably arose. The importance of marine plants can be seen not only in the production of organic materials, but also in their various ecological roles, serving as nurseries and habitats and as a direct source of food. It should not be surprising that the study, management, and expansion of marine plant communities have become international issues. The types of marine plants and their communities will be examined in terms of structure and function. Human use and abuse of these communities will also be generally considered (Chap. 5) and examined for each marine plant community (Chaps. 8 to 12).

MARINE PLANTS AND THEIR CLASSIFICATION

Many present-day classifications group organisms into five kingdoms as first proposed by Whittaker (1969) and expanded by Margulis and Schwartz (1988):

1

the Monera, Protist, Plantae, Fungi, and Animalia. The marine organisms included in this text are those that release oxygen from water during photosynthesis and contain the key pigment, chlorophyll *a*. These organisms are found in the Kingdoms Monera (Cyanobacteria and Prochlorophyta), Protist (algae), and Plantae (ferns, flowering plants).

Kingdom Monera

Of the roughly 100,000 species of bacteria, probably less than 2000 occur in the marine environment. However, this group plays a number of significant roles in energy transfer, mineral cycles, and organic turnover. In contrast to the other four kingdoms, members of the Monera have a prokaryotic cytology in which the cells lack membrane-bound organelles such as chloroplasts, mitochondria, nuclei, and complex flagella. The genetic material, deoxyribonucleic acid (DNA), is not organized into large complex chromosomes as in eukaryotic cells. Further, the cell walls of bacteria consist of peptidoglycans in which the carbohydrate polymers are interconnected with short chains of amino acids.

Bacteria can be subdivided into two major groups or subkingdoms, Eubacteria and the more primitive Archeobacteria. Eubacteria include the only bacteria that contain chlorophyll *a* and are placed in the divisions Cyanobacteria or blue-green algae and Prochlorophyta (Chloroxybacteria). Cyanobacteria are of special interest as they play a major role in nitrogen fixation as well as being primary producers in the marine environment. They will be considered in Chap. 7 on microalgae.

Kingdom Protist

As presently organized, the kingdom includes a wide variety of "animals" (protozoa), "plants" (algae), and "fungi" (slime molds). They are grouped together due to their simple structure and for the most part are unicellular species. In earlier classifications, the algae, which are plantlike, photosynthetic, and contain chlorophyll *a*, were placed in the Kingdom Plantae. Presently, most classifications have removed algae from the Plant Kingdom (e.g., Raven et al., 1992) due to their simple construction (a majority being unicellular), the variety of accessory pigments used in photosynthesis, the lack of protecting tissue around gametic cells, and their highly varied life histories. However, analyses of ribosomal RNA and gene DNA sequences indicate that red algae and green plants are closely related (Ragan and Butell, 1995). With the exception of the more protozoanlike divisions (Euglenophyta, Pyrrhophyta, and Cryptophyta) and some classes of Chrysophyta, macroalgae and related species may be better aligned within the Kingdom Plantae.

Macro- and microalgae are the dominant photosynthetic organism in the marine environment. They are viewed as "primitive" photosynthetic organisms because of their simple construction and reproduction. However, algal fossils are known since the pre-Cambrian (Proterozoic; see Table 2-1) 590 million years before present (m ybp) for macroalgae (1300 m ybp) and microalgae (1900 m ybp; van den Hoek et al., 1995). A comparison of five recent phycology texts demonstrates a wide variation in algal classification (Bold and Wynne, 1985; South and Whittick, 1987; Lee, 1989; Sze, 1986; van den Hoek et al., 1995). This is because the authors emphasize different algal features and the immense diversity of algae (Norton et al., 1996). In a review of algal biodiversity, Norton et al. note that the number of undescribed algal species may exceed the known ones (ca. 41,431) by a factor of 4 to 8.

Algae can be separated based on types of photosynthetic pigments, storage products, chloroplast structure, inclusions in the cell (e.g., pyrenoids and trichocysts) chemistry and structure of the cell wall, construction of the flagella, type of cell division, and life history. Detailed structural information on algae can be found in the classic volumes of Oltmanns (1922a, 1922b, 1923), Fritsch (1935, 1945), and Bourrelly (1966, 1968, 1970). In this text, a classification similar to that of Bold and Wynne (1985) is followed with minor exceptions, and seven divisions of algae are considered. The reader is referred to van den Hoek et al. (1995) for a classification in which cytological (flagellar structure and mitosis) and molecular (DNA and RNA sequences) were used. There are three well-defined divisions of macroalgae or seaweeds (Chlorophyta, Phaeophyta, and Rhodophyta; Chap. 7) and four divisions of predominately microalgae that have marine representatives (Chrysophyta, Pyrrhophyta, Euglenophyta, and Cryptophyta; Chap. 6). The percent of marine algal species varies greatly between divisions, as shown in Table 1-1.

Kingdom Plantae

Of the various divisions of "true" plants, only 0.085% of the estimated 300,000 species of Angiosperms are found in marine environments (Table 1-1). There are a few marine species of mosses, lower vascular plants (ferns), and no marine gymnosperms (Table 1-1). The seagrasses (Chap. 11), mangroves (Chap. 10), and salt marsh plants (Chap. 9) form three major marine angiosperms communities in coastal waters. Further, the number of species are few when compared to terrestrial flowering plants. All of these plant communities have come under serious stress due to anthropogenic activities.

Seagrasses are the only truly submerged angiosperms of the marine environment. Although of recent origin in the oceans (ca. 65 to 40 million years before present), seagrasses are recognized as important primary producers in coastal communities and can be used to determine the environmental "health" of coastal and estuary ecosystems. The two types of tidal marsh plant communities, the "grasslandlike" salt marshes and the dense, coastal mangrove forests appear at first glance to have little in common. However, these two coastal communities show strong parallels in both their ecological roles and in the abiotic factors that control their development.

TABLE 1-1. Chlorophyll-Bearing Plants

Division/Class	Common Name (chlorophylls)	Percent Marine	Identified Species
Cyanobacteria			
Cyanophyta	Blue-green (a)	8	2,000
Prochlorophyta	(a, b)	75	4
Algae[a]			
Rhodophyta	Red algae (a)	98	4,000
Chlorophyta	Green algae (a, b)	13	14,720
Euglenophyta	Euglenoids (a, b)	3	900
Phaeophyta	Brown algae (a, c)	99	1,500
Chrysophyta	(a, c)		
Chrysophyceae	Golden-brown algae	20	1,000
Xanthophyceae	Yellow-brown algae	15	600
Bacillariophyceae	Diatoms	50	10,000
Pyrmnesiophyceae	Haptonemonads	50	300
Eustigmatophyceae		8	12
Raphidophyceae	Chloromonads	50	15
Pyrrhophyta	Dinoflagellates (a, c)	90	2,000
Cryptophyta	Cryptomonads (a, c)	50	200
Bryophyta[b]			
	Mosses, Liverworts (a, b)	0	25,000
Vascular plants[b]			
	(a, b)		
Psilophyta	Wisk ferns	0	3
Sphenophyta	Horse tails	0	15
Lycophyta	Club mosses	0	1,000
Pterophyta	Ferns	0.1	12,000
Gymnospermophyta	Cone plants	0	722
Angiospermophyta	Flowering plants	0.09	285,000

[a]Number of algal species; after Norton et al. (1996).
[b]Number of species; after Raven et al., 1992.

BOTANICAL NOMENCLATURE

Organisms are classified within a kingdom into *divisions* (Table 1-1) similar to the phyla of the animal kingdom. Each division can be subdivided into one or more *classes* that contain one or more *orders*. *Families* make up the next subgroup, and within each family, the various *genera* and their *species* occur. Species are named according to the binomial system of nomenclature proposed by Linnaeus in his *Species Plantarum* published in 1753. The scientific name, which is in Latin, is a *binomial*: It contains the genus and species, as well as the name(s) of the persons who described the plant. Typically, the name is a descriptive term for the organism.

Ulva lactuca Classification

The common green alga, sea lettuce, has the generic name of *Ulva*, the Latin name for marsh plant, and the species name of *lactuca*, a descriptive term meaning "lettuce." The plant is classified as follows:

Taxonomic Unit	Name	Ending of Unit
Division	Chlorophyta	-phyta
Class	Ulvophyceae	-phyceae
Subclass	Ulvoideae	-oideae
Order	Ulvales	-ales
Family	Ulvaceae	-aceae
Name (genus and species)	*Ulva lactuca* Linneaus	

The plant's name is italicized because it is in Latin and the author's name of that species is included. If a species is transferred to another genus, the original author's name for the species is retained and the person who made the transfer is cited as well. For example, the tropical green seaweed *Caulerpa prolifera* (Forsskäl) Lamouroux indicates that Forsskäl is the author of the species *prolifera* and that Lamouroux transferred the species from another genus (*Fucus*) to *Caulerpa*.

Species Descriptions

All descriptions of new plants or descriptions of new taxonomic units (divisions, classes, orders, families) are governed by the rules of the International Code of Botanical Nomenclature, with amendments and revisions occurring every five years at the International Botanical Congress. All new genera and species or lower taxa must have a Latin description that is diagnostic if the name is to be valid. An example is the erection of a new variety of the calcified, green seaweed *Halimeda lacrimosa* Howe called var. *globosa* Dawes and Humm as follows:

> Plantae usque and 10 cm altae, densae, plus minusve globosae; segmenta pro parte majore globosa, rare obovoidea vel pyriformia, 5–10 mm diam., cava. utriculi peripherales 35–40 μm diam., 80–100 mm longi; membrane utriculares apicales 7–9 μm crassae, post calcificatae remanentes cohaerentes. Utriculi subcorticales 100–130 μm diam., omnes ferentes 16–17 utriculi peripherales. Fila littos centrales conjugentes in 3s ve. 4s, rare 2s vel 6s ad nodia. Sporangia ignota.

The translation of the Latin description describing the new variety *globosa* Dawes and Humm of *Halimeda lacrimosa* Howe is

Plants forming dense masses to a height of 10 cm; the segments mostly spherical in shape, rarely obovoid or pyriform, 5–10 mm in diameter, hollow, the peripheral utricles 35–45 μm in diameter, 80–100 μm long, the apical walls 7–9 μm thick remaining coherent after decalcification. Subcortical utricles 100–300 μm in diameter, each bearing 16–17 peripheral utricles. Filaments of the central strand fusing at the nodes in threes and fours, rarely in twos and sixes. Sporangia unknown.

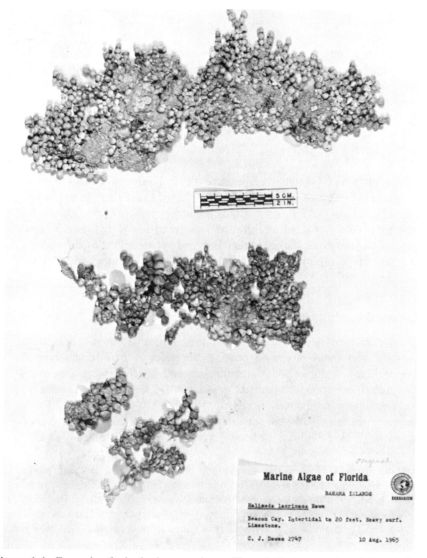

Figure 1-1. Example of a herbarium specimen. The calcified green alga *Halimeda lacrimosa* Howe was dried and glued to the herbarium sheet and a label placed in the lower right corner.

Marine Algae of Florida

HERBARIUM

Bostrychia binderi Harvey

Collected in Cockroach Bay, Tampa Bay, West coast.
27°41'N, 82°30'W.

26 July 1996.

Growing on black mangrove pneumatophores by boat ramp.

Collector: C.J.Dawes.
Identified: C.J. Dawes.

Collection Number: 27102.

Figure 1-2. A herbarium label for *Bostrychia binderi*. The label contains the collection location, date, ecological characteristics, collector name, and collection number.

Type Specimen

A *type* specimen, from which the species description was made, is placed in a herbarium (storage site of pressed plants) and its location given (Fig. 1-1). Further, a herbarium label giving all the pertinent information (species name, location, date of collection, habitat, name of collector, and identifier) is mounted with the pressed specimen (Fig. 1-2).

MARINE PLANT ENVIRONMENTS

Marine plants are limited to the *euphotic* zone or upper regions of oceans where submarine irradiance is sufficient for photosynthesis. This is true of both *benthic* (attached) and *planktonic* (free-floating) species (Fig. 1-3). The light-limiting depth of a species is called the *compensation point*, or the intensity of irradiance where photosynthesis and respiration rates are balanced. Thus, benthic or bottom-dwelling forms are limited to the shallower areas of the continental shelves (usually to about 150 m). The *phytoplankton* or free-floating photosynthetic species will be limited to the upper illuminated oceanic waters (usually to about 200 m). Planktonic plants are generally more abundant along the coasts and in the shallower regions of the continental shelves and areas where nutrient-rich water occurs. The increased abundance of coastal phytoplankton reflects the higher levels of nutrients eroded from the continents, ressuspended from the benthic substrata, and circulated from deep nutrient-rich seawater (upwelling).

Exposed Coasts

Vertical subdivisions of rocky coastal environments can be separated on the basis of light penetration, the euphotic region where submarine light supports

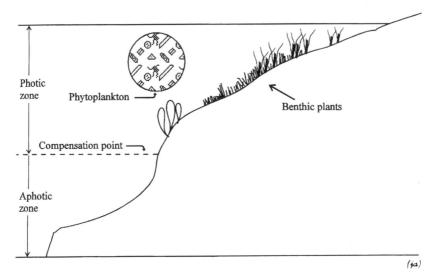

Figure 1-3. Distribution of marine plants with relation to light penetration. Phytoplankton and benthic plants are limited to the photic zone.

plant growth and the aphotic depths where light does not penetrate. This text follows the advice of Lüning (1990) in that *zones* indicate vertical divisions and *regions* are used in the context of biogeographical distributions. Vertical zonation can be considered in terms of biological distributions or physical areas based on water levels. The most commonly used biological zonations (Table 1-2) include the "classical" system (Lüning, 1990), the "universal" system (Stephenson and Stephenson, 1972), and rocky system (Lewis, 1964). Stephenson's universal system included a *supralittoral fringe* (upper limit of *Littorina* and covered only by extreme high-water spring tides), the *midlittoral*

TABLE 1-2. A Comparison of Intertidal Zone Classifications

Tidal[a]	Biological	Classical	Universal	Lewis
Above spray zone	Terrestrial lichens and plants	—	Supralittoral	Maritime
Spray	Marine lichens	Supralittoral	Supralittoral Fringe	Littoral Fringe
Intertidal	Barnacles	Eulittoral	Midlittoral	Eulittoral
Intertidal Fringe	Kelps	Sublittoral Fringe	Infralittoral Fringe	Sublittoral

Source: Modified from Lüning (1990).

[a]Tidal levels do not correspond exactly to biological or other classifications due to the effect of waves and climatic factors.

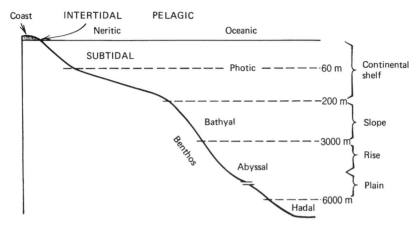

Figure 1-4. Profile of pelagic and benthic zones of oceans. Pelagic regions include neritic or waters lying over continental shelves and oceanic waters. Benthic zones include the coast (maritime), intertidal, and continental shelf (subtidal), and slope (bathyal), abyssal, and hadal depths.

zone that lies between the upper limit of barnacles and upper limit of *Laminaria*, and the *infralittoral fringe* that extends from the midlittoral zone to the extreme low-water spring tide (see Fig. 8-3). Stephenson's midlittoral zone is further divided into *upper*, *middle*, and *lower* subzones. Lewis (1964) modified the universal classification to three regions called the *littoral fringe*, *eulittoral zone*, and *sublittoral zone* (Table 1-2; Fig. 8-4). This modified classification has been used by Morton (1991) in his study of zonation in the Bay of Fundy, Nova Scotia, Canada. Lüning (1990) follows a modified version of the classical system with three divisions in the euphotic zone: the *supralittoral* (spray zone), *eulittoral* (the zone that shows periodic or aperiodic exposure), and *sublittoral* (the zone that is either never exposed or occasionally emersed but wetted by waves). Because the zones are based on the distribution of key organisms, it can be difficult to relate a biological classification of one habitat to a second one. In this text, the euphotic zone is divided into five physical zones: the maritime, spray, intertidal, subtidal fringe, and subtidal (Fig. 1-4). The subtidal region of benthic plants is limited to the euphotic zone. The aphotic zones, where no plants occur, include the Continental Slope, Continental Rise, Abyssal Plain, and Hadal zones (Chap. 2).

Each area of the coastal euphotic zones has unique abiotic and biotic features that influence the plant communities found there. The *maritime* zone lies above the high-tide mark and is essentially terrestrial but under oceanic influence being subjected to varying degrees of wave spray, mist, and salt carried by the wind. Maritime vegetation can be dominated by *halophytic* plants (tolerant of salt and lack of water). The maritime communities may be found growing on sand dunes, as coastal prairie, chaparral, or forests. Maritime vegetation may show the effect of salt pruning where the seaward branches are killed back and the plants appear to be leaning away from the coast (Fig. 1-5).

Figure 1-5. Maritime vegetation shows adaptations to the effects of waves and wind. Sand dunes on the low-energy west coast of Florida are stabilized by sea oats (*Uniola paniculata*, arrow) and a variety of halophytic dicot bushes.

The width and extension of a *spray* (supratidal) zone is dependent on wave activity. If wave action is high (e.g., exposed cliffs of southern California; Fig. 1-6), the upper portion of the spray zone will extend beyond the highest tide mark (*dilation*) due to frequent wetting. Plants in the supratidal zone must tolerate long periods of desiccation and extremes in climatic factors. Plants typical of the spray zone (Fig. 1-6, zone 1) include lichens and blue-green algae. In colder waters, red algae (*Porphyra, Bangia*), small green algae (*Codiolum, Blidingia,* and *Urospora*), and large brown algae (*Pelvetia, Fucus*) may occur in the spray zone if wave activity is high. In the tropics, this zone, even if wetted by high wave activity, will primarily have blue-green algae due to intense sunlight and high air and water temperatures.

The *intertidal* zone is regularly covered and uncovered by the tides and may range from a few centimeters (Baltic Sea in Europe; Patos Lagoon in Brazil) to 15 m (Bay of Fundy, Nova Scotia) in vertical height. In regions where the intertidal zone is limited, aperiodic fluctuations due to wind patterns (Patos Lagoon) may be more important although erratic. The upper limit of intertidal plants is usually controlled by abiotic factors that result in desiccation due to higher temperatures, intense sun, and limited wetting. In contrast, the lower limit of an intertidal species are more often under the influence of biotic factors such as competition or grazing. In temperate waters of North America, the upper limits of the intertidal zone are usually delineated by barnacles (Fig. 1-6, zone 2), below which *Porphyra* will grow (Fig. 1-6, zone 3) along with larger brown algae such as *Fucus, Pelvetia,* and *Ascophyllum* may occur. The lower limit of

Figure 1-6. Intertidal zonation of a cliff exposed to moderate-energy waves in southern California. Four zones exposed during a spring low tide include (1) a spray zone with blue-green and small green filamentous algae, (2) an upper intertidal zone delineated by barnacles and bordered by the red alga *Porphyra*, (3) a lower intertidal zone containing frondose red algae and fucoids, and (4) a subtidal fringe containing kelps and a red algal turf.

such a community may be marked by an upper fringe of larger brown algae such as the kelp *Laminaria saccharina* and *L. digitata* (Fig. 1-6, zone 4). In subtropical and tropical regions, the upper area can show various levels of color (West Summerland Key, Florida; Stephenson and Stephenson, 1972) due to encrusting and endolithic (penetrating the limestone) algae. The dominant plants in tropical intertidal regions are again small blue-green, green, and red algae that usually are found under ledges or in solution holes in the substrata.

A wide variety of macroalgae can occur in the intertidal zone (Fig. 1-6, zones 2 and 3) of temperate rocky shores, and microalgae, particularly blue-green species, dominate hard substrata in the tropics. The lower portion of the intertidal region is called the *subtidal fringe* (Fig. 1-6, zone 4), which is exposed to the air infrequently, perhaps only during low wave activity. The algae growing at the subtidal fringe have more features in common with species found below than species of the intertidal zone. Such species can tolerate only short periods of exposure and may demonstrate adaptations to high wave energy. In temperate (cool-water) communities, the dominant macroalgae may be large brown (*Alaria, Laminaria*) or red (*Palmaria, Mastocarpus*) seaweeds. In tropical regions, brown (*Padina, Sargassum*) and red (*Laurencia, Pterocladia*) macroalgae may occur.

Abiotic factors that are important to marine plant distribution in the intertidal zone include tidal ranges, tidal periodicity, substratum, wave action, and climatic features, such as air temperature and precipitation. Of these, the substrata is most critical as a quick comparison of a sandy beach (unconsolidated sand) and a rocky shore (consolidated or hard substratum) will demonstrate. Chapter 2 deals with abiotic factors in terms of importance to marine plants and the chapters dealing with marine plants (Chaps. 8 to 12) evaluate the influence of various factors on these communities.

The euphotic zone of the continental shelf that is not uncovered by tides is called the *subtidal* zone (Fig. 1-4). Subtidal plants can extend from the lowest tide mark to the edge of the continental shelf if there is sufficient irradiance. In shallow subtidal areas (to 50-m depth), benthic algae can form extensive communities such as the giant kelp forests of the west coast of North America if a hard substrata is available. At greater depths, light becomes limiting and algal diversity and abundance are limited to shade-tolerant forms.

Submerged aquatic vegetation (SAV) includes not only macroalgae, but also seagrasses, which are flowering plants. Subtidal plant communities form important habitats and are the primary source of food for many organisms. Thus, the biotic factors of predation (grazing) and competition play a major role in the distribution of SAV. The chapters on macroalgae (Chap. 8) and seagrasses (Chap. 11) will cover the adaptations and ecological roles of these subtidal plants.

The use of modern diving gear, such as SCUBA, and the increased availability of research vessels and submersibles (Fig. 1-7) have allowed marine botanists to explore the deep-water floras of the world such as in Bermuda (Searles and Schneider, 1987). Such studies have demonstrated that brown, green, and red algae can grow at depths of 88 to 268 m where irradiance is only 0.009 μmol photons m^{-2}s^{-1} (Littler and Littler, 1984).

Estuaries

Estuarine communities, where freshwater mixes with oceanic water, are some of the most productive and diverse of coastal communities. An *estuary* is an inlet of the sea that extends into a river or stream basin (Figs. 1-8 and 3-3) and is therefore a transition zone between freshwater and marine water worlds. Moving from a river through an estuary to the sea demonstrates the changes in depth, energy (currents, turbulence), water clarity, salinity, sediments (organic content, redox potential), and biota. Abiotic features in estuaries vary, depending on the degree of protection from water motion (waves, tidal currents), the quantity of freshwater input, and circulation patterns including residence time of the water, depth, and the salinity gradient. Further, abiotic factors change both temporally and spatially, so that a wide variety of habitats exist in estu-

Figure 1-7. The Johnson-Sea Link II submersible. Submarines extend direct observation and SCUBA collection (to 50 m) to the end of the continental shelf (courtesy of Harbor Branch Foundation, Fort Pierce, Florida).

Figure 1-8. Aerial view of a tidal estuary in Queensland, Australia. Complex interactions occur between the fringing mangroves (f), the subtidal seagrasses (s), and mud flats (m) within this small tropical estuarine ecosystem (courtesy of R. A. Davis, University of South Florida, Tampa).

aries. Geomorphological features influence the abiotic conditions of an estuary and can be used to define four types: drowned river valleys, bar-built estuaries, fjords, and deltaic formations (Ketchum, 1983). One of the more dramatic events in estuaries with large basins and narrow exits is a strong current that produces intense tidal rapids and results in unique plant and animal habitats. An example of tidal rapids is found in the Great Bay Estuary of New Hampshire.

Some organisms live out their entire life in an estuary (clams, oysters, scallops) and others use estuaries as migration routes (anadromous salmon, catadromous eel). Estuarine food webs are complex and dynamic and may be dominated by grazing and detrital components. Estuaries may contain salt marsh, mangrove, and seagrass communities and support some of the world's greatest fisheries for oysters, clams, shrimp, crabs, and fishes (Committee on Biological Diversity, 1995). Estuarine production is high per unit area and may be "the most productive natural ecosystem of the world" (Day et al., 1989). The biological complexity and nutrient cycling in estuaries have been compared to that of tropical rain forests because of the wide variation of habitats found there. Most humans live within 100 km of bays and estuaries worldwide; thus, human impact on estuarine plant communities has been severe. Anthropogenic influences including dredge filling, removal of tidal and benthic communities, and pollution have taken a large toll on estuaries. Because of their biological importance, reviews of estuaries worldwide (Ketchum, 1983) and detailed examination of how they function (Day et al., 1989) have been published. Chapter 2 will consider the abiotic features of estuaries in relation to their support of marine plant communities. Chapters 9 and 10 deal with predominately estuarine plant communities: salt marshes and mangrove forests.

Oceans

Although the continental slope and rise and the ocean basin are devoid of photosynthetic plants, these deeper waters are usually nutrient-rich. Deep, cold oceanic waters provide nutrients when they are circulated to the nutrient-poor upper water through *upwelling* or vertical water currents. The coastal or *neritic* waters are usually more nutrient-rich and typically have a light yellow to brown coloring. The color is due to yellow *gelbstoff*, which is composed of dissolved organic and humic materials from continental runoff and benthic sediments (Lüning, 1990). *Oceanic* water found over the continental slope and ocean basins is usually nutrient-poor and clear blue in color. Because of the lack of nutrients, oceanic waters typically have low productivity and have been called "wet deserts."

The ocean waters can also be divided into *photic* and *aphotic* zones based on the penetration of light (Fig. 1-3). Microalgae or *phytoplankton* are the dominant types of plants in oceanic waters and these are reviewed in Chapter 7. Microalgae are depth-limited by their need for light. If they sink below the photic zone, there will be insufficient irradiance for photosynthesis and the cells will use up their reserves and die. As might be expected, species of microalgae differ in their abilities to utilize low irradiances. Similar to macroalgae, the compensation point of phytoplankton will vary from species to species. Oceanic waters are usually transparent due to the lower levels of nutrients, plankton, and suspended material and thus have a deeper photic zone (ca. 150 m) than neritic waters (ca. 80 to 100 m).

CHAPTER 2

Abiotic Factors

The abiotic factors influencing marine plants are similar to those of terrestrial environments with the exception of the saline aquatic medium in which they live. Although water is denser than air, its water molecules move at high speeds (ca. 580 km h^{-1} at room temperature), and dissolved substances (oxygen, carbon dioxide, nutrients, and minerals) diffuse slowly through it to the plant cells. Further, seawater is also more benign to marine organisms than freshwater, having similar osmolarities and carrying a variety of elements essential to life. This is in contrast to the corrosive aspect of seawater, as demonstrated by its effect on metal parts of ships.

Yale Dawson (1966), who wrote the first text dealing with marine botany, presented a list of abiotic factors to be considered when studying a marine plant community. The factors were grouped into *physical* (light, substratum, temperature, relative humidity, rain, and pressure), *chemical* (salinity, oxygen, nutrients, carbon dioxide, pH, pollution), and *dynamic factors* (waves, currents, tides). Reviews of the effects of abiotic factors on seaweeds are given in texts by Lüning (1990) and Lobban and Harrison (1994); on salt marsh and mangrove plants by Day et al. (1989), Adam (1990), and Robertson and Alongi (1992); and on seagrasses by McRoy and McMillan (1977), Phillips and McRoy (1980), and Larkum et al. (1989a). In this chapter, three major abiotic factors are described (*geological*, *physical*, and *chemical*), and supporting information is also given in the specific chapters dealing with phytoplankton (Chap. 6) and various macrophyte communities (Chaps. 8 to 12).

GEOLOGICAL FACTORS

The Oceans

Approximately 71% of Earth's surface is covered by seawater and can be divided into the 5 major and 20 minor ocean basins (Sverdrup et al., 1964; Davis, 1986; Fig. 2-1). The Atlantic, Pacific, and Indian oceans are the three largest basins, covering 321.1 × 10^6 km^2 of Earth. These three account for 89% of all the oceans and have a mean depth of 4057 m. The lip of each basin corresponds to the continental shelf with the main part of the basin being the

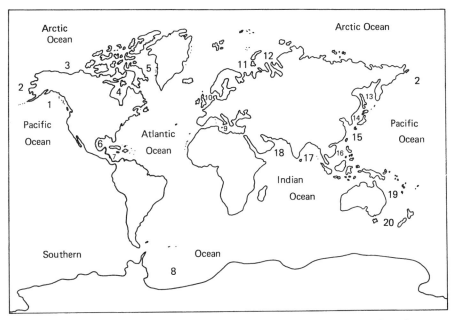

Figure 2-1. Map of the world's oceans showing the five major (named) and twenty minor (numbered) oceans. The minor oceans are the (1) Gulf of Alaska, (2) Bering Sea, (3) Beaufort Sea, (4) Hudson Bay, (5) Baffin Bay, (6) Gulf of Mexico, (7) Caribbean Sea, (8) Weddell Sea, (9) Mediterranean Sea, (10) North Sea, (11) Barents Sea, (12) Kara Sea, (13) Okhotsk Sea, (14) Sea of Japan, (15) Yellow Sea, (16) South China Sea, (17) Bay of Bengal, (18) Arabian Sea, (19) Coral Sea, and (20) Tasman Sea.

ocean proper where the average depth is about 3700 m. Overall, the oceans are shallow when compared to Earth's diameter, with the average depth being equivalent to the thickness of a coat of paint on a globe with a 25-cm diameter.

The surface of Earth below the oceans has a topography similar to that on land, as seen in a hypsographic curve (Fig. 2-2) with the largest portions being between 600 and 3500 m. The submerged portion of a continental block includes the *continental margin* or *shelf* (to 200 m), *slope* and *rise* (200 to 3000 m; Fig. 1-4). Beyond the continental rise, the ocean floor, which is called the *abyssal plain* (3000 to 6000 m), comprises 33 to 75% of the basin and contains abyssal hills, ridges, and trenches. Because depth distribution of benthic plants is limited by light penetration, the vertical limits within this text will be the euphotic zone (Fig. 1-4), which can range from less than 5 m in turbid coastal waters such as Helogland, Germany, to 280 m in oceanic waters of the Bahamas (Lüning, 1990).

Location

A coordinate system is used to define horizontal (latitude and longitude) and vertical (elevation or depth) positions on Earth's surface. Latitude and longi-

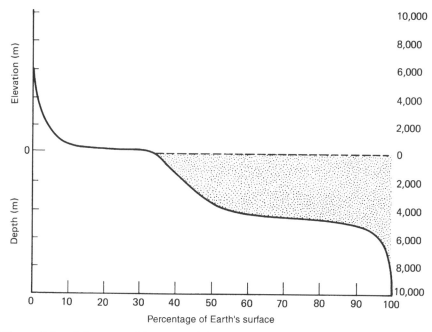

Figure 2-2. A hypsographic curve giving the percent of Earth's surface above or below an elevation or depth. The mean elevation of land is sea level (0 m), and the mean depth of the sea is 3729 m.

tude are horizontal, angular coordinates that divide Earth's surface. *Latitude* is defined in degrees from the equator (north or south) and expressed as lines parallel to the equator. *Longitude* is defined as the angular distance from the meridian plane running (east or west) through the Royal Observatory at Greenwich, England, and expressed as lines running at right angles to the equator. For example, the Marine Station of the University of the Philippines is at Bolinao, Pangasinan, on Luzon Island, with coordinates of 16° 24.08′ N latitude and 119° 54.28′ E longitude and a lighthouse on the southern tip of Anclote Key, on the west coast of Florida, is at 28° 10′ 10″ N and 82° 50′ 42″ W. In both examples, the degrees, minutes, and seconds (or tenths of minutes) are used to indicate distance from the equator (north or south latitude) and from the Greenwich Royal Observatory (west or east longitude). Location of a site can be done using Global Positioning Satellites (GPSs), LORAN locators, and with geodetic charts (App. A).

Geological History and Plate Tectonics

The age of Earth can be divided into four eras, each of these divided into periods, and then into epochs (Table 2-1). Simple marine life probably existed in Precambrian times ca. 3 billion ybp (years before present), and by the Cam-

TABLE 2-1. **Geologic Time Scale with Age in Millions of Years before Present (ybp). Events Include Separation of Continents and Generalized Climate**

Period	Epoch	Events
Cenozoic Era (65)		
Quaternary (2)	Pleistocene	Glacial events. Humans evolve.
Tertiary (65)	Pliocene (5)	Desert formation, cool.
		Central American uplift.
	Miocene (25)	Grasslands, moderate climate.
	Oligocene (38)	Modern flowering plants.
		South America and Anarctica.
	Eocene (55)	Grasslands begin, tropical.
		Australia and Antarctica.
	Paleocene (65)	Early primates, cool climate.
		Large inland seas disappear.
Mesozoic Era (248)		
Cretaceous (144)		Angiosperms appear, dinosaurs disappear.
		Tropical; Africa and South America separate.
Jurassic (213)		Gymnosperms (cycads) appear, mild climate.
		Continents and low, large inland seas present.
Triassic (248)		Gymnosperm and fern forests, first dinosaurs.
		Super continent with large arid areas.
Paleozoic Era (590)		
Permain (286)		Origin of cycads, conifers. Glaciation, arid.
Carboniferous (360)		Amphibians diversify, forests appear.
		Warm climate, swampy low land.
Devonian (408)		Fishes appear, land plants diversify.
		Large seas, local mountains.
Silurian (438)		Extinction event, first fossil plants.
		Mild climate, flat continents.
Ordovician (505)		Extinction event, first land plants, fungi.
		Climate mild, shallow seas.
Cambrian (590)		Faunal diversification.
		Mild climate, large seas.
Precambrian Era (4500)		

Origin of life (3.5 billion), eukaryotes (1.5 billion), and multicellular animals (700). Formation crust and beginning of continental movement.

Source: Modified from Raven et al. 1992.

brian or Ordovician Period, the oceans were chemically similar to those of today. By contrast, geography and climate of Earth have undergone several major changes, including extensive glaciation in the Precambrian, Permian, and Quaternary periods, as well as the most recent separation of the present land masses around 180 to 200 million ybp.

As noted by Davis (1986), "the most significant advance in earth sciences during the last half of the twentieth century is the establishment of *plate tectonics*." Although ignored for many years, the concept of *continental drift*, first proposed by Alfred Wegener in 1912 is now a basic tenet in Earth sciences. Continental drift has caused the movement of large fragments of Earth's crust; formation, separation, and rejoining of continents; and modification of ocean basins (Davis, 1986). During the early Mesozoic Period (about 200 million ybp), a single super continent existed called Pangaea (Fig. 2-3). The breakup of Pangaea occurred along two main fractures, between the Americas and Europe-Africa and between Africa-India and Australia-Antarctica. The separation of Europe and North America about 190 million ybp produced the North Atlantic Ocean. The South Atlantic Ocean came into existence about 125 million ybp

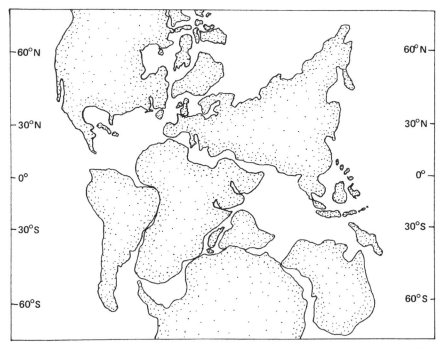

Figure 2-3. The single continent Pangaea around the mid-Cretaceous (110 million ybp). Gondwanaland is still intact in the Southern Hemisphere, whereas the continents in the Northern Hemisphere are in various stages of separation and the Atlantic Ocean is beginning to form.

with the separation of South America and Africa. The Pacific Ocean is what remained of the predrift (Pangaean Sea).

The importance of continental drift to the evolution of marine plants is evident in two theories used to understand biogeographic distribution. One theory relies on the effect of continental drift and the other emphasizes the importance of temperature (Lüning, 1990). Speciation across oceans and within biogeographic floras probably resulted from the effects of both processes. The *vicariance hypothesis* proposes that continental drift resulted in the separation of populations, which ultimately resulted in the differentiation of new species (speciation). For example, McCoy and Heck (1976) proposed that speciation of corals, seagrasses, and mangroves in the Caribbean and Pacific oceans occurred because of a combination of continental drift and isolation, using the vicariance theory. Another theory regarding species evolution and biogeographic distribution is the *dispersal hypothesis*, which includes the concept of *centers of origin*. The second theory proposes that species have evolved through dispersal from a primary or central population and because of isolation differentiated into new species. The isolating factor commonly cited in the dispersal hypothesis is temperature, which results in biogeographical floras, while the vicariance hypothesis most commonly relies on the effect of separation by oceans.

Classification of Coasts

Shorelines of continents can be identified based on whether the substrata are *consolidated* (hard) or *unconsolidated* (loose sediment), *young* (primary, youthful) or *mature* (older); in addition, the levels of energy of the coast are important (high or low wave activity). Rocky or consolidated coasts are usually primary landforms (youthful) and subject to high wave energy and subsequent erosion. Rocky coasts can be divided into their geographical landforms, including coastal cliffs (steep, vertical; Fig. 8-1), shore platforms (horizontal or sloping surfaces; Fig. 2-4) and limestone (porous, irregular; Fig. 2-5) coasts. The different types of rocky coasts and levels of wave energy greatly influence the types of *lithophytic* (rocky) plant communities. As wave energy increases, lithophytic communities expand upward (*dilation*) beyond the high-tide mark due to effects of spray (see Fig. 8-4).

Unconsolidated coasts or beaches, those found on barrier islands and sandy shores, do not support extensive populations of marine macrophytes as found on rocky shores. Waves are the fundamental force operating on a beach, making these habitats some of the most variable of coastal landforms that occur on both primary (youthful) and secondary (mature) coasts. A beach can be divided into three zones: *backshore*, *foreshore*, and *nearshore* (Fig. 2-6). The backshore extends landward from the highest tide mark to the dunes or cliffs, the foreshore is the intertidal zone of the beach, and the nearshore is the zone from the lowest tide level seaward. Beaches subject to high wave energy tend to have a steeper gradient on the foreshore and a coarse particle or grain size (Table 2-2).

The abundance and distribution of intertidal and subtidal organisms can be

Figure 2-4. A rock platform coast on Kangaroo Island, Victoria, Australia. The granite substrata is not porous or pitted, and the seaweeds are exposed to intense surf (from Dawes, 1995; with permission of Academic Press).

greatly reduced where sand deposition occurs on rocky outcrops. For example, Daly and Mathieson (1977) found that such habitats in New Hampshire are dominated by opportunistic annuals (e.g., the green alga *Enteromorpha intestinalis*). *Psammophytic* or sand-dwelling plants were common and mostly perennial as seen with the wiry, turf-forming red alga *Ahnfeltia plicata*. Psammophytic seaweed species showed special adaptations, including rapid life histories and sand-resistant crustose phases that can withstand seasonal sand burial and erosion.

Estuaries

Major reviews of estuaries include Ketchum (1983), Knox (1986), and Day et al. (1989). *Estuaries* are semienclosed coastal bodies of water having a free connection with the open sea and being diluted by freshwater from land drainage (Fig. 1-8). Such habitats can be divided into a marine region (lower) with a direct connection to the sea, a middle region where salt water and freshwater mix, and an upper or fluvial region dominated by freshwater but subjected to tidal action (Ketchum, 1983; Day et al., 1989). Two characteristics of estuar-

Figure 2-5.   The intertidal zones on a limestone reef of West Summerland Key, Florida. The four zones are the (1) white to black spray zone, (2) upper yellow intertidal zone, (3) lower gold to brown intertidal zone, and (4) subtidal fringe partially exposed by the surf (from Dawes, 1995; with permission of Academic Press).

ies are that they are young and ephemeral. Estuaries are geologically young because they were formed when the sea reached its present level about 5000 ybp (Day et al., 1989). They are geologically ephemeral because they are "sediment sinks" or depositional environments that have short life expectancies in terms of hundreds to thousands rather than millions of years. Estuaries may be classified on their geomorphology, salinity and density, and type of circulation (Ketchum, 1983); each type greatly influences the distribution of salt marshes, mangals, seagrass beds, and macroalgal diversity.

Geomorphic estuarine types include drowned river valleys, bar-built estuaries, fjords, and deltaic formations. Because of the rise in sea level since the last ice age, many estuaries are *drowned river valleys* (Great Bay Estuary, New Hampshire). Other estuaries in flat coastal plains were created through longshore transport of sediment that formed *bars* across the mouth of a river (Fig. 1-8). Bar-built estuaries usually have a shallow, restricted opening to the sea (Sarasota Bay, Florida). *Fjords* occur in more mountainous regions where glaciers cut deep gorges that become flooded due to rising sea level (Somes Sound, Maine; Norway). *Deltaic* estuaries occur where rivers carry

TABLE 2-2. The Relationship between Particle Size
(Wentworth Grain-Size Sale) and Beach Slope

Particle Size	Name	Average Slope
Gravel		
64–256 mm	Boulders	<24°
4–64 mm	Cobbles	24°
2–4 mm	Pebbles	17°
1–2 mm	Granules	11°
Sand		
0.5–1 mm	Very coarse	9°
250–500 μm	Coarse	7°
125–250 μm	Medium	5°
62–125 μm	Fine	3°
2–62 μm	Very fine	1°
Mud		
>2–2 μm	Silt/Clay	0°

Figure 2-6.   Beach zonation on a barrier island off Virginia. The four intertidal zones include (1) a maritime dune region, (2) a backshore, (3) a foreshore impacted by the waves, and (4) a nearshore that is rarely exposed (from Dawes, 1995; with permission of Academic Press).

large amounts of sediment to form extensive deltas (Mississippi River, Louisiana).

Classification of estuaries based on *salinity stratification* includes well-mixed to weakly stratified, to strongly stratified types, as well as fjord and arrested salt wedge types. The first three types of salinity regimes occur in drowned river valleys and depend on the level of river water flow. The significance of salinity stratification is that the upper freshwater layer usually has higher oxygen levels than the lower seawater; thus, the more highly stratified the estuarine water column, the lower the dissolved oxygen in the bottom seawater. *Well-mixed* estuaries show little difference in salinity between the surface and bottom water (Long Island Sound, New York). Estuaries showing *weak stratification* have a limited undiluted layer of seawater near the bottom. There can be a net flux of surface freshwater seaward and of bottom seawater landward with a gradient between the two ends (Southhampton River, England). *Strongly stratified* estuaries exhibit a net flux of seawater landward and freshwater seaward with a 2 to 10 ppt (parts per thousand) difference in salinity between surface and bottom water (Columbia River, Washington). A *fjord* estuary typically has three distinct zones of water: an upper brackish layer, an intermediate layer of increasing salinity, and a bottom layer of stable high salinity. Estuaries with large river flows may exhibit an undiluted surface freshwater layer and a similar bottom *saltwater wedge*; these are called arrested salt wedge estuaries (Ishikari River, Japan).

Estuaries classified by *tidal currents* recognize the influence of tides, which is the basis for the word estuary (L. *aestus* = tide). Estuaries with large openings can have *standing-wave tides* where the tidal amplitude is similar throughout the basin and occurs at the same time. Larger estuaries with restricted openings, such as the Great Bay Estuary, experience *progressive wave tides*, which show increased delay further into the basin. These estuaries experience tidal rapids where there are narrow channels. A *mixed tidal wave* occurs in elongated estuaries with the standing tidal wave near the mouth and an increasing progressive wave at the upper end.

An important feature of estuaries and estuarine water is the fluctuation of abiotic factors, seasonally, tidally, as well as periodically. Major factors that influence the ecology of estuaries include seasonal variation in temperature, the degree of protection from wave energy, the type and rate sedimentation, and the amount and characteristics of freshwater input. Usually, there is a wide fluctuation in salinities and temperatures, and a strong influence of terrestrial nutrients and sediments; thus, environmental conditions of estuaries can be quite diverse. Temperature in a temperate estuary (Great Bay) will range from -2 to $27°C$ and in a subtropical estuary (Tampa Bay) from 12 to $33°C$. Salinities in the same two bays can range from less than 5 to 35 ppt. In addition, sediment loads along with dissolved nutrients can be very high in estuaries compared to the surrounding oceanic waters. These ranges in abiotic factors occur during tidal changes and between the oceanic and freshwater portions of an estuary. If evaporation is equal to or greater than the freshwater source, as found in arid regions like Baja California (Pacific Mexico), the net result can be an increase in salt content above the open coast and the water will be *hypersaline*.

Two major marine angiosperm communities form tidal communities in estuaries, namely, salt marshes (Chap. 9) and mangroves (Chap. 10). Both of these communities show their greatest development in estuaries where they are protected from high levels of water motion, they are supplied with sediment, and experience freshwater input. The high levels of organic matter in estuaries can support large populations of bacteria. Bacterial activities can result in depressed levels of oxygen and thus greatly influence the types of benthic organisms (*epifauna* and *infauna*, and benthic plants) that form intertidal and subtidal communities. Probably the greatest diversity of animals that live within (infauna) substratra occur in estuaries due the unconsolidated substrata. Because the sediment load of estuaries results in low water transparency, benthic plants are usually limited to shallow depths (0.5 to 3 m). The lack of benthic plants, high bacterial activity, and low levels of dissolved oxygen result in many specialized adaptations of estuarine animals. All of the biota show adaptations to changes in salinity (Adam, 1990).

PHYSICAL FACTORS: LIGHT

Penetration and Absorbance

The interaction of light in aquatic ecosystems has been reviewed by Kirk (1994). Of the Sun's electromagnetic spectrum that reaches sea level (Fig. 2-7), about one-half is visible light (390 to 760 nm) and ultraviolet (UV; 290 to 390 nm) and the other half is infrared (IR; 760 to 3000 nm). In terms of energy content, 3% of the impinging wave energy (λ) is UV, 52% is visible, and 45% is IR. The wavelengths are reflected at the water's surface, absorbed (conversion of radiant energy), and scattered (deviated). The effect of water results in vertical attenuation of the UV, visible, and IR spectra, which can be expressed as *transmittance* (T) in Eq. 2-1:

$$T = I_2/I_1 \qquad\qquad (2\text{-}1)$$

where I_1 = irradiance at depth 1, and I_2 = irradiance at a depth 1 m lower.

An intertidal community on a clear day near the equator could experience a photosynthetic active radiation (PAR; 400 to 700 nm) of 2500 μM photons m^{-2} s^{-1}, and a deep-water (140-m) population of the tropical green alga *Halimeda incrassata* in the Eniwetok Lagoon would have a maximum of 25 μmol photons m^{-2} s^{-1}. The normal depth distribution of seaweeds is around 200 m in highly transparent oceanic water where the percent transmission is 0.01%. Thus far, the greatest depth of a macroalga was recorded at 268 m in the Bahamas where the irradiance was 0.009 μM photons m^{-2} s^{-1} (Littler and Littler, 1984).

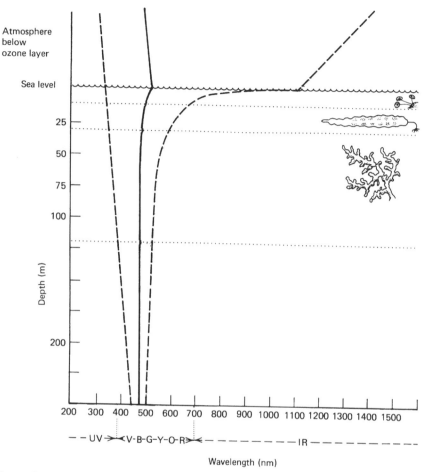

Figure 2-7. The penetration of sunlight through the atmosphere and seas of Earth. Much of the ultraviolet radiation is removed by the ozone layer and most of the infrared light is removed at sea level. The solid line shows the wavelength of maximum intensity at increasing depths, and the dashed lines are the boundaries for 95% of available solar energy.

As outlined in Figure 2-8, there are two *optical* classes of seawater, including *oceanic* water characterized by Types I to III and *coastal* water characterized by Types 1 to 9 (Jerlov, 1976). Water is colorless, so that color is due to the available wavelengths that penetrate and are reflected from it. The transmittance maximum in oceanic waters is in the blue part of the spectrum. Coastal waters reflect slightly more brown to yellow, due to the scattering and selective absorption by *gelbstoff* (yellow substances, gilvin), which is composed of dissolved organic and humic materials from continental runoff and benthic sediments (Lüning, 1990). Thus, oceanic waters appear blue at the surface and are

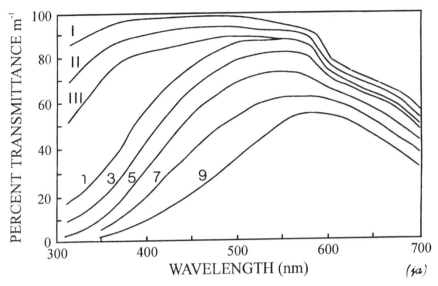

Figure 2-8. A classification of optical water types shows that the distribution of light spectra varies with the types of oceanic (I, II, III) and coastal (1, 3, 7, 9) waters. (After Jerlov, 1976; with permission of Elsevier Science Publ.).

increasingly green with depth, and coastal waters appear more brown to yellow at the surface. The distribution of benthic plants is limited to the euphotic zone with crustose algae extending to ca. 0.001 to 1% of surface irradiance (Lüning, 1990).

Measurement of Light

Light can be measured either as the available energy or the number of quanta (photons for visible and UV λ) impinging on a unit surface area. *Irradiance* (W m^{-2}) is a measure of energy and *photon fluency* (μmol photons m^{-2} s^{-1}) is a measure of the rate of quanta hitting a unit area. Photon fluency units have replaced the einstein (1 E = 1 mol of photons = 6.02×10^{23} photons) in the literature. Photon fluency measurement is usually done with a quantum meter that measures the total energy of photosynthetically active radiation (PAR = 400 to 700 nm) in μmole photons. Irradiance (μW cm^{-2}) can be measured using a radiometer or a quantum meter. Lüning (1990) gives the following relationships in Eq. 2-2 and Li-Cor Inc. (1985) in Eq. 2-3:

$$1 \text{ W m}^{-2} \approx 2.50 \times 10^{18} \text{ phot. m}^{-2} \text{ s}^{-1}$$
$$\approx 4.2 \text{ } \mu\text{mol phot. m}^{-2} \text{ s}^{-1} \tag{2-2}$$
$$\text{klux} \times A = \mu\text{mol phot. m}^{-2} \text{ s}^{-1} \tag{2-3}$$

where klux = 92.9 fc, and A = source of irradiance (daylight = 18, metal halide = 14, sodium or mercury = 14, white fluorescence = 12, incandescence = 20).

PHYSICAL FACTORS: TEMPERATURE

Temperature is the most fundamental abiotic factor for organisms as it affects all levels of biological organization from molecular, cellular, organismal to communities. High temperatures result in denaturation of proteins and damage to enzymes and membranes. Low temperatures can cause disruption of the lipids and proteins in membranes and mechanical damage to the cell through the production of ice crystals. Probably the strongest demonstration of the importance of temperature can be seen in the geographic distribution of marine floras. Mangrove communities cannot tolerate regular frosts and are replaced by salt marshes as one moves from tropical to temperate latitudes. Latitudinal differences of seaweed vegetation are of great interest and are dealt with in Chapter 4 as well as a detailed review by Lüning (1990).

Heat is energy of molecular motion. It can be expressed as calories (1 cal = heat required to raise 1 g of water 1°C) and is transferred via radiation, convection, and conduction. Temperature is a measure of heat and it is expressed in °C on the Celsius scale or K on the Kelvin scale). At -273.15°C or 0 K, all molecular motion should cease. Temperature can be measured by calibrating the expansion of fluids (e.g., Hg in a glass thermometer) and by determining the levels of electrical resistance (conductivity meter).

PHYSICAL FACTORS: WATER MOVEMENT

The effects of water movement can be seen in the adaptation of marine plants to waves, currents, and tides. The first two aspects are critical in terms of mechanical and chemical stress. Plants must adapt to the damaging aspects of waves and currents and to the slow diffusion rates of nutrients and gases in water.

Boundary Layer

Gases and ions diffuse 10,000 times slower in water than in air. The slowness of diffusion in water is evidenced by the need to stir water in growth studies and during measurement of oxygen. Thus, water movement (turbulence, waves, currents) is an important factor in increasing the chance of a plant to come into contact with dissolved substances. Further, there is a *boundary layer* of slowly moving seawater that surrounds submerged plants. Nutrients, oxygen, and carbon dioxide all must diffuse through that layer to be taken up by submerged plants. Submerged plants have responded to the boundary layer through morphological and structural adaptations, which will be reviewed in the chapters on macroalgae (Chap. 8) and seagrasses (Chap. 11).

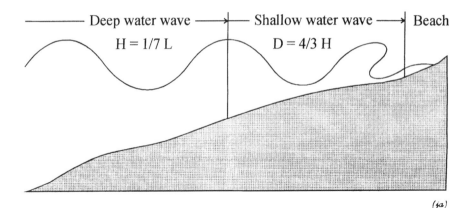

Deep water wave ⟶ ⊢— Shallow water wave ⟶⊣ Beach

H = 1/7 L   D = 4/3 H

(ja)

Figure 2-9. A wave encountering shallow water. The wave will break when water depth equals 4/3 of its height due to slowing of the orbital speed of water molecules near the substrata.

Waves

The generation of waves is caused by wind (meteorological), gravitational pull (tides), earthquakes and landslides (geological). Waves result from the deflection of wind as it blows over water; changes in atmospheric pressure also supply the energy to produce air movements (see what follows). The *height* (H), *period* (T), and *length* (L) of waves depend on the velocity and duration of the wind, the distance the wave has to travel, and the water *depth* (D). The ideal wave in deep water ($D > 1/2L$) changes as it encounters the drag of the substratum in shallow water (Fig. 2-9). Because of drag at the base of the wave in shallow water, the orbits of water molecules at the top of the wave are moving faster than the wave velocity (*celerity*). The different orbital speeds cause the wave to steepen and break. The wave will break when $D = (4/3)H$ or when $H = 1/7(L)$.

Wave energy can be measured with a dynamometer (kg cm^{-2}), which is a calibrated spring attached to a *drogue* or float that is dragged with the wave (Fig. 2-10). The dynamometer (Jones and Demetropolus, 1968) is attached to the substratum and the drogue is pulled with each passing wave. The energy available in a breaking wave can be expressed in force (newtons; N = kg m s^{-1}, where m = mass), pressure or force per unit area (pascals; Pa = N m^{-2}) or mechanical energy (joules; J = N-m). Wave height and frequency can be measured with a sighting level or a surveying instrument by calibration against a rod embedded in the surf.

Tides

The periodic rise and fall of the sea level due to the gravitational attractions of the Sun and Moon are gravitational waves or *tides*. Being much closer, the Moon's gravitational pull is about two times stronger than the Sun's (46% of

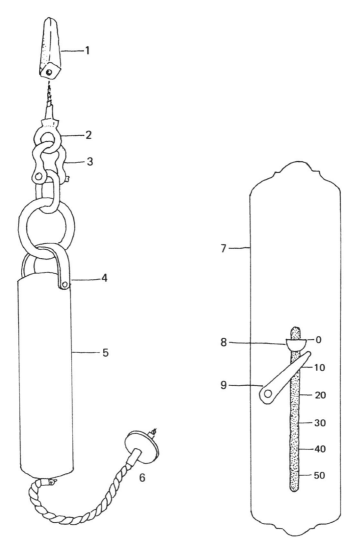

Figure 2-10. A spring dynamometer used to measure wave force. A hardwood plug (1) is driven into a rock and the dynamometer (5) attached by an eyescrew (2), connecting link (3), and split pin (4). The dynamometer is a calibrated spring balance (7) with a blocker (8) that stops the pointer (9) from returning to zero after a wave pulls the drogue (6) (after Jones and Demetropolus, 1968).

Moon's effect) and thus the tidal pattern repeats itself on a 24-h, 50-min cycle in larger ocean basins. Gravitational pull occurs on the side of Earth facing the Moon and centrifugal forces occur on the opposite side due to Earth's rotation (Fig. 2-11).

The shapes of the ocean basins increase or reduce the gravitational effects

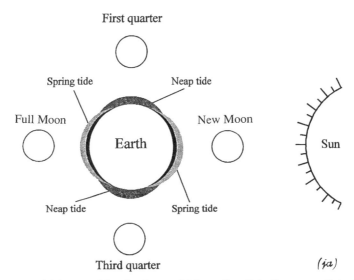

Figure 2-11. Tidal control by the Sun and Moon. Parallel alignment of the Sun and Moon results in a spring tide every 28 days. Neap tides are produced when the Sun and Moon are at right angles to one another.

of the Sun and Moon, which influence the type of tide and its amplitude (Fig. 2-12). Tidal regularity is similar to the sloshing of water in a bowl where advancing waves may amplify or cancel out receding ones. Many coasts have *semidiurnal* tides (two high and two low tides every 24 h, 50 min) that are *equal* (Balboa) or *unequal* in size (San Francisco), whereas others have *diurnal* tides with a single high and a single low each day (Pakhoi). Other areas show *mixed* tides, where tides vary from semidiurnal to a single high or a single low in the 24-h, 50-min period (Manila). When Earth, Sun, and Moon are in alignment (Fig. 2-11), the gravitational effects are combined and very high and low (*spring*) tides result (at new and full Moon). When the three solar bodies are aligned in a right angle, the weakest, *neap* tides are formed. *Mean sea level* (*MSL*) is the average of spring and neap tides.

Currents

The rotation of Earth and the atmospheric circulation combine to produce the *wind-driven* horizontal currents of the surface waters of the oceans. The deep-water portion of the ocean (continental slope and rise, abyssal plain) also has currents that are driven by differences in salinity and temperature (*thermohaline*). The horizontal currents are caused by prevailing wind patterns (i.e., westerlies, trade, and polar winds) that result in a series of *gyres* (Fig. 2-13).

Air at the equator and warm latitudes becomes heated, expands, and then rises. After expansion and rising, a low-pressure area is produced into which

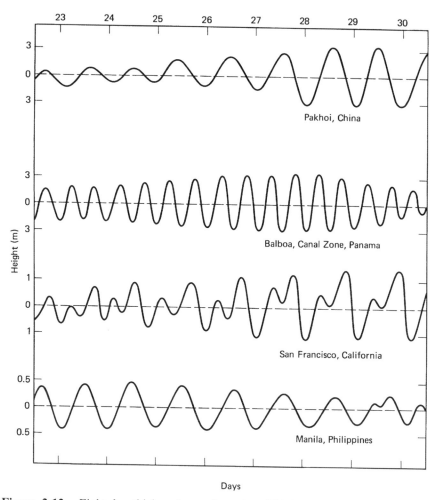

Figure 2-12. Eight-day tidal cycles at four sites. Tidal amplitude (range) and frequency (period) vary, depending on basin size, prevailing winds, and currents.

cooler air moves. Rising air is carried north or south from the equator where it eventually cools, contracts, and sinks, creating a high-pressure area. The deflection of air masses to the right in the Northern Hemisphere and to the left in the Southern Hemisphere is due to Coriolis force from the spinning of the planet. The curving movement of the air masses results in the westerlies and trade winds (*prevailing winds*) in both hemispheres, which are separated by regions without prevaling winds (*doldrums*).

The easterly trade winds produce the equatorial currents, common to all oceans. Because of the land masses in the Northern Hemisphere, the equatorial currents are deflected north to become the western boundary currents, which are usually the strongest currents of a circulation pattern. At about 40° to 50°

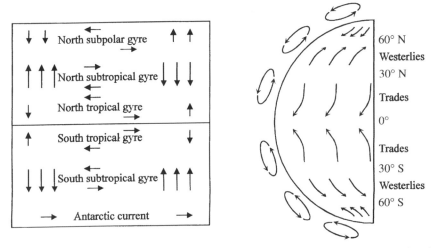

Figure 2-13. Idealized wind patterns or atmospheric gyres of the world. The two hemispheres have similar wind patterns that are modified because of the placement of their continents.

N latitude, the water is deflected eastward, forming the westerly trade winds, and then south after meeting the eastern side of the continents (Fig. 2-14). The Sargasso Sea can be used as an example. The sea is essentially a large eddy confined within the following currents: the North Equatorial Drift (moves to the west and northward into the Caribbean and Gulf of Mexico), the Gulf Stream (moves to the north along eastern North America, the North Atlantic Current (moves to the east over the North Atlantic Ocean), and the Canaries Current (moves to the south along the western European coast). It appears that the Sargasso Sea biota, including the two pelagic species of *Sargassum* (*S. fluitans*, *S. natans*), has a long geological history. The sea dates back 17 million ybp to a Carpathian section of the Tethys Sea that was on the edge of Pangaea (Lüning, 1990).

Vertical currents (*upwellings*, *downwellings*) can result from a number of factors. *Thermohaloclines* are caused by changes in seawater density due to differences in salinity and temperature in two bodies of water that meet. *Offshore winds* will push surface waters away from a coast allowing deep water to rise. Deep water may be pulled to the surface through interactions between currents that pass close to one another (*divergences*), when nearby surface water is pulled along with a strong current (*wake streams*), and when rapid currents pass over shallower or rough benthos (*turbulence*) as seen in tidal rapids. In all cases, cooler, nutrient-rich deeper water, if brought into the euphotic zone will support increased phytoplankton production, which in turn supports zooplankton and fish. Thus, upwellings sites can be regions of high primary and secondary productivity.

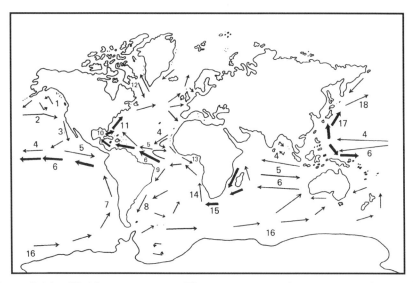

Figure 2-14. World ocean currents. The strongest oceanic currents are shown with bold arrows and conform to the equatorial and western boundaries of the oceans. The major currents are (1) Alaskan, (2) North Pacific, (3) California, (4) North Equatorial, (5) Eastern, (6) Southern Equatorial, (7) Peru, (8) Falkland, (9) Brazil, (10) Florida, (11) Gulf Stream, (12) Labrador, (13) Guiana, (14) Benguela, (15) Agulhas, (16) Western Wind Drift, (17) Kuroshio, and (18) Oyashio.

CHEMICAL FACTORS

Seawater

Pure water, an apparently simple compound, exists in all three states (solid, liquid, gas) on Earth and has some remarkably complex properties. Because of the polarization of water molecules, they act like small magnets and have strong hydrogen bonds (bonding energy of 109 to 111 kcal mol^{-1}). Consequently, pure water freezes at 0, not $-150°C$, and boils at 100, not $-100°C$. The addition of salt to water alters it's physical properties, as the ionic bonds between the dissolved salt and water molecules must be overcome in order to freeze or boil. The addition of salt results in a raising of osmotic pressure, boiling point, density, and electrical conductance and a lowering of vapor pressure and freezing point (Sverdrup et al., 1964; Davis, 1986).

The composition of oceanic seawater is uniform throughout the world due to the presence of 12 major elements that account for about 99.9% of all the dissolved solids (Table 2-3). Almost all of the natural elements (73/89) are found in seawater with concentrations of less than 1 part per million (ppm). The major gases in seawater are nitrogen, oxygen, and carbon dioxide of which the last two have great significance in the biological processes of respiration and photosynthesis.

TABLE 2-3. Major Components (Elements) of Seawater

Component	Form	Concentration (ppt or g kg^{-1})	Percent of solids
Chloride	Cl^-	19.000	55.01
Sodium	Na^+	10.500	30.40
Sulfate	SO_4^{-2}	2.650	7.67
Magnesium	Mg^{+2}	1.350	3.91
Calcium	Ca^{+2}	0.400	1.61
Potassium	K^+	0.380	1.10
Bicarbonate	HCO_3^-	0.140	0.41
Bromide	Br^-	0.065	0.19
Boric acid	H_3BO_3	0.026	0.08
Carbonate	CO_3^{-2}	0.018	0.01
Strontium	Sr^{+2}	0.008	—
Fluoride	F^-	0.001	—
Total		34.538	99.94

Source: After Davis (1986) and Milne (1995).

Salinity

A kilogram of average seawater contains 34.7 g of salts, which can be expressed as 34.7 ppt (gm kg^{-1}, o/oo) salinity. Thus, the definition of *salinity* is the weight of solids obtained by drying 1 kg of water under standard conditions. More recently, oceanographers have replaced this definition with one that is based on the conductivity of a sample of seawater compared to a standard solution of KCl in which the mass fraction of KCl is 32.4356×10^{-3}, the temperature 15°C, and the pressure 1 atm (Parsons, 1982). However, use of conductivity is only valid in oceanic, not estuarine waters due to changes in the ratios of chemical constituents. A history of the development of salinity and standard seawater is presented by Culkin and Smed (1979).

Seawater is considered to be *oceanic* if its salinities are between 32 and 38 ppt, *hypersaline* if above 38 ppt, and *estuarine* if it fluctuates between 1 and 32 ppt. Based on the salinity classification of Redeke (1933), seaweeds can be classified according to their tolerance to seawater and may be *euryhaline* or *euhaline* (30 to 40 ppt), *polyhaline* (18 to 30 ppt), *mesohaline* (3 to 18 ppt), or *oligohaline* (0.5 to 3 ppt). The term *brackish* has been used in a variety of ways (see Remane and Schlieper, 1971), but is most commonly used to describe low salinities (0.5 to 5 ppt) by limnologists or low, stable salinities by oceanographers. However, the term for many scientists includes waters showing salinities ranging from oliohaline through polyhaline (Remane and Schlieper, 1971). In this text, brackish will be used as described for the Baltic Sea (Wallentinus, 1991). Wallentinus calls the Baltic Sea "one of the worlds largest brackish water areas" where stable, low (mesohaline = 5 to 10 ppt) salinities occur in most of the basin.

Salinity can be determined through density (*hydrometer* and *pycnometer*), resistance to an electrical current (*conductivity meter* or *salinometer*), the bending of light from air to saltwater (*refractometer*), and by determining the concentration of chlorine through titration with $AgNO_3$ (*chlorinity*). The hydrometer is a Cartesian diver that measures the specific gravity and the pycnometer is a volumetric flask that determines the weight of the salts in a standard volume; both require a standard temperature and pressure. A refractometer is a hand-held unit that measures the refractive index of a medium. Light passing from air to pure water has a refractive index of 1.3330 and that passing through 35-ppt seawater has a refractive index of 1.3394. The conductivity meter measures the current flowing across a Wheatstone (electrical) bridge at a known temperature and pressure; the more ions, the higher the electrical conduction. The conductivity–chlorinity relationship in seawater can be seen in Eq. 2-4:

$$Cl\ (ppt) = -0.7324R_{15} \tag{2-4}$$

where R_{15} is the ratio of the electrical conductivity of a given sample to one where salinity = 35 ppt with both samples held at 1 atm and 15°C.

Chlorinity is the international method (Knudsen Procedure) for salinity determination using the relationship shown in Eq. 2-5. To determine salinity, the chloride content of a known volume of water at 20°C is determined by titration with silver nitrate to form silver chloride, a white precipitate (Eq. 2-6).

$$Salinity = 0.03 + (1.805 \times chlorinity) \tag{2-5}$$

$$AgNO_3 + NaCl \leftrightarrow AgCl + NaNO_3 \tag{2-6}$$

The standard Knudsen Procedure for determining salinity requires special glassware and Standard (Copenhagen) Seawater and usually its accuracy is far more than needed for marine botany studies (Strickland and Parsons, 1968; Dawes, 1981). A less accurate, simplified field titration technique for chlorinity is given in App. A.

Oxygen in Seawater

In the euphotic zone, seawater oxygen concentration is about 0.9% (depending on temperature, partial pressure of the gas, salinity, and biological activity) versus 21% in the atmosphere. The two major sources of oxygen in seawater are the atmosphere and plants. The latter are involved in the oxygen cycle through their photosynthesis and respiratory activities (Eq. 2-7).

$$\begin{aligned} &\text{photosynthesis} \rightarrow \\ 6CO_2 + 6H_2O &\leftrightarrow C_6H_{12}O_6 + 6O_2 \\ &\leftarrow \text{respiration} \end{aligned} \tag{2-7}$$

Levels of oxygen vary throughout marine waters. Polar seas contain about twice the amount of oxygen as tropical waters due to their lower water temperature and salinities. Benthic communities below the photic zone tend to experience lower concentrations of oxygen (0.2 to 0.3%) than surface communities. Thus, infaunal organisms are likely to experience limited oxygen supplies and show adaptations to these conditions. Oxygen measurements are of particular value in studies of tidal communities and estuaries, which are areas of high biological activity and enrichment due to the impact of terrestrial sediments and runoff.

Determination of oxygen levels in seawater is usually by titration (Winkler technique) or by measurement of oxygen diffusion (oxygen meter). Details of both techniques are given in App. A along with a third procedure that allows measurement of pressure changes during photosynthesis or respiration (manometric technique). The Winkler titration procedure is a three-step process where a tetravalent compound of $2MnO(OH)_2$ is produced in relation to the amount of dissolved oxygen present in water. The compound is acidified in the presence of potassium iodide and the amount of liberated iodide that results is chemically equivalent to the amount of bound oxygen. The free iodide is measured via titration with sodium thiosulfate to sodium iodide and the amount of dissolved oxygen is calculated.

To carry out a Winkler procedure, standard biological oxygen demand bottles (300-mL BOD bottles) are filled to overflowing with seawater without allowing turbulence; 1 mL of manganous sulfate ($MnSO_4$; 480 g L^{-1}) and 1 mL of alkaline potassium iodide (Kl + KOH; NaOH, 500 g, 500 mL^{-1}; Kl, 300g, 450 mL^{-1}) are added (Dawes, 1988). The two 1-mL solutions are added below the surface, and the BOD bottle is stoppered and gently inverted. Any dissolved oxygen rapidly oxidizes an equivalent amount of divalent manganese to basic hydroxides of higher valency states, as shown in Eqs. 2-8 and 2-9.

$$MnSO_4 + 2KOH \rightarrow K_2SO_4 + Mn(OH)_2 \qquad (2\text{-}8)$$

$$2Mn(OH_2) + O_2 \rightarrow 2MnO(OH)_2 \quad \text{(brown precipitate)} \qquad (2\text{-}9)$$

The brown-colored manganese compound [$MnO(OH)_2$] is allowed to settle in the bottle, shaken a second time, and allowed to settle again. The compound is stable over a number of days if stored in a dark cool place. After the precipitate has settled, concentrated sulfuric acid is added to form manganese sulfate, which reacts with the potassium iodide (KI), as shown in Eqs. 2-10 and 2-11:

$$MnO(OH)_2 + 2H_2SO_4 \rightarrow 3H_2O + Mn(SO_4)_2 \qquad (2\text{-}10)$$

$$Mn(SO_4)_2 + 2KI \rightarrow MnSO_4 + K_2SO_4 + I_2 \qquad (2\text{-}11)$$

The liberated iodine is chemically equivalent to the amount of dissolved oxygen in the sample and is titrated with a standardized sodium thiosulfate solution ($Na_2S_2O_3$). The latter substance is oxidized to tetrathionate, removing molecular iodine and causing the solution to clear, as shown in Eq. 2-12:

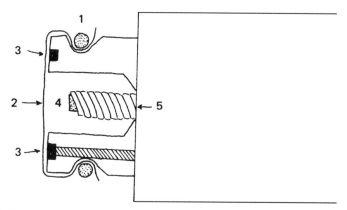

Figure 2-15. An oxygen probe. The probe is covered by a semipermeable membrane (2) held in place by a rubber or metal O-ring (1) and contains a concentrated KCl solution (4). A weak current is passed through the rhodium or gold cathode outer ring (3) and silver anode (5), resulting in accumulation of oxygen at the cathode and a change in current.

$$2Na_2S_2O_3 + I_2 \rightarrow Na_2S_4O_6 + 2NaI$$

A starch indicator solution (ca. 4%, aqueous) is added (ca. 1 mL) before all the iodine is bound, turning the solution blue due to the reaction between the free iodine and starch. Continued removal of iodine by titration with sodium thiosulfate results in the easily seen clearing of the solution as the blue color disappears.

Oxygen also can be measured using a dissolved oxygen analyzer (Thomas, 1988) with probes (sensors) that have a membrane-enclosed polarographic (Clark-type) electrodes (Fig. I-9; see App. A). The probe contains a gold or rhodium cathode and a silver anode that is bathed in a solution of buffered potassium chloride (Fig. 2-15). The probe is either an *active* (e.g., Yellow Springs Instrument Co., Ohio) or *passive* (e.g., Orbisphere Laboratory, Geneva) type. The active type energizes the probe even when turned off, and as such, the probe is highly stable. The passive type only applies voltage to the probe when switched on and thus requires more frequent recalibration. A thin teflon membrane separates the probe tip from the seawater. Oxygen diffuses through the membrane, resulting in a current between the anode and cathode that is proportional to the percent saturation of oxygen in the sample. The current between the cathode and anode is amplified by the oxygen meter after correction for temperature and pressure. Because the oxygen is consumed at the sensor tip, the seawater must be constantly mixed by a magnetic stirrer with a magnet in the BOD bottle or a probe with a built-in stirrer.

Carbon Dioxide in Seawater

This gas is the product of respiration, the substrate in photosynthesis (see Eq. 2-7), and an important factor controlling the pH of seawater. The interactions of CO_2 and seawater are summed up in reversible Eqs. 2-13 and 2-14.

$$CO_2 + H_2O \leftrightarrow H_2CO_3 \leftrightarrow H^+ + HCO_3^- \leftrightarrow 2H^+ + CO_3^{-2} \quad (2\text{-}13)$$

$$ \underset{\text{Carbonic acid}}{} \quad \underset{\text{Bicarbonate}}{} \quad \underset{\text{Carbonate}}{}$$

$$CO_2 + H_2O + CO_3^{-2} \leftrightarrow 2HCO_3^- \qquad\qquad\qquad (2\text{-}14)$$

$$\underset{\text{Bicarbonate}}{}$$

About 99% of the CO_2 absorbed from the air is converted to bicarbonate or carbonate and remains in these forms. Because of the reservoir, 50 to 60 times more CO_2 is present in seawater than in air. In photosynthesis, CO_2 is removed and the HCO_3^- ion is converted to H_2CO_3 and ultimately to CO_2 (Eq. 2-13). The H^+ ions are also used and replaced so that photosynthesis causes a slight reduction of CO_2, a larger decrease in bicarbonate ions, a small decrease in H^+, and a slight increase in carbonate ions. Respiration produces a reversal of these changes.

Ionic Concentrations in Seawater

Although the concentration of the hydrogen ion (H^+) influences the acidity of seawater, the pH of seawater remains basic, ranging between 7.4 and 8.4 (average is 7.8) due to carbonate buffering (Eqs. 2-13 and 2-14). The pH of seawater rises when CO_2 is removed via photosynthesis and drops when CO_2 is added via respiration. Thus, changes in pH can be used to measure rates of photosynthesis, although it is more common to measure changes in oxygen levels. Removal of CO_2 also can be accomplished by production of calcium carbonate ($CaCO_3$) via the process of *calcification* (Eq. 2-15).

$$Ca^{+2} + 2HCO_3^- \leftrightarrow Ca(HCO_3)_2 \leftrightarrow CaCO_3 + H_2CO_3 \qquad (2\text{-}15)$$

$$\phantom{Ca^{+2} +} \underset{\text{Bicarbonate}}{} \quad \underset{\substack{\text{Calcium} \\ \text{bicarbonate}}}{} \quad \underset{\substack{\text{Calcium} \\ \text{carbonate}}}{} \quad \underset{\substack{\text{Carbonic} \\ \text{acid}}}{}$$

The calcium carbonate will dissolve if the H^+ concentration is high (acidic) or if the CO_3^{-2} concentration is low. In the photic zone, the process of CO_2 removal via photosynthesis can help maintain a high pH so that both biotic and abiotic calcification can occur.

Nutrients

Twenty-one *essential elements* are required by all plants (DeBoer, 1981). Of these, four are *primary elements* necessary for plant growth, namely, oxygen, carbon, nitrogen, and phosphorus, which have concentrations of 857,000, 28, 0.5, 0.07 mg L^{-1}, respectively, in seawater (Davis, 1986). In addition, silicon (3 mg L^{-1} in seawater) is a critical element for diatom culture, because it forms SiO_2, which is part of its cell walls. In contrast to flowering plants, which make their own vitamins, most seaweeds are *auxotropic*, requiring some organic compounds. For example, in culturing many seaweeds such as the red alga *Eucheuma*, three vitamins are recommended in trace amounts: B_{12} (cyanocobalamin), thiamine, and biotin (Dawes and Koch, 1991).

In addition to being primary elements, nitrogen, and phosphorus can be *limiting* nutrients to marine plants. The low concentrations of nitrogen, phosphorus, and silicon in oceanic waters will influence phytoplankton growth. In coastal waters, due to terrestrial runoff and mixing with benthic sediments, higher concentrations of these nutrients may be present. In addition to standard laboratory analyses (Strickland and Parsons, 1968; Parsons et al., 1989), field kits are available to measure the levels of these nutrients in seawater.

Nitrogenous compounds commonly found in seawater include *nitrate* (NO_3; oceanic: 0.1 to 43, neritic: 1 to 600 μg atoms L^{-1}), *nitrite* (NO_2; oceanic: 0.01 to 3.5; neritic: 0.1 to 50 μg atoms L^{-1}), and *ammonia* (NH_3; oceanic: 0.35 to 3.5, neritic: 5 to 50 μg atoms L^{-1}). Sources of these compounds can be found in the nitrogen cycle (Fig. 2-16), with all components (fixation, denitrification, ammonification, nitrification) occurring in the marine environment (Milne, 1995). Studies on the utilization of nitrogen indicate that macroalgae and microalgae are efficient in taking up both ammonia and nitrate. Uptake and use of nitrate and ammonia are briefly discussed in Chap. 4. Determination of nitrate can be done by cadmium reduction of nitrate to nitrite and diazotizing, which is accomplished by coupling with a sulfanilamide to produce a highly colored azo dye. Nitrite is first determined without cadmium reduction and then subtracted from the final analysis. A more direct technique is the ultraviolet spectrophotometric screening method (A.P.H.A., 1989). Ammonium can be extracted from the sediment using 2 M KCl (Bremmner, 1965) and then can be oxidized to nitrite using Nessler's reagent (mercuric iodide), which is then diazotized with sulfanilamide. Alternatively, ammonium can be colormetrically analyzed (Kempers and Zweers, 1986).

Phosphorus is present in seawater as $H_2PO_4^-$ and HPO_4^{-2}, with a mean concentration of 0.07 ppm; even so, levels range from 0 to 0.003 mg atoms L^{-1} at the surface to 0.09 mg atoms L^{-1} in deeper waters. The phosphorus cycle (Fig. 2-17) can be separated into addition, removal, and uptake (concentration) phases. Uptake and use of inorganic phosphate by marine plants are briefly discussed in Chap. 4. There are temporal (seasonal, diurnal) changes in phosphate levels in the oceans with highest concentrations occurring during the winter in temperate waters and at night when plants are not photosynthesizing. About 14 million metric tons of phosphorus are carried into the sea through erosion and land runoff annually (Davis, 1986). The most commonly measured form is orthophosphate because plants typically absorb and incorporate this form, making phosphorus available to other organisms. In addition to dissolved phosphate

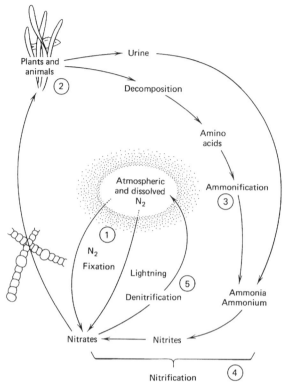

Figure 2-16.   The nitrogen cycle. The five components of the nitrogen cycle are (1) ammonia production via nitrogen fixation and conversion to nitrates (nitrification), (2) incorporation of ammonia and nitrates by plants and animals, (3) release of nitrogenous compounds through decomposition and excretion and conversion to the ammonia ion and ammonium, (4) oxidation of ammonia to nitrites and nitrates, and (5) denitrification and release of molecular nitrogen. Whereas nitrogen fixation and use of nitrates and ammonium require energy, the remaining steps (3 to 5) release energy.

in the water column and in pore water of substrata, bound phosphate of sediments can be extracted using a sodium bicarbonate solution (Olsen and Sommers, 1982). The dissolved orthophosphate then can be reacted with an acidified molybdate solution to form a phosphomolybdate-heteropoly acid, which is reduced to an intense blue-colored phosphomolybdate and measured using a spectrophotometer.

Silicon is critical to cell wall formation in diatoms, which form the single most important group of primary producers in oceanic waters. There are wide ranges in concentration of the element, from 0 to 0.5 mg L^{-1} in clear oceanic waters to 8.4 mg L^{-1} in neritic waters. The source in coastal waters consists of clays, where undissolved SiO_2 and dissolved $Si(OH)_4$ exist. The silicon cycle can be divided into active and inactive pools (Fig. 2-18) with the dissolved

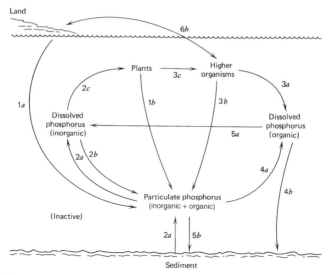

Figure 2-17.   The phosphorus cycle. Three major components make up the phosphorus cycle: (a) addition, (b) removal, and (c) uptake. Phosphorus is obtained from terrestrial sediments (1a) or organisms (6b), either organic (4a) or inorganic (2a). Phosphorus is converted from organic to inorganic material (5a) and taken up by plants (2c) that are consumed by animals (3c). Decomposition of plants (1b) and animals (3a, 3b) releases dissolved and particulate phosphorus that can return to the sediment via precipitation (4b, 5b). Removal of phosphorus from the sea is done by feeding of terrestrial animals (6b).

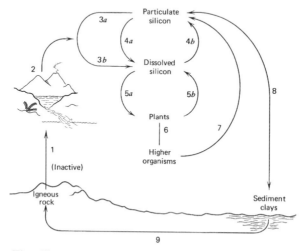

Figure 2-18.   The silicon cycle. Silicon is released from terrestrial and marine sediments (2, 8) and removed via burial (9) and geological events (1). Particulate (3a) and dissolved (3b) silicon in seawater are converted back (4a) and forth (4b). Plants use (5a) or release (5b) dissolved silicon. Animals obtain silicon through grazing (6) and release it via decomposition (7).

compound obtained from weathering of igneous rock and erosion of clays. Measurement of soluble forms of silicon is carried out by reaction of orthosilicate acid (H_4SiO_4) with molybdate to form the 1 : 12 silicomolybdate acid molecule. The absorbance can be measured with a spectrophotometer of the yellow product. Absorbance of the blue product after reduction with Metol (*p*-methyl amino pheno sulfate) also can be measured and both techniques are described.

CHAPTER 3

Biotic Factors

The chapter gives an overview of marine plant relationships with their environment, which is called *ecology*. The overview is a requirement to understanding the biology of marine plants because they do not live as isolated units. Consideration will be given to basic tenets of marine ecology including levels of organization (populations to ecosystems), processes of community development (succession, energy transfer), strategies (evolutionary, plant responses), biological interactions (forms of symbiosis, competition, predation), and growth (rates and responses).

MARINE ECOLOGY

Ecological Units

Photosynthetic marine organisms range from unicellular eukaryotic algae and Cyanobacteria to macrophytic seaweeds and flowering plants. Marine plants can be grouped into ecological units ranging from populations, to communities, to ecosystems. Such *populations* consist of a group of individuals of a species that occupies a specific area and can be evaluated in terms of age structure, survival, germination, and death. Chapman (1979) suggested that when studying populations of a species, three types of parameters should be measured: size or density (number of individuals per area), age distribution (juvenile vs. mature individuals), and spatial patterns of individuals. One example of a survivorship study with the giant kelp, *Macrocystis pyrifera*, showed that it had a low survival rate, with only 5 out of 387 germlings surviving after nine months (Fig. 3-1; Rosenthal et al., 1974). Usually, population studies include an evaluation of physiological and life history data, and extensive field measurements.

Size class distribution studies of standing seaweed populations can help determine the longevity patterns, as described by Knight and Parke (1931). All the individuals in an *annual* population will have a life history of less than one year and an example might be the North Atlantic brown seaweed *Leathesia differmis*. Species that are *pseudoperennial* include plants that pass a portion of each year in a reduced perennating form such as the tropical brown seaweed *Sargassum filipendula*. A *biannual* develops over a two-year period, such as

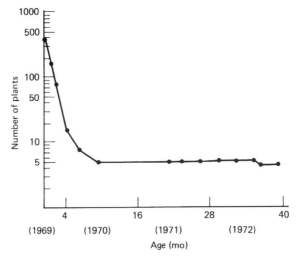

Figure 3-1. Survivorship curve for young *Macrocystis pyrifera* sporophytes. Based on a 5.7-year study, it was found that over 95% of all sporophytes never reached one year in age, beyond which survival was 100% (after Rosenthal et al., 1974).

Fucus distichus, and a *perennial* is a species that will survive more than two years, such as *M. pyrifera*. An example of a size class study can be seen in a two-year study of populations of the red seaweed *Eucheuma isiforme* (Dawes et al., 1974). In that study, a deep water (Anclote Key) and a shallow Florida Keys (Bahia Honda) population showed distinct annual growth patterns, in contrast to another Florida Keys (Molassas Key) population that was a perennial.

Communities are populations of species that coexist and interact with one another and live in a particular area. Thus, communities are biotic in nature and include species interactions. Community structure can be studied in terms of the dominant species, numerical relationships, floristic features, and their specific abiotic characteristics. The primary requirement of any quantitative study of a community is an adequate sampling procedure to ensure representation of all populations within the community. Some common sampling procedures (transect, quadrat) are presented in App. A.

Dayton (1975a) described kelp-dominated communities that occur at different depths on Amchitka Island in Alaska (Fig. 3-2). He found a three-layer canopy formed by the large brown seaweeds. The kelp *Alaria fistulosa* (reaching 22 m in height) formed the uppermost layer. Below this was a second layer formed by species of *Laminaria*, then a subcanopy of *Agarum clathratum* (as *A. cribosum*), and, finally, small red and green algae that formed a shade-tolerant turf community directly on the substrata.

A single species (beds of the giant kelp *Macrocystis pyrifera*) or group of species (such as a mangrove community) can dominate and, thus, characterize a community. However, it may be difficult to determine the boundaries due to the

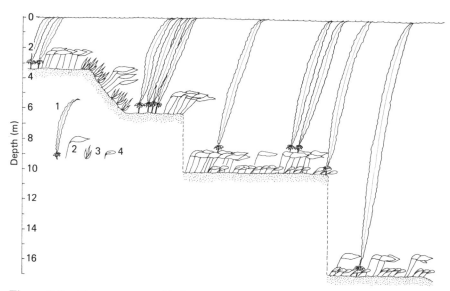

Figure 3-2.   Kelp canopy profile formed at three different depths off Amchitka Island, Alaska. *Alaria fistulosa* (1) is the largest species and forms the primary canopy, *Laminaria* spp. (2) a secondary layer, and *L. longipes* and *Agarum clathratum* (3) the third canopy. A turf community of red and green algae (4) forms a shade-tolerant benthic layer (after Dayton, 1975a).

patchy nature of the dominant marine plants. Thus, many studies of communities employ statistical (e.g., Sokal and Rohlf, 1981) analyses of species diversity and determine standing stock or biomass, or relative dominance using quadrats or transect sampling procedures (App. A). For example, Hoffman and Dawes (1997), using statistical programs to analyze species distribution, reported that salterns were part of salt marshes (Chap. 9) and mangrove communities (Chap. 10) and were not distinct from them. The significance of the study was the conclusion that salterns should be protected as part of a tidal community continuum and not be destroyed or altered.

Phytosociological methods, using floristic classifications, are effective means of classifying vegetational units, such as intertidal seaweed populations. Phytosociology identifies communities using the dominant types of vegetation (species and morphology). The method primarily incorporates two types of numerical measurement: the cover-abundance and sociability of each species (see App. A).

Each species within a community has a specific habitat or *niche*, which includes all environmental and physiological aspects of the organism's interaction with the environment. Grinnell (1917) described the ecological or environmental niche as the ultimate distributional unit of a species or subspecies. Three types of niches have been described by Vandermeer (1972): *fundamental* (ideal one without competition), *partial* (with partial competition), and *realized* (actual site and extent of occupation). All three types can be found in the marine environment. A fundamental niche might be a bed of turtle grass (*Thalassia testudinum*) within which there are no other species. A partial niche could be the edge of the bed where shoal grass (*Halodule wrightii*) interacts and a realized niche would be a stable mixture of the two species.

Ecosystems are the largest ecological unit and include all communities in a region along with associated abiotic characteristics. Tansley first introduced this term in 1930, emphasizing the interaction between organisms and the abiotic factors of an area. Avoiding extraordinary events (e.g., meteorites, earthquakes, volcanic activity), an ecosystem is thought to be a self-contained unit. The organisms within such a system do not depend, for the most part, on events outside the ecosystem. It is within the level of an ecosystem that energy flow (fixed carbon and other organic compounds) and various mineral cycles (nitrogen, phosphorus, silicon, sulfur) occur.

Tampa Bay, Florida, is an example of an extensive estuarine ecosystem. The bay includes salt marsh and mangrove forests, riverine, seagrass, and hard and soft bottom communities. An aerial photo of Cockroach Bay (Fig. 3-3), a subsidary of Tampa Bay, shows some of the dominant communities. The ecosystem concept, developed in terrestrial ecology, is not as useful in open water systems but quite functional in studies of coral reefs and coastal communities. In coastal and estuarine ecosystems, there is a close proximity between terrestrial abiotic

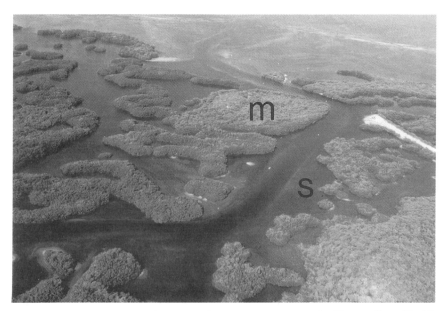

Figure 3-3. Aerial view of a mangal-seagrass ecosystem in Tampa Bay, Florida. Numerous overwash mangals (m) and seagrass beds (s) are separated by tidal channels.

and biological factors (e.g., land runoff, soil nutrients) that can influence them. For example, it is difficult to completely separate the effects of terrestrial and marine influences on the Tampa Bay ecosystem. The surrounding watershed (i.e., rivers and wetlands) and terrestrial runoff all influence Tampa Bay. Further, the boundaries of marine systems are also difficult to determine because of the possibility of import and export. The submerged seagrass and coastal tidal marsh communities can export a major part of their biomass to the continental shelf.

Food Webs and Chains

A series of *trophic* or energy-transfer levels can be found within an ecosystem such as Cobscott Bay (Gulf of Maine) (Fig. 3-4), which is explained in Chap. 9. The development of simulation models to better understand energy flow has become more common, with examples available for most marine plant systems. The energy-transfer levels involve a set of feeding relationships or *food chains* if linear (single direction and set of levels) or *food webs* if complex (organisms can interact at different levels). *Autotrophs* or *primary producers* are the self-supporting, photosynthetic (plants, bacteria) and chemosynthetic (bacteria) organisms producing the biological energy of food webs and chains (e.g., *Spartina*, algae; Fig. 3-4). Photosynthetic autotrophs convert the raw energy of sunlight, carbon dioxide, and water in photosynthesis into organic compounds. In a similar way, chemo-autotrophs use the energy of inorganic chemical bonds to produce organic compounds in chemosynthesis. The autotrophs must interact with and also show adaptations for organisms that "prey" on them. *Heterotrophs*, which consist of both *primary* (herbivores) and *secondary* (carnivores) forms, depend on the autotrophic production of organic matter. Primary consumers range from phagotropic unicellular dinoflagellates that feed on phytoplankton, to large sea urchins and manatees that graze on seagrasses, to insects that eat mangrove leaves. Secondary consumers include carnivores that feed on the herbivores, such as sea otters that eat sea urchins and many fish, to birds that eat insects.

Two other groups of heterotrophic organisms play a significant role in the development of tropic levels. *Decomposers* (saprophytic fungi and bacteria) liberate nutrients by breaking down organic compounds to simpler forms that are then available to the autotrophs and detritivores. *Detritivores*, including species of starfish, mullet, and snails, feed on decomposing plant or animal material. In estuaries, these two trophic groups are of great importance in making organic material available for consumers and nutrients for the plants.

Biomass and Productivity

Two important measurable aspects of marine plant communities are biomass and productivity. *Biomass* is the "standing stock" or mass of a species or community and is usually expressed in terms of weight (fresh or dry) per unit area

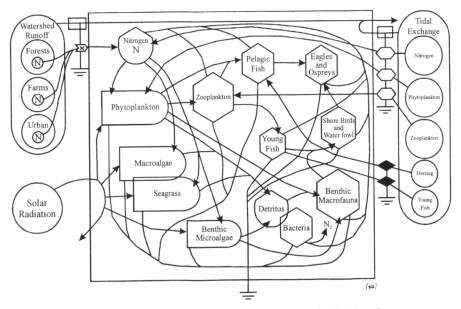

Figure 3-4. An energy-flow model of Cobscook Bay, Gulf of Maine. Input, conversions, and outputs are modeled in order to determine the significance of each component the habitat's food web (after Kidman, 1995; with permission from Dr. D. Campbell, U.S. Environmental Pollution Agency, Atlanta).

(gm^{-2}). For example, the biomass of a subtidal, temperate kelp community is about 400 to 1600 g dry wt m^{-2} (Mann and Chapman, 1975), whereas that of a seagrass meadow ranges from 100 to 1200 g dwt m^{-2}. It is useful to express biomass in terms of *ash-free* dry weight. Ash-free dry weights are obtained by oxidizing the biomass in a muffle oven (4 h at 500°C), which does not remove the calcium carbonate, but does consume the organic matter. The ash weight is then subtracted from the total dry weight to obtain the weight of organic material. *Productivity* is the rate of production, usually expressed in terms of mg of oxygen or carbon fixed per unit time. The net primary productivity of a tropical rain forest (500 to 1800 g C m^{-2} y^{-1}) might be then be compared with a subtidal bed of the brown seaweed *Sargassum* sp. (2500 g C m^{-2} y^{-1}) or the seagrass *Thalassia testudinum* (330 g C m^{-2} y^{-1}).

The morphology of seaweeds has been used to predict the productivity (Table 3-1) of coral reef macroalgae (Littler et al., 1983). Thus, a *functional-form* model (Litter and Littler, 1980) has been used to evaluate the photosynthetic, nutrient uptake and structural aspects of seaweeds in relation to grazer protection. External and internal plant structure will influence photosynthetic efficiency, ability to take up nutrients, and resistance to predation. For example, a thick, tough fucoid such as *Fucus* will be less likely to be eaten. But, due to its thick morphology *Fucus* will show a lower photosynthetic and nutrient uptake

TABLE 3-1. **Macroalgal Morphologies and Functional Relationships**

Morphology	Construction	Examples
Blades	Thin, 1-few cells thick, soft	*Ulva, Dictyota*
Filaments	Filamentous, few cells, thick, delicate, soft	*Ceramium, Cladophora*
Coarsely branched	Erect, many cells thick, fleshy to wiry	*Gracilaria, Codium, Gelidium*
Thick or leathery	Branches or blades thick, corticated, rubbery	*Laminaria, Fucus*
Flexible, calcareous	Calcified but jointed, hard surface	*Corallina, Halimeda*
Crustose	Calcified or not, encrusting, prostrate	*Petrocelis, Clathromorphum*

Note: Arrow indicates highest productivity.
Source: Modified from Littler et al. (1983).

efficiency than a thin bladed seaweed like *Ulva*. Similar features have been used to evaluate the successional stage of plants (Table 3-2), where *Fucus* can be considered to be a late-successional form and *Ulva* would be an opportunistic stage in the development of a community.

ECOLOGICAL STRATEGIES

Succession

Most communities are not static, but constantly changing. Thus, *succession*, or predictable shifts in the dominant species of a community, occurs over time. The concept of succession was developed in terrestrial ecology and identifies a series of developmental stages, which include *pioneer* (early stage, opportunistic), *intermediate*, and *stable* (mature, late stage) forms that make up the final *climax* (mature) community. Three types of successional models have been suggested (Connell and Slatyer, 1977). The *facilitation* model proposes that pioneer species modify their habitat so that it is more suitable for latter stages. The *tolerance* model suggests that later stages simply overgrow or replace earlier species, and the *inhibition* model describes cases where early stages prevent establishment of later stages. All three types of succession can occur, depending on the abiotic conditions within a community.

Early-stage or pioneering plants are usually annuals, which show rapid growth and high rates of reproduction. Pioneers reestablish rapidly and invest little in plant structure having simple morphologies and being thin or delicate forms. Pioneers may be considered the "weeds" of the marine environment with species of the green seaweed *Ulva* being excellent examples. At the other end of the successional continuum are the climax or mature species. Stable, predictable environmental conditions are a prerequisite for the establishment of

TABLE 3-2. Characteristics, Costs, and Benefits of Pioneer and Late
Successional Seaweeds

Early-State (Pioneer) Species	Late-Stage (Mature) Species
Characteristics	
1. Not seasonally controlled reproduction	Seasonally controlled reproduction
2. Colonize disturbed areas	Replace earlier stages
3. Annuals, short life histories	Complex, long life histories
4. Simple construction, high productivity and surface area to volume ratio	Complex construction, low productivity and surface area to volume ratio
6. Escape grazing via rapid growth	Slow growth, chemical and structural grazing defenses
Costs	
1. High mortality, limited time for reproduction	Slow growth, reproduction is delayed
2. Outcompeted for light, crowded out for substrata	Reduced nutrient uptake due to low SA:V ratio
3. Easily damaged and eaten	Energy diverted to structure
4. Sensitive to abiotic factors: waves desiccation	Require specialized niches
Benefits	
1. Highly productive	Long-term reproductive phase
2. Rapid invasion of new areas	Competitive for space, light
3. Rapid uptake of nutrients	Resistant to grazing
4. Not specialized, colonize different habitats	Complex life histories allow survival in various habitats

Source: Modified from Littler and Littler (1980).

climax species, which are usually composed of perennials. A common feature
of climax communities is that this stage will support higher species diversity,
with correspondingly fewer individuals per species.

   Disturbances result in community perturbation (Rykiel, 1985) and consist of
two types. *Disasters* occur with sufficient frequency for a single organism to
experience them (e.g., storms) and *catastrophes* are only experienced by occasional generations. What is a disaster for a perennial may be a catastrophe for an
annual. Because of disturbances, the successional progression can be disrupted
or the climax community damaged and the process restarts at some earlier stage.
The three successional models described before can be found in recoveries of
marine plant communities after a disturbance. A inhibition model where pioneer plants prevent succession was demonstrated by Sousa (1979). The dominant, intertidal red alga *Gigartina canaliculata* required three years to recover in

southern California, but it could be prevented from establishing by the pioneer green alga *Ulva lactuca*. Another inhibitional example was described by Kennelly (1987), who found that a dictyotalean algal turf prevented the establishment of a "climax" kelp *Ecklonia* when the latter was removed. A facilitation model, where pioneers modify the substratum making the environment more suitable, was demonstrated by Williams (1990). The green, calcified tropical algae *Halimeda* and *Udotea* so modified the soft substratum in St. Croix subtidal sand flats that first *Syringodium filiforme* and then *Thalassia testudinum* could establish. The tolerance model, in which later-stage species colonize and grow over the earlier stages, is best seen in replacement studies. Mangrove seedlings can establish in salt marsh communities and shade out the original marsh. Also the giant kelp can reestablish and shade out the other benthic plants, without any previous stages in succession.

Succession in Marine Communities

Studies of physical disturbance suggest that succession does occur in seaweed, salt marsh, mangrove, seagrass, and coral reef communities (see Chaps. 8 to 12). In an early study of seaweed succession, Doty (1957) followed algal recolonization after a lava flow in Hawaii. A series of successional stages of microalgae and macroalgae occurred on the sterile, newly formed rock, with a climax community, similar to that found on 100-year-old laval flows, being established within 10 years.

Another example of seaweed succession was reported for Torch Bay, Alaska (Duggins, 1980). One year after the removal of the kelp forest and continuous removal of grazers, a mixed canopy of kelps (i.e., *Nereocystis luetkeana* and *Alaria fistulosa*) formed, and an understory of *Costaria costata* and *Laminaria dentigera*, developed. During the second and third years, continuous stands of *L. setchellii* and *L. groenlandica* developed and the community had returned to its original composition. The urchin grazing was suggested as a major factor inhibiting the natural succession of the kelps.

The regular zonation of flowering plants within intertidal mangrove and salt marsh communities has also been considered to be an example of succession from marine to terrestrial (or perhaps the other direction?) habitats (Chaps. 9 and 10), that is, because the community, with its characteristic species and zonation, recovers after a disturbance. Succession in seagrass communities has been documented in Australia (Shepherd and Womersley, 1981; Birch and Birch, 1984) and the Caribbean (Williams, 1990). For example, in the U.S. Virgin Islands, rhizophytic green algae (*Halimeda, Penicillus, Udotea, Caulerpa*) are early colonizers, stabilizing and adding organic matter to the sediments. The seagrasses *Syringodium filiforme* and *Thalassia testudinum* then grew into the area.

Coral reef communities also show algal succession mediated by herbivorous fish and urchins (Berner, 1990; Glynn, 1990; Hixon and Brostoff, 1996; Chap. 12). Thus, if herbivorous fish were excluded (by caging), algal succession on a Hawaiian coral reef crest progresses through three stages within a year. An early dominance of small green and brown filamentous algae (*Enteromorpha, Ecotocarpus*) initially occurred followed by thin and finely branched red algae (*Centroceras, Taenioma*), and then foliose and coarsely branched algae dominated. If damselfish were allowed to graze within the cages, the process of algal succession was slowed, and outside the cages, grazing was so intense that all erect algae were removed.

Modeling Marine Plant Strategies

A number of models have been proposed to clarify the responses of plants to their environment, including succession, the r and K continuum (Pianka, 1970), the CSR model (Grime, 1977), and the habitat template (Taylor et al., 1990). Such models can help to explain the sequence and complexity of a marine plant community, but they may oversimplify the patterns and neglect some of the biotic and abiotic factors. The r-K *model* is taken from a graphic presentation of plant succession, where the r phase is a rapidly rising population growth and their high rate of utilizing the habitat's resources. The K phase occurs when population growth has ceased and all available materials are being used. The model thus emphasizes features of early-stage (*r*-strategists or pioneers) and late-stage (K-strategists or mature) plants (Table 3-1). The former do not use up environmental supplies, whereas the latter may reach the carrying capacity or utilize all available supplies. As noted in Table 3-1, perennial algae can be regarded as K-strategists, as they make maximal use of environmental resources and incorporate a large portion of their carbon production into their plant body.

The *CSR model* (Fig. 3-5; Grime, 1977), assumes that plants have evolved in terms of *competition* (C), *stress* (S), and disturbance (R = ruderal). Stressful conditions such as high salinity, suboptimal temperatures, or shading can restrict the production of maximum biomass. Stressful conditions can be viewed as "external constraints, limiting resource acquisition, growth or reproduction of organisms" (Grime, 1977). Others consider stress as the "gradient between ideal conditions and the ultimate limits of survival" (Grigg and Dollar, 1990). Stress-adapted plants tend to be perennial. In a review, Davison and Pearson (1996) considered two types of stresses; they also emphasized that Grime's concept is too limited. *Limitation stress* can reduce growth due to inadequate resources such as light and nutrients. *Disruptive stress* causes cellular damage and requires allocation of resources to prevent or repair damage (e.g., freezing and desiccation).

Plants adapted to disturbances (i.e., storms, wave energy, herbivory) often experience partial or total loss of biomass. Grime (1977) identifies disturbance-tolerant plants as ruderal or opportunistic (weedy) species that are similar to the r-strategists. Examples of seaweeds would include the green algae *Ulva* or *Enteromorpha*, which occur after oil spills and sewage outfalls. However, there are also stress-tolerant perennials such as psammnophytic seaweeds like *Mastocarpus stellatus* and *Fucus vesiculosus*, which tolerate months of sand cover in New England coasts. Where there is low stress and disturbance, plants may have to compete more intensely in order to survive. Such competitive plants would be

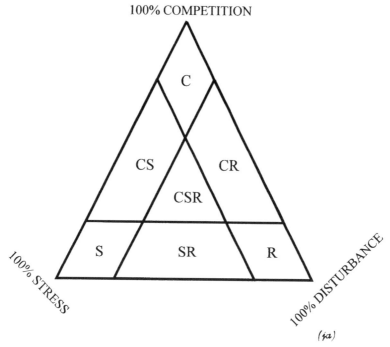

Figure 3-5. The triangular competition-stress-ruderal (CSR) model of plant strategies. The model uses three strategies of plants, namely, competition (C), stress tolerance (S), and disturbance tolerance (ruderal, R). Combinations of strategies are also shown: stress-tolerant competitor (CS), ruderal competitor (CR), stress-tolerant ruderal (SR), and (CRS) competitive stress-tolerant ruderal (after Grime, 1977; with permission of University of Chicago Press).

comparable to K-strategists. One example of a community exhibiting a high level of competition for space and light is a kelp-dominated forest. There are no "pure" C, S, or R plants, but rather species having a mix of the three features.

The *habitat template* model (Taylor et al., 1990) utilizes r-K selection to identify early- and late-stage, resource-using species and imposes on this concept resource impoverishment (I) or carrying capacity of the habitat (Fig. 3-6). The r and K features of a species (Table 3-1) can be moved along the *I*, or *impoverishment*, gradient from high to low carrying capacity of the environment resulting in four "extreme" types of strategies (rather than three of the CSR model). Unlike the CSR model, competition intensity can remain high, even in stressful or disturbed communities like xerophytic flowering plants in deserts. An example in the marine environment may be seagrass beds that are subjected to high water motion and growing in a low-nutrient environment as found in depressions in coral reef flats. As impoverishment increases (lower carrying capacity, K, right side of Fig. 3-6), plant biomass decreases. As noted in Chap. 10, mangrove zonation is probably better explained with a template rather than a triangular CRS model.

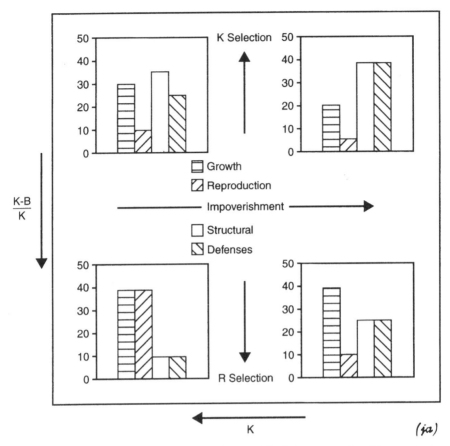

Figure 3-6. The four-way Habitat Template model of plant strategies. Four strategies of plants (r, K, r-l, K-l) are presented in terms of relative allocation of energy for growth, reproduction, structural development, and defense against grazing. The x axis shows the relative carrying capacity of the habitat (K), and the y axis shows the mean annual distance below K that the vegetation is maintained after some stress (after Taylor et al., 1990; with permission, Oikos).

BIOLOGICAL INTERACTIONS

Diverse biotic interactions may occur within the subtidal and intertidal zones, including plant-plant, plant-animal, and plant-microbe. Symbiotic relationships (i.e., parasitism, mutualism, epiphytism, mimicry, commensalism) occur throughout marine plant communities just as the more evident biological interactions of predation and competition. Plant-animal interactions are probably the best-known examples (John et al., 1992), perhaps due to the dominance of zoological studies of intertidal communities.

Symbiosis

The range of organismic symbiotic relationships range from parasitism to mutualism (Goff, 1983). *Parasitic* relationships, where one organism obtains nutrients at the detriment of another, are particularly common within red algal species (Chap. 6), whereas few occur with brown or green algae. Examples of *mutualistic* relationships, where both organisms benefit, are abundant in the marine environment. Examples of marine plants include microalgae (zooxanthellae in corals and foraminifera, algal and fungal components of marine lichens), macroalgae (endophytic fungi in fucoids, endosymbiotic bacteria in coenocytic green algae), and flowering plants (ants that live on and protect mangroves). *Commensalistic* relationships, which occur when one organism benefits without affecting the other, are uncommon between marine plants, unless *mimicry* (the imitation of another) and *epiphytism* (growing on another plant) are included. An example of marine plant mimicry is the red seaweed *Eucheuma arnoldii*, which grows among and resembles branches of hermatypic corals.

The wide variety of epiphytic algae is best compared with the abundance of epiphytic vascular and nonvascular plants in tropical rain forests where there is also "standing room only" (Ducker and Knox, 1984). Algal epiphytes are very common (see also "Biofouling") and demonstrate a solution to space problems within intertidal and subtidal zones. Because they are so common, epiphytism is probably the most extensively studied form of marine plant commensalism (Ducker and Knox, 1984). Linskens (1963) differentiates epiphytes based on spatial and functional relationships with their host (*basiphyte*). Structural types of epiphytes include *holoepiphytes*, which are attached to the outer layer of the basiphyte (e.g., seagrass epiphytes) and *amphiepihytes*, which penetrate the tissues (e.g., *Polysiphonia lanosa* on *Ascophyllum nodosum*). The preceding example is also called a *hemiparasite*. Functionally, most epiphytes show an *indifferent* type of attachment in which neither partner is affected. In a few cases, a *mutualistic* relationship occurs where the epiphyte and basiphyte exchange nutrients (Ducker and Knox, 1984). Epiphytes occur on seagrasses, macroalgae, mangroves, and salt marsh plants.

Although the majority of algal epiphytes have no specific basiphyte (e.g., ectocarpoid species), there are examples where specificity occurs between epiphyte and host. Some examples include the relationship between the Southern Hemisphere brown algae *Notheia anomala* on *Hormosira banksia* and the Northern Hemisphere brown algae *Elachistia furcicola* on *A. nodosum*. Specificity of algal epiphytes also extends to seagrasses with the red alga *Smithora naiadum* occurring only on the seagrasses *Zostera marina* and *Phyllospadix scouteri* (Harlin, 1980). Further, *S. naiadum* and its seagrass basiphytes show active movement of ^{32}P and photosynthetic products (carbohydrates).

Many examples of epiphytism appear to be commensalistic, but with closer examination, it is evident that the basiphyte can suffer. Thus, epiphytic algae growing on healthy portions of seagrasses or macroalgae can reduce host photosynthesis through shading and by nutrient removal. In such cases, epiphytism can be damaging to the host and thus be classified under parasitism. However, basiphytes can "defend" against epiphytism (Ducker and Knox, 1984) as described for macroalgal epiphytes in Chap. 8. Epiphytes of salt marsh plants (Chap. 9) and mangroves (Chap. 10) have been less well studied when compared to studies on seagrass epiphytes (Chap. 12; Borowitzka and Lethbridge, 1989).

Competition

An active demand by two or more organisms for a resource that is potentially limiting is called *competition*. Denley and Dayton (1985) describe two types: exploitative and interference competition. In *exploitative competition*, organisms compete for a limiting resource, without being directly antagonistic to one another. One example is a filamentous algal epiphyte that is more effective in absorbing nutrients (i.e., from the water) than a larger, fleshy algae. *Interference competition* may not relate directly to a limiting resource. Thus, an alga may overgrow another or release toxic chemicals; the latter example is a form of *allelopathy*. Competition occurs in all marine plant communities.

The zonation of intertidal algae (hard-substrate communities) is usually controlled by a variety of abiotic and biotic factors. Upward extension (*dilation*) is usually limited by abiotic factors such as temperature or desiccation, whereas biotic factors, including competition and predation, limit expansion into the lower zone. Within the subtidal zone, competition for space and light control will cause patchy distributions of species. For example, the surf grass *Phyllospadix torreyi* dominates the lower part of the southern California intertidal zone and desiccation prevents its upward expansion (Stewart, 1989). By contrast, the more desiccation-tolerant calcarious turf, dominated by the red alga *Corallina pinnatifolia*, occurs above the seagrass. If the *P. torreyi* is removed, the algal turf will extend down into the bare zone until it is outcompeted by the more slowly growing seagrass rhizomes. The preceding interaction is an example of interference competition. A second example of interference competition has been shown in Caribbean subtidal seagrass meadows (Williams, 1987). That is, turtle grass (*Thalassia testudinum*) will outcompete manatee grass (*Syringodium filiforme*) through shading and efficiency of nutrient removal from the sediment. A third example occurs with the exotic green alga *Caulerpa taxifolia*, which has invaded the Mediterranean and is outcompeting the seagrass *Posidonia oceanica* (Villele and Verlaque, 1995).

Soft-sediment intertidal communities (mangroves, salt marshes) show in inverse control in zonation of plants when compared with hard-substrate intertidal macroalgal communities, where biotic factors (competition, predation) determine lower boundaries and abiotic factors (desiccation, salinity) determine the upper boundaries. In soft-sediment habitats, abiotic factors such as wave activity and degree of flooding limit the downward expansion of mangroves and salt marsh plants, while biotic factors like competition with glycophytes limit their upward expansion (Ball, 1980; Bertness and Ellison, 1987). The difference in the control of zonation may reflect the limitations of marine angiosperms to a submerged existence as well as their ability to tolerate sea water.

In southern California kelp beds (*Macrocystis pyrifera*), competition for light occurs with understory algae adapted to lower light levels. After canopy removal, the understory communities can change. The removal of a two-layered kelp canopy (the upper being the giant kelp and the lower being the elkhorn kelp *Pterygophora californica*) resulted in rapid recruitment of new kelp plants, with only moderate changes in the understory red algal flora (Reed, 1990). The effect of shading demonstrates exploitative competition where juvenile kelp sporophytes could only develop if the overstory kelps were removed.

Predation

Grazing patterns have been described for mesoherbivors, suspension feeders, molluscs, crustaceans, sea urchins, and herbivorous fishes (John and Lawson, 1990; John et al., 1992) as well as turtles and manatees. Although the assumption is sometimes made that coevolution has occurred between plants and their herbivores (John et al., 1992), it is not always true (Hay and Fenical, 1988). By contrast, the assumption that most plant communities have herbivores does seem to be true. Carpenter (1986) describes three types of grazers: fish having a large foraging range, but occurring in low density; urchins with moderate foraging range and found in higher densities; and mesograzers such as copepods and amphipods that exhibit small foraging ranges and occur in high densities. Mesograzers appear to play a role throughout the oceans affecting algal succession, survival, overgrazing, and detrital feeding. Even so, this group of herbivores is poorly understood (Brawley, 1992). Warm, tropical waters support a variety of herbivorous fish, particularly around coral reefs (Horn, 1989), so that seaweeds and seagrasses must have adaptations to defend against predation (see Chap. 12). In contrast, there are fewer species of herbivorous fish in temperate and cold water habitats.

Shifts in predators (secondary consumers) of herbivores can result in "chain" reactions if a higher consumer is removed from a community. For example, overfishing of barracuda, a secondary consumer in tropical regions, can result in overgrazing of algae and crustose corals by herbivorous fish. Bird predators, including gulls (*Larus glaucescens*) and oystercatchers (*Haematopus bachmani*), when in large populations, can remove significant populations (45 to 59%) of the purple sea urchin *Strongylocentrotus purpuratus*. Removal of the purple urchin from the lower intertidal zone of Tatoosh Island, Washington State, caused a sixfold increase in algal diversity and a 24-fold increase in algal cover (Wootton, 1995).

Grazing can differentially damage or remove early or late successional species (Farrell, 1991). In the *inhibition model* of succession (Connell and Slatyer, 1977), pioneer species can be heavily damaged by grazers, resulting in enhanced succession. An example would be grazers that preferentially feed on early successional species. For example, when the periwinkle *Littorina littorea* was added to an upper New England tide pool, it removed the green alga

Enteromorpha sp. but did not affect larger individuals of the red alga *Chondrus crispus*, so that the red alga became the tide pool dominant (Lubchenco, 1978). In the *facilitation model*, the rate of succession during the early stages can be slowed due to grazing. For example, removal of the purple urchin (*Strongylocentrotus*) *purpuratus* from an intertidal community causes a mixture of algal species that could not develop otherwise. In the *tolerance model*, the effect of grazing on or competition by early-stage plants may not play a significant role in the rate of succession of later stages.

Extreme grazing of seaweeds (Elner and Vadas, 1990) and seagrasses (Camp et al., 1973) is not that common. In most subtidal communities, herbivores coexist with the plants. Although moderate grazing pressure will usually support higher species diversity, it can also cause algal overgrowth. Intense grazing will reduce diversity and cause a domination by algal crusts that are not easily eaten (Hackney et al., 1989). Many sessile organisms, such as corals, are dependent on grazers to prevent overgrowth by filamentous and fleshy algae, and this is very evident in coral reefs (Chap. 12).

The effects of grazing also can be difficult to interpret, as shown in a review of the recurring devastation of the kelp-macroalgal beds by the sea urchin *Strongylocentrotus droebachiensis* in Nova Scotia (Elner and Vadas, 1990). Elner and Vadas found that one cause of urchin overgrazing was due to shifts from high (urchin decline) to low (urchin expansion) water temperature over a 15-to-29-year cycle. Previous studies had proposed the overharvesting of the lobster, a predator of the urchins, resulting in an increase in grazing pressure on the macroalgae. Another reinterpretation of overgrazing of macroalgae by urchins helped explain overgrazing of the giant kelp *Macrocystis pyrifera* by the purple urchin *Strongylocentrotus purpuratus* and red urchin *S. franciscanus*. Earlier studies (Wilson et al., 1977) suggested that a combination of hunting of sea otters, a predator of the urchins, and enrichment through domestic pollution allowed increased urchin numbers. It had been thought that the sea otter, by preying on the urchin, kept them from overgrazing kelps. Sea otters were hunted for their fur and to reduce their impacts on abalone populations. It was also believed that the higher organic content of coastal waters, due to sewage pollution, enhanced urchin larval survival. Schiel and Foster (1986) proposed that the sea otters were not important for urchin removal, and Tegner (1980) demonstrated that abalone was a competitive herbivore for drift seaweed. The latter has been overharvested as well as sheephead fish, which is a predator of urchins. Without competition or predation, urchins overpopulated and grazed the bases of kelps after stripping the other algae from the substrata.

Plants have evolved *defenses* (escapes) against grazing (Hay, 1981a; Littler and Littler, 1988; Duffy and Hay, 1990), and crustose algae and corals may depend on grazers for enhanced growth (Slocum, 1980). Escapes from grazing can be grouped into temporal, spatial, structural, and chemical categories. *Temporal* escapes include rapid growth, rapid recruitment, and life histories involving an annual or heteromorphic partner where the macroscopic plant either disappears or is a not-easily grazed crust. *Spatial* escapes include refuge habitats (intertidal, crevices), as well as association with toxic, stinging, carnivorous, or territorial animals. *Structural* escapes include morphological adaptations, such as tough, massive branches or small dense turfs that minimize access. Differences in texture may also prevent grazing, including tough, cartilaginous, or calcified branches. Mimicry, in which one species resembles another, and the presence of constituents that limit food value (e.g., calcification) can also be included under structural defenses (Berner, 1990). *Chemical* defenses are probably the most frequently cited adaptations to prevent grazing (Hay and Fenical, 1988), particularly for tropical green (*Caulerpa, Halimeda*) and brown (*Dictyota, Lobophora*) seaweeds (Hay and Fenical, 1988; Chap. 12) and temperate kelps (*Lessonia* Martinez, 1996). However, few of the secondary metabolites found in seaweeds have been shown to be effective antiherbivory compounds because careful behavioral studies are required. One example is the green alga *Neomeris annulata*, which uses a chemical defense against the sea urchin *Diadema savignyi* (Lumbang and Paul, 1996). The pantropical species produces at least three brominated sesquiterpenes, each of which deters feeding by the sea urchin *Diadema savignyi* even when present in lower than natural concentrations. Another example is the higher phenol content in subtidal populations of *Lessonia migrescens* resulting in a higher resistance to grazing than intertidal plants (Martinez, 1996).

Biofouling

Although epiphytism has a biological connotation, *biofouling* usually refers to marine organisms growing on man-made structures; thus, it has an anthropogenic emphasis (see review of Evans and Hoagland, 1986). Biofouling can result in new, productive communities as seen with oil platforms throughout the world. The importance of these man-made reefs is evident by the increase in coastal and deep-water fish around them due to the fouling communities of invertebrates and macroalgae. Succession of fouling communities on the oil platforms in the North Sea indicates that *Enteromorpha* species rapidly colonize new surfaces and the community "matures" with kelps such as *Alaria esculenta* and several species of *Laminaria*.

Because of the commercial importance of biofouling, studies have been done to determine the adhesive products of seaweeds (Falkner, 1977) and their antimicrobiological effects (Henriquez et al., 1979). Studies on controlling fouling have been extended to seaweed mariculture, in particular, maintaining clean tanks, piping, and pumps. For example, studies on tank culture of the red alga *Gracilaria conferta* (Friedlander, 1992) indicated that control of the abiotic factors could reduce fouling by green (*Ulva lactuca*) and brown (*Ectocarpus confervoides*) epiphytes. This included modifying light quality (to a green wavelength) and intensity (to low intensity: 172 mmol photons m^{-2} s^{-1}), keeping salinity above 20 ppt, and maintaining a lower pH than normal seawater (7.0, to maintain CO_2). Other fouling studies have focused on the mode of attachment of *Enteromorpha* (Evans and Christie, 1970) and *Ulva* (Braten, 1975).

CHAPTER 4

Physiological Ecology

Physiological ecology is the study of mechanisms that allow organisms to respond to their environment. An understanding of such processes in marine plants will aid in predicting their survival in a changing environment. By monitoring physiological responses, it is possible to determine if marine plants can *acclimate* or adjust through phenotypic responses to new conditions and what characteristics are *adaptive*, that is, the tolerance and genetic limits.

Ecophysiology also bridges the gap between molecular and organismal biology. Suborganismal processes such as osmoregulation of an estuarine alga to changes in salinity are adaptive and can be demonstrated through study of the alga's physiology. Probably the most common types of ecophysiological studies are time-course experiments, in which the physiological responses to changes in abiotic conditions are enumerated, for example, measurement of photosynthesis or respiration versus changes in salinity, temperature, irradiance, or water movement. The linking of carbon gain (loss) illustrates how the strategic needs of plants are met (Mooney, 1976).

This chapter reviews the physiological and morphological responses of marine plants to several abiotic factors introduced in Chap. 2. Accounts of the ecophysiology of seaweeds (Lüning, 1990; Lobban and Harrison, 1994), mangroves (Tomlinson, 1986), seagrasses (Larkum et al., 1989a), and salt marsh plants (Long and Mason, 1983; Adam, 1990) are available, and a classical source on physiological adaptations of marine organisms can be found in the edited series by Kinne (1970, 1971, 1976).

IRRADIANCE AND PHOTOSYNTHESIS

Life as it is known on Earth primarily evolved because of the interaction between photosynthetic organisms and light. The primary exception to this pattern concerns chemosynthetic bacteria. Without light, photosynthesis would not occur, molecular oxygen would not be available, and at least 90% of all energy-containing molecules would disappear. Presently, there is still much to learn about the interaction between marine plants and light.

Light and Marine Plants

Both the quality and quantity of visible light are important to photosynthesis and change during transmittance through seawater. Thus, submerged marine plants must respond to both changes in light intensity and spectral quality. Submerged seaweeds, seagrasses, and benthic intertidal algae in tidal marshes can tolerate a range of photon influence levels; even so, they will bleach under high light levels and cease to grow under reduced light.

Hellebust (1970) considered functional and structural responses of light on submerged plants. Functional responses include photoacclimation through pigment production; changes in photosynthetic rates, as shown by sun and shade responses; reorientation of chloroplasts, phototactic and phototrophic responses; and initiation of reproduction due to photoperiod signals to short- and long-day plants. Examples of functional responses of seaweeds are given in what follows and can be seen in the responses of a thin-blade green algae such as *Ulva* (Titlyanov et al., 1987). The plant responded to low irradiance by increasing pigment content, photosynthetic enzymes, and electron-transport components. Seaweeds respond to changes in light quality as well; under red light, there is an accumulation of carbohydrates, and under blue light, protein synthesis, respiration, and enzyme activity are enhanced (Dring, 1988).

Structural responses include changes in morphology and cytological structure. Species of the green algal genus *Caulerpa* will become etiolated, producing elongated, narrow photosynthetic branches when grown under low light. The green alga *Bryopsis*, when grown in weak light, only produced prostate filaments, whereas under high light, it produced typical feathery erect branches (Hellebust, 1970). The fronds of the intertidal brown *Fucus* narrow if the plant is exposed to high light. Cytological responses to light are evident in deepwater species such as the green alga *Tydemania* that becomes a "black box" by distributing chloroplasts throughout its coenocytic body (Gilmartin, 1966). Chloroplasts will move to cell walls that are perpendicular to incoming light in brown seaweeds to reduce the effect of high irradiances (Nultsch et al., 1981).

In contrast to the less complex anatomy of seaweeds, the leaf of a flowering plant is constructed of vascular, parenchymatous (mesophyll), and epidermal tissues (Fig. 11-6). Thus, leaves of marine vascular plants are excellent organs to demonstrate adaptations to light. For example, seagrasses contain most of their chloroplasts in the epidermal cells, similar to the outer cells of seaweeds, to allow more efficient use of submarine illumination (Fig. 11-7). In contrast, leaves of mangrove and salt marsh plants may contain a number of "hypodermal" layers of cells to reduce or reflect the intense solar irradiation (Figs. 9-7, 10-7, and 10-8). Upper, sun leaves of mangroves usually are two to three times thicker than the shade leaves found within the canopy on the same tree. In addition to adaptations to irradiance, the leaves of tidal marsh vascular plants exhibit *xeromorphy*, that is, adaptation to water and nutrient deficiencies (Chaps. 9 and 10).

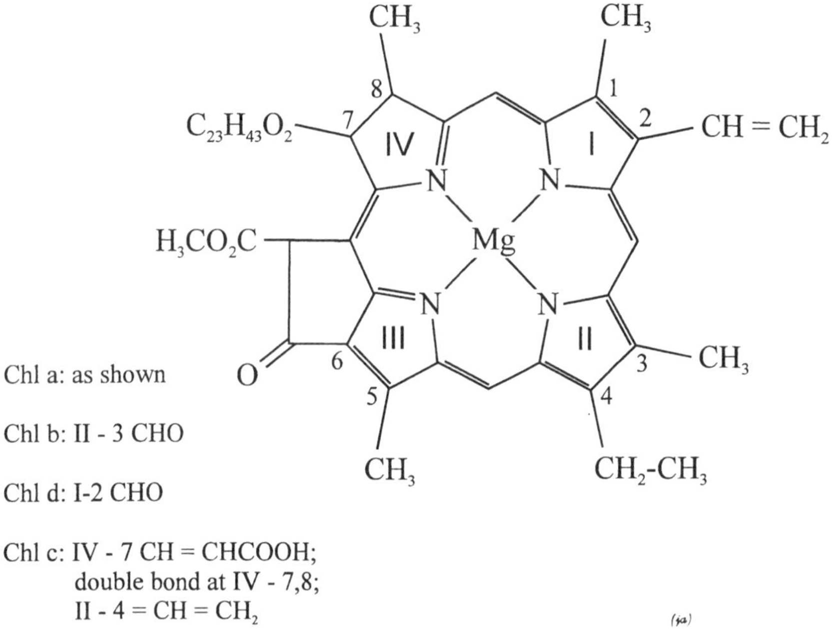
CH3
CH3
C23H43O2
CH = CH2
H3CO2C
Mg
IV
I
III
II
N
O
CH3
CH3
CH2-CH3
1
2
3
4
5
6
7
8
Chl a: as shown
Chl b: II - 3 CHO
Chl d: I-2 CHO
Chl c: IV - 7 CH = CHCOOH;
double bond at IV - 7,8;
II - 4 = CH = CH2

Figure 4-1. The molecular structure of chlorophylls *a* and *b*. Chlorophyll *a* differs from *b* by having CH—CH_3 on ring II instead of — CHO. Chlorophyll c_1 has —C_2H_5 at R (4-2b) and c_2 has —CH=CH_2 (after Dawes, 1981).

Pigments and Light-Harvesting Antennas

The photoreceptor pigments of the Chlorophyta and Angiospermophyta (chlorophylls *a* and *b*) suggest a unidirectional evolution from primitive green algae to flowering plants and rRNA sequences support this idea (van den Hoek et al., 1995). In contrast, other divisions of macroalgae (Chap. 6) and microalgae (Chap. 7) show at least two directions (chlorophylls *a*, *c* and fucoxanthin; chlorophyll *a* and phycobilins; Table 4-1). Because of the diversity of photosynthetic pigments in algae, they have been used to distinguish divisions since Harvey (1841). The pigments can be grouped into chlorophylls, carotenoids (carotenes and xanthophylls), and phycobilins (biliproteins). Because the pigments have specific absorption and fluorescent spectra, they can be identified after extraction.

All oxygen-producing photosynthetic plants contain chlorophyll *a*, with chlorophyll *b* occurring in the green algae, Euglenophyta, Prochlorophyta, and flowering plants, and chlorophyll *c* occurring in the "brown"-colored algae (Table 4-1). Chlorophylls are cyclic tetrapyrroles having a magnesium atom and a phytol tail (Fig. 4-1). The absorption differences between chlorophylls *a*, *b*, and *c* are associated with the side chains of the molecules. Chlorophylls are easily extracted using lipid solvents, such as chloroform or acetone. Cartenoids,

TABLE 4-1. Photosynthetic Pigments Common to Divisions and Classes of Algae[a]

Division or Class	Chlorophylls			Carotenes		Xanthophylls								Phycobilins		
	a	b	c	α	β	Z	N	L	V	F	D^1	D^2	S	PC	PE	AP
Cyanophyta	+				+	+								+	+	+
Prochlorophyta	+	+			+	+										
Rhodophyta	+			+	+	+		+						+	+	+
Chrysophyta																
Chrysophyceae	+		+	+	+	+	+	+	+	+	+	+				
Xanthophyceae	+		+		+		+				+	+				
Eustigmophyceae	+		+		+		+		+							
Bacillariophyceae	+		+		+		+			+	+	+				
Raphidophyceae	+		+		+	+	+			+	+	+				
Prymnesiophyceae	+		+	+	+		+		+	+	+	+				
Phaeophyceae	+		+		+	+	+		+	+	+	+				
Cryptophyta	+		+	+						+	+			+	+	
Pyrrophyta	+		+	+	+		+				+	+				
Euglenophyta	+	+			+		+		+							
Chorophyta	+	+		+	+	+	+	+	+				+			

Source: Dawes (1981); Kirk (1994); and van den Hoek et al. (1995).

[a]Abbreviations: Z = zeaxanthin; N = neoxanthin; L = lutein; V = violaxanthin; F = fucoxanthin; D^1 = diatoxanthin; D^2 = diadinoxanthin; S = siphono-xanthin; PC = phycocyanin; PE = phycoerythrin, AP = allophycocyanin.

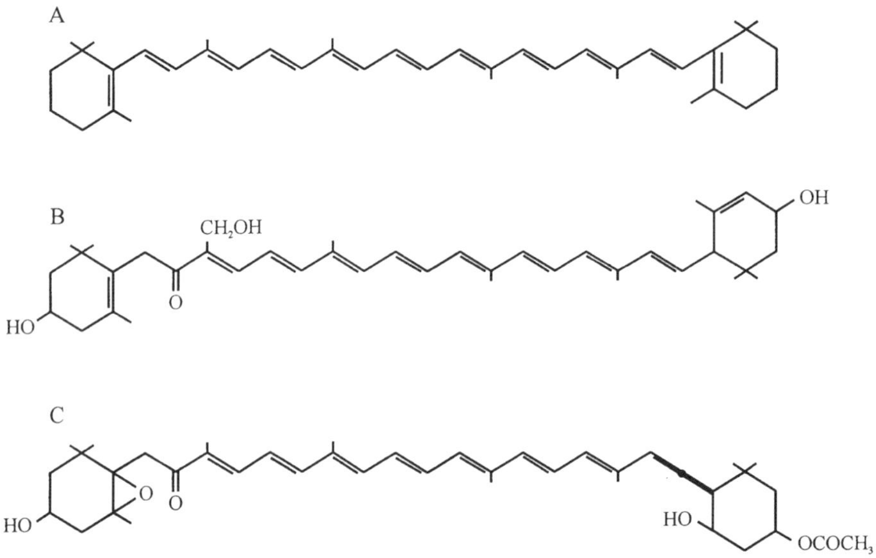

Figure 4-2. Examples of three cartenoid pigments: β carotene (A), siphonoxanthin (B), and fucoxanthin (C) (modified from Kirk, 1994; with permission of Cambridge University Press).

which can be divided into carotenes and xanthophylls, are C_{40} isoprenoid hydrocarbon chains (Fig. 4-2). Carotene pigments are unsaturated [β carotene; Fig. 4-2(A)] and the xanthophylls have some degree of saturation with oxygen [siphonoxanthin, fucoxanthin; Figs. 4-2(B) and 4-2(C)]. Cartenoids can be separated using lipid solvents such as diethyl and petroleum ether, n-propanol, as described by Hellebust and Craigie (1978).

Pigments occur in the chloroplast and are organized as pigment-protein complexes in the thylakoids. These complexes function as light-harvesting "antennae" (Lobban and Harrison, 1994). The basic component of these antennae are two types of cartenoid- and chlorophyll-containing complexes, Photosystems I (P_{700}) and II (P_{680}). In the former, each subunit contains 20 to 45 chlorophyll a molecules and 20 to 45 β carotene molecules and the latter contains about 40 chlorophyll a molecules and an unknown amount of β carotene. The two complexes are associated with a variety of light-harvesting pigments, including chlorophylls a and b, lutein, violaxanthin, and neoxanthin in flowering plants, with different combinations in various groups of algae.

The third group of light-harvesting pigments, the phycobilins, occur in the prokaryotic Cyanophyta and eukaryotic Rhodophyta and Cryptophyta [Figs. 4-3(A) and 4-3(B)]. The phycobilins reflect red (phycoerythrobilin group) and blue (phycocyanobilin group) light and are soluble in water, due to their covalent binding with protein. Thus, the term *biliproteins*. Phycobilins occur in

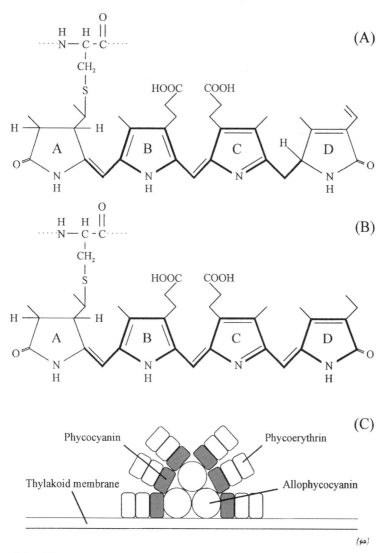

Figure 4-3. The structure of two phycobilins, phycoerythrin (A) and phycocyanin (B), and a diagramic view (C) of a hemidiscoidal phycobilisome (modified from Kirk, 1994; with permission of Cambridge University Press).

35-nm diameter particles called *phycobilisomes* [Fig. 4-3(C)], which are attached to the surface rather than within thylakoid membranes (Figs. 6-28 and 7-1).

The organization and morphology of chloroplasts plus the types and amount of pigment influence the light-harvesting ability of a plant. The number, size, and distribution of chloroplasts will determine the efficiency of light acquisition by plants as do structural adaptations of photosynthetic tissues. For example, the leaves of mangrove trees contain layers of cells (hypodermis, hydrocytes, crystoliths) to filter out the intense tropical sunlight. Further, the chloroplasts are aligned vertically to incoming light in the photosynthetic tissue, the mesophyll, of mangrove leaves to further reduce the effect of high irradiances. In contrast, seagrass blades concentrate their plastids in epidermal cells in order to more efficiently capture submarine irradiance that penetrates the water column.

Hay (1986) modified the concept of Littler et al.'s (1980) *functional-form* groups (Chap. 3, Table 3-1) in seaweeds to include their light-harvesting ability and proposed four basic types: (1) flat, opaque blade and crustose plants (*Laminaria, Lithothamnion*); (2) multilayered, transparent plants (*Ulva*); (3) multilayered, branching forms (*Gracilaria*); and (4) multilayered blade species (*Sargassum*). The advantage of multiple tissue layers in a plant is that it can become a "black box." Thus, even with a limited number of light-harvesting pigments per cell, a thicker, multilayered plant will trap almost all of the submarine illumination that is encountered.

However, the light-harvesting categories are not as easily applied as one might think because other factors such as nutrients, competition, and grazing will influence the plant's morphology. In addition to light, a seaweed's morphology is adapted for nutrient acquisition and tolerance of water movement. Ramus (1978) found that the thin, transparent morphology of *Ulva* was less efficient in absorbing submarine illumination than the thick, multilayered body of *Codium* although both green seaweeds had the same pigments. Increased pigments in *Ulva*, due to nutrient uptake, increased its ability to absorb light, but not to the level of the thicker *Codium*.

Responses to Light Quality and Quantity

The action spectra (photosynthetic activity) of the various pigments usually parallel their absorption spectra as shown in the green alga *Ulva* [Fig. 4-4(A)], the brown alga *Laminaria* [Fig. 4-4(B)], and the red seaweed *Porphyra* [Fig. 4-4(C)]. Specific pigments are most effective if the alga is exposed to the complementary wavelengths. For thin species of Chlorophyta, there is a "green window," which means the green light is reflected and not available for photosynthesis. The "window" can be removed by having a thicker morphology, making the alga optically black (black-box effect), which results in almost total absorbance with the green alga *Codium* (Ramus, 1978).

Engelmann (1884) demonstrated in a simple, but elegant experiment that red and blue light differentially triggered photosynthesis in algae, an early, accurate example of action spectra. He used aerophilous bacteria that would accumulate at sites where oxygen was produced in photosynthesis on the filaments of the green alga *Cladophora* and the red alga *Callithamnion*. Engelmann also proposed the idea of *phylogenetic chromatic adaptation* to explain the vertical distribution of seaweeds and their adaptation to light quality. He argued that red seaweeds occurred in deeper waters because they evolved phycoerythrin that absorbs in the blue-green range of visible light, which make up the dominant

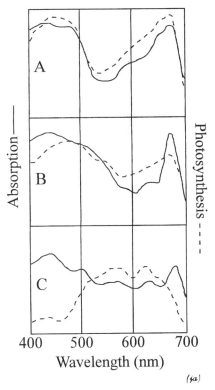

(ya)

Figure 4-4. The action spectra of three thin-bladed macroalgae. Compare the relationship between the action (photosynthesis) and absorption spectra (pigments) in *Ulva* (A), a young (thin) blade of *Laminaria* (B), and *Porphyra* (C) (modified from Lüning, 1990; with permission of John Wiley & Sons, Publ.).

wavelengths available at depth. Among others, Oltmanns (1892) noted that red algae also occur in shallow water and green and brown seaweeds can be found in deeper water. He believed that *intensity adaptation* through increase in total pigments or thickness of the plant body was more important than chromatic adaptation. It has now become apparent that both morphological adaptations and pigment concentration (Hay, 1986; Ramus, 1978) are effective in trapping available submarine illumination. Hence, the original idea of phylogenetic chromatic adaptation has been rejected (Saffo, 1987).

Even so, there are a number of examples that show shifts in pigment in response to light quantity and quality. Members of the Cyanophyta show *ontogenetic chromatic acclimation* or shifts in types of pigments when exposed to different wavelengths (Prezelin and Boczar, 1986). They reported that species of *Tolypothrix* change from blue-green to red when cultured under red and blue light, respectively. Ontogenetic chromatic acclimation has also been demonstrated in such diverse green seaweeds as *Cladophora*, *Valonia*, and *Codium*

(Yokohoma et al., 1977) and *Caulerpa* (Reichert and Dawes, 1986). Deep-water forms, representing four different orders of green algae, all produce the xanthophyll siphonxanthin with an absorption peak around 540 nm, the blue-green range found in deep water. Only *Caulerpa racemosa* from deep-water plants (37 m) had detectable levels of siphonoxanthin compared with a shallow-water population from 1 m. However, after being cultured under low irradiance, the shallow-water form also produced detectable levels of siphonoxanthin.

Another response to light quality can be seen in the effects of *UV radiation* on seaweeds and seagrasses. Seaweeds can cope with increases in UV through protection and synthesis of UV-absorbing compounds. For example, growing tips of *Gracilaria chilensis* increase their UV absorption within a day of exposure (Molina and Montecino, 1996). In contrast, the green alga *Ulva expansa* showed a significant decrease in growth if exposed to either high levels of solar PAR with or without additional UV-B (290 to 320 nm). Thus, further ozone depletion in the upper atmosphere resulting in increases in UV-B radiation would negatively affect this alga (Grobe and Murphy, 1994). Antarctic seaweeds, including the red alga *Palmaria decipiens* and the green algae *Enteromorpha bulbosa* and *Prasiola crispa*, contain pigments that absorb UV at different wavelengths (295, 309, 322, 337 nm). The latter species, which occurs on high, exposed rocks, showed a decline in photosynthetic rates and loss of chlorophyll after being exposed to an enhanced ratio of UV-B to visible light (Post and Larkum, 1993). The study indicates that even a terrestrial Antarctic alga can be stressed by UV-B, so that the presence of UV-absorbing compounds are only effective for specific wavelengths. Thus, much of the natural UV spectra will not be absorbed by the pigments and can damage the photosynthetic apparatus, as shown by the dinoflagellate *Prorocentrum micans* (Lesser, 1996). Blue light also plays a role in a plant's ability to tolerate UV radiation. Sporophytes of the brown alga *Laminaria hyperborea* recovered after being UV irradiated (253 nm) if exposed to blue light, but not if exposed to red or green light (Han and Kain, 1992). The recovery of the brown seaweed after blue-light exposure is typical of what happens when terrestrial plants are exposed to UV. The basis for this reversal is not well understood. It is thought that the decline in photosynthesis after exposure to UV is caused by a decrease in chlorophyll and Rubisco activities.

Sun and Shade Plants

Plants not only *photoacclimate* to light quality, but also to light intensity (Falkowski and LaRoche, 1991). Because (the rate of) photosynthesis is dependent on the level of illumination, there is a relationship between the two, as demonstrated by *P-I curves* [photosynthesis-irradiance curves; Kirk, 1994; Fig. 4-5(A)]. Such curves also offer insights into a plant's tolerance to irradiance, as well as its ability to acclimate and general physiological state. Thus, *P-I* curves have been used extensively to evaluate the physiological tolerances of marine algae.

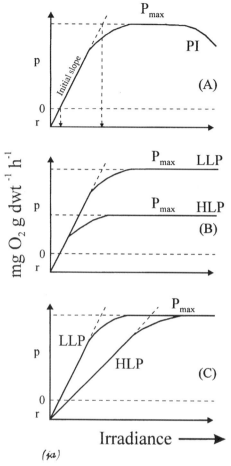

Figure 4-5. Photosynthesis-irradiance (*P-I*) curves of plants showing their acclimation to light. The *P-I* curve (A) illustrates the compensation (I_c) and saturation (I_k) irradiances, maximum rate of photosynthesis (P_{max}), photosynthetic efficiency (alpha), and photoinhibition (PI) where r = respiration and p = photosynthesis. Acclimation to changing irradiance by high (HLP) and low (LLP) light plants may be carried out by changing the number (B) or size (C) of photosynthetic antennae, with this response being evident in the shape of the *P-I* curves (4-5B and 4-5C modified from Ramus, 1981; with permission of Blackwell Scientific).

In the dark, there is no photosynthesis, and like all aerobic organisms, plants consume O_2 and release CO_2 due to respiration. With a gradual increase in irradiance from zero (dark), O_2 production begins causing the measured consumption of O_2 to decrease. At a specific level of irradiance, the production and consumption of O_2 is equal and is the *compensation irradiance* (I_c). Beyond I_c, increased irradiance results in the net production of O_2, which usually occurs in a linear fashion until the maximum rate of photosynthesis or *photosynthetic capacity* (P_{max}) is reached. The linear slope of the *P-I* curve, which originates in the dark and increases to P_{max}, is called *alpha* (α). It is an expression of the photosynthetic efficiency of the plant. The steeper the slope, the more efficient that plant is to increasing levels of low irradiance. Because the photosynthetic response is not linear near P_{max}, the irradiance at which P_{max} is reached or the *saturating irradiance* (I_k) is extrapolated. Determination of I_k is done by extending the linear part of alpha through the horizontal extension of P_{max}. The saturating irradiance is determined from where these two lines cross. With further increases in irradiance, P_{max} may remain level, drop slightly, or show a rapid decrease. A large decline in P_{max} due to excessive irradiance indicates *photoinhibition*, a process that involves damage to components of the photosystem, most commonly Photosystem II (Raven and Sprent, 1989).

Based on the responses to a *P-I* curve, a plant can be classified as either a *sun* or *shade* plant. Whereas the dominant tidal marsh plants (mangroves, salt marsh) are typically sun plants, seagrasses show some shade adaptation. Seaweeds may also be either sun (i.e., intertidal) or shade adapted (i.e., deepwater) species. Sun plants have a low photosynthetic efficiency (the slope of α is gradual), high I_c and I_k, and photoinhibition occurs only at a high irradiance, if at all. A shade plant shows the reverse pattern. Many plants can *photoacclimate* to changes in irradiance, usually by changing pigment concentrations. A shift from high to low irradiances results in a loss of energy (Falkowski and LaRoche, 1991), so that there is a diversion of macromolecule biosynthesis from lipids to carbohydrates or proteins used to produce light-harvesting pigments, and resulting in an increase in photosynthetic efficiency. A shift to high light intensity normally results in a pigment dilution through growth and removal. The *P-I* curves also can suggest how pigments are assembled into light-harvesting antennae (Ramus, 1981). Thus, if an alga acclimated to low light is exposed to higher irradiances, it can either change the *number* [Fig. 4-5(B)] or *size* [Fig. 4-5(C)] of its light traps. A shift to a higher P_{max} without a change in α indicates that an alga has changed the *number* of antenna "traps" without changing their size. A shift in α to a more efficient (steeper) rate indicates that an alga has changed the *size* but not the number of antenna traps.

Light as a Signal

In addition to serving as the source of energy for photosynthesis, light also serves as an environmental signal for plant development (Lüning, 1990). Such responses can be *photoperiodic* (responses to daylength), *photomorphogeneic* (responses to spectral range), or *phototropic* (movement toward or away from light).

Photoperiodic responses in plants include germination, changes in growth forms (morphological alterations), or life history stages and initiation of reproductive structures. Most of these responses involve temperature as well as light triggers. Three types of plants are recognized with regard to day length, *short-*

day (respond to photoperiods less than a critical day length), *long-day* (respond to photoperiod more than a critical day length), and *day-neutral* (day length not critical). Examples of short-day seaweeds are more common than long-day plants. A good example is the life history of the green alga *Monostroma grevillei*, which includes haploid (N) gamete-producing monostromatic blades and a diploid *"Codiolum"* stage that produces haploid zoospores via meiosis. The zygotes germinate into the diploid plant in the summer that will undergo meiosis in the winter, triggered by the short-day signal. The haploid zoospores that are the product of meiosis grow directly into the blade plant in the spring.

Photomorphogenic responses in seaweeds usually involve the effect of light quality. For example, blue or ultraviolet (UV) light will induce the formation of reproductive caps in the green alga *Acetabularia*, whereas red or green light inhibits growth (Schmid, 1984). Blue or UV light triggers the production of sperm and eggs in the filamentous gametophyte of the Laminariales (Lüning and Neushul, 1978). Phototrophic responses include the induction of polarity in germinating *Fucus* zygotes (Buggeln, 1981) and circadian rhythms in the green alga *Caulerpa prolifera* (Dawes and Barilotti, 1969). *Circadian rhythms* are regular biological rhythms of growth or activity that occur on or about a 24-hour basis. In the example of *C. prolifera*, chloroplasts migrate during the night from the tips of blades to the rhizome and return in the morning. The chloroplast migration pattern will continue for days, even if the plant is held in continuous darkness or under low light levels. Other examples can be found in the review of Sweeney (1977).

Carbon Fixation

All plants use CO_2 as their inorganic carbon substrate in photosynthesis. A major factor for submerged plants such as seaweeds and seagrasses is the rate of diffusion of CO_2, which is rapid in air but very slow in water (10^4 times slower). The existence of a slow-moving or motionless boundary layer around submerged plants further restricts the rate of gas and nutrient exchange. Thus, some seaweeds are only found in areas of high water motion, whereas others have adapted to grow in protected, low-energy habitats (Neushul, 1972). On the positive side, CO_2 is part of the carbonate buffer system and usually occurs in higher concentrations in seawater than in air (Eqs. 2-13 and 2-14). Some seaweeds can take up the bicarbonate ion (HCO_3^- directly from seawater using the enzyme carbonic anhydrase that will rapidly covert HCO_3^- (36 mM CO_2 min^{-1}) to CO_2 (Surif and Raven, 1989).

All plants utilize the C_3 (Calvin) cycle in carbon fixation, regardless of whether the C_4 (Hatch-Slack) pathway also exists (Hopkins, 1993). It appears that seaweeds only have the C_3 pathway, in which the enzyme ribose-1,5-bisphosphate carboxylase oxygenase (*RuBisCO*) combines one molecule of CO_2 with a molecule of 5-carbon ribose-1-5-bisphosphate. The result is two 3-carbon molecules of 3-phosphoglycerate, which ultimately become part of a 6-carbon sugar such as glyceraldehyde 6-phosphate (Eq. 4-1).

$$
\begin{array}{l}
CH_2\!-\!O\!-\!P^* \\
| \\
C\!=\!O \\
| \\
CHOH \\
| \\
CHOH \\
| \\
CH_2\!-\!O\!-\!P^*
\end{array}
\quad +\quad CO_2 \;+\; H_2O \;\longrightarrow\; 2 \;\times\;
\begin{array}{l}
CH_2\!-\!O\!-\!P^* \\
| \\
CHOH \\
| \\
C\!=\!O \\
| \\
O^-
\end{array}
\qquad (4\text{-}1)
$$

(RuBP carboxylase)

Ribulose   +   (Enzyme) $\longrightarrow$ 3-Phosphoglycerate

1-5 Bisphosphate

Because RuBisCO is also an oxygenase (will react with oxygen), photosynthesis in C_3 plants is always accompanied by *photorespiration*, a light-dependent process that consumes O_2 and releases CO_2 (Eq. 4-2). The uptake of oxygen results in production of one molecule of 3-phosphoglycerate and one of phosphoglycolate via RuBisCo-oxygenase. The 2-carbon phosphoglycolate molecule is then respired.

$$
\begin{array}{l}
CH_2\!-\!O\!-\!P^* \\
| \\
C\!=\!O \\
| \\
CHOH \\
| \\
CHOH \\
| \\
CH_2\!-\!O\!-\!P^*
\end{array}
\quad +\quad O_2 \;\longrightarrow\;
\begin{array}{l}
\text{1PGA} \\[4pt]
\text{(3-Phosphoglycerate)} \\[4pt]
+ \\[4pt]
COO^- \\
| \\
CH_2\!-\!O\!-\!P^*
\end{array}
\qquad (4\text{-}2)
$$

(Ribulose)   (RuBisCO)   (Phosphoglycolate)

5-C sugar   +   (Enzyme) $\longrightarrow$ Products

Photorespiration results in a *net loss* of fixed carbon that is not coupled with production of high-energy bonds as found in a molecule of adenosine triphosphate (*ATP*), so there is no energy gain. Because most marine plants are C_3 plants, it can be assumed that photorespiration occurs. As O_2 levels rise in a cell, RuBisCO switches from being a carboxylase (carbon-fixing) to an oxygenase (oxidizing) enzyme. Photorespiration occurs in seaweeds as seen in the effects of inhibitors on the red alga *Hypnea musciformis* (Fig. 4-6). However, the process can be suppressed in seaweeds due to their effective carbon-concentrating systems (Kerby and Raven, 1985) that maintain high levels of CO_2 in the cells. Factors that increase the rate of photorespiration include increased levels of oxygen, light, and temperature whereas increased levels of CO_2 lower the rate.

Some salt marsh grasses (e.g., *Spartina*) are C_4 plants, whereas none of the mangroves are. In the C_4 pathway, the enzyme phosphoenolpyruvate (*PEP*) carboxylase combines a molecule of CO_2 with a 3-carbon molecule of phospho-

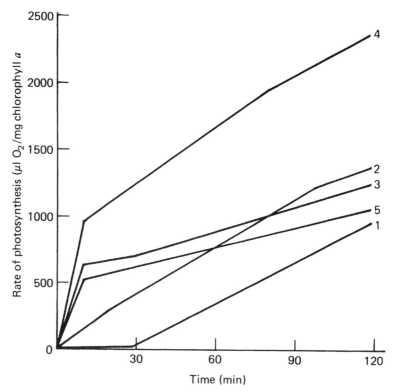

Figure 4-6. Photorespiration in the tropical red *Hypnea musciformis*. The five curves show photosynthetic responses of the red alga to increasing concentrations of potassium glycidate (PG), an inhibitor of photorespiration: (1) no inhibitor, (2) 2.9 mM PG, (3) 14.3 mM PG, (4) 28.6 mM PG, and (5) 57.1 mM PG). Oxygen production increases with increasing levels of PG until the highest concentration causes cellular damage.

enolpyruvate to form a 4-carbon sugar oxaloacetate (Oxalo-A). Oxaloacetate then can be reduced to malate or converted to aspartate (Eq. 4-3) via the addition of an amino group ($-NH_2$).

$$
\begin{array}{l}
CH_2 \\
| \\
CH_2-O-P^* \\
| \\
C=O \\
| \\
O^- + HCO_3^- + P_i \longrightarrow \text{(PEP)}
\end{array}
\qquad
\begin{array}{l}
COO^- \\
| \\
CH_2 \\
| \\
C=O \\
| \\
COO^-
\end{array}
\qquad
\begin{array}{l}
COO^- \\
| \\
CH_2 \\
| \\
CH-OH \quad \text{(or)} \\
| \\
COO^-
\end{array}
\qquad
\begin{array}{l}
COO^- \\
| \\
CH-H_3N^+ \\
| \\
CH-OH \\
| \\
COO^-
\end{array}
\qquad (4\text{-}3)
$$

Phosphoenolpyruvate (enzyme)   Oxalo-A   Malate   (or)   Aspartate

The significance of the C_4 (Hatch-Slack) pathway is that CO_2 is more efficiently trapped and made available for the C_3 cycle. Aspartate and malate are broken down to release CO_2, either at another time (temporal separation in the succulent dicot, *Sesuvium*) or in a different part of the leaf (spatial separation in *Spartina*). Because PEP carboxylase will not function as an oxygenase like RuBisCO, photorespiration will not occur. Further, the affinity of PEP carboxylase for CO_2 is higher than RuBisCO; thus, as levels of CO_2 decline and O_2 rises, the production of malate and aspartate continues.

TEMPERATURE

If light can be considered the single most important abiotic factor for life due to its involvement in photosynthesis, then temperature must be the most fundamental factor for determining the distribution of organisms. Temperature affects chemical reactions (metabolic rates) and thus all processes of an organism. Plants are essentially *poikiotherms* as they do not regulate their temperature and must adapt to their environment. Because of this pattern, temperature tolerances should be considered at the cellular (enzymatic) physiological (photosynthesis, respiration) and organismal (growth, reproduction) levels.

Temperature and Marine Plants

The effects of temperature have been reviewed for blue-green algae (Oppenhiemer, 1970), algae (Gessner, 1970), and seaweeds (Lüning, 1990; Lobban and Harrison, 1994). Such plants can be *eurythermal* (tolerant of wide temperature ranges) or *stenothermal* (intolerant of wide temperature ranges). Most Arctic, Antarctic, and temperate intertidal plants tolerate low to freezing temperatures and recover rapidly when exposed to high temperatures. In contrast, tropical seaweeds are usually intolerant to about 10°C. The red alga *Chondrus crispus*, which grows at a low intertidal zone, could only tolerate −20°C for three hours, whereas another red seaweed *Mastocarpus stellatus*, which extends into the mid-intertidal zone can tolerate −20°C for longer periods (Dudgeon et al., 1989). Structural responses of marine plants to temperature include differences in size, as shown by the large morphologies of intertidal seaweeds in cold temperate habitats. By contrast, few intertidal seaweeds of any size will be found in tropical habitats where intense sunlight and high temperatures are common (e.g., *Turbinaria*, *Sargassum*). Functional responses to temperature can be demonstrated through changes in respiration and photosynthesis (see "Metabolic Processes") as well as alterations in growth and reproduction.

Metabolic Processes

The importance of thermal effects can be seen at the molecular level. The most common example is the Q_{10} effect, which is the ratio between the metabolic rate at a given temperature t and the metabolic rate at $t + 10$. The Q_{10} ratios for enzymatically controlled reactions vary between 1.1 to 5.3 with an average of 2.0. Thus, within the tolerances of the organism's enzymes, the metabolic rate will double with every 10°C rise in temperature. Both V_{max} (maximum velocity of catalytic activity) and K_m (Michaelis constant) in the Michaelis-Menten response (Eq. 4-4) are affected by temperature.

$$v = V_{max}[S]/(K_m + [S]) \tag{4-4}$$

where v = initial rate of the catalytic activity, and $[S]$ = substrate concentration.

The effect of Q_{10} can be seen in the photosynthetic responses of the brown alga *Ascophyllum nodosum* collected in the winter (Fig. 4-7, J) and summer

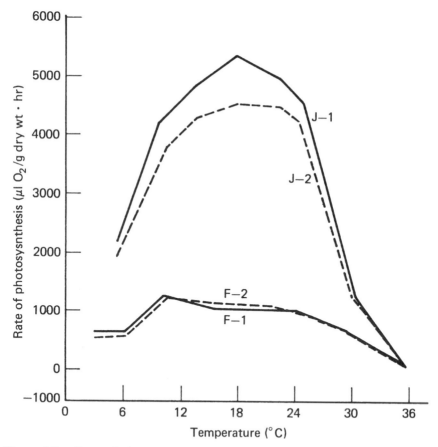

Figure 4-7. Seasonal photosynthetic temperature responses by the temperate brown alga *Ascophyllum nodosum*. The attached (solid line) and entangled salt marsh ecad *scorpioides* (dashed line) both have identical photosynthetic responses in July (J-1, J-2), peaking at 18°C, whereas in February (F-1, F-2) they peak at 10°C (after Chock and Mathieson, 1978).

(Fig. 4-7, F) from an estuary and open coast habitats. Although plants exhibited different levels of response, the photosynthetic rates of summer plants doubled for each 10°C rise in temperature between 6 and 18°C, that is, their range of tolerance. Higher temperatures resulted in decreased photosynthesis, indicating that enzymes were adversely affected (Fig. 4-7).

Heat and Cold Tolerance

Enzymatic rates will depend on the *level* of available enzyme as well as the *tolerance range* (kinetics) of an enzyme. A plant exposed to increasing or decreasing ranges in temperature will be *stressed*. As described in Chap. 3, stress is usually exposure to continuous suboptimal conditions that restricts plant productivity (Grime, 1977). Measurement of stress can include the monitoring of enzymatic activity, physiological processes, or growth and reproductive responses.

Heat tolerance is based on a plant's ability to stabilize protein synthesis (such as the production of heat-shock proteins), whereas cold tolerance is possible by increasing the level of enzymes in the cell or the production of modified enzymes (e.g., *isozymes*) that function at different temperature ranges. The latter approach occurs in cold-water seaweeds, which usually have temperature optima above environmental conditions (Lüning, 1990). Freezing tolerance is well known for intertidal algae, which avoid the damaging effect of cellular ice crystal formation with antifreezelike substances such as DMSP (β-dimethyl-sulphonioprionate; Karsten et al., 1989). Acclimation to temperature changes in brown algae has been shown for enzymes of *Laminaria saccharina* (Davison and Davison, 1987) and seasonal photosynthetic acclimation of *Ascophyllum nodosum* (Chock and Mathieson, 1978; Fig. 4-7). Acclimation is also evident for *A. nodosum* respiration, which increases less with enhanced temperatures in summer ($Q_{10} = 1.5$) than during winter ($Q_{10} = 2.0$; Kanwisher, 1966).

Interpretation of photosynthetic and respiratory temperature responses must be used with caution as they are short-term and not the same as growth. For example, the highest photosynthetic rates in the tropical red alga *Eucheuma denticulatum* were recorded at 30 and 35°C (Fig. 4-8). However, when the same samples were returned to 25°C, the photosynthetic responses were lower than those of plants held continuously at 25°C, which was the optimum temperature (Dawes, 1979). The decline after exposure to higher temperatures suggests that enzymatic degradation probably occurred at the higher temperatures but were not evident over the short measurement period.

Growth and Reproduction

The average optimal temperatures for growth of seaweeds are 0 to 10°C for polar species, 10 to 15°C for cold-temperature species, 10 to 20°C for warm-temperate species, and 15 to 30°C for warm temperate to tropical species (Lüning, 1990). Studies using growth as a monitor of temperature tolerance

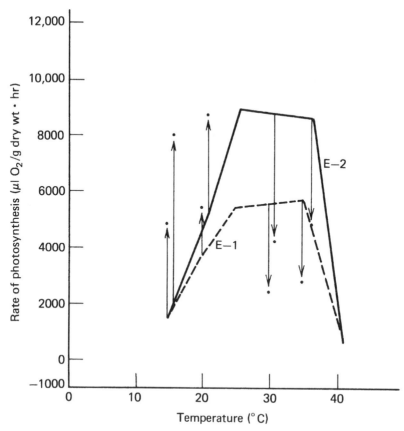

Figure 4-8.   Thermal photosynthetic damage in the tropical red alga *Eucheuma*. Two Philippine species, *E. kappaphycus* (E-1, dashed line) and *E. denticulatum* (E-2, solid line), showed identical photosynthetic responses to increasing temperatures. After exposure to 30 to 40°C, their photosynthetic responses declined when returned to 25°C, suggesting thermal damage (after Dawes, 1979).

in marine plants have followed the concept of Gessner (1970) in which structural and functional responses of plants are considered. Gessner (1970) and Lüning (1990) also point out the continuum among the optimal temperature range where reproduction (*reproduction limits*) occurs, the suboptimal temperatures where only growth occurs (*growth limits*), and the extreme temperatures where growth ceases and death occurs (*lethal limits*). Such ranges can be very narrow for seaweeds. For example, the red alga *Chondrus crispus* had optimal sporeling growth at 21 to 24°C, and growth ceased at 26°C (Prince and Kingsbury, 1973). By contrast, the seagrass *Ruppia maritima* from North Carolina and Florida showed broad temperature tolerances, with both populations exhibiting the same growth and photosynthetic patterns when exposed to 14, 22, and 30°C (Koch and Dawes, 1991).

Growth response has also been used to understand the biogeographic distribution of macroalgae (see what follows). For example, four amphi-Atlantic tropical macroalgae (*Dictyopteris delicatula*, *Ceratodictyon intricatum*, *Ernodesmis verticellata*, and *Lophocladia trichoclados*) showed no ecotypic differentiation between tropical (Canary Island) and temperate (Mediterranean) isolates when grown between 20 and 30°C (Pakker et al., 1996). In contrast, four other macroalgae (*Microdictyon boergesenii*, *M. tenuis*, *Wurdemannia minnata*, and *Valonia utricularis*) showed marked differences in growth at low temperatures for tropical compared to temperate isolates (Pakker and Breeman, 1996). The temperature tolerances shown in the two studies indicated that the first four species originated from tropical populations and the last four are descendants of the eastern Mediterranean after the Pleistocene Glaciation.

Ice and Marine Plants

The mechanical problem of ice formation in cold temperate intertidal communities is another aspect of temperature. "Ice rafting" of salt marsh peat and vegetation to tidal flats is a common sight during the late winter in New England (see Fig. 9-14; Hardwick-Witman, 1984). Large pieces of ice will rip out patches of the salt marsh and drift away; these patches may end up on tidal flats or beaches where the salt marsh vegetation reestablishes or dies. Ice scouring of seaweeds in the shallow subtidal and intertidal regions is another example of mechanical damage by ice. Thus, in the same Great Bay Estuary, Mathieson et al. (1982) estimated that 136 metric tonnes dry weight of *Ascophyllum nodosum* can be removed by ice rafting each year, resulting in the ecad *scorpioides*, as described for drift algae (Chap. 8). In Newfoundland, the removal of the canopy of climax kelp *Alaria esculenta* by pack ice can enhance submarine illumination on rocky substrata and increase the diversity of opportunistic seaweeds (Keats et al., 1985). In the tideless northern Baltic Sea, the upper limit of *Fucus vesiculosus* appears to be controlled by ice scraping (Kiirkki and Ruuskanen, 1996). Residual holdfasts can regenerate new plants, but because of the limited growth period, they usually remain vegetative. However, some of the larger plants will sink below the ice pack and survive to reproduce during the next summer.

Biogeography

The geographical distribution of marine plants as mediated by temperature can be examined at two levels, local and latitudinal. Locally, the occurrence of seaweeds within intertidal habitats reflects their tolerance to ranges of air temperature and, hence, desiccation. By contrast, subtidal species usually show a narrower tolerance to temperature because of the limited fluctuation in water temperature. Ecotypic differentiation to temperature has been demonstrated in species that have wide latitudinal ranges, such as the brown filamentous alga *Ectocarpus siliculosus* (Bolton, 1983), but not for the red seaweed *Chondrus crispus* (Lüning et al., 1987) nor the seagrass *Ruppia maritima*, as described before (Koch and Dawes, 1991).

Latitudinal and longitudinal distribution of seaweeds, seagrasses, and tidal marsh plants is the result of past migration and changes in coastlines, including the separation of continents. As pointed out in Chap. 2, theories used to explain the distribution of marine plants tend to fall into two categories, vicariance and dispersal biogeography. The former considers phyletic relationships in terms of historical vicarious geology such as the separation of continents. The dispersal theory relies more on the development of new species due to spreading and then development of barriers. The present distribution of marine plants is probably due to a combination of both processes.

Of all physical factors, temperature plays the largest role in delineating floristic differences of marine plants at different latitudes. Many studies have outlined the tolerances of seaweeds to temperature and their latitudinal distribution, including Biebl (1970), van den Hoek (1984), Lüning and Freshwater (1988), Lüning (1990), and Bischoff and Wiencke (1995). The last study compared strains of the amphi-equatorial green alga *Urospora penicilliformis* and explained temperature tolerances based on periods of geological isolation. Setchell emphasized that mean water temperatures could be used to evaluate the distribution of seaweeds (1920a) and seagrasses (1920b). Thus, he proposed that 5 or 10°C *isotherms* (boundaries of mean water temperatures averaged over years) could be used to identify biological boundaries and these have proven useful to separate latitudinal floras of marine plants. Den Hartog (1970) and Larkum and den Hartog (1989) outlined the latitudinal distribution of the 12 seagrass genera, proposing that water temperature is the controlling factor. In a similar approach Chapman (1977) reviewed the latitudinal distribution of mangals and salt marshes (Fig. 9-6) and noted that mangroves are limited to regions without prolonged frosts.

Michanek (1979) and Van den Hoek (1984) identified seven biogeographical regions in the world's oceans, with each having distinct seaweed floras. Lüning (1990) presented the same seven regions (e.g., Fig. 4-9) in his review of seaweed floras. All of these floras are based on regions that Briggs (1974) established in his text on biogeography of marine fauna. The seven major grouping are as follows:

1. Arctic group (one region)
2. Cold temperate group of Northern Hemisphere (three regions)
3. Warm temperate group of Northern Hemisphere (three regions)
4. Tropical group (four regions)
5. Warm temperate group of Southern Hemisphere (five regions)
6. Cold temperate group of Southern Hemisphere (five regions)
7. Antarctic group (one region)

Photoperiod or day length varies latitudinally and also plays a significant role

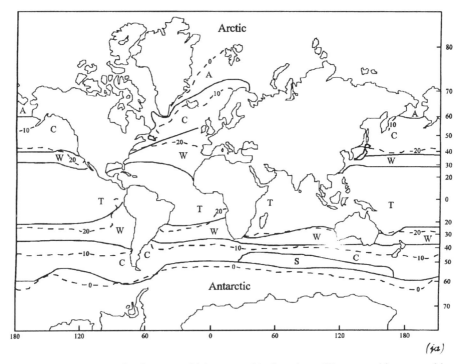

Figure 4-9. August isotherms and biogeographical regions. The seven biogeographical regions for August isotherms (0, 5, 10, 15, 20, 25°C) include summer (Northern Hemisphere) and winter (Southern Hemisphere): Arctic, Antarctic, cold temperate (C), warm temperate (W), and tropical (T) (modified from Lüning, 1990; with permission from John Wiley & Sons, Publ.).

in controlling the life histories of marine plants. Photoperiod is usually coupled with temperature (cf. Lüning, 1990). An example can be found in the reproductive responses of the tetrasporic stage of *Asparagopsis armata* (previously identified as a distinct plant, *Falkenbergia rufolanosa*). Cultivars collected from Australia, Italy, and Ireland were subjected to different photoperiods, temperatures, and nutrients (Guiry and Dawes, 1992). Italian cultivars showed 100% tetraspore production at 17 to 21°C and under 9 to 10 h. By contrast, short-day Irish plants required 17°C for spore production. Thus, the Italian and Irish strains exhibited temperature and photoperiod regimes comparable to their original habitats.

SALINITY

Salinity can influence the local distribution of seaweeds and seagrasses, and to a lesser degree salt marsh and mangrove plants. The first two types of plants are *obligate halophytes* as they require saltwater for germination and growth. However, some seaweeds such as the estuarine red alga *Bostrychia binderi* will grow in "pure" spring water that contains high levels of calcium and CO_2 (Dawes and McIntosh, 1981). In contrast, many salt marsh and mangrove plants are *facultative halophytes* because they can usually germinate and grow in freshwater. Tidal flowering plants occur in estuarine and oceanic habitats and are outcompeted by freshwater species in the latter's habitat due to their slower growth. The slow growth of salt marsh and mangrove plants is, in part, due to their adaptations to prevent water loss and limit salt uptake. Such adaptations are highly important for growth in marine systems, but require extensive physiological and structural development.

Osmoacclimation

All marine plants must adjust to the salinity in their environment (Chap. 2). Estuarine species are said to be *euryhaline* because they can acclimate to fairly wide shifts in salinity (e.g., 5 to 30 ppt within a tidal change). By contrast, *stenohaline* species can only acclimate to narrow ranges of salinity. Seawater with a salinity of 35 ppt has an osmotic potential of approximately −2.0 MPa (1 MPa or megapascal = 10 bars = 10 atmospheres of pressure; 1 bar = 760 mm Hg). To survive, a plant must maintain an internal water potential of −2.0 mPa or less. Osmotic potentials in coastal and oceanic seaweeds can range from −2.6 to −3.0 MPa, with their cells being equal (isotonic) or slightly higher (hypertonic) than ambient conditions. They lose water to the surrounding seawater if their protoplast has a lower osmotic pressure than the surrounding medium (hypotonic).

The water potential (osmotic pressure) in a plant cell must change with alterations in salinity (*osmoregulation*) and this process consists of two phases (Wyn Jones and Gorham, 1983). In the first 30 min of an increase (or decrease) in salinity, the seaweed shifts its internal ion concentrations (K^+, Na^+, Cl^-), which results in a rapid flux to increase (or decrease) cell *turgor pressure*. The latter is the internal osmotic pressure pressing against the cell wall or membrane. Changes in water potential and increases in ions can be damaging to enzymes of the cell. Thus, such ionic fluctuations are more prevalent in algae with large vacuoles (e.g., the red alga *Griffithsia*) where the potentially toxic ions can be separated from the protoplasmic enzymes. After this initial alteration (i.e., the next few hours), the macroalgae will produce (or remove) *compatible solutes*, which are organic compounds that are soluble, have no net charge, and do not inhibit cellular activities. Carbohydrates, amino acids (proline), quaternary ammonium compounds (betaines), and β-dimethylsulfoniopropionate (DMSP) are all used as compatible solutes. The carbohydrates will differ with brown algae (mannitol), red algae (e.g., Floridoside), and green algae (sucrose), and the other compatible solutes can be found in many seaweeds (proline).

The first part of the two-phase process was shown for *Caloglossa leprieurii* (Mostaert et al., 1995) and 17 other seaweeds (Kirst and Bisson, 1979). Such species maintain their turgor pressure over a wide range of salinities (10 to 35 ppt) by changing their concentrations of one or more ions (i.e., Na^+, K^+, or Cl^-) with Cl^- being primarily involved in the ion pump for turgor regulation. The second phase was shown by Kirst (1979) for the green phytoflagellate *Platymonas subcordiformis* and by Mostaert et al. (1995) for the red alga *Caloglossa leprieurii*. The initial transient increase for ion levels are used to "bridge the concentration gap" until sufficient levels of the photosynthate mannitol was synthesized, demonstrating the importance of this compound as a compatible solute and osmolite.

Seagrasses appear to osmoregulate in a similar fashion to seaweeds (Tyerman, 1989), with their surrounding leaf sheaths protecting the most sensitive meristematic tissues near the leaf base (Chap. 11). Epidermal cells of seagrass blades appear to be the primary site of osmoregulation. Salinity fluctuations initially result in changes of the ions Na^+ and Cl^- (probably in the vacuole) concentrations, with K^+ remaining constant in *Halophila ovalis* and *Zostera capricorni*. Proline, glycinebetaine, and other compatible solutes occur in the cytoplasm of *Halodule wrightii*, *Thalassia testudinum*, and *Ruppia maritima*, suggesting that these compounds are used in later stages of osmoregulation.

Emergent flowering tidal plants show at least three types of salinity adaptations, including exudation, exclusion, and shedding, which are not unique to any single species (Tomlinson, 1986). *Exuders*, such as the black mangrove *Avicennia germinans* and the cord grass *Spartina alterniflora*, allow a certain portion of the salt to move up the xylem to their leaves. The salt is concentrated in their leaves and the brine exudates through hydathodes and salt glands, which are modified glands (Chaps. 9 and 10; Fig. 10-12). *Excluders*, such as the red mangrove *Rhizophora mangle*, prevent most of the salt in sediment pore water from entering their roots via ion pumps. *Shedders* include many salt marsh plants and are characterized by concentrating salt in cell walls of their leaves and stems. Because these plants tend to die back each year, the above-ground portions containing the salt are shed at the end of a growth season. More will be said about salt tolerance in the chapters dealing with marine flowering plants (Chaps. 9 to 11).

Photosynthesis and Growth

Photosynthetic and respiratory rates are also influenced by salinity. For example, both populations of the red alga *Hypnea musciformis* collected from a mangrove estuary and an oceanic site showed broad tolerances and similar optima during the winter but not in the summer (Dawes et al., 1976; Fig. 4-10). The distinct summer responses reflect the lower salinity of the estuary (due to summer rain) that caused a broader acclimation of that population than the oceanic plants.

Growth studies will also demonstrate if a species can acclimate to changes in salinity (euryhaline) or cannot respond (stenohaline), being a *haloecotype*. Isolates of the brown alga *Ectocarpus siliculosus* were taken from three estuarine habitats differing in salinity (Russell and Bolton, 1975). After three years of

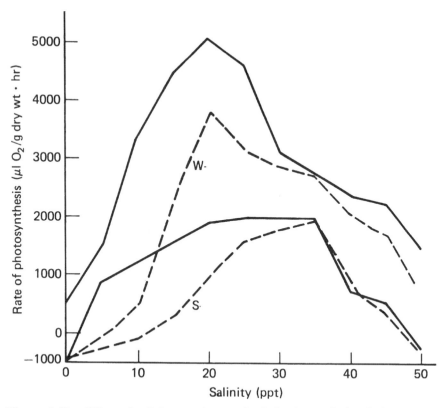

Figure 4-10. Effect of salinity on phytosynthesis in the tropical red alga *Hypnea muscifomis*. Plants collected from a mangrove estuary (solid line) and exposed coast (dashed line) both have broad tolerances to salinity with maxima at 20 ppt in winter (W) and 36 ppt in summer (S) (after Dawes et al., 1976).

culture in 34-ppt salinity, they were transferred to a series of different salinities, where they showed optimal growth in the salinity of their original habitat. The data suggest that each population was a distinct haloecotype. In another study, Russell (1988a) demonstrated the evolution of low-salinity seaweed ecotypes in the Baltic Sea and showed that it included both differential salinity tolerances and anatomical traits.

WATER MOVEMENT

Waves and water currents (tidal and wind) affect marine plants in special ways, including mechanical stress and changes in nutrient acquisition. The form and function of seaweeds also reflect adaptation to water movement (e.g., filamentous, foliose, and massive; Table 3-1).

The Boundary Layer

Gases and ions diffuse 10,000 times slower in water than in air. The slowness of diffusion in water is evidenced by the need to stir water in growth studies and during measurement of oxygen. All submerged surfaces have a *boundary layer* of slowly moving seawater through which nutrients, oxygen, and carbon dioxide must diffuse. Koch (1994) demonstrated that the boundary layer in the seagrasses *Thalassia testudinum* and *Cymodocea nodosa* can limit photosynthetic rates due to the slow diffusion of CO_2. She measured photosynthetic rates in CO_2 saturated seawater and reported a Michaelis-Menten response with increasing rates of water flow up to saturation. Algal turf communities on coral reefs showed higher (up to 20%) photosynthetic rates when water currents were oscillatory rather than unidirectional, suggesting that the boundary layer was reduced by the effect of turbulence (Carpenter and Williams, 1996). Many seaweeds, including encrusting algae and microscopic germlings, live within this boundary layer either throughout their life or during a portion of their existence (Neushul, 1972; Fig. 8-9). Any increase in turbulence (mixing, flow rate, aeration) will reduce the thickness of the boundary layer. Thus, some plants have evolved the ability to grow in areas of high water movement (tidal rapids, high-energy intertidal zone). Other species have developed surface structures such as spines (*Macrocystis*), outgrowths (*Laminaria*), holes (*Agarum*), or fine hairs (*Codium*) that increase surface area and turbulence in slow-moving water.

Waves

Plants are affected by waves directly in a number of ways. *Mechanical stress* will result in drag, which removes plant parts, and impact, which causes shearing of parts or entire plants from the substratum. By carrying sediment, water will *erode* and *abrade* plants. An increase in water movement will *reduce* the boundary layer around submerged plants. The hydrodynamic effects of waves on plants are a significant aspect of water movement on high-energy shores, although the dispersion of erosive materials such as sand can destroy subtidal or intertidal communities. Lewis (1964) emphasized the importance of wave action on zonation of rocky shores, and Seapy and Littler (1979) compared seaweed communities of adjacent exposed and protected sites (Chap. 8). The latter study showed that sites sheltered from high wave activity had lower diversity (fewer species) but higher biomass and primary productivity. The high productivity was due to a predominance of more delicate blade forms of seaweeds, which have a higher surface-to-volume ratio in sheltered communities than the crustose, ropelike, or thick (tough) plant bodies found in the exposed site. Denny (1988) developed a "structure-exposure index" (S-E index), which describes the properties of the organism, not its environment. The S-E index incorporates all factors that can influence the organism including grazing. As a seaweed grows, it can change with age and change from one category to another.

Tides

The zonation of intertidal organisms is primarily influenced by tides in both lithophytic and tidal communities. Intertidal zonation is considered to be primarily under the influence of tidal levels with other abiotic factors (rainfall, sunlight, air temperature) playing a secondary role. The frequency and amplitude (height) of the "tidal environment" (Lewis, 1964) are the controlling factors and play a critical role on coasts having consolidated substrata (Chap. 8). On some coasts, *critical tidal levels* (Doty, 1946) appear to restrict some seaweeds to specific zones. In a study of intertidal seaweeds on the west coast of North America, Doty identified eight possible tidal levels that coincided with dominant marine organisms (Fig. 4-11). However, other factors can complicate intertidal zonation, especially wave activity and air temperature.

Tidal currents will influence subtidal plant communities as well. In confined bodies of water (Puget Sound, Washington, and Great Bay Estuary, New Hampshire), the erosive, high-energy tidal currents can limit the type of submerged vegetation. Mathieson et al. (1977) found that the shearing force was so high in some of the Great Bay tidal rapids that the diversity and size of many seaweeds were significantly reduced. Further, species characteristic of high-energy, open coastal environments dominated.

Desiccation

Exposure of the intertidal seaweeds to air will cause stress due to dehydration, increased concentration of solutes in cells, and inability to take up nutrients, which are characteristics of *desiccation*. When exposed to air during low tide, intertidal species essentially experience a terrestrial environment, which includes exposure to high temperatures and desiccation. Bell (1993) found that the rate of aerial photosynthesis decreased with increasing desiccation in the intertidal red alga *Mastocarpus papillatus*. Further, upon emersion, the apical tips showed a rapid recovery in photosynthesis if desiccated at 15 to 25°C versus the damaging effect of exposure to 30°C.

Water loss is directly related to the surface area and volume of the plant. Thus, saccate algae such as the intertidal red seaweed *Halosaccion americanum* can supply water from its thallus cavity (Oates, 1986). Intertidal species such as fucoids of rocky shores and macroalgal epiphytes of in tidal communities not only "tolerate" desiccation, but their photosynthesis may be enhanced. Oates and Murray (1983) found that the fucoid alga *Hesperophycus* was able to carry on photosynthesis longer in air than its neighbor *Pelvetia*, which grows at a lower intertidal zone (shorter exposures to air). Quadir et al. (1979) found that the brown alga *Fucus distichus* had higher photosynthetic rates during periods of exposure than when submersed. Similar higher photosynthetic responses to exposure are known for intertidal salt marsh algae such as the filamentous red alga *Bostrychia binderi* (Dawes et al., 1978). *Porphyra linearis* occurs in the uppermost intertidal zone in the North Atlantic and Mediterranean coasts and is

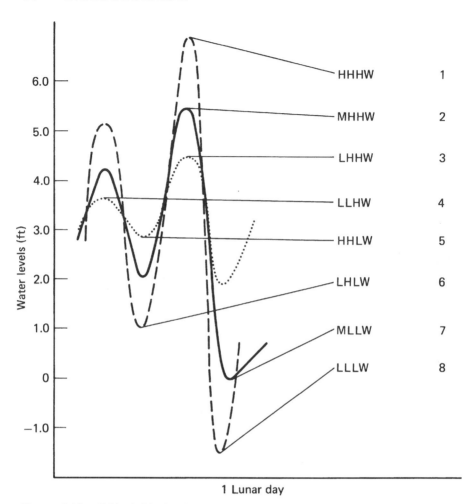

Figure 4-11.   Critical tide levels. Intertidal zonation of marine algae is particularly sensitive to tidal levels, as shown by Doty's (1946) scheme for the Oregon coast where highest-high high water (HHHW) reached 6.5 ft (2.5 m) and lowest-low low water was −1.2 ft (−0.6 m). The spray zone extends above #1 and the subtidal zone occurs below #8. The average monthly exposure in each zone was correlated with the organisms present. For example, zone 2 had marine lichens and a mean exposure of 120 hours; zones 4 and 5 supported the brown fucoid *Pelvetia*, with a mean exposure of 6 to 9 hours; and zones 7 and 8 supported the red alga *Prionitis*, which was almost always submerged.

able to regain photosynthetic activity after three weeks of drying to 5% water content (Lipkin et al., 1993). The tolerance of *P. linearis* to desiccation corresponded to the period between successive spring tides when it would be covered by seawater. *Porphyra capensis*, another high intertidal species in South

TABLE 4-2. **Essential Elements in Seaweeds**

Element	Examples of Function	Examples of Use
Nitrogen	Proteins, cell structure	Amino acids
Phosphorus	Energy transfer, structure	DNA, ATP
Potassium	Protein stability, pH control	As ions
Calcium	Enzymes, structure, ions	$CaCO_3$, alginate
Magnesium	Enzymes, ion transport	Chlorophyll
Sulfur	Enzymes, structure	Carrageenan
Iron	Respiratory enzymes	Cytochromes
Maganese	Photosystem II e^- transport	Chloroplast membrane
Copper	Electron transport	Plastocyanin
Zinc	Enzymes	Carbonic anhydrase
Molybdenum	Nitrate reduction	Nitrate reductase
Sodium	Water balance	Ion pump
Colbalt	Vitamin B_{12}	B_{12}
Chlorine	Halogenated metabolites	Antigrazing products
Bromine	Halogenated metabolites	Antigrazing products
Iodine	Halogenated metabolites	Antigrazing products

Sources: After DeBoer (1981); Lobban and Harrison (1994).

Africa, can photosynthesize either during submersion or emersion. The plant is covered with a mucilage sheath that helps to resist desiccation (Levitt and Bolton, 1991).

NUTRIENTS

Ten of the 21 essential nutrients required by plants are required by all algae: carbon, hydrogen, oxygen, nitrogen, phosphorus, magnesium, copper, manganese, zinc, and molybdemum (DeBoer, 1981; Table 4-2). In laboratory culture, some vitamins (B_{12}, thiamine, biotin) are also routinely added for seaweed growth (Dawes and Koch, 1991) but they do not appear to be universally required by seaweeds. Lobban and Harrison (1994) give a detailed review of the importance of nutrients to seaweeds, and culture media are presented by Stein (1973).

Uptake and Kinetics

Uptake of nutrients may not require expenditure of energy. For example, after diffusion through the boundary layer, essential elements can be taken up through simple *adsorption* by an algal cell wall or by *passive transport* of nonelectrolytes such as CO_2, NH_3, O_2, and N_2. Passive transport is simply diffusion from high to lower concentrations (source to sink). *Facilitated diffusion* is a more rapid version of passive transport down an electrochemical gradient, where any expenditure of energy is indirect. That is, the removal of elements via conversion into other compounds results in a "sink condition." *Active transport* results when ions (e.g., NO_3^-, $PO_4^{-3}NO_3^-$, PO_4^{-3}) are transported across the cell membrane against an electrochemical gradient (e.g., ion pumps). It is more rapid and requires energy.

Both facilitated diffusion and active transport demonstrate Michaelis-Menton saturation kinetics (i.e., Eq. 4-4). In the process, phosphate energy bonds in ATP (adenosine triphosphate) are used by ATPases to move ions across the cell membrane using proton (H^+) pumps. The uptake kinetics of ammonium and phosphate have been described for a number of seaweeds because of the limited availability and importance of these nutrients for growth (Lobban and Harrison, 1994). The uptake kinetics of nutrients can change, depending on concentrations in the cell. In the red seaweed *Gracilaria tikvahiae*, uptake of NH_4^+ [Fig. 4-12(A)] is linear, demonstrating active transport (Friedlander and Dawes, 1985). In contrast, uptake of PO_4^{3-} by the same alga [Fig. 4-12(B)] had two saturation (facilitated) phases (0 to 0.2, 0 to 2.0 μM) and one linear (active) phase (0 to 11 μM).

Nitrogen, usually in the form of nitrate, is taken up and stored by a cell or converted to cellular nitrite (via nitrate reductase) and ultimately to ammonia

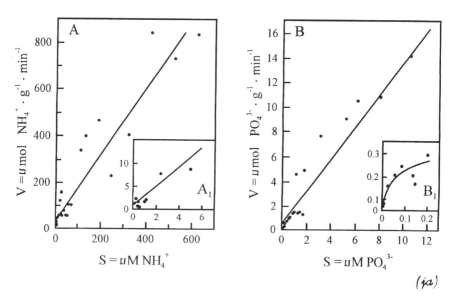

(ja)

Figure 4-12. Uptake kinetics of ammonium and phosphate by *Gracilaria tikvahiae*. Pulse feeding experiments were used to measure uptake of all three concentrations of ammonium (A: 0.02, 0.2, 2.0 mM) and the lowest alone (A1: 0.02 mM). A parallel experiment used three concentrations of PO4 (B: 0.002, 0.02, 0.2 mM) and the lowest alone (B1: 0.002 mM). Uptake of ammonium was linear, regardless of concentration, whereas that of phosphate varied according to substrate concentration (modified from Friedlander and Dawes, 1985; with permission of the Journal of Phycology).

(via nitrite reductase). The conversion does not require ATP as the process is a decrease in free energy. Ammonia is usually toxic to all organisms at concentrations above 1.0 mM (e.g., photosynthetic reactions are stopped). Thus, ammonia must be rapidly incorporated, and this is done by the enzyme glutamine synthetase in the plastid and cytoplasm. Ammonia is assimilated into the amide position (via ATP) of glutamate to form glutamine via the enzyme glutamine synthetase. The second step is the transfer of the amide nitrogen to 2-oxoglutarate to produce glutamate. Ammonia assimilation takes place in the chloroplast, where ATP and reduced ferredoxin are available. Thus, ammonia assimilation is actually part of a photosynthetic process in plant cells, with a major source of ammonia coming from respiration, where two molecules of glycerine are converted to one molecule each of serine, ammonia, and carbon dioxide.

Phosphorus is acquired primarily as orthophosphate ions and to a lesser extent via inorganic polyphosphates and organo-phosphor compounds. Phosphomonoesterases (phosphotases) on the cell surface are required to cleave the ester linkages joining phosphate groups to organic compounds. Once in the cell, inorganic phosphate is stored in vacuoles or polyphosphate vesicles (Lundberg et al., 1989) or incorporated into phosphorlated compounds (Chopin et al., 1990). Storage in polyphosphate bodies does not occur in vascular plants but is common among algae (Lundberg et al., 1989).

Controlling Factors

As in other metabolic processes, uptake of nutrients is affected by a various abiotic (temperature, salinity, irradiance, water movement) and biotic (morphology, cell size) factors. Nutrient-uptake studies have been the subject of numerous experiments with regard to the mariculture of economically important seaweeds such as *Gracilaria* (see Table 4 in Dawes, 1987). Hanisak (1987) reviewed the relationship among water temperature, flow rates, aeration, pH, nutrients, and carbon fixation based on tank culture studies of *G. tikvahiae*. If maximum uptake of CO_2, nitrate, and phosphate is to occur, optimal ranges for each abiotic factor must also be present in the habitat.

In her studies on seagrasses and water currents, Koch (1993) showed that currents up to a critical level (0.25 cm s^{-1} for *Thalassia testudinum*, 0.64 cm s^{-1} for *Cymodocea nodosa*) resulted in improved CO_2 diffusion and increased photosynthesis. Above critical current levels, carbon fixation was limited by enzymatic kinetics. The different critical current levels for the two seagrass species may be due to different blade morphologies that affect friction velocities and turbulence. The latter will affect the thickness of the boundary layer. In the giant kelp *Macrocystis pyrifera*, increased irradiance and water current can enhance nitrate uptake but not ammonium (Wheeler, 1985).

CHAPTER 5

Human Affairs and Marine Plants

Hugo Grotius in his volume *Mare Liberum* (1609) considered that the seas could not be spoiled and were free for the use and benefit of all. The free use of the seas has been the prevailing philosophy of governments up to the middle of the twentieth century when environmentalists, the fishing industry, and managers of bays and estuaries began to identify serious losses and increasing levels of pollution. Covering over two-thirds of Earth, the oceans and coastal waters have been viewed as sources of food, energy, waste removal, and transportation. The most heavily impacted marine plant communities have been estuarine and nearshore coastal systems, which are close to human habitation. Almost all texts dealing with marine plant communities offer examples of human impacts. In addition, there are books that focus on algae and the influence of humans (Jackson, 1968; Lembi and Waaland, 1989). Human influences on marine plant communities can be grouped into direct impacts including mechanical damage, pollution, and biological and indirect effects such as global warming and changes in sea level. The positive side in human affairs dealt with in App. B are the uses of marine plants that with proper management can be renewable resources for farming, food, and industry.

DIRECT DAMAGE: PHYSICAL

Mechanical damage in marine plant communities is most often due to the physical effects of construction. Long-term effects of industrialization and urbanization on marine plant communities are an active topic of research and presently they are intensely regulated in most developed countries. *Dredge-fill* operations for the building of ports and harbors, dredging of channels, and general development of coastal communities have resulted in extensive losses of tidal and seagrass communities. For example, in Tampa Bay, 81% (30,273 ha) of seagrasses and 44% (12,857 ha) of mangrove and salt marshe land have been lost due to construction and resultant turbidity from runoff and pollution (Lewis and Estevez, 1988). Coastal construction has been directed toward the transport of vessels (shipping channels, harbors) and the protection of coastal property (jetties, sea-walls), as well as improvement of tourist areas (renourishment of beaches). The direct effects of construction result in removal of marine plant

Figure 5-1. Storm damage to a groin and wooden seawall near Sarasota Bay, Florida (courtesy of R. A. Davis, University of South Florida, Tampa).

habitats and thus the loss of communities and their filtering, stabilizing, and protective roles.

The effects of storms on developed beaches without the typical coastal dune vegetation are well known. Thus, *seawalls* (parallel to the shore) have been constructed to protect residential and industrial property, and *groins* (perpendicular to shore) are established to reduce beach erosion. Seawalls are intended to protect property from direct wave erosion and groins are supposed to trap sediments moving along the shore and thus increase the width of beaches. Both seawalls and groins usually prove ineffective against storms even in low-energy coasts (Fig. 5-1) due to occasional storms. Unless massive barriers are made, wave impact and erosion will require constant repair and replacement of both types of protective structures. By contrast, sand dune vegetation [Figs. 1-5 and 5-5(A)] and coastal plant communities can absorb the effect of water motion.

Within the last 40 years, a code of restrictions has been initiated on marine construction at the national, state, and local levels within the United States. Similar regulatory efforts were previously initiated in Great Britain and are now in effect in most other countries. Unfortunately, circumvention of these regulations occurs frequently. Historical comparisons of large construction projects in coastal and estuarine habitats are more easily documented than small, incremental modifications to bridges, roads, residential land, and other developments. Long-term losses of marine vegetation can be easily determined by comparing old and recent aerial photographs. For example, development of bridges, roads, and dredge-fill operations has been documented in Tampa Bay's National Estuarine Program review. By contrast, small modifications due to private homes or small businesses (marinas, hotels, parks), can go unnoticed unless regular aerial reviews are conducted. Even so, these smaller effects can result in serious modifications to bay productivity over time.

Shallow-water marine plant and animal communities also can be damaged and stressed due to boats and docks. The increased use of power boats will cause *propeller cuts* in shallow seagrass beds. Motor boat scars can be seen in all parts of the world where seagrass and algal communities occur in the shallow bays and estuaries (Fig. 5-2). With the increase in the number of private motor boats, the problem of seagrass bed destruction has become very serious. In Florida alone, more than 70,040 ha of the 2,700,00 ha of seagrass meadows show scarring due to propellers (Sargent et al., 1994). The problem in propeller cuts is the slow rate of recovery after the seagrass beds are damaged. For example, the rate of recovery for turtle grass cut by a propeller requires 4.1 to 7.5 years (Dawes et al., 1997). Other types of boat damage can be seen in anchor damage to coral reefs and seagrass beds. Anchorage at many coral reef areas worldwide is now forbidden, the boats tying up to permanently anchored floats.

Figure 5-2. Propeller cuts in shallow subtidal seagrass beds of Tampa Bay. The aerial photo of a channel between mangals shows the prop cuts (arrows) through the beds (courtesy of Dr. Nick Ehringer, Hillsborough Community College).

DIRECT DAMAGE: POLLUTION

Another stress on natural communities is seen with the addition of nutrients and other toxic materials or *pollution*. Thus, a wide variety of pollutants that impact communities include thermal effluents, oil spills, sewage and garbage, organic chemicals, metals, and debris. Some of these pollutants are *toxic* (e.g., high temperatures, pesticides, herbicides, chemicals, components of oil spills), whereas others enhance plant growth being *eutrophic* in nature (sewage, garbage, some components of oil spills). Comparative historical studies are needed to determine if serious damage has occurred due to such pollutants, and it is usually more difficult to determine the effects from pollution than from physical damage. In some cases, reviews of marine plant floras may not demonstrate any serious loss. For example, Edwards (1975) compared four seaweed collections made between 1793 and 1971 from one of the most polluted shores in Great Britain, County Durham, and found only a 17% reduction in species over almost two hundred years.

Thermal Pollution

Increases in seawater temperature can occur when industrial complexes discharge water, causing thermal stress in the coastal communities. As temperature is one of the most fundamental abiotic factors for all organisms, it is not surprising that an increase of 1°C can result in a 10% decrease in macroalgal species (Devinny, 1980). Thermal stress is particularly prevalent where ambient seawater temperatures are near the upper limits for marine plants, for example, in subtropical and tropical areas. Studies on the effect of thermal effluents have shown decreased diversity and biomass for macroalgae in Korea (Kim et al., 1992) and New England (Schneider, 1981). Thermal loading in the salt marshes of Maine resulted in a 48% decrease in biomass during the first year of discharge, and a further decrease of 40% occurred in year two (Vadas et al., 1976).

Oil Spills

Crude oil (petroleum) is a mixture of about 10,000 types of hydrocarbons, with about 20% being paraffins (aliphatics), 55% naphthenes (cycloalkanes), and 20% aromatics (benzenes). The last contains many toxic compounds (Preston, 1988). In areas such as Santa Barbara, California, natural seeps release oil, asphalt, and tar into the sea from fractures; these hydrocarbons are used as sources of energy by bacteria. Major problems can occur when massive amounts of petroleum are suddenly released. Oil spills from supertankers such as the *Torrey Canyon* in 1967 (139 million liters), the *Castillo de Bellver* in 1983 (293 million liters), and the *Exxon Valdez* in 1989 (40 million liters) were dwarfed by the deliberate dumping in the Persian Gulf by Iraqi troops of 1,110 million liters in 1991. The Kuwait crude oil was transported by currents south along the northeastern shoreline of Saudi Arabia where shallow subtidal seagrass and intertidal salt marsh and limited *Avicennia marina* communities occur. The seagrass communities along the coast of Saudi Arabia did not show any damage (Kenworthy et al., 1993); in contrast, intertidal communities showed losses of algal mats (Al-Thukair and Al-Hinai, 1993) and damage to salt marsh and mangroves (Boer, 1993). Such events can cause both short- and long-term damage to coastal and benthic communities.

Because oil spills include releases of both heavy and light compounds, they effect plants through the uptake of lighter toxic material and by coating the plant with the heavier components. In addition, there can be synergistic interactions between the oil coat and exposure to bright sunlight, as shown for red and black mangrove seedlings (Proffit et al., 1995). Uptake of the toxic benzene-containing compounds causes damage to cellular metabolism of epifaunal and infaunal inhabitants of oil-polluted communities. Coating of seagrass blades, salt marsh, and mangrove stems and roots by heavier components can prevent gas exchange. The roots and stems of mangroves have openings called *lenticles* (Chap. 10), which allow gas exchange for roots growing in anaerobic environments. If sealed, due to tar or viscous oil, the roots can die. For example, a minor spill of crude oil (Bunker #2) at the mouth of Tampa Bay coated the black mangrove pneumatophores so that their lenticles were blocked. There was no initial effect, but after one year, some trees dies. Other trees showed increased lenticle growth, proliferation of branched pneumatophores, and adventitious roots, which may be a response to limited availability of oxygen, as seen in the Persian Gulf oil spill (Boer, 1993).

The effects of a crude oil spill depends on the type of coastal community. An oil spill on an exposed rocky coast will have a more limited effect in contrast to a long residence time (10 years or more) in unconsolidated, protected coasts with tidal communities and seagrass beds. Because oil does not penetrate hard substrata, recovery by macroalgae is usually within two to three years. Most intertidal macroalgal communities were immediately destroyed after the *Torrey Canyon* oil spill on the southwestern tip of England. However, within two months, pioneer (r-selected) green algae (e.g., *Enteromorpha, Ulva*) covered the rocks. Within a year, the green algal-barnacle monoculture was succeeded by brown and red seaweeds, and within five years, the community had returned to prespill conditions. In another study, Foster et al. (1971) found that after a massive oil spill off Santa Barbara, California, there was an immediate dieback of the rocky intertidal marine algae. However, recovery of the seaweed community was complete within two years. Oil spills in the tropical waters of Puerto Rico resulted in losses of mangrove forests that had not recovered after 15 years. Similarly, damage to coral reefs near Eilat, Israel, in the Red Sea allowed the colonization of macroalgae on the dead coral skeletons. In the first case, oil contaminants remained in the unconsolidated sediments, and in the second, the slow-growing corals were killed, so that a macroalgal community dominated.

Domestic Wastes

Whereas nutrification is an increase in nutrients that results in a visible community shift, *eutrophication* refers to the input of either dissolved nutrients or organic matter that results in excessive oxygen demand. Of all the aspects of pollution, the addition of garbage and sewage are most commonly cited as sources of eutrophication. Barges daily haul compacted garbage from New York City that is dumped into the ocean just a few kilometers offshore from the harbor. San Diego generates over 2000 tons of feces, 9.3 million liters of urine every 24 h, which is discharged into the Pacific Ocean through a 4-km-long pipe after only primary treatment (Milne, 1995). Effluents from wood pulp mills contain large quantities of waste organic matter, including tannins and lignins and toxic compounds such as hydrogen sulfide, methyl mercaptans, and wood resins. Agricultural runoff into rivers and streams can contribute to enhanced levels of nutrients in estuaries.

Nutrient-rich sediments resulting from deforestation, dredge-filling, and channelization are carried by rivers to the coast. Because many rivers are channelized, the nutrient loads that would have been dispersed over flood plains are now carried directly into the sea. For example, the intensive channelization of the Mississippi and contributing rivers has resulted in thousands of tons of nutrients being carried to the northern Gulf of Mexico (Davis, 1986). The result is an erosion of the Mississippi Delta due to lack of sediments and the formation of a 50+ km^2 "dead zone" off the coast of Alabama. The latter has been formed from nutrient-rich sediments discharged from the Mississippi River.

Just as fertilizers support growth of terrestrial crops (and weeds), domestic wastes support blooms of macroalgal and phytoplankton. The effects of eutrophication are many and include (1) decreases in dissolved oxygen that can reach *anaerobic* (without oxygen) levels, (2) shifts in species so that early (r) strategists dominate and diversity declines, and (3) degradation of the habitat from decreased light transmittance due to decreases in turbidity and increases in phytoplankton growth. The decline of seagrasses in Chesapeake Bay (Kemp et al., 1983), Tampa Bay (Lewis and Estevez, 1988; Johansson, 1991), the Dutch Wadden Sea (Giesen et al., 1990), and along the Florida Keys (Lapointe et al., 1994) are some examples of increased turbidity due to enhanced phytoplankton growth, noted in the first three studies just cited, and increased macroalgal and seagrass epiphytes, noted by Lapointe et al. (1994). Dredging can contribute to *nutrification* (an increase in nutrients) by suspending nutrient-rich sediments and lowering water transparency via release of particulate material in the water column.

Eutrophication from complex domestic wastes can result in higher biological oxygen demand (BOD), a reduction in species diversity, plankton blooms (Chap. 7) that reduce water transparency, and a shift to rapid macroalgal colonizers (e.g., green algae). Littler and Murray (1974) found that macroalgae growing near a sewage outfall in southern California had higher net production and were small-growth forms with simple, short life histories. In other words, they were dominated by pioneering plant species. As a general rule, the more *oligotrophic* (lacking nutrients) the habitat, the greater the effect of organic enrichment on the community. There can be exceptions to the preceding statement as seen in Western Australia, where there is a lack of phytoplankton and seagrass epiphytes, so that the effects of increased nutrients did not change the seagrass community (D. I. Walker, personal communication). The decline of coral reefs in Hawaii and the Florida Keys (Sorokin, 1993) and documented *harmful macroalgal blooms* (Chap. 8; Bach and Josselyn, 1978; Merrill and Fletcher, 1991; Valiella et al., 1992; Maze et al., 1993; Lapointe, in press) are examples of the effects of nutrification. Eutrophication and nutrification have resulted in massive macroalgal blooms of *Cladophora vagabunda* and *Gracilaria tikvahiae* in Waguoit Bay, Massachusetts (Valiella et al., 1992), of *Cladophora prolifera* in Bermuda (Bach and Josselyn, 1978; Lapointe and O'Connell, 1989), and a "green tide" of *Ulva lactuca* in the Venice Lagoon and North France (Merrill and Fletcher, 1991; Maze et al., 1993). The green tide problem has required a costly ($6.1 million) yearly removal off the coast of France and Venice in the northern Mediterranean.

The frequent blooms of the brown filamentous alga *Pilayella littoralis* in Nahant Bay, Massachusetts (Wilce et al., 1982), do not appear to be caused by anthropogenic nutrification, but rather by water circulation and nutrient-rich upwellings (Donald Cheney, personal communication). The brown seaweed will develop intense biomass seasonally and then is washed ashore, where it rots on public beaches. Lapointe (in press) has described macroalgal blooms on coral reefs off West Palm Beach, Florida (*Codium isthmocladum*; Fig. 5-3), and Discovery Bay, Jamaica (diverse multilayered algal community including *Chaetomorpha linum*, *Lobophora variegata*, and *Sargassum polyceratium*). The two macroalgal blooms are supported by a bottom-up increase of nutrients rather than the loss of grazers. The bloom of *Codium isthmocladum* on deep reefs (20 to 30 m) appears to be linked to nutrient-rich water from urbanization, sugar cane farming, and channelization. The multilayered algal bloom off Discovery Bay is also supported by eutrophic groundwater seeping from fractures in the underlying porous limestone.

Marine algae are used as *phytometers* or *bioindicators* of polluted conditions (Levine, 1984; Haglund et al., 1996) because they are sessile permanent monitors of an area, easily collected, and tend to accumulate a variety of compounds. Native populations of seaweeds can be used to monitor levels of chemical elements released from mining (Vasquez and Guerra, 1996), and growth inhibition of laboratory cultivars of *Gracilaria* is effective in toxicity assessments (Haglund et al., 1996). The kelp forests in the British Isles have been monitored via a model of their energy web in order to predict eutrophic conditions (Bellamy et al., 1968). Some seaweeds such as the brown rockweed *Ascophyllum nodosum* show incremental growth, each internode cor-

Figure 5-3. The native green alga *Codium isthmocladum* has overgrown the coral reefs that occur 20 to 30 m off West Palm Beach, Florida (courtesy of B. E. Lapointe, Harbor Branch Oceanographic Foundation).

relating with a year's growth. Thus, the different internodes of *A. nodosum* allows historical tracing of water quality or environmental conditions (Levine, 1984).

Synthetic Hydrocarbons

Pesticides, herbicides, PCBs, antifouling compounds, and components of pulp mill effluents are some examples of synthetic hydrocarbons. Although all of these substances are toxic, they usually occur in such low concentrations in seawater that they do not pose a serious problem. Even so, *biomagnification* of these compounds can cause their accumulation in food chains. The compounds are concentrated at each level, from phytoplankton to various zooplankton, filter feeding invertebrates including shell fish, fish, and finally to birds and man. At sufficiently high concentrations, these compounds are toxic to birds, fish, and humans.

Pesticides include the degradable organic-phosphate compounds such as parathion and persistent chlorinated hydrocarbons like DDT. The long-term effects of DDT are known to cause thin egg shells in brown pelicans, which results in the death of the unhatched chicks or crushing of the shell during incubation by the mother. The causes of reduced photosynthetic rates in macroalgae and phytoplankton by DDT are less understood. *Herbicides* (e.g., Paraquat,

Diquat) are intensively used along freshwater streams and in lakes to reduce weed cover; hence, they may enter diverse estuarine and coastal communities. Spore germination and growth plus photosynthesis of macroalgae are all negatively affected by herbicides such as Paraquat and phenocarboxylic acid derivatives (3 AT).

A group of hydrocarbons called *PCBs* (*p*olychlorinated *b*iphenyl hydrocarbons) are similar to DDT in that they are resistant to degradation and persist for long periods within diverse substrata. Because they are highly stable compounds, as well as being toxic to most organisms, they are banned from production in many countries. Although PCBs may lower photosynthesis in phytoplankton, little else is known about their effects on marine plants. *Antibiofouling* organotins such as triphenyltin (TPT) and tributyltin (TBT) are extensively used as biocides on pilings and boat hulls. As biocides, these compounds are very toxic to macroalgae, apparently affecting mitochondria development and function (Evans and Hoagland, 1986). Unfortunately, because of their long life span, biocides can remain in the sediment of harbors and bays for many years causing a variety of impacts on marine plants and their associated fauna.

Heavy and Radioactive Metals

Heavy metals are normal constituents in seawater, occur in low concentrations, and, in some cases, are required by marine plants in trace amounts. Metals such as mercury, cadmium, lead, arsenic, copper, and chromium occur in industrial and domestic wastes and they can contaminate seawater and coastal areas. Mercury has a variety of industrial uses and so it is common in sewage. Lead contaminates are most abundant in runoff from roads and come from the exhaust of combustion engines that use leaded gasoline. Due to their low solubility, mercury and lead are not found in high concentrations in seawater, but they can reach higher levels in organisms via biomagnification. Heavy metals are passively adsorbed (Fuge and James, 1974) as well as taken up actively by macroalgae. The use of macroalgae as scavengers of nutrients from primary treated sewage water is very effective. However, with the possibility of uptake of heavy metals from the sewage, the algae may not be useable as fodder or as fertilizers.

As outlined by Rai and colleagues, the general order of toxicity of heavy metals to seaweeds is as follows: Hg > Cu : Cd > Ag > Pb > Zn (Rai et al., 1981). Although the direct effects of many heavy metals are poorly understood, various studies have shown that excessive accumulation (3 to 5 ppm) of mercury, lead, zinc, and copper can cause changes in cell appearance, growth, and metabolism of phytoplankton (see Round, 1981). Edwards (1972) reported that 0.01 ppm copper in seawater depressed the growth of red algae such as *Callithamnion*. Mercury inhibits enzyme functions resulting in cessation of growth, inhibition of photosynthesis, reduction of pigment levels, and increased cell permeability. The level of toxicity of diverse heavy metals is also related to their physical-chemical state. If they are bound to organic particles, their toxicity can be reduced. If in the form of a hydroxide, oxide, carbonate, or phosphate, they do not stay in solution because of the high pH of normal seawater. Hence, the dissolved concentrations of most heavy metals remain low.

Radioactive isotopes are created mostly by the operation of nuclear reactors (power plants, nuclear-powered vessels), but they also occur in medical materials and in military weapons. Marine plants can be exposed to radioactivity both from natural (e.g., ^{40}K) and man-made sources (e.g., ^{90}Sr). The stability or half-life of radioactive substances can result in long-term (long half-life) or limited (short half-life) exposure. The uptake of radioactive isotopes by marine plants can occur by adsorption, as shown in *Porphyra* with ^{106}Ru, and by active uptake, as shown by Bachman and Odum (1960) for marine benthic algae with ^{65}Zn. However, little information is available regarding their effects on such plants. Even so, there is strong evidence documenting the problem of bioaccumulation of radioactive isotopes within diverse food chains (Robertson, 1972).

Debris

The amount of nonbiodegradable garbage (debris) found in mangrove forests, salt marshes, and beaches demonstrates a lack of concern for our marine environment (Fig. 5-4). Types of debris include plastic bags and bottles, metallic objects, glass, and diverse materials such as fishing gear. Plastics are slow to degrade and are used extensively to make containers, monofilament fishing lines, polypropylene rope and coverings. Plastic debris now occurs from the Arctic Circle to tropical atolls and tends to concentrate in estuaries. In the Bay of Biscay and in Seine Bay, an average of 14 pieces per subtidal hectare were collected off the continental shelf (Galgani et al., 1995) with up to 95% being plastic bags. However, most debris, being plastic, will float. Debris is usually not a direct problem for marine plant communities and may even offer new substrates for epiphytic and macroalgae. However, debris can cause physical damage to an intertidal community by covering substrata and if of sufficient size will remove plants during storms.

BIOLOGICAL DAMAGE

The harvesting of marine animals has increased throughout the past three centuries, causing a serious decline in fish and invertebrate (shell fish, shrimp, lobster) yields. In part, this is due to overfishing, causing a decrease in the average adult size and volume of fish and invertebrate catches. The decline of these economically important animals is also due to a loss of coastal marine plant communities, which are used as nurseries and habitats for commercial and supporting organisms. The destruction of salt marshes, mangroves, seagrass, and seaweed habitats probably is reflected in declining harvests of commercial fish and shell fish as well as wide fluctuations in population size. Harris (1996) is

Figure 5-4. Miscellaneous debris collected within three minutes in a mangal on the west coast of Florida.

questioning what the state of seaweed and faunal communities will be in the Gulf of Maine due to the intensive harvesting of the sea urchin *Strongylcentrotus drobachiensis* and subsequent invasions of other species. He has observed what appears to be an unstable progression of urchin barrens to algae, to mussels (*Mytilus edulis*), to sea stars (*Asterias* spp.), to algae over a two-year study in the subtidal regions of southwest Gulf of Maine. Complicating factors in Harris's 20-year review is a rise in sea water temperature (18 to 20°C) and invasions of an exotic green alga (*Codium fragile*), colonial tunicates (*Botrylloides diegensis* and *Diplosoma* spp.), and a bryozoan (*Membranipora membranacea*) into urchin barrens.

Another problem that results from overfishing is the disruption of the food chain. Overgrazing may be linked to an imbalance in predator-prey populations (Chap. 3, "Predation"). Such imbalances have caused overgrazing of seaweed communities by urchins (Tegner, 1980; Schiel and Foster, 1986) when predators have been removed or overgrowth of corals by algae when grazers are in limited supply (Glynn, 1990). The subtidal kelp-dominated communities have been replaced by lower productive coralline "barrens" in many areas of the northwest Atlantic Coast in the Gulf of Maine (Vadas and Steneck, 1988). The barren areas appear to be linked in part to increases in sea-urchin populations that will die back due to disease during periods of higher water temperature (Chap. 3, "Predation"; Elner and Vadas, 1990). However, the increases in urchin populations is also linked to the lack of their natural fish predators because of overexploitation (Vadas and Steneck, 1995). The authors point out that overfishing along the western North Atlantic over the past three centuries has not only resulted in closing of fishing regions, but in a semipermanent change in the coastal seaweed communities.

Harvesting and Aquaculture

Use of nets to collect shrimp, mullet, and other fish in shallow coastal waters has been previously cited as destroying the shallow seagrass beds and smaller animals that play a role in the food chain for commercial fish. Laws banning the use of nets in coastal waters have become more common in the United States (e.g., Florida Net Ban of 1995) with the intention of limiting damage to the seagrass and macroalgal communities and protecting other organisms of the food chains. Fishery managers hope that improvement in habitats, increases in primary consumers within food chains, and reduced damage to subtidal communities will increase the number of secondary consumers. Such an expectation is open to question and the loss of coastal wetlands and subtidal habitats may already be so excessive that a return to previous fish and invertebrate productivity is impossible.

Aquaculture plays an ever increasing role in production of shrimp, fish, and seaweeds throughout the world (Chap. 6). However, in the past 50 years, large areas of coastal wetlands have been destroyed to produce shrimp, fish, and algal ponds, especially in the tropics (Fortes, 1988, 1991). For example, shrimp pond production and logging in Ecuador have resulted in a loss of 28,500 ha of mangrove forest or 14% of the total mangrove area (Blanchard and Prado, 1995). Mangrove destruction in Ecuador is occurring at an annual rate of 3,742 ha y^{-1} (Clirsen, 1992). Throughout the tropics, mangrove communities are in danger of being destroyed in order to develop shrimp, fish, or algal farms (Linden and Jernelov, 1980). Similarly, 100 years ago, vast areas of salt marshes of in the southeastern United States were converted to rice farms through dikes and diversion of freshwater. The good news is that most of these salt marshes have recovered, although the original dikes are still visible in Georgia and South Carolina.

Exotic Species and Biological Pollution

Establishment of introduced plants in new environments (*exotics*) is not as well documented for marine as for terrestrial communities. Criteria for identifying introduced species may include one or more of the following (Ribera and Boudouresque, 1995): (1) The species is new to the region. (2) There is a geographic discontinuity between the species known range and the new population. (3) The new location is highly localized. (4) A reasonable migration pattern can be established. (5) The species will dominate its new site by occupying niches originally held by natives. (6) The site of introduction is nearby. (7) The species is genetically identical to its original, geographically discontinuous range. In most examples, the exotic will suppress resident plants through niche replacement and eventually alter the community structure and function

(Thompson et al., 1987). Gray (1986) could find no common set of genetic attributes that would characterize exotics and suggested that the new habitat may "break genetic molds" after introduction.

Exotic microalgae and planktonic species (Bolach, 1993) and *macroalgae* (Ribera and Boudouresque, 1995) have been documented. Exotic seaweeds are known in the Mediterranean (60 taxa: 40 red, 12 brown, and 8 green algae), the European Atlantic (26 taxa: 18 red, 4 brown, and 4 green), South Australia (19 taxa: 4 red, 8 brown, and 7 green), but less so elsewhere. Ribera and Bourouresque offer seven ways in which exotic seaweeds may be introduced: (1) attachment of ship's hulls (*Antithamnion algeriense* into the Mediterranean), (2) removal of ballast water from a ship nearby land (many microalgae such as the diatom *Biddulphia sinensis* into Europe), (3) fishing bait (*Fucus spiralis* into the French Mediterranean), (4) scientific research (*Kappaphycus alvarezii* into Hawaii), (5) aquaculture (deliberate introduction of *Macrocystis pyrifera* into China), (6) aquaria (possibly *Caulerpa taxifolia* into the Mediterranean), and (7) migrations through man-made connections like the Suez Canal (*Solieria dura* into the eastern Mediterranean).

One of the most intensely documented examples of an exotic seaweed (over 300 articles) is the fucoid alga *Sargassum muticum* (Critchley et al., 1990). The species was introduced on oyster spat from Japan to the Pacific coast of North America and the Atlantic coast of Europe. Currently, it has extended along much of the Pacific coast of North America, as well and into Europe to the Mediterranean. The alga spreads by both vegetative propagues and zygotes and forms dense canopies that shade out local plants. Other seaweeds that are exotic to European coasts include the brown algae *Undaria pinnatifida* and *Laminaria japonica*, both of which were probably carried (i.e., their spores) in ballast waters of ships (Rueness, 1989) or escape from cultivation sites (Floc'h et al., 1996). The former is now a permanent member of the Atlantic flora with a preference for settlement on artificial structures (Floc'h et al., 1996).

A number of exotic marine green algae are also known. The coenocytic alga *Codium fragile* subspecies *tomentosoides* was introduced in 1956 from western Europe to Long Island, New York. It now covers large areas of rocky shores and commercial oyster beds in Long Island Sound, resulting in loss of habitat and oyster yields. The alga has spread south to New Jersey and north to Nova Scotia (Carlton and Scanlon, 1985). Recently, it has established on New Zealand rocky shores (Trowbridge, 1995), where native herbivores exert little grazing pressure, although there are at least four gastropods, two echinoids, a generalists snail (*Turbo smaragdus*), and two specialist ascoglossans (*Placida dendritica* and *Elysia maoria*). *Caulerpa taxifolia*, another coenocytic green alga, has over 191 documents dealing with its invasion and replacement of seagrass communities in the Mediterranean (Boudouresque et al., 1996). It was first discovered off the coast of Monaco in 1984 and either was introduced from the Monaco Aquarium (Meinesz and Boourouresque, 1996) or is an invader from the Red Sea (Chisholm et al., 1995). Regardless, the species has a broad temperature tolerance, grows well in eutrophic water, and has significantly replaced seagrass meadows of *Posidonia oceanica* as well as algal beds (Meinesz et al., 1993). By 1994, the plant had covered an area of 1000 to 2000 ha and was increasing by factors of 2 to 10 annually (Boudouresque et al., 1995). Typical of exotics, the replacement of *P. oceanica* by *C. taxifolia* has resulted in a significant loss in the diversity of native algae (Verlaque and Fritayre, 1995). Caulerpa's biomass averages 500 g dwt m^{-2}. It will form extensive meadows in 9 m and is apparently adapted to oligotrophic waters (Meinesz et al., 1995; Delgado et al., 1996).

Exotic *salt marsh* species are less well known. In salt marshes of Europe, *Spartina anglica* is a polyploid of the *S. townsendii* complex (*S. alterniflora* × *S. maritima*; Guenegou et al., 1988) and has replaced the natural species and vegetatively filled-in bays that originally were open (Gray et al., 1991). *Spartina anglica* is an aggressive species, which was intentionally introduced to increase salt marsh coverage in New Zealand and China. In a similar fashion, *S. alterniflora*, a native of the east coast of North America, has been introduced into San Francisco Bay in order to improve salt marsh productivity (Callaway and Josselyn, 1992). Another example of a salt marsh exotic is the purple loosestrife (*Lythrum salicaria*). It is a perennial species that can invade low salinity regions of salt marshes (Thompson et al., 1987). It is even a more serious exotic of freshwater wetlands throughout the United States. *Lythrum* is typical of any exotics whose distribution and spread have been enhanced by the lack of natural enemies, as well as its ability to inhabit disturbed areas. Purple loosestrife is tolerant of salinities up to about 5 ppt and can form large monotypic stands. Studies of control agents present in Europe, the source of purple loosestrife, suggest that three insects, a root-mining weevil (*Hylobius transversovittatus*) and leaf-eating beetles (e.g., *Balerucella calmariensis*), can be used in the United States (Malecki et al., 1993).

A subtropical exotic, brazilian pepper (*Schinus terebinthifolius*), demonstrates a similar problem as the temperate loosestrife. Brazilian pepper has replaced large parts of the ecotone vegetation between the mangrove forest and terrestrial vegetation in subtropical and tropical regions of Florida. Because the tree can tolerate low levels of salt, it has replaced *Chonocarpus erecta* (Buttonwood) and even *Laguncularia racemosa* (white mangrove) along the west coast of Florida. Current research is being directed to possible biotic vectors that would attack the tree.

INDIRECT DAMAGE

Of the various anthropogenic influences that can affect marine plant communities, two global atmospheric modifications seem to be increasing, increases in carbon dioxide gas and a depletion of the ozone layer. The former may cause major problems of *global warming*, and the latter may result in increased *ultraviolet-B* (UV-B) radiation.

Global Warming and Sea Levels

The escape of solar infrared radiation (IR) from Earth's surface is slowed by atmospheric CO_2 and other rare gases including methane. Without CO_2, Earth's surface would cool to $-18°C$. The present concentration of atmospheric CO_2 is 350 ppm and this reflects an increase of about 70 ppm since the industrial revolution began around 1850 (Davis, 1986). The increase in CO_2 concentration is due to burning of fossil fuels and the removal of forests. The result is a slight rise +0.5°C) in the average global temperature over the last 100 years (Milne, 1995). The increase in atmospheric CO_2 and subsequent rise in global temperature is called the *global warming*, while the trapping of infrared is the *greenhouse effect*. With the continuation of forest removal and burning of fossil fuels, atmospheric CO_2 concentrations could reach 600 ppm by the year 2080 and an average global temperature increase of 1.5 to 4.5°C (Davis, 1986). Warming of the atmosphere would cause increased respiration and breakdown of organic compounds by bacteria and fungi and thus more CO_2 would be released. A positive feedback (higher temperatures, more CO_2 released) can then result in further increases in temperature. It is thought that even a slight rise in sea temperatures would result in major shifts of marine species, including changes in phytoplankton diversity and extensions of coral reefs extend to higher latitudes. Such changes would cause major shifts in the biogeography of marine plant communities, as seen in an historical study of invertebrate abundance at Monterey, California (Barry et al., 1995). The abundances of 45 invertebrates for 1931–1933 and 1993–1994 showed increases in eight of nine southern species and decreases of five of eight northern species.

An increase in air and sea temperatures could result in a differential warming of the polar caps when compared to tropical oceanic regions. The polar caps would warm because continental air heats faster than air over tropical oceans. Thus, in addition to a rise in seawater temperature, a melting of the huge reserve of fresh-water at the poles will increase the sea level. Presently, sea levels are rising and it appears the rate is increasing, supporting the greenhouse effect theory. The present-day absolute or *eustatic* rise of the oceans is estimated at 1.2 mm/y or 12 cm/century (Gornitz et al., 1982) and projected to increase from 30 to 100 cm/century (Davis, 1986; Milne, 1995). The *apparent* (local) changes in sea level in North America vary from 0.1 cm (Alameda, California) to 4.0 cm (Sandy Hook, New Jersey) over the past 40 years (Stevenson et al., 1986). The variations in apparent changes of sea level reflect not only eustatic changes, but also land subsidence, sediment supply, wave energy and erosion, and changes in rainfall patterns.

If the rate of increase in sea level continues, this would almost certainly result in flooding of coastal wetlands. Globally, a rise in sea level of 33 cm would result in a landward advance of saltwater 30 m (Milne, 1995). However, in low-lying areas having limited terrestrial relief, a 33-cm rise could result in seawater extending many kilometers from the coast. It is difficult to predict whether specific wetlands can survive increased sea-level rises. For example, 4 out of 15 salt marshes of the eastern United States and Gulf Coast had accretion rates that were lower than the present local rise in sea level (Stevenson et al., 1986). Shifts in accretion may be due to natural changes in sediment supply, tidal creek patterns, and anthropogenic influences. The large marshes of the southeastern United States may have expanded due to deforestation and farming, which increased sediment load due to erosion in the 1800s (Stevenson et al., 1988). Ironically, recent reforestation, reduction of farming, dams, and flood control of rivers may lower the sediment load and cause reduction of the same marshes.

Glaciation and deglaciation events have occurred regularly over the past few million years, resulting in coastline shifts worldwide. The present situation of high sea levels and extensive estuaries has existed for about 10 to 20% of the past few million years. For example, the sea level was 100 m lower about 15,000 ybp (years before present) and similar to present-day levels about 35,000 ybp (Davis, 1986). Thus, fluctuations in sea level have been rapid on an historical basis, suggesting that the tidal communities can survive. Geological profiles of existing salt marshes and studies in marshes affected with land subsidence (e.g., San Francisco Bay) indicate survival is possible if the rate of sea-level changes or land subsidence is slow (Callaway and Josselyn, 1992). However, if it is rapid, as in Louisiana salt marshes (Chap. 9) or Pacific atoll mangroves (Chap. 10), permanent flooding will kill the plants as sedimentation is too slow to allow buildup of the marsh. It appears that the key for survival of salt marsh and mangrove forests lies primarily in the rate of sedimentation versus the increase in sea level (Ellison and Farnsworth, 1996).

In addition to changes in sea level, global warming would reflect an increase in atmospheric CO_2 levels (Davis, 1986) and thus increased availability to submerged plants. Much speculation is being made as to what responses microalgae and macroalgae would have to an increase in available CO_2. The two major groups of marine macrophytes are seaweeds (macroalgae) that evolved in the oceans about 1 billion ybp and seagrasses that probably originated about 90 million ybp (Chap. 1). Seagrasses had to adapt to media where the diffusivity of CO_2 (Chap. 2) is many orders of magnitude lower than in air, and they evolved when the pH of seawater was lower and CO_2/HCO_3^- ratios were higher in Cretaceous seas than today (Beer and Koch, 1996). Thus, seagrasses today are at a "disadvantage" compared with seaweeds in their ability to obtain the photosynthetic substrate under high pH and high CO_2/HCO_3^- concentration ratios of today's oceans (but see Beer, 1996). Beer and Koch (1996) suggest that should global CO_2 levels again rise, nearshore seagrasses may proliferate and outcompete macroalgae.

Ozone and UV-B

At sea level, about 50% of the 900 J $m^{-2}s^{-1}$ of the sun's total irradiance consists of infrared wavelengths, while 5% is UV and the remainder visible (PAR) irradiation (Salisbury and Ross, 1978). The ultraviolet spectrum consists of UV-A (320 to 400 nm) equal to 4% of total PAR and UV-B (280 to 320 nm) or about 0.8% of PAR, while UV-C (190 to 280 nm) is absorbed in the stratospheric (10 to 50 km above sea level) ozone (O_3) layer (Vincent and Roy, 1993). UV-B can damage DNA and immune systems and can cause sunburn, skin cancer, and cataracts in humans but is mostly blocked by the ozone layer. Reduction of stratospheric O_3 would increase the intensity of UV-B at sea level (Milne, 1995). Thus, there has been international concern regarding the production of chlorine containing compounds called chlorofluorocarbons (CFC; e.g., Freon) that are used as refrigerants and propellants (aerosol cans). Being highly stable, CFSs will pass through the lower troposphere into the stratosphere, where UV radiation strips off the Cl^- ions that then react with and destroy O_3.

Increased UV penetration may cause enhanced cellular and photosynthetic damage for diverse organisms (Vincent and Roy, 1993) and, in particular, seaweeds, phytoplankton, and seagrasses (Larkum and Wood, 1993). The effect of increased UV-B on phytoplankton is particularly important, since they inhabit surface oceanic waters. If phytoplankton photosynthesis were significantly diminished by UV-B radiation, a serious loss of CO_2 fixation capability in the world's oceans would result. Inhibition of photosynthesis also occurs if seaweeds and phytoplankton are exposed to UV-B. In the Precambrian and early Paleozoic periods, the ozone layer had not developed and primitive seaweeds probably encountered higher levels of UV-B (Davis, 1986). Probably seaweeds were restricted to greater depths in the sea where the UV impact was lessened. As O_2 and O_3 increased in the atmosphere, seaweeds colonized shallower zones.

Protective mechanisms by seaweeds and phytoplankton to lessen the impacts of UV irradiance include avoidance, quenching, and screening (Vincent and Roy, 1993). Some organisms avoid UV wavelengths by growing at greater depths (i.e., deep-water seaweeds) or migrating downward during the day (i.e., phytoplankton) in the ocean. Seaweed cartenoids (e.g., β carotene) will neutralize (quench) singlet state oxygen radicals and will protect photosystem II reaction centers by releasing energy at the molecular level (Tefler et al., 1991). Mangrove and salt marsh plants may screen UV wavelengths via their epidermis, which absorbs about 95% of all UV light so that it does not reach the lower photosynthetic mesophyll cells. The brown alga *Ecklonia radiata* produces mycosporine-like amino acids (MAAs) that absorb in the 310–360 nm range (Marchant et al., 1991); these compounds increase during the Arctic spring (Wood, 1987). Similar studies on Antarctic seaweeds, which are exposed to higher levels of UV-B, have found that upper intertidal summer species of green algae contain high levels of various MAAs (Post and Larkum, 1993). In a comparative study, Maegawa and colleagues (1993) reported that deep-water red seaweeds are intolerant to full solar radiation (1–5% UV), while shallow-water species are more tolerant. Thus, the presence of UV-absorbing pigments may be a determining factor in vertical distribution of red algae.

MANAGEMENT AND RESTORATION

Until the 1950s, coastal wetlands were considered to be wastelands that could be "improved" by dredging, draining, or filling (Mitsch and Gosselink, 1993). Seagrass beds, being subtidal and therefore not visible, were not even considered as important habitats. The loss of a large percentage of these communities has resulted in significant decreases in water quality, and marine animals and terrestrial wildlife (e.g., Lewis and Estevez, 1988). The recent recognition of the importance of these habitats has caused increased regulatory, restoration, and management activities (see Chaps. 8 to 12).

Management

When discussing management of marine plant communities, two terms should be considered. Foremost, *conservation* is the protection and preservation of species, communities, or habitats, and *management* is the process of controlling the use and the maintenance of these communities. Management usually involves alteration and restoration of habitats for use by humans. Unfortunately, management practices can shift easily from maintenance or enhancement to exploitation of the resources, which is often beyond the carrying capacity of the habitat. An example of this is the "protection" of mangrove forests in Ecuador, where needs of shrimp farms override the maintenance of coastal wetlands (Blanchard and Prado, 1995). In the United States, management of coastal communities has shifted from simply laws and enforcement to the development of objectives for environmental protection, recreation, aesthetics, and production of renewable resources. With the recognition of the ecological importance of tidal wetlands and seagrass habitats, protection has expanded into restoration of damaged communities.

One national program within the United States is the National Estuarine Program (NEP), which has been established in many of the major estuaries throughout the country. The NEP coordinates the initial stages of estuarine protection and enhancement, with the goal of turning over management to local governments after 5 to 10 years. Similar efforts have been carried out in Great Britain and the Netherlands and to a lesser degree by countries bordering the Mediterranean. Managers have moved from traditional monitoring of water quality (e.g., water transparency, turbidity, nutrients) to studies of marine plant habitats as phytometers. An example of such management approaches is the Great Bay Estuary report (New Hampshire, Maine), which includes assessments of primary and secondary producing organisms as well as standard hydrological, geomorphological, and pollution data (Short, 1992). Such a management approach allows a broader assessment of the health of an estuary, as exemplified within Chesapeake Bay (Dennison et al., 1993). Submersed aquatic vegetation (SAV) was used to assess the health of Chesapeake Bay, including species of freshwater aquatics and seagrasses.

Legal Support

The "No Net Loss" (NNL) concept is probably the most significant wetlands policy established in the United States during the 1990s (Mitsch and Gosselink, 1993). The National Wetland Policy Forum ensured that wetlands should not continue to decrease and, where feasible, increase. A serious problem with the NNL is the legal definition of a wetland and this has continued to be a serious hinderance for its retention. Internationally, the Third Law of the Sea Convention in 1982 established the 12-mile (19.2-km) territorial sea and 200-mile (322.5-km) zone of exclusive control, which has allowed better control and regulation of coastal environments.

Restoration, Mitigation, and Creation

The recognition that salt marshes and mangrove swamps (Chaps. 9 and 10), coral reef (Chap. 12), and seagrass (Chap. 11) communities are valuable, highly productive habitats has resulted in attempts to expand and improve them (e.g., Lewis, 1982; Short, 1992). Rehabilitation or *restoration* of damaged communities and *creation* of new communities are two methods of producing more marine plant habitats. Presently, *mitigation*, the creation of new communities to replace ones that are destroyed, is used extensively where development (roads, bridges) will result in damage to a wetland. The success of wetland plantings is also linked to the development of nurseries and the improvement of planting techniques. For example, it is now possible to purchase salt marsh and mangrove propagules from nurseries so that existing communities are not damaged for donor stock. Probably the greatest success (survival) has been in the restoration of salt marshes because these communities extend from Florida to Canada (Broome et al., 1988) and thus numerous planting studies have been carried out. Techniques for planting salt marsh plants, mangroves, and seagrasses are published for many parts of the world. For example, plantings of dune vegetation in the Netherlands are highly successful [Fig. 5-5(a)]. In Florida, salt marsh [Fig. 5-5(b)] and mangrove restoration (Fig. 5-6) has also proven successful and detailed techniques for planting of tidal halophytes are available (Barnett and Crewz, 1990). Although planting techniques have been developed for seagrasses (Fonseca, 1989, 1994; Chap. 11), cultivars are from donor beds or from drift seeds. There is a serious need for the development of seagrass nurseries.

Figure 5-5. Restoration of coastal plant communities. Dune vegetation is established in the Netherlands (A) and a salt marsh (B) is restored in Florida (Fig. 5-5A), is courtesy of R. A. Davis, University of South Florida, Tampa).

Figure 5-6. Mitigation of mangals damaged by dredge spoil. Red mangrove seedlings planted 2.5 years ago near Tampa Bay, Florida, reached 0.5 m in height, demonstrating the need for land preparation to ensure proper tidal flushing (courtesy of R. Lewis, Lewis Environmental Service).

CHAPTER 6

Macroalgae

INTRODUCTION TO THE SEAWEEDS

Algae (alga, singular) can be defined as "photosynthetic, nonvascular plants that contain chlorophyll *a* and have simple reproductive structures" (Trainor, 1978). Thus, this definition includes both prokaryotic (Cyanophyta) and eukaryotic forms, as shown in Chap. 1. The biodiversity of algae is large but difficult to determine, mainly because of the limited biogeographic inventories worldwide (Norton et al., 1996). Bold and Wynne (1985) noted that the morphological diversity and cytology of algae make them difficult to clearly define. Macroalgae, which are primary found in the Divisions Chlorophyta (green algae), Phaeophyta (brown algae), and Rhodophyta (red algae), are commonly called seaweeds because of their size, multicellular construction, and attachment to firm substrata. The three major divisions of seaweeds contain taxa that have more fundamental (e.g., cytological, chemical, life histories) differences between one another than with the vascular plants. The differences between divisions of seaweeds are evident when comparing the photosynthetic pigments, reserve foods, cell wall, mitosis, flagellar construction, morphology, and life histories.

Dring (1982) has listed 900 Chlorophycean, 997 Phaeophycean, and 2540 Rhodophycean marine species worldwide. Thus, there are fewer marine macroalgal than microalgal species (Chap. 7). Within the Chlorophyta alone, the majority of species are unicellular or microscopic filamentous forms, whereas in the brown and red algal divisions, almost all the species range from filamentous forms to large thalloid plants. A number of texts that deal with seaweed biology and taxonomy include Bold and Wynne (1985), Sze (1986), South and Whittick (1987), Lee (1989), and van den Hoek et al. (1995).

Seaweeds probably evolved in the late Precambrian, or 900 to 600 million ybp (years before present); see Table 2-1 and van den Hoek et al. (1995) for further details. Fossils of calcified species, similar to present-day genera of seaweeds, can be found as early as 600 million ybp. It is interesting that seagrasses, which evolved around 100 m ybp in the Cretaceous, have not "displaced" seaweeds as the dominant marine vegetation. Thus, although primitive in terms of plant body structure, algae have dominated the benthic marine communities since the end of the Precambrian Period.

A word about seaweed construction and growth pattern is needed to clarify some of the terms used in this chapter. Foremost, macroalgal growth can occur by the division of one or more *apical cells* (cells at the tip of a branch; e.g., the brown alga *Dictyota*), which can form an *apical meristem* (a group of apical cells; e.g., most red algae; see *Eucheuma*). By contrast, seaweed growth may occur throughout the plant and not be restricted to branch tips, and this is called *diffuse growth* (no defined area of cell division; e.g., the brown alga *Hummia*) or *intercalary growth* (defined areas of cell division; e.g., the brown alga *Laminaria*). *Trichothallic growth* is a form of intercalary cell division that produces a hair or filament in addition to the production of new branch tissue (e.g., the brown alga *Sporochnus*). Trichothallic growth results in the production of tufts of hairs at the tip of each branch that may be long lasting or only temporary. Macroalgal construction can range from *filamentous* (one or more rows of cells; e.g., the green alga *Cladophora*), to *foliose* (flattened or membranous blades; e.g., the red alga *Halymenia*), to *tubular* (cylindrical or terete; e.g., *Eucheuma*). Tubular construction may be *multiaxial* (central core or *medulla* being filamentous; e.g., *Eucheuma*) or *pseudoparenchymatous* (central core cells cuboidal or spherical in shape but still in filaments; e.g., *Hummi*). Finally, branches can exhibit *parenchymatous* construction, where the cells are isodiametric and not filamentous as found in *Dictyota*.

DIVISION CHLOROPHYTA

The "green algae" are dominated by unicellular, freshwater species. Of the 16,800 known species (Norton et al., 1996), about 10% occur in marine environments and are mostly macroalgae. A number of the green seaweeds are used as direct food (seavegetables; App. B). Higher plants probably evolved from green algae based on similar photosynthetic pigments, reserve food, cell-wall chemistry, flagellar structure, and cell division. Variations in cell division, flagellar structure, and cell-wall features have resulted in a number of taxonomic reorganizations of the Chlorophyta (van den Hoek et al., 1995). The present text uses the three-class system presented by Bold and Wynne (1985) and Kumar (1990).

Cytology

Chlorophylls *a* and *b*, which also occur in higher plants, provide the typical "green plant" coloration found in most green algae. In addition, green algae contain β carotene and the xanthophylls lutein, zeaxanthin, which also occur in higher plants, and some members also contain violaxanthin, neoxanthin, siphonein, and siphonoxanthin (Table 4-1; van den Hoek et al.,

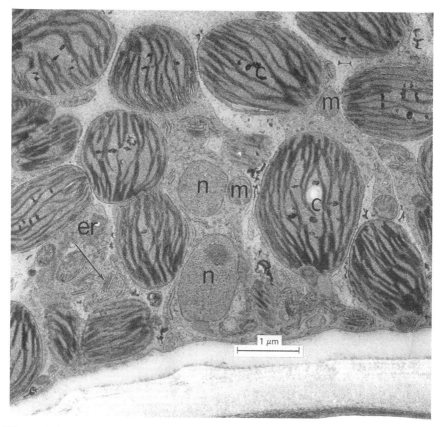

Figure 6-1. An electron micrograph of the cytoplasm of the tropical green alga *Caulerpa verticillata*. Visible cell organelles include chloroplasts (c) with thylakoids in bands of 3 to 5 and containing starch and plastoglobules, mitochondria (m), nuclei (n), endoplasmic reticulum (er), and Golgi bodies (arrow).

1995). Siphonoxanthin plays a role in the acclimation of deep-water species to the blue-green spectral quality of submarine illumination (Chap. 4).

The cell structure is eukaryotic and most green algae are uninucleated. Even so, two orders of marine chlorophytes are *multinucleated*, that is, each of the cells contain many nuclei (e.g., Siphonocladales, Cladophorales). Other orders contain *coenocytic* species, which are multinucleated and single-cell organisms, that is, there are no cell walls or cell membranes separating the nuclei of the entire plant (e.g., Caulerpales; Dawes and Rhamstine, 1967). Chloroplast thylakoids are grouped into bands 3 to 5 (Fig. 6-1) and can vary in morphology from cup-shaped, discoid, reticulate, to laminate. *Pyrenoids* or amylase-containing protein bodies occur in the chloroplasts of some green algae (e.g., Cladophorales). Reserve food occurs as starch (amylose and amylopectin), which is chemically similar to that in flowering plants.

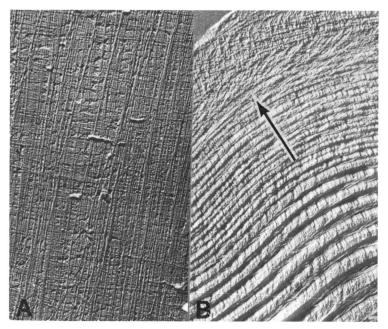

Figure 6-2. The cell wall of the green alga *Apjohnia laetevirens* from south Australia. A surface replica (A) and etched cross-section (B) show the crystalline cellulose microfibrils in alternating layers (Dawes, 1969). An arrow indicates the outer wall. Unit marks equal 1 μm.

The structure and composition of green algal cell walls range from cellulose microfibrils, which are typical of flowering plants (Ulvales), to highly crystalline cellulose (Siphonocadales, Cladophorales), to polymers of xylan and mannan, which also form microfibrils (Caulerpales, Dasycladales, Codiales). Mixtures of cellulose, xylan, and mannan polymers are also known for members of the Derbesiales (Kloareg and Quatrano, 1988). Microfibrils formed from different polymers are structurally arranged in random to highly organized layers (*Apjohnia*, Fig. 6-2; Dawes, 1966). Several tropical green algae (Codiales, Dasycladales) have calcified walls consisting of the aragonite form of calcium carbonate crystals. Some calcified species (*Halimeda*) account for a significant portion of the primary production and calcium carbonate sediments in tropical waters (Colinvaux, 1974).

Cell division in plant cells can be divided into *karyokinesis* (division of the chromosomes and nucleus) and *cytokinesis* (division of the cytoplasm or cell), as outlined in Figure 6-3. Two types of nuclear division [Fig. 6-3(A)] occur in green algae, including *closed* (intranuclear) karyokinesis, where the nuclear envelope does not break down, and *open karyokinesis*, where the nuclear envelope disappears. The latter type of nuclear division is characteristic of flowering plants. The process of cytokinesis is closely linked with karyokinesis in uni-

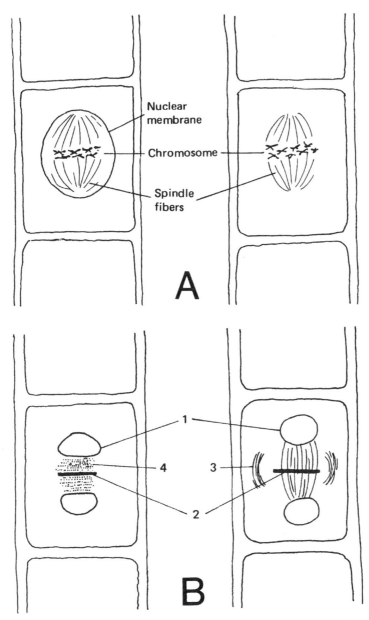

Figure 6-3.   Two forms of karyokinesis and cytokinesis. The breakdown of the nuclear membrane in open karyokinesis (right cell) is contrasted with closed (left cell) mitosis (A). The phragmoplast (3) consisting of microtubules arranged at right angles to the cell plate (2) in cytokinesis (right cell) is contrasted with the phycoplast (4) and its microtubules that run parallel (left cell) to the cell plate (B). The reforming daughter nuclei (1) are seen at the two poles in each dividing cell.

nucleate cells, but not in multinucleate or coenocytic species. Cytokinesis [Fig. 6-3(B)] in cells exhibiting closed karyokinesis occurs via a set of spindle fibers or *phycoplast*, which is arranged parallel to the developing cross wall of the cell. In open karyokinesis, the spindle fibers are at right angles to the developing cell wall, with the structure being called a *phragomoplast*. A new cell wall can be formed in two ways. In multicellular green algae, it develops from a *cell plate*, which is the result of deposition of wall material from Golgi vesicles along the phycoplast or within the phragomplast. In many unicellular species, a *cleavage furrow* forms by the ingrowth of the cell membrane that pinches the cell in half.

Green seaweeds typically produce motile asexual spores (*zoospores*), or *gametes*. The "typical" motile cells of green algae have a pair of apically inserted flagella of equal length (*isokontan*) that lack hairs (i.e., *whiplash* flagella). However, there are many variations on this theme, which are useful in separating green algae (van den Hoek et al., 1995). The flagella of green algae have the typical eukaryotic construction, with nine peripheral pairs of microtubules surrounding two central, single microtubules (axoneme). The latter are connected to a basal body, forming a complex connection to striated fibers, a possible capping plate (*Batophora*), and a cytoskeleton (Lobban and Harrison, 1994). A stigma or eyespot, which consists of orange or red carotenoid pigments within a chloroplast, can also be associated with the flagellar root system. The construction of the flagellar root system and type of cell division have been used to identify classes of green algae (van den Hoek et al., 1995).

Life Histories

Drew (1955) described algal *life histories* as a "recurring sequence of somatic and nuclear phases." The sequence, therefore, is an alternation of haploid (gametophytic) and diploid (sporophytic) phases, although the alternation need not be regular. Three basic patterns of life histories are exhibited by algae (Bold and Wynne, 1985; Fig. 6-4) with all three types found in the Chlorophyta. The *haplontic* life history is one in which the dominant plant is haploid (1N) with the zygote being the only diploid (2N) stage [i.e., zygotic meiosis; *Chlaymdomonas*, Fig. 6-4(A)]. Haplontic life histories are unknown in green seaweeds, although they are common in many unicellular species. In a *diplontic* life history, the diploid phase is dominant, with gametes being the only haploid phase [i.e., gametic meiosis; Fig. 6-4(B)]. Species in the Dasycladales, Caulerpales, Siphonocladales, and Ulvales (in part) show diplontic life histories. *Haplodiplont* (diplobiontic) life histories include free-living diploid and haploid plants and they exhibit sporic meiosis; they may either look alike [*isomorphic*, Fig. 6-4(C)] or be different [*heteromorphic*, Fig. 6-4(D)]. Species of the Ulvales (in part), Cladophorales, and Bryopsidales either show isomorphic or heteromorphic life histories.

As noted by Lobban and Harrison (1994), seaweed life histories are a "continuous interaction between the plants and their biotic and abiotic environments." Triggers such as changes in temperature, day length, and tidal cycles

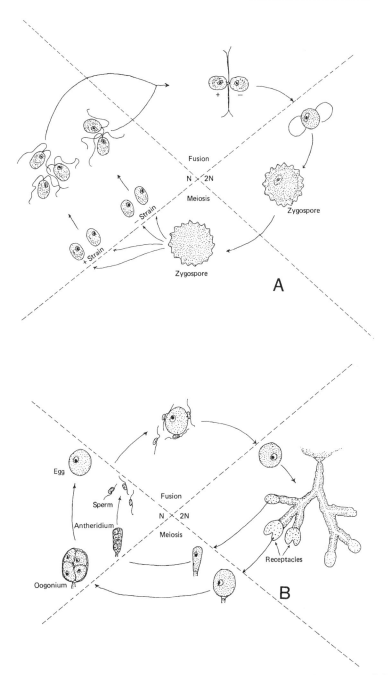

Figure 6-4. Life history diagrams of haplontic (A: *Chlamydomonas*), diplontic (B: *Fucus*), isomorphic haplo-diplontic (C: *Ulva*), and heteromorphic haplo-diplontic (D: *Laminaria*) algae.

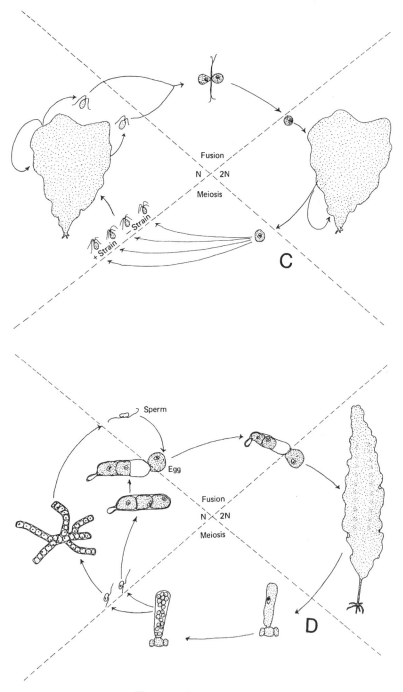

Figure 6-4. (*Continued*)

can result in a switch from vegetative to reproductive modes and thus a shift in the life history. Day length (and temperature) can serve as a signal that induces seaweeds to undergo reproduction (Chap. 4). Thus, many algal life histories are controlled by abiotic factors. The heteromorphic life history of the foliose green alga *Monostroma* is determined by short-day responses in which the diploid, unicellular "Codiolum" phase is triggered to undergo meiosis, producing zoospores that grow into a haploid foliose phase (Lüning, 1990).

Why have a haplodiplontic life history? Some studies have shown that there are advantages to having two free-living phases in a haplodiplontic life history as in the diploid crustose phase of the brown alga *Scytosiphon lomentaria* (Littler and Littler, 1980). The crusotose "Ralphsia" diploid plant is resistant to grazing when compared to the erect, tubular gametophyte in *S. lomentaria*. In the case of the Antarctic brown alga *Desmarestia menziesii*, the gametophytes have different photosynthetic characteristics than the sporophytes (Gomez and Wiencke, 1996). The young sporophytes and heteromorphic, filamentous gametophytes are shade-adapted and can survive (grow) as subcanopy species. In contrast, its mature, large-bladed sporophyte is a sun-adapted canopy plant. In the case of the intertidal red alga *Mastocarpus papillatus*, its diploid (*Petrocelis*) phase can avoid grazing because it is a perennial crust. In contrast, the bladed haploid phase occurs when grazers are uncommon (Slocum, 1980).

Taxonomy

The classification of Chlorophyta has been an area of considerable debate since the early 1970s, when a series of studies using the electron microscope demonstrated distinctive forms of cell division (karyokinesis and cytokinesis; Mattox and Stewart, 1984), flagellar construction and root architecture, and, more recently, nucleic acid sequences within chloroplasts and mitochondria (RNA) plus nuclei (DNA). For example, in reviews of the Chlorophyta, the number of classes has increased from one (Blackman and Tanseley, 1903), to three (Bold and Wynne, 1985; Kumar, 1990), and to ten (van den Hoek et al., 1988; 1995). Three classes of green algae are recognized here, the Chlorophyceae containing the majority of species, plus the Prasinophyceae and Charophyceae.

The *Prasinophyceae* are characterized by unicellular, motile green algae that have the following five features: (1) cell walls covered by one or more layers of fibrillar scales consisting of organic compounds; (2) flagella attached in a depression or grooved region of the cell and being covered with scales and hairs; (3) flagellar roots with a complex basal body; (4) a single bowl-shaped chloroplast with a central pyrenoid and surrounding starch grains; and (5) specialized ejectosomes (i.e., mucocysts or trichocysts; see Chap. 7) in some taxa. Four orders are placed in this class (van den Hoek et al., 1995). One marine example is the genus *Pyraminomas*, which is a pear-shaped unicell with four flagella (quadriflagellated) having scales and being inserted in an apical depression.

The *Charophyceae* (stoneworts) contain a single order with five genera. The complex morphology of these species includes a well-developed apical meri-

TABLE 6-1. Key to the Six Orders of the Chlorophyceae That Primarily Contain Marine Algae

1 Sheetlike or tubular, with cells being uninucleate	*Ulvales*
1 Filamentous or siphonous, with simple to complex morphology . .	2
2 Filamentous	3
2 Siphonous tubes, either singular or interwoven	4
3 Chloroplast netlike or discoid and connected	*Cladophorales*
3 Chloroplast uniform, with holes (perforate)	*Acrosiphoniales*
4 Branches in whorls, plants radial	*Dasycladales*
4 Plants not radially symmetrical	5
5 Segregative (internal cleavage) cell division	*Siphonocladales*
5 No segregative cell division, coenocytic plants	*Caulerpales*

stem, plus distinctive "leaf" and "stem" cells. A few species occur in brackish water, but none are marine. Stoneworts are mentioned here because they may be a side branch in the evolution of vascular plants (van den Hoek et al., 1988; Graham, 1996).

Six of the 15 orders of the *Chlorophyceae*, which are organized according to Bold and Wynne (1985), have marine species. The orders can be separated based on chloroplast, cell arrangement, and morphology using Table 6-1. The *Ulvales* can be separated into five families containing biseriate (Percursariaceae) to polyseriate (Schizomeraceae) filaments and monostromatic (Prasiolaceae, Monostromataceae) or distromatic blades or tubes (Ulvaceae). All of these families have laminate, parietal chloroplasts with pyrenoids and uninucleate cells, and one family, Prasiolaceae, has axial or stellate chloroplasts. All families but the Schizomeraceae have marine species. Species of the foliose *Ulva* and the tubular *Enteromorpha* (Fig. 6-5) can be found in cold temperate to tropical waters, including estuarine and oceanic waters. For example, the green alga taxon *Ulva lactuca* is known to occur from Newfoundland to the Bahamas. Sexual reproduction in different green alga taxa ranges from *isogamous* (identical + and – gametes), to *anisogamous* (motile gametes of different sizes), and to *oogamous* (nonmotile egg and motile sperm). Haplontic, diplontic, and haplodiplontic life histories are known for different species in the Ulvales.

The *Cladophorales* contain two families, the Cladophoraceae, which are filamentous species, and the Anadyomenaceae, whose filaments are fused together to form delicate blades. Three genera of the Cladophoraceae are common to marine habitats: *Rhizoclonium* (more delicate, unbranched filaments, producing rhizoids), *Chaetomorpha* (coarse, unbranched filaments), and *Cladophora* (branching filaments). The monotypic family Anadyomenaceae is known for the delicate, brilliant-green blades, which consist of anastomosed filaments, as in *Anadyomene stellata* (Fig. 6-6). All species in this order are multicellular, with each cell having many (multinucleated) nuclei. The parietal chloroplasts are constructed in the form of a net (reticulated plastic) or occur as segmented discs.

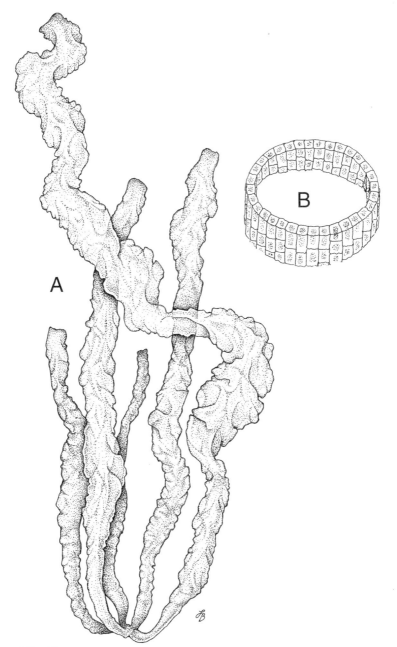

Figure 6-5. The morphology (A) of the cosmopolitan green alga *Enteromorpha intestinalis* and cross-section (B) of the thallus showing the tubular construction that is 1 to 2 cm in diameter.

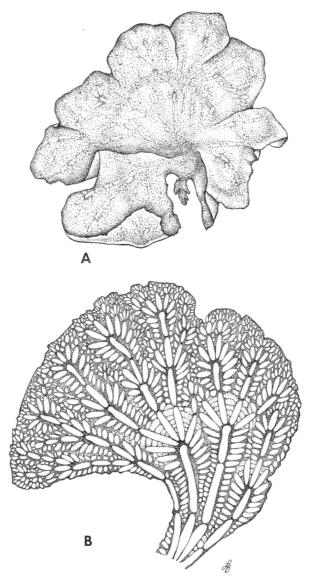

Figure 6-6. The tropical green alga *Anadyomeme stellata* is a leafy member (A) of the order Cladophorales whose filaments fuse to form elegant "veined" blades 3 to 6 cm in diameter (B).

Life histories are usually isomorphic haplodiplontic and the gametophytic phase produces biflagellated isogametes whereas the sporophyte produces quadriflagellated zoospores.

The monotypic order *Acrosiphoniales* contains the family Acrosiphoniaceae with three cold-water marine genera: *Urospora* (unbranched filaments), *Spongomorpha* (branched, uninucleate filaments), and *Acrosiphonia* (branched, multinucleate filaments). The species, unlike those of the Cladophorales, have a single, perforated chloroplast and typically a heteromorphic haplodiplontic life history in which the haploid gametophyte alternates with a unicellular sporophyte. The unicellular sporophyte was previously described as a separate genus (*Codiolum*). *Urospora* is a worldwide, predominantly cold-water genus having unbranched filaments, multinucleated barrel-shaped cells (up to 150 μm in length and 80 μm in diameter), and a netlike chloroplast. Cell walls in this order tend to be composed of noncellulosic microfibrils such as xylan.

The *Siphonocladales* contains three families: the Siphonocladaceae (filamentous construction), Boodleaceae (netlike or bladelike construction of anastomosing filaments), and Valoniaceae (plants consisting of an aggregation of vesicles). The order consists of tropical marine species that have multicellular filamentous construction, each cell being multinucleated. Cytokinesis is by *segregative cell division*, in which the cytoplasm divides into protoplasmic portions of varying size, each of which rounds up and produces an enveloping membrane. The segregative units can expand outward from the parent cell to exogenously produce irregular-shaped branches (*Siphonocladus*) or rhizoids for attachment (*Valonia*; Fig. 6-7). The segregative units also can enlarge within the original cell to form a type of pseudoparenchymatous (basically filamentous) tissue

Figure 6-7. The multinucleated tropical green alga *Valonia macrophysa* consists of dark-green cells, the largest of which reaches 2.0 cm in diameter.

(*Dictyosphaeria*). *Valonia ventricosa* forms the largest multinucleated cell of all plants, reaching 10 cm in diameter; even so, the internal segregative cell division can result in thousands of cells, some of which form basal rhizoids. The life history of *Dictyosphaeria cavernosa* is isomorphic haplodiplontic, with the gametophyte producing isogametes.

The *Caulerpales* have a siphonous or coenocytic construction. Four of the six families contain marine genera, which, unlike most seaweeds, can form psammophytic communities (growing on unconsolidated sediment; see Chap. 8) in tropical and subtropical waters. The Bryopsidaceae, including *Bryopsis* (Fig. 6-8) and *Derbesia*, have heteromorphic haplodiplontic life histories. The Caulerpaceae is a monotypic family with over 73 tropical species of *Caulerpa*, which are separated by their distinct morphologies. All species possess a rhizome that produces erect "blades" and rhizoids that penetrate soft sediments [Fig. 6-9(A)]. The genus is characterized by internal *trabeculae*, which are ingrowths of the cell wall [Fig. 6-9(B)]. Species of *Caulerpa* have both chloroplasts and starch-bearing leucoplasts and lack any cross walls; hence, the plant exhibits a coenocytic construction (Dawes and Rhamstine, 1967). Wound healing is critical in coenocytes due to the lack of cross walls and is via production of a carbohydrate wound plug (Dawes and Goddard, 1978; Menzel, 1980)

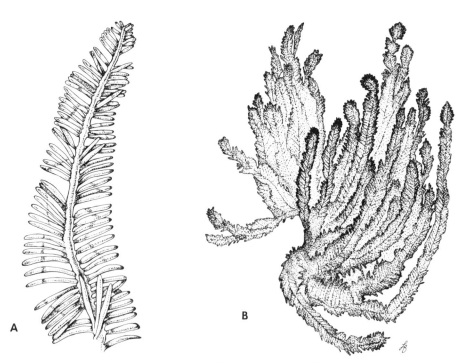

A

B

Figure 6-8.   The featherlike branches (2 to 3 cm tall) of the warm-water green alga *Bryopsis pennata* (A) form dark-green mats (B) in tide pools.

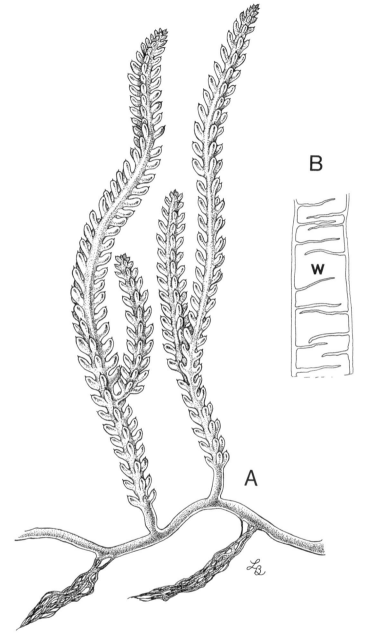

Figure 6-9.   *Caulerpa cupressoides*. The coenocytic green algal genus *Caulerpa* contains over 100 tropical species, including *C. cupressoides* with compressed (2.5-mm-dia.) branches (A). All species have internal wall struts (W), which are also called trabeculae (B).

unlike the proteinaceous wound plug of *Bryopsis* (Burr and West, 1971). The life history of *Caulerpa* is diplontic, with anisogamous gametes being released from tubes or papillae on the blades. The Codiaceae [e.g., *Codium isthmocladium*; Fig. 6-10(A)] contains genera with colorless interior coenocytic filaments called *siphons*, which are interwoven to form a multiaxially constructed thallus. The surface is formed by *utricles* [Fig. 6-10(B)], which are the tips of the

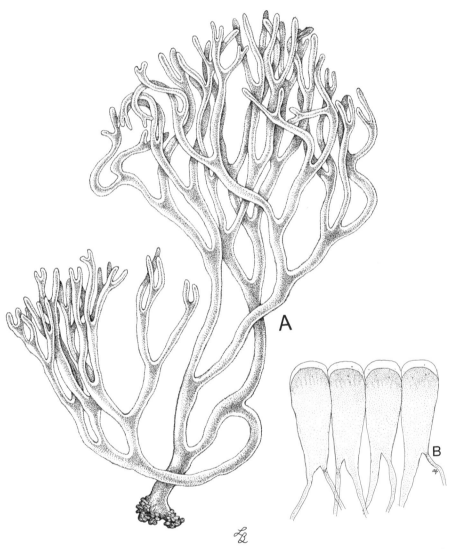

Figure 6-10. The complex coenocytic green alga *Codium isthmocladum* reaches 1 to 2 dm in length (A) and can overgrow coral reefs (Chap. 5). The plant's surface is constructed of utricles whose outer wall can be thickened as in *C taylorii* (B).

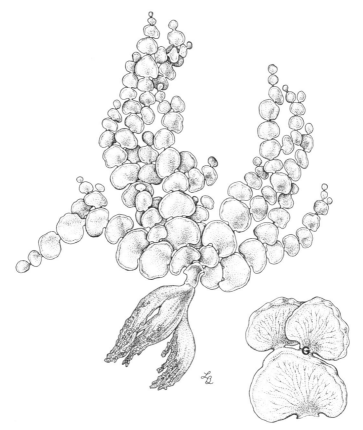

Figure 6-11. *Halimeda discoidea* is a tropical siphonaceous green alga whose siphons are interwoven into calcified segments and noncalcified geniculae (G).

exterior siphons and are fused along their side walls. Like *Caulerpa*, members of the Codiaceae are psammophytic, producing basal rhizoids that anchor into soft substrata. Their life history is diplontic and they produce anisogametes. The Udoteaceae contains over 100 species of tropical or subtropical siphonous algae, many of which are calcified. All species produce rhizoids that permit attachment in unconsolidated substrata. Calcified genera include *Halimeda* (Fig. 6-11), *Udotea*, and *Penicillus*, and noncalcified genera include *Chlorodesmis*, *Avrainvillea*, and *Cladocephalus*. Species of the Udoteaceae can dominate tropical psammophytic communities, with *Halimeda* being responsible for up to 90% of the sediment in some atolls.

The *Dasycladales* contain eight tropical genera that are placed in two families (Dasycladaceae, Acetabulariaceae). Members of the order are characterized by whorled branching and superficial calcification. At least 50 fossil genera are known as far back as the Ordovician Period (Table 2-1). The life history

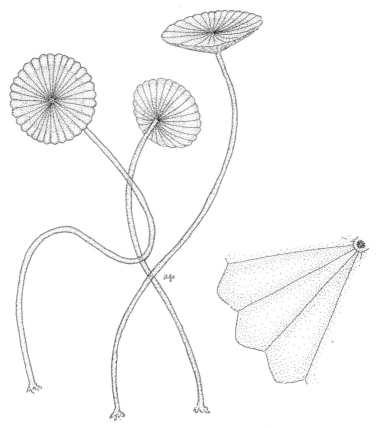

Figure 6-12. *Acetabularia crenulata* is a peltate tropical green alga consisting of a delicate, calcified stalk and a whorl of branches (rays) forming the cap.

is diplontic and isogametes are produced in cysts that are released from the branches. Whereas *Acetabularia* (Fig. 6-12) remains uninucleate until fertile, *Cymopolia* is vegetatively multinucleate. Because of its large primary nucleus and easily handled cell, *Acetabularia* has been used in morphological studies where the nucleus of one species is transferred to another.

DIVISION PHAEOPHYTA

The brown algae are placed in a single class, Phaeophyceae, and contain about 265 genera with 1500 to 2000 species (Norton et al., 1996) that are almost exclusively marine. Only five or six genera have species recorded from freshwater habitats (Bold and Wynne, 1985; van den Hoek et al., 1995). Brown algae, which are primarily dominant in temperate areas, range from small filamentous forms (*Ectocarpus*), to massive intertidal rockweeds (*Ascophyllum*, *Fucus*), to subtidal large kelps (*Macrocystis, Nereocystis*). Some tropical brown seaweeds, such as the dictyotalian taxon *Lobophora variegata*, will grow at depths of 100 m. Most of brown algae are *lithophytes*, which require stable hard substrata for attachment, and a number of the filamentous, smaller species are epiphytes. A few, such as *Sargassum fluitans* and *S. natans* of the Sargasso Sea, occur only as drift populations, whereas others like *Pilayella littoralis* can form extensive drift populations that contaminate beaches within Boston Bay (Wilce et al., 1982). A number of species are economically important such as the kelps, *Macrocystis* and *Laminaria*, and the rockweeds, *Ascophyllum* and *Fucus*, of temperate latitudes and tropical species of *Sargassum*. These are harvested as sources of the phycocolloid alginic acid and also used as cattle fodder and supplements to fertilizers (App. B).

Cytology

Brown algae contain chlorophylls *a* and *c* (c_1, c_2, c_3), β carotene, fucoxanthin, and neofucoxanthin, as well as other cartenoids (Table 4-1). Typically, their eukaryotic cells are uninucleate, except for some of the medullary cells in *Laminaria* and *Durvillea*. Chloroplast thylakoids are arranged in bands of 3, with a girdling lamella just inside the double plastid membrane (Fig. 6-13). Pyrenoids are found in a variety of brown algal orders (Ectocarpales, Dictyotales, Laminariales). They are stalked and occur within the double membrane of the chloroplast. Plastids are covered by the chloroplast endoplasmic reticulum (CER), which is continuous with the outer nuclear membrane. The CER provides a close relationship among the chloroplasts, the endoplasmic reticulum, and the nucleus. Physodes (fucosan granules) are common, particularly in the cells of intertidal species, and may function in filtering sunlight (Ragan, 1976), serve as antifoulants (Sieburth and Conover, 1965), or contribute to wound plug development (Fagerberg and Dawes, 1976). A common component of physodes are various types of tannins; of special interest are the phlorotannins (polymers of phloroglucinol), which may discourage grazing (Ragan and Glombitza, 1986). The release of volatile brominated methanes by seaweeds (10^4 tonnes y^{-1}), which are common in brown algae, may play a role in the destruction of the ozone $_3$) layer above the Arctic Ocean (Wever, 1988).

The reserve food in brown algae is a β-1-3 linked glucan (*laminarin* or *chrysolaminarin*), with some β1-6 linkages. It may account for 2 to 34% of the plant's dry weight. Mannitol, a low-molecular sugar-alcohol, is also present and is thought to serve both as a reserve food and as an osmoticant (Chap. 4). All brown algae examined have cellulose as a structural component, their microfibrils being arranged in alternating lamellae (Dawes et al., 1961). Plasmodesmata are common and usually penetrate the cell wall in well-defined pit fields (Fig. 6-14). The cell wall also contains alginic acid (D-mannuronic and L-glucuronic acid), which probably functions in structural (as a cementing agent) and ion-exchange roles, is extracted as a phycocolloid (App. B). Fucoidan, of which L-fucose (2-deoxy-L-mannose) is the primary component, is a water-soluble extract from brown algae.

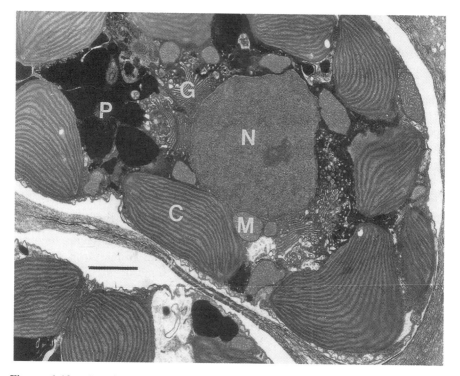

Figure 6-13. An electron micrograph of the epidermal cell of the tropical fucoid brown alga *Sargassum filipendula*. The central nucleus (N), surrounded by Golgi bodies (G), physoid bodies (P), and mitochondria (M); peripheral chloroplasts (C) with thylakoids in bands of 3, and electron-dense physodes are present. Unit mark equals 1 μm.

Mitosis begins with the duplication of two centrioles. It is a form of closed karyokinesis [Fig. 6-3(a)] where the spindle fibers penetrate the nuclear envelope. Breakdown of the nuclear envelope only occurs as the chromosomes migrate to the two nuclear poles (anaphase) and it is then re-formed around the daughter nuclei. Cytokinesis is carried out by an inward furrowing of the cell membrane, while the new wall is deposited on the cleavage furrow. All brown algae produce motile flagellated cells, which either function as zoospores or gametes. Motile cells are *heterokonts*, that is, the flagella are unequal in length and morphology. The two flagella are laterally inserted into the ellipsoid to a dorsiventrally flattened monad. The shorter basally oriented flagellum is smooth (acronematic), whereas the longer anterior-oriented flagellum is hairy (pleuronematic) with appendages (mastigonemes) arranged in two rows. Two exceptions are the spermatozoids of the Dictyotales, which have a single anterior pleuronematic flagellum, and the Fucales, which have a short, thickened, anterior pleuronematic flagellum and a posterior acronematic proboscis.

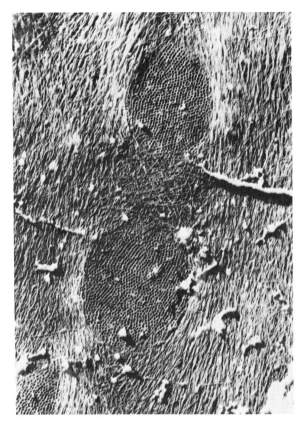

Figure 6-14. The cell wall of the tropical brown alga *Dictyota dichotoma* showing two pit fields and cellulose microfibrils arranged in lamellae (Dawes et al., 1961).

Although uniflagellated, there is a residual basal body in the dictyotalean sperm suggesting evolution from an ancestral heterokont.

Life Histories

With the exception of diplontic life histories in the Fucales and Durvillaeales, the typical brown algal life history is haplodiplontic and is either isomorphic or heteromorphic. The gametophytic generation is usually reduced in most heteromorphic life histories, but there are exceptions. The Fucales have a diplontic life history with the sporophyte producing eggs and sperm. Reproduction occurs in two types of structures, namely, *plurilocular* and *unilocular sporanga*. In the former, a single motile cell is produced per locule via mitosis, and these sporangia can function as gametangia (sexual) or as zoosporangia (asexual). Unilocular sporangia are enlarged single-cell structures in which meiosis usually occurs. The four haploid cells produced in the unilocular sporangia via meiosis, or multiples of 4 after subsequent mitotic events, may be released as nonmotile spores (Dictyotales) or mitotically divide further to produce meio-zoospores.

Taxonomy

The brown algae are usually recognized as a distinct division, as presented in this text. However, others place the class Phaeophyceae in the Chromophyta, Chrysophyta, or Heterokontophyta. The amalgamation with other "brown algae" recognizes the similarity in pigments and flagellar structure. Based on new information on life histories, growth patterns, and plant construction, the number of orders containing brown algae has increased from 11 (Dawson, 1966), to 12 (Dawes, 1981), to 13 (Bold and Wynne, 1985), to 16 (van den Hoek et al., 1995). The key to 14 orders presented in Table 6-2 does not include two questionable orders, the Syringodermatales and Ascoseirales.

The *Ectocarpales*, which consist of the family Ectocarpaceae and about 30 genera, are thought to contain the most primitive species of brown algae. The uniseriately branched filaments have diffuse growth via intercalary cell division. Most life histories are isomorphic, fertilization being isogamous to anisogamous. *Ectocarpus* (Fig. 6-15) is a common example; its cells contain one to a few branching, ribbon-shaped chloroplasts. Plurilocular sporangia (Fig. 6-15B) are found on both the haploid and diploid plants; the motile cells of the former may function as either zoospores or gametes, whereas on the latter plant, they can only function in asexual reproduction. Unilocular sporongia (Fig. 6-15A) are common only to diploid plants. The zygote grows into a sporophyte on which unilocular sporangia can develop and produce 1 N zoospores after meiosis. These haploid zoospores grow into gametophytes that normally produce only plurilocular sporangia, whereas mitosis will result in either asexual zoospores or gametes. Thus, both phases of the ectocarpoid life history can perpetuate itself through plurilocular production of asexual zoospores. Muller (1972, 1977) has shown that *E. siliculosus* can have small sporophytes compared to the gametophytes, as well as complex variations of the typical isomorphic haplodiplontic life history.

The *Ralfsiales* contain three families with genera that have a crustose morphology. Some previous members of this order, such as *Ralfsia pacifica*, are the microthalloid diploid phase of larger fleshy species found in the Scytosiphonales. The species have a basal layer of radiating filaments, with each cell producing an erect branch forming the tightly compacted pseudoparenchymatous crust. The life history is not well known for many of the crustose forms placed in this order, and some "species" may be diploid phases of other scytosiphonean algae. Anisogametes have been seen in *Neoderma tingitana*.

The *Sphacelariales* consist of 10 genera that are distributed from temperate to tropical waters. Growth is by a prominent apical cell and the life history is an isomorphic haplodiplontic type. Thus, some classifications align this order with the Dictyotales (van den Hoek et al., 1995). Plants are small filamentous tufts that are multiseriate in construction but do not form pseudo-

TABLE 6-2. A Dichotomous Key to 14 Orders of Brown Algae

1	Life cycle diplontic, lacking a gametophytic phase..............	2
1	Life cycle haplodiplontic, with gametophytic and sporophytic phases.......................	3
	2 Growth by means of a distinct apical cell...................	*Fucales*
	2 Growth is diffuse, without apical cells.....................	*Durvillaeles*
3	Filamentous or pseudparenchymatous construction..............	4
3	Parenchymatous or multiseriate construction, at least in one phase of life history	8
	4 Life history isomorphic to slightly heteromorphic...........	5
	4 Life history heteromorphic, the sporophyte dominant........	6
5	Filamentous morphology, usually having more than one plastid per cell, pyrenoids present...........................	*Ectocarpales*
5	Crustose morphology, pseudoparenchymaous, one plastid per cell, pyrenoids absent......................................	*Ralfsiales*
	6 Oogamous sexual reproduction (eggs and sperm)...........	7
	6 Isogamous sexual reproduction	*Chordariales*
7	Growth trichothallic, each axis ending in one filament (uniaxial construction)...	*Desmarestiales*
7	Growth considered to be trichothallic, each axis ending with a tuft of filaments	*Sporochnales*
	8 Isomorphic life history (except for *Cutleria*), at least one phase showing trichothallic growth	9
	8 Heteromorphic life history, no trichothallic growth, one phase having parenchymatous construction, the other filamentous or pseudoparenchymatous....................................	12
9	Trichothallic growth	10
9	Apical cell growth..	11
	10 Multiseriate construction, uniseriate apical regions, multiseriate basal regions, forming only quadrinucleate monospores on diploid plants.............................	*Tilopteridales*
	10 Parenchymatous construction at least in one phase, only unilocular sporangia on the diploid plant...................	*Cutleriales*
11	Plants erect, flattened, four or eight nonmotile spores formed per unilocular sporangium, oogamous reproduction.................	*Dictyotales*
11	Plants erect, terete, many motile spores per unilocular sporangium, isogamous to oogamous reproduction..............	*Sphacelariales*
	12 Vegetative cells with one platelike chloroplast and conspicuous pyrenoid, larger gametophytes bearing only plurilocular sporangia	*Scytosiphonales*
	12 Vegetative cells with many chloroplasts, with or without pyrenoids, larger plant bearing unilocular sporangia.........	13
13	Growth apical or diffuse, isogamous or anisogamous reproduction...	*Dictyosiphonales*
13	Growth intercalary with a localized meristem and superficial merstematic layer (meristoderm), oogamous reproduction.......	*Laminariales*

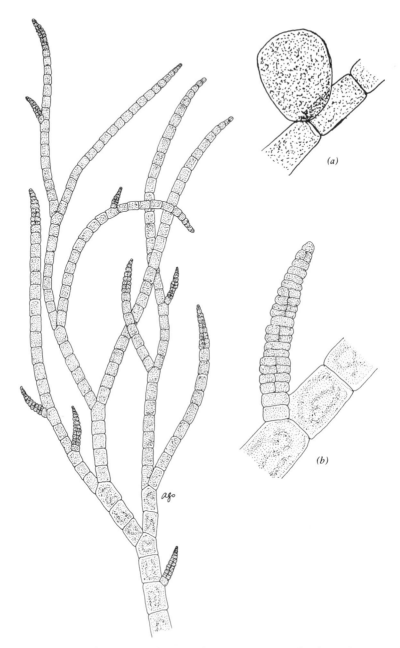

Figure 6-15. The filamentous brown alga *Ectocarpus siliculosus* bears conical plurilocular sporangia. Diploid plants produce unilocular sporangia (a); plurilocular sporangia (b) are found on both diploid and haploid plants. The main axes are 40 to 60 μm in diameter.

parenchymatous tissue (*Sphacelaria*). Gametic morphology ranges from isogamous (*Cladostephus*), to anisogamous (*Sphacelaria*), to oogamous (some species of *Halopteris*). The development of specialized branches or *propagulae* in asexual reproduction is a feature of *Sphacelaria*.

The *Tiliopteridales* contain a few genera characterized by filamentous construction and muliseriate in the lower regions of an otherwise uniseriate filament. Thallus growth is trichothallic and the plants may contain monosporangia (large spherical cells). Monosporangia of gametophytes contain a single large nucleus, whereas those on the sporophytes have four nuclei (quadrinucleate). In the North Atlantic species *Haplospora globosa*, the gametophytes produce eggs and sperm and the life history is thought to be isomorphic.

The *Cutleriales* contain two genera, *Cutleria* and *Zanardinia*, both exhibiting anisogamy. The former, found in the Gulf of California, exhibits an alternation of heteromorphic phases in which the microthalli are crustose sporophytes (i.e., Aglazonia stage). The small sporophyte of *Cutleria* is parenchymatous and the macothallic gametophyte is fan-shaped and has trichothallic growth. *Zanardinia* exhibits an alternation of isomorphic phases, with both showing trichothallic growth and parenchymatous construction.

The *Dictyotales* contain 16 genera in a single family, the Dictyotaceae. The group is pantropical to warm temperate with isomorphic haplodiplontic life histories. The species are flattened dichotomously branched blades [See *Dictyopteris*, Fig. 6-16(A)] with one or more apical cells [Fig. 6-16(B)] and having parenchymatous construction being two to many cells thick [Fig. 6-16(C)]. Typically, the gametophytes are dioecious and sexual reproduction is oogamous. Female gametophytes produce a single egg in each oogonium with the latter occurring on surficial sori. Male gametophytes produce pale-colored sori of plurilocular sporangia, with each cell releasing a single sperm. As noted previously, the sperm have a single pleuronematic flagellum. Typically, the diploid sporophyte produces four haploid, nonmotile spores via meiosis; each of these grow into a gametophyte with sex segregation often occurring. *Dictyota dichotoma* resembles *Dictyopteris delicatula* [whose blades have midribs; Fig. 6-16(A)], and is probably the most widespread species of the order. Its dichotomous blades grow by a single apical cell [Fig. 6-16(B)], is paranchymatous, and three cells thick [Fig. 6-16(C)].

The *Chordariales* are a large, diverse order containing six or more families. Members are thought to be primitive and possibly related to the Ectocarpales, species that are uniseriate filaments having intercalary growth. Life histories in the Chordariales are usually heteromorphic with alternations between a haploid microthallus and a diploid macrothallus. Sporophytes range from discoid filamentous epiphytes to pseudoparenchymatous plants. The zygote may produce a microdiploid thallus called the *plethysmothallus*, which can grow directly in the macrothallic sporophyte or asexually reproduce itself through production of plurilocular sporangia and zoospores. *Cladosiphon*, a member of the family Chordariaceae, has a loose, pseudoparenchymatous sporophyte that alternates with a small filamentous gametophyte.

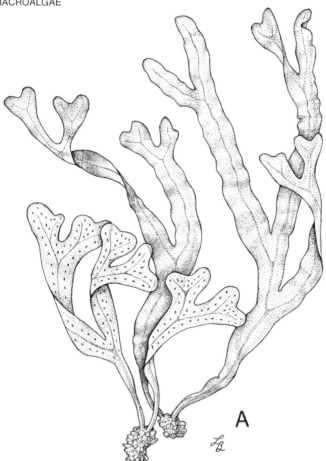

Figure 6-16. Members of the tropical brown algal order Dictyotales have flattened branches (A: *Dictyopteris delicatula*), grow by apical meristems of single (B: *Dictyota dichotoma*), or multiple cells, and are parenchymatous in construction (C: *D. dichotoma*).

Members of the *Sporochnales* are placed in two families with six genera that mostly occur in the tropical waters of the Southern Hemisphere. *Sporochnus* and *Neria* occur in the Gulf of Mexico and extend into the deeper waters off North Carolina. The life history is heteromorphic with a macroscopic sporophyte and microscopic gametetophytes that produce eggs and sperm. Trichothallic growth results in turfs of hairs at branch tips.

The *Desmarestiales* have a worldwide distribution in temperate waters, with *Desmarestia* being an important component of the Antarctic flora. Members of this order show parallel evolution in morphology in the Antarctic when compared with the kelps of the Northern Hemisphere and Arctic waters. The single family contains two genera, *Desmarestia* and *Himanthothallus*, both exhibiting

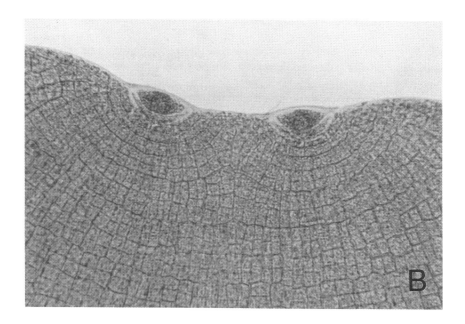

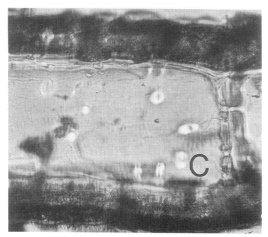

Figure 6-16. (*Continued*)

trichothallic growth and pseudoparenchymatous construction. The life history is heteromorphic, with the microthallic gametophytes exhibiting oogamous sexual reproduction. Sulfuric or malic acid can occur in cell vacuoles (pH 0.8 to 1.8), with concentrations being up to 0.44 N in some species. The acids will bleach out other algae if left in close contact.

The *Dictyosiphonales* contain four families. The different genera have heteromorphic haplodiplontic life histories, with their sporophyte being macroscopic and exhibiting parenchymatous construction and diffuse (intercalary) growth. The life histories can be flexible; in some cases, there is an intermediate, pleismothallic stage that gives rise to the macrothallic sporophyte. Some species are small, as seen in the genus *Elachistia*, which is an epiphyte on various seaweeds (*Ascophyllum, Laminaria, Chondrus*). The gametes are isogamous. *Hummia onusta* (Fig. 6-17) is a species created out of two other species, *Stictyosiphon subsimplex* being the sporophyte [Fig. 6-17(A)] and *Myriotrichia onusta* the gametophyte [Fig. 6-17(C)]. The combination of two former genera demonstrates the diverse morphologies found in this order. Whereas the sporophyte is a large parenchymatous [Fig. 6-17(B)] branching plant with apical growth, the gametophyte is a small epiphyte having uniseriate filaments [Fig. 6-17(D)] with a discoid filamentous base, diffuse growth, and only producing plurilocular sporangia.

The *Scytosiphonales* represent a small order containing two families, Chnoosporaceae and Scytosiphonaceae. The life history is diplohaplontic in which the sporophyte is reduced. The species were removed from the Dictyosiphonales based on the presence of a pyrenoid and a single plastid per cell, plus only having plurilocular structures on the macroscopic gametophyte. A *Ralfsia*-like sporophytic stage of *Scytosiphon* produces unilocular sporangia that are apomeitic, whereas the tubular, erect gametophyte produces plurilocular sporangia and anisogametes. The interplay of environmental controls on reproduction in *Scytosiphon* is discussed by Bold and Wynne (1985) and van den Hoek et al. (1995). Four North Atlantic genera include *Colpomenia, Hydroclathrus, Rosenvingea*, and *Petalonia*, of which the first three genera have species that occur in the Caribbean.

The *Laminariales*, commonly called "kelps," include the largest and most complex brown algae. They can dominate the lower intertidal and subtidal zones of temperate to Arctic latitudes and are primarily confined to the Northern Hemisphere. A number of the species are harvested for phycocolloids (e.g., alginic acid) and fodder (App. B). All of the species have obligate heteromorphic life histories with a macrothallic sporophyte and microthallic gameteophyte and oogamous sexual reproduction. Spectral quality, day length, and water temperature all play a role in controlling zoospore and gamete production (Dring, 1988).

The sporophytes of kelps are parenchymatous in construction and lengthen by an intercalary meristem (*transition zone*) that is found at the base of the blade. Kelps grow in diameter via a superficial *meristoderm*. Most species are perennial, with one specimen of *Pterogophora* known to be 17 years old. Kelps have a highly differentiated morphology (Fig. 6-18). Most species have a holdfast (attachment organ), stipe (the stem), and lamina (blades). The anatomy of kelps is the most complex of all seaweeds, with tissue differentiation includ-

Figure 6-17.   The tropical brown *Hummia onusta* combines two previously existing species. The sporophyte *Stictyosipon subsimplex* (A) is an epiphyte on seagrass blades and has parenchymatous construction (B), and the gametophyte *Myriotrichia subcorymbosa* (C) is a small filamentous epiphyte that produces plurilocular sporangia (D).

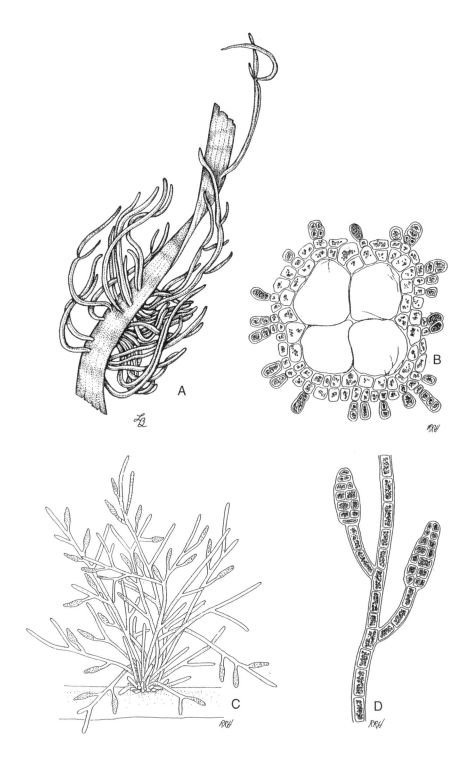

141

Figure 6-18. *Macrocystis pyrifera.* Michael Neushul is examining the holdfast of a giant kelp washed ashore at Point Dune, Southern California.

ing an epidermis, outer and inner cortex, and a central medulla that contains sievelike cells called *trumpet hypae*. The latter function in conduction, similar to sieve cells in higher plants (Schmitz and Srivastava, 1980).

   The four families are separated based on thallus morphology, blade differentiation, and location of sporangial sori that are either on distinct *sporophylls* or vegetative blades. The family Chordaceae, which is monotypic (*Chorda*), has no distinct stipe and blade and has unilocular sporangia over the entire plant. Members of the Laminariaceae have single blades, or if multiple, they are not produced by splitting of the transition zone at the base of the blades. Examples of this family include *Laminaria*, *Agarum*, and *Costaria*. The Lessoniaceae contain the largest known kelps, including *Macrocystis* (Fig. 6-18) and *Nereocystis*. The family is characterized by having longitudinal divisions that extend into the intercalary meristem of the blades. For example, in *Macrocystis*, the thallus consists of a stipe and lateral blades with enlarged bases (*pneumatocysts*) that were produced by the intercalary merisetem. The individual blades with the pneumatocysts were split off (overtopped) from the primary blade containing the transition zone. The pneumatocysts function as floats and vary from a single one for each of the blades (*Macrocystis*) to a single, large float with blades extending from it (*Nereocystis*). The sea palm *Postelsia* is a member of this family, but it lacks pneumatocysts [Fig. 8-4(A)]. The plant is an erect, low intertidal species of high-energy coasts of the west coast of North America. The Alariaceae is characterized by a *Laminaria*-like blade and the presence of small, basal spore-bearing blades or *sporophylls* that proliferate from the transition zone. The blades are not divided by splits into the intercalary meristem. Genera include *Alaria* in the North Atlantic and *Egregia* and *Eisenia* in the northwestern Pacific. *Egregia* has a primary blade that is a long, compressed structure producing lateral blades and small pneumatocysts.

The *Fucales* have a diplontic life history and exhibit a pronounced oogamous reproduction with a unique sperm, as previously described. Growth is from one or more apical cells. Plant construction is also complex with a holdfast, stipe, blades, and floats. Most of the genera are cold-water species with their center of distribution in the Southern Hemisphere. However, *Sargassum* exhibits a pantropical distribution. A number of genera such as *Ascophyllum, Fucus Pelvetia,* and *Sargassum* have ecads that result from being unattached and entrained within estuaries ("Drift Seaweeds and Blooms," Chap. 8). One of the best known examples of drift macroalgae can be found in the North Atlantic gyre called the Sargasso Sea, where two unattached, floating species of *Sargassum* occur.

Sexual reproduction within the Fucales occurs in fertile tips (*Fucus*) or specialized branches (*Sargassum*) called *receptacles*. Sporangia called oogonia (egg producing) and antheridia (sperm producing) are found in cavities called *conceptacles* (Fig. 6-19) and may be the equivalent of unilocular sporangia (where meiosis occurs). The number of eggs within an oogonium is characteristic of

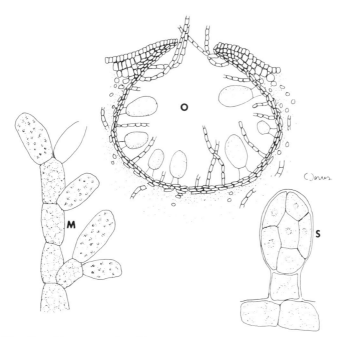

Figure 6-19. Fucoid conceptacles. The conceptacle of the temperate brown and dioecious alga *Fucus vesiculosus* bears oogonia (O), an antheridial branch (M), and a single oogonium (S).

different genera with *Fucus* having eight, *Ascophyllum* four, *Pelvetia* two, and *Hesperophycus*, *Sargassum*, and *Cystoseria* one. Usually, each antheridia produces 64 uniflagellated sperm.

Four families are recognized in the Fucales. The Fucaceae has a flattened morphology and a four-sided apical cell. Its members include *Fucus* and *Ascophyllum* in the North Atlantic and *Hesperophycus* and *Pelvetia* in the North Pacific. *Pelvetia* also occurs in the eastern North Atlantic. Members of the Sargassaceae, which contains tropical genera such as *Sargassum* and *Turbinaria* (Fig. 6-20), have a radial organization due to the three-sided apical cell. Floats (pneumatocysts, bladders) are common on many species, and lateral branches occur in the axes of subtending leaves, similar to flowering plants. The Cystoseiraceae contains about 16 genera; it is similar to the Sargassaceae except that the branches do not arise in leaf axils. The Hormosiraceae is monotypic and restricted to the Southern Hemisphere. Unlike the other genera, *Hormosira* consists of hollow, globose segments.

The monotypic order *Durvillaeales* was created for the genus *Durvillaea*, which has diffuse, rather than apical growth, but has many features of the Fucales. Species produce a massive holdfast, stipe, and blade and are only found in the Southern Hemisphere. As with *Fucus*, *Durvillaea* has a diplontic life history and produces conceptacles containing oogonia and antheridia. One of the four species, *D. antarctica*, forms large stands in the lower intertidal and upper subtidal exposed sites in cold waters of New Zealand and South America.

DIVISION RHODOPHYTA

According to Norton et al. (1996), there are 4000 to 6000 red algal species, although some estimates range from 2500 to 20,000. By far, the majority of species are marine with about 3% (150 species from 20 genera) being freshwater (Sheath, 1984). Features of red algae include eukaryotic cells, a complete lack of flagellar structures, food reserves of floridan starch, which is an amylopectin (α 1-4 main chain, β 1-6 side chain glucans), the presence of phycoblins, chloroplasts without stacked thylakoids, and no external endoplasmic reticulum (Fig. 6-24). There are over 300 economically important species of red algae (App. B) that are used as a direct food source (sea vegetables) and commercial colloidal extracts (App. B). Aside from several texts in phycology, the review by Cole and Sheath (1990) is an excellent source of data on the Rhodophyta.

The apparent lack of flagella and flagellar root systems in red algae, as well as an early fossil record of calcified forms (Cambrian, 590 million ybp), suggests an early evolutionary separation from other eukaryotic organisms (Bold and Wynne, 1985). However, more recent rRNA (28S) data suggest that the Rhodophyta evolved well after the evolution of flagella and probably came from ancestral green algae (Cole and Sheath, 1990; van den Hoek et al., 1995). If the latter evidence is true, then ancestral red algae probably had flagella that were subsequently lost.

Figure 6-20. *Turbinaria turbinata*. The tropical fucoid has reproductive branches (receptacles) with sunken conceptacles; its "leaves" are triangular appendages. The plant reaches 0.5 m in height.

Cytology

In addition to chlorophyll *a*, the red algae contain α and β carotenes, and the xanthophylls zeaxanthin and lutein (Table 4-1). Of special interest are the phycobiliproteins or *phycobilin* pigments including r-phycocyanin, r-phycoerythrin, c-phycocyanin, and allophycocyanin. Phycobilins are water-soluble compounds organized into structures called *phycobilisomes*, which occur on the surface of

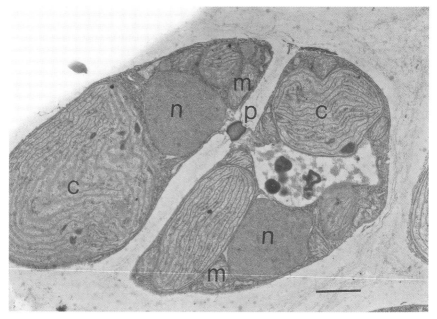

Figure 6-21. An electron micrograph of the epidermal cells of the tropical red alga *Hypnea musciformis* connected by a pit connection (p). Each cell contains a nucleus (n), chloroplasts (c), and mitochondria (m). Unit mark equals 1 μm.

the chloroplast thylakoids (see *Porphyridium*; Fig. 6-25). The red algal chloroplast is constructed of a typical double-membrane chloroplast envelope. Thylakoids are separated and not grouped into bands; hence, the chloroplast has a distinct appearance in ultrathin sections (Fig. 6-21). In more advanced red algae, a *boundary* or girdle thylakoid will parallel the inner chloroplast membrane (Fig. 6-21).

Pyrenoids are present in the chloroplasts of primitive red algae and are not associated with the floridan starch. The latter substance occurs in the cytoplasm rather than the chloroplast. Floridian starch is insoluble in boiling water and appears refractive using light microscopy. The reserve food is a branched amylopectin consisting of α (1-4) glucans with β 1-6 side glucan side chains. Thus, this starch, unlike that of green algae, is not contained within the chloroplast, and due to its highly branched nature is insoluble in boiling water.

Mitosis is closed [Fig. 6-3(A)] with the nuclear membrane remaining through karyokinesis and the telophase spindle being persistent. An electron dense area (using transmission electron microscopy) called a *polar body* occurs at each pole during karyokinesis. Spindle fibers also penetrate the nuclear envelope at each pole but do not radiate from the polar bodies. Cytokinesis occurs during late telophase by furrowing of the plasmalemma where constituents are then deposited to form the new cross wall.

The cell wall has an inner layer of randomly arranged microfibrils (Fig. 6-22; Dawes et al., 1961) as well as an outer amorphous layer that may contain sulfated galactan polymers. Some of the latter are economically important phycocolloids including agar, carrageenan, funoran, and furcellarin (App. B). Microfibrils are composed of cellulose polymers except in two primitive red algae, *Porphyra* and *Bangia*, where the polymers are of xylose and mannose. Calcification is characteristic of cell walls of members within the Corallinales, with its crystalline form being calcite instead of the aragonite of calcified green algae. The process of calcification is linked to carbon fixation, with uptake of CO_2 from cell walls resulting in an increase in pH and an increase CO_3^{-2} ions during photosynthesis. The increase in carbonate results in a precipitation of $CaCO_3$. Coralline red algae play a critical role in coral reef development via their primary productivity, cementation of coral reef rubble, and serving as a source of sediment (Chap. 12).

Figure 6-22. The cell wall of red alga *Ceramium* sp. contains cellulose microfibrils having a random arrangement. The thin regions (arrow) are pit fields traversed by plasmodesmata. The polystyrene balls are 0.814 μm in diameter (Dawes et al., 1961).

A distinctive feature of red algal cell walls are *pit plugs* (Fig. 6-22), which are visible under the light microscope. The plugs are neither a "pit" nor a "connection," but consist of distinctive proteinaceous material that can appear refractive (Pueschel, 1989). Pit plug formation occurs through incomplete cytokinesis when the annual ingrowth of the new wall ceases with an intercellular cytoplasmic connection remaining. Condensation of vesicular material results in plug formation, with only the cytoplasmic membrane remaining continuous between sister cells. The construction of pit connections (cap layers, consistency of the plug core) have been used as a taxonomic trait (Pueschel, 1989) and seven types have been described (van den Hoek et al., 1995). Whereas *primary* pit connections result from the equal division of two sister cells, *secondary* pit connections are formed through mutual contact of neighboring cells.

Life Histories

Asexual reproduction is common throughout the Rhodophyta. Indeed, it may be the only mode of reproduction for many members of the subclass Bangiophycidae (e.g., order Porphyridiales). Guiry and Irvine (1989) list 10 types of spores produced by red algae (Fig. 6-24). Specialized *monospores* (asexual reproductive cell from a monosporangium) and *paraspores* (possibly asexual cells from a parasporangium) may occur in some species. Sexual reproduction is known for a few members of the Bangiophycidae and almost all species studied in the subclass Florideophycidae. The typical life history, which is *triphasic*, includes three phases, with two being diploid and one haploid. Several red algae exhibit permutations on the triphasic theme. Hence, the reader is refered to various phycological texts (Bold and Wynne, 1985; South and Whittick, 1987; van den Hoek et al. 1995) for details. The life history presented in what follows is characteristic of about 70% of the red algae studied (Bold and Wynne, 1985).

Sexual reproduction in red algae is a type of oogamy involving a nonmotile "sperm" or *spermatium* from a male plant and a "egg" or *carpogonium* on the female gametophyte. A typical triphasic life history for the red alga *Eucheuma* is shown in Figure 6-23. The free-living diploid plant is called the *tetrasporophyte*, which, when reproductive, produces specialized cells (*tetraspore mother cells*) that undergo meiosis, producing four haploid *tetraspores*. The pattern of division within the tetrasporangia varies among orders, being cruciate to zonate to tetrahedral (see Fig. 6-24), as well as irregularly cruciate or zonate (Guiry and Irvine, 1989). On release, the haploid tetraspores germinate and grow into free-living *male* and *female gametophytes*, which can be isomorphic (*Eucheuma*; Fig. 6-23) or heteromorphic (*Bonnemaisonia*). The male gametophyte produces *spermatangial* cells, usually at the end of branches, that produce *spermatia*. The female gametophyte produces *carpogonia* on vegetative branches, which have an elongated hairlike extension called a *trichogyne*, the receptive site for spermatia during fertilization. The position of the carpogonium (on vegetative or specialized branches) and the number of cells of the carpogonial branch are taxonomic features use to separate orders and families in the red algae.

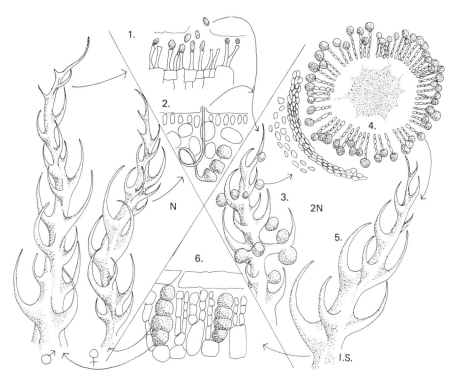

Figure 6-23. The life history of the tropical red alga *Eucheuma isiforme*. The fleshy carrageenophyte (see Fig. 6-31) has a triphasic life history. Male gametophytes produce nonmotile spermatia (1) whose nucleus fuses with the carpogonium (2) to produce diploid carposporophytes (4) that in turn form cystocarps (3) on the female gametophyte. The carposporophytes release carpospores that grow into the diploid tetrasporophytes (5) that are isomorphic with the gametophytes. The tetrasporophytes produce haploid tetraspores (6) that grow into male and female gametophytes.

Fertilization occurs when a spermatium attaches to the trichogyne of the carpogogonium, and its nucleus fuses with the carpogonial nucleus. In most red algae, the zygote usually undergoes a series of mitotic divisions that produce a mass of diploid cells, the *carposporophyte*. The latter stage is often considered to be small parasitic generation of the female gametophyte. In many species, it consists of densely packed *gonimonoblast* filaments that divide to produce *carpospores*. The entire structure may be enclosed in a vaselike protective encasement of branches or *pericarp* that is part of the female gametophyte. The carposporophyte plus the pericarp, which may be visible to the naked eye, are called a *cystocarp*. In *Eucheuma* (Fig. 6-23), the initial group of diploid cells will fuse to produce a large *fusion cell* from which gonimoblast filaments form. Fusion cells are not common to most red algae.

Diploid nuclei from the initial fusion of a spermatium and carpogonium may

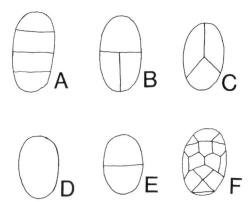

Figure 6-24. Examples of sporangial division. Tetrasporangia may be divided zonately (A), cruciately (B), or tetrahedrally (C). Three specialized sporangia include monosporangia (D), bisporangia (E), and polysporangia (F).

be transferred to other receptive units called *auxiliary cells*; these may either be very near (*procarp condition*) or distant (*nonprocarp condition*) from the carpogonium. If transfer of the diploid nucleus is nonprocarpial, then a *connecting filament* (ooblast filament) grows from the fertilized carpogonium to auxiliary cells, as is found in *Eucheuma*. Thus, from a single fertilization, a large number of carposporophytes and ultimately carpospores can be produced. On release, the carpospores will germinate and grow into new free-living tetrasporophytes. In summary, the triphasic life history of *Eucheuma* includes haploid male and female gametophytes, a parasitic diploid carposporophyte, and a diploid, free-living tetrasporophyte that is isomorphic with the gametophytes (Fig. 6-26).

Two questions are often asked regarding the red algal life history. Why is there such an elaborate life history and what is the role of this elaborate fertilization process? The triphasic life history might be an evolutionary "compensation" related to the total lack of flagellated cells, zoospores, or gametes (Searles, 1980). Certainly, the absence of any flagellated stage must be one of the most unique characteristics of this division of eukaryotic plants in the Kingdom Protoctista.

Taxonomy

As noted by Woelkerling (1990), there are many different treatments of red algae with the number of orders ranging from 10 to 18 and one or two classes or subclasses. In this text, two subclasses (Bangiphycidae and Florideophycidae) and one class (Rhodophyceae) are detailed (Bold and Wynne, 1985; South and Wittick, 1987; but see van den Hoek et al., 1995). The former subclass is small, containing about 1% of the genera and most of the six freshwater red algal genera.

The *Bangiphycidae* are characterized by six primary traits: (1) uninucleate cells; (2) a single stellate, central plastid; (3) intercalary (diffuse) cell division; (4) an absence of pit connections or if present they lack a cap; (5) the usual lack of sexual reproduction (except for species of *Porphyra*, *Bangia*, and *Rhodochaete*); and (6) being simple unicellular or multicellular forms. When sexual reproduction is present, there is no carposporophyte nor a triphasic life history. This subclass contains three orders with marine species: the Porphyridiales, Compsopogonales, and Bangiales.

The *Prophyridiales* consist of unicellular, colonial, to pseudofilamentous forms that are placed in 18 genera. Sexual reproduction is unknown. Probably the most studied species of this order is the unicellular *Porphyridium aerugineum* (family Porphyridiaceae), which occurs in freshwater, soil, and marine habitats. The globose cells have a single large stellate chloroplast with a central pyrenoid (Fig. 6-25). *Goniotrichum alsidii* (family Goniotrichaceae) occurs as

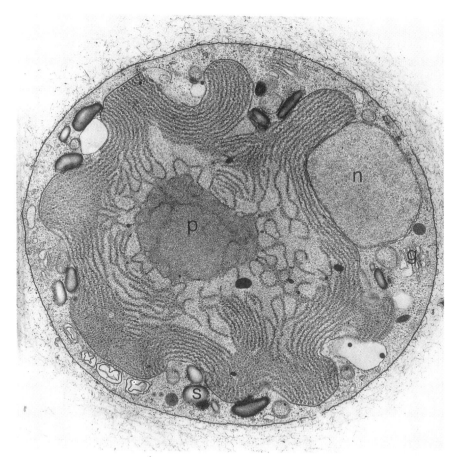

Figure 6-25. An electron micrograph of the edaphic red alga *Porphyridium aerugineum*. The central, stellate chloroplast contains a central pyrenoid (p); starch grains (s), golgi bodies (g), and a nucleus (n) are at the periphery of the cell. The cell is about 20 μm in diameter (courtesy of Beth Gantt, Smithsonian Institution).

an epiphyte on other algae and displays the pseudofilamentous, branching aspect typical of the order.

The *Compsopogonales*, which contain three families, produce monosporangia that undergo unequal division to produce monospores (Fig. 6-24). The family Erythropeltidaceae contains three marine genera. These include *Erythrocladia*, small epiphytic pads of radiating filament; *Erythrotrichia*, which is either unbranched or branched uni- to multiseriate filaments; and *Smithora*, which is a small monostromatic blade found specifically on the seagrasses *Zostera* and *Phyllospadix* (Chap. 11).

The *Bangiales* are a monotypic order that contains two relatively common marine genera, *Bangia*, with a filamentous to solid cylinder construction, and *Porphyra*, which is a bladed species. Because of its economic importance as a food in Japan, Korea, China, and elsewhere (App. B), *Porphyra* has been intensively studied. The heteromorphic life history of *Porphyra* (Fig. B-3) involves a haploid stage consisting of a monostromatic to distromatic blade and a diploid filamentous (*Conchocelis rosea*) phase that bores into shells (Fig. B-4).

The subclass *Florideophycidae* has been subjected to major revisions since 1980 (van den Hoek et al., 1995), because of enhanced information regarding life histories (Cole and Sheath, 1990), pit plug morphology (Pueschel, 1989), and molecular phylogentic studies (Freshwater et al., 1994). The main features of this subclass include five features: (1) occurrence of multinucleated cells in many species; (2) presence of pit plugs; (3) presence of several to many discoid chloroplasts in a cell; (4) cell division is typically apical; (5) only multicellular species are present; and (6) sexual reproduction is common (Bold and Wynne, 1985). The classification used in this text does not recognize five other orders (Achrochaetiales, Batachospermales, Hildenbrandiales, Gracilariales, Ahnfeltiales) as found in van den Hoek et al., 1995). A key to nine orders of the Florideophycidae containing marine species is modified from Dawes (1981) and Bold and Wynne (1985) and presented in Table 6-3.

The *Palmariales* contain genera separated from the Rhodymeniales because of their unique life histories. That is, isomorphic male gametophytes and tetrasporophytes are macroscopic, whereas the female gametophytes are small (a 0.1-mm dia. disc with 1-mm blades). After fertilization, the tetrasporphyte develops directly on the female plant (parasitic?), producing tetrasporangia on stalk cells. *Palmaria palmata*, formerly known as *Rhodymenia palmata*, is a common foliose plant in the North Atlantic and Pacific oceans, having a circumboreal distribution. It is commonly called *dulse* along the maritime coasts of Canada and Maine in the North Atlantic and is eaten dry as a snack or in cooked meals (App. B).

The *Nemaliales* have been reduced in stature after the removal and elevation of the families Bonnemaisonaceae and Batrachospermaceae to ordinal level (van den Hoek et al., 1995). The remaining species still exhibit a wide range of life histories, sexual reproduction, and morphology. Most probably, this order will be further modified in the future. Of the four remaining families, the Acrochaetiaceae (Acrochaetiales; van den Hoek et al., 1995) show filamentous uniaxial development and monospore formation in *Audouinella* (Fig.

TABLE 6-3. A Dichotomous Key to Nine Orders of the Florideophycidae

1 Carposporophytic phase absent..............................	*Palmariales*
1 Carposporophytic phase present..............................	2
2 Carposporophyte develops directly from fertilized carpogonium or subtending cell, auxiliary cells absent.........	3
2 Carposporophyte develops from an auxiliary cell.............	5
3 Nutritive chains of cells fuse with gonimoblast filaments; plants have alternation of isomorphic generations....................	*Gelidiales*
3 Nutritive chains of cells lacking, plants have alternation of heteromorphic life history.....................................	4
4 Growth uniaxial, pericarp well developed, and vesicle cells present..	*Bonnemaisoniales*
4 Growth uniaxial to multiaxial, pericarp absent or very limited, and vesicle cells absent.....................................	*Nemaliales*
5 Auxiliary cells produced after fertilization and from the supporting cell...	*Ceramiales*
5 Auxiliary cells present prior to fertilization....................	6
6 Auxiliary cell an intercalary part of a vegetative filament, in normal pattern of branching.................................	*Gigartinales*
6 Auxiliary cell not intercalary in vegetative filament.............	7
7 Auxiliary cell in an accessory (nonvegetative filament, on the supporting cell (procarpial) of the carpogonial filament or remote from the carpogonial branch (nonprocarpial)..........	8
7 Auxiliary cell that is the terminal cell of a two-celled filament borne on the supporting cell of the carpogonial branch (procarpial)...	*Rhodymeniales*
8 Zonate tetrasporangia simultaneously cleaved, intercalary meristem present, reproductive structures in conceptcales, cell wall impregnated with calcite calcium carbonate...........	*Corallinales*
8 Tetrasporangia, if zonately divided not simultaneously cleaved; no intercalary meristem, conceptacles absent; if calcified, aragonite calcium carbonate on surface.......................	*Cryptonemiales*

6-26), but not in *Rhodochorton*. Different taxa can attach by either a single cell or disc of cells, and growth is *heterotrichous*, as there is differential development of the erect and prostrate filamentous portions. Members of the family Helminthocladiaceae include the tropical genera *Liagora* and *Helminthocladia* [Fig. 6-27(A)]. Species of the former genus are lightly to heavily calcified (aragonite; on the surface only). The latter genus is pantropical in distribution and gelatinous in texture, consisting of interwoven filaments. The life histories in this family can be triphasic and heteromorphic with their carpospores forming an "*Audouinella*" stage that can reproduce both by monospores (asexual) and tetraspores. The latter spores can grow into a protonemalike (prostrate filaments) structure that produces erect "buds." The buds then develop into the macrothallic gametophytes that produce carpogonial [Fig. 6-27(B)] and spermatangial [Fig. 6-27(C)] branches.

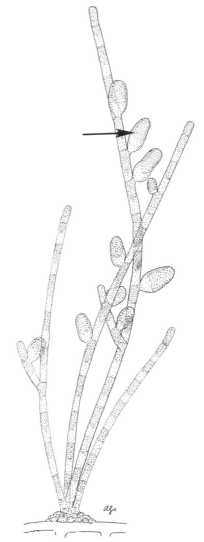

Figure 6-26. The tropical epiphytic red alga *Audouinella hypnea* consists of filaments (8 to 10 mm dia.) that develop from a filamentous base and produce monosporangia (arrow).

Figure 6-27. The gelatinous, tropical red alga *Helminthocladia clavadosii* is a deep-water species that reaches 4 dm in length and has progressive branching (A). Cortical tissue contains carpogonial branches (B) with a trichogyne and spermatangial branches (C), with chains of spermatia (arrow).

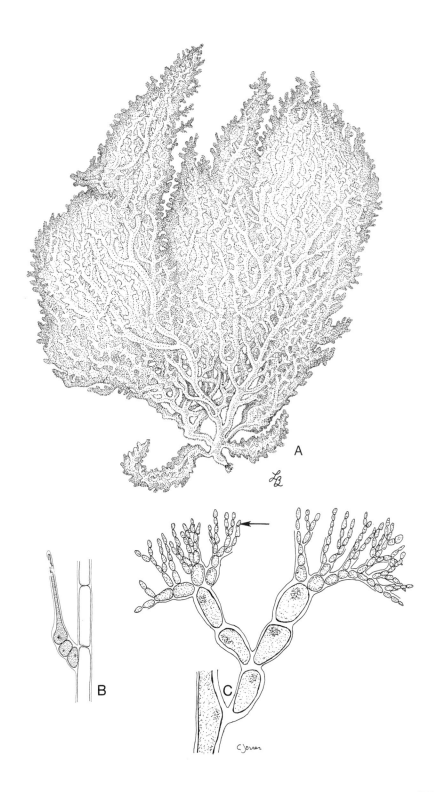

Members of the *Gelidiales* are usually placed in two families, the Gelidiaceae and Wurdemanniaceae. The life history is a typical triphasic one, with the gametophytes and tetrasporophytes being isomorphic. In some species, only the tetrasporic phase is known, which suggests a "modified" life history. Species in this order lack an auxiliary cell, but after fertilization, connecting filaments will branch and fuse with specialized *nutritive cells*. Unlike auxiliary cells, these secondary fusions do not result in carposporophytes. The genera yield the economically important phycocolloid, agar (Fig. B-8), and some are grown as sea vegetables (App. B). *Gelidium* is a rather polymorphic genus. It ranges from terete to compressed branches [Fig. 6-28(A)] with a dense, central medulla that obscures the uniaxial nature of the plant. Small, thick-walled filaments (*rhizines*) are present in the medullary tissue [Fig. 6-28(B)] and their presence and position have been used to distinguish genera. The rhizines are specialized filaments that develop at the apex and extend to the base of the plant and may function in structural support.

The order *Bonnemaisoniales* was erected to contain heteromorphic members of the Nemalionales. That is, *Bonnemaisonia hamifera* was found to be the gametophytic phase of a tetraspore-bearing plant previously identified as *Trailliella intricata*. Similarly, *Asparagopsis armata* is the flesh gametophyte, and *Falkenbergia rufolanosa* is its free-living filamentous tetrasporophyte. The tetrasporophytes of both *B. hamifera* and *A. armata* were originally placed in the Ceramiales because they were either uniseriate, branched filaments (*T. intricata*) or were composed of three cells in cross-section (*F. ruflanosa*). Tetraspore production in *A. armata* is induced by a short day and moderate temperatures (Guiry and Dawes, 1992), and this is the case for *B. hamifera* as well (Lüning, 1990).

The *Cryptonemiales* are a large order with 12 families with over 100 genera, but more recently have been reduced through the elevation of a number of families to ordinal level. Some classifications combine this order with the Gigartinales (see what follows; van den Hoek et al., 1995) in part because of the parallel morphological patterns found in the two orders. The two orders are primarily separated by the auxiliary cell, which is on a specialized branch in the Cryptonemiales, whereas in the Gigartinatles, it occurs on a vegetative branch. The Cryptonemiaceae contain a variety of temperate and tropical genera, including *Halymenia*, *Corynomorphya*, *Cryptonemia*, and *Gratelopuia*. *Halymenia* species are usually foliose [Fig. 6-29(A)] and large (to 0.5 m), with a delicate multiaxial construction of slender medullary filaments [Fig. 6-29(B)] that can interconnect to form "ganglia." Unlike the multiaxial members of the Cryptonemiaceae, the Gloiosiphoniaceae are uniaxial and appear to have heteromorphic life histories with a crustose tetrasporic phase and erect, fleshy gametophytes. Some authors (e.g., van den Hoek et al., 1995) have raised this family to ordinal status.

"Coralline algae," or members of the *Corallinales*, are impregnated with calcite calcium carbonate (Bosence, 1991). They occur in both temperate and tropical waters throughout the world. Reproductive structures occur in pits or *concepticals* and growth is by both intercalary (diffuse) and apical meristems.

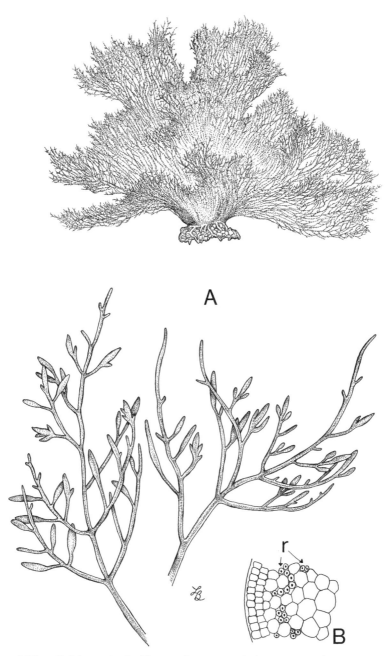

Figure 6-28. *Gelidum crinale*. The small, agar-producing red alga forms wiry tufts 2 to 5 cm in height (A) and has terete branches. Cross-sections of mature branches show thick-walled cells (rhizines = r) in the outer region of the medulla (B).

Again, this order was previously delineated as a family of the Cryptonemiales, being characterized by its zonately divided tetrasporangia (Fig. 6-24) and the carpogonial branches functioning as auxiliary cells. The *articulated* species have noncalcified *genicula* or joints between the calcified segments [*Amphiroa*; Fig. 6-30(A)]. *Nonarticulated* species lack genicula and are crustose or with erect, nonjointed branches [*Lithothamnion*; Fig. 6-30(B)]. Although only 35 genera, species occur throughout the world and play a conspicuous role in the development of coral reefs and the formation of calcium carbonate sediments.

The *Gigartinales* contain the highest number (28) of families in the Rhodophyta. Most probably a number of families will be raised to ordinal status based on nucleic acid sequences (van den Hoek et al., 1995). In addition to exhibiting typical triphasic life histories, members of this order produce an auxiliary cell on normal vegetative branches. A number of genera are harvested (and farmed) for the phycocolloid, carrageenan (App. B, Fig. B-9). For example, the family Solieriaceae contains a number of tropical carrageenophytic genera, including *Solieria*, *Eucheuma* (Figs. 6-23 and 6-31), and *Kappaphycus*. The family is characterized by the fusion cell that occurs on fertilization and some species have a filamentous medulla. The Gracilariaceae, which are characterized by multiaxial construction with the medullary cells being parenchymatous, have six genera of which *Gracilaira* (Fig. 6-32), with over 100 species, has the largest number. Currently, it is the leading source of agar in the world, having replaced members of the Gelidiales. Some species of this order, such as *Mastocarpus stellatus* (formerly, *Gigartina stellata*; in the family Gigartinaceae), have a heteromorphic life history, consisting of a tetrasporic crust that was called *Petrocelis cruenta* and a foliose erect gametophyte.

The *Rhodymeniales* contain about 40 genera that are placed in three families. The species are characterized by multiaxial growth and triphasic isomorphic life histories. The female gametophytes show a procarp arrangement with three- to four-cell carpogonial branches and adjacent two-cell auxiliary cell branches. *Champia salicorniodes* (Fig. 6-33) has a four-cell carpogonial branch and a two-cell auxiliary cell branch, both of which arise from the same supporting cell in a procarpial arrangement. Many taxa have a hollow or partially hollow construction, with terete to foliose morphologies. *Chrysymenia* is a member of the Rhodymeniaceae that lacks central medullary filaments, whereas *Champia* in the family Champiaceae has hollow segments with internal longitudinal filaments.

The *Ceramiales* order is the largest and most clearly defined of the

---

Figure 6-29. The tropical red alga *Halymenia floresia* consists of highly divided soft blades (A) that reach 4 dm in length. The loose medulla (B) contains stellate ganglialike cells (arrows) that are about 50 mm in diameter, whereas the cortex has tetrahedrally divided tetrasporangia (t).

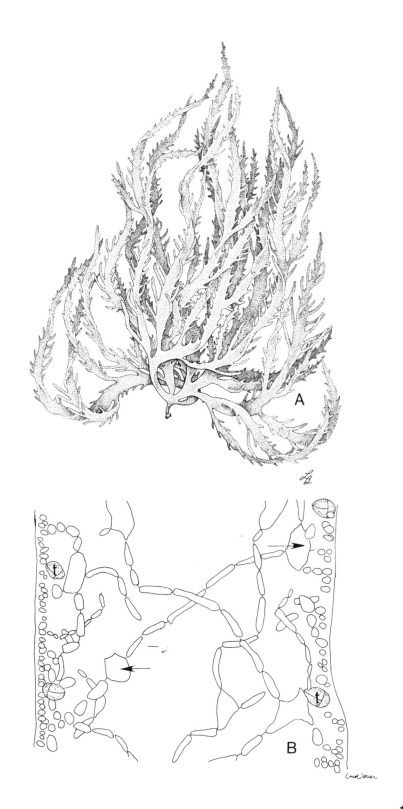

159

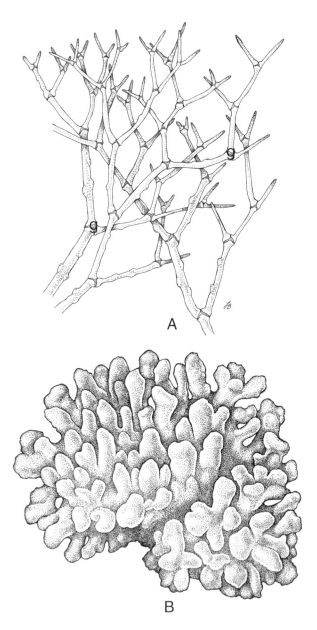

Figure 6-30. Crustose and articulated tropical coralline red algae. *Amphoria fragilissima* (A) is an articulated coralline that reaches 2 to 4 cm and has calcified segments separated by noncalcified geniculae (g). By contrast, *Porolithon antillarum* (B) is a crustose, nonarticulated coralline whose erect branches (to 1 cm tall) lack geniculae.

Figure 6-31. *Eucheuma isiforme* is a tropical carrageenan-producing red alga that forms large, orange-yellow bushes 0.5 m tall.

Rhodophyta. There are four families, with each having an auxiliary cell that develops directly from the supporting cell of a carpogonial filament *after* fertilization. The Ceramiaceae, with more than 100 genera, contain many delicate forms; typically, they are filamentous and uniseriate in morphology and their carposporophyte exposed (the pericarp-covering carposporophyte is limited). One example is *Callithamnion*, a small, delicate uniseriate alga that has alternate branching (Fig. 6-34). The cells in many members of this family are multinucleate, and pit connections are easily visible under the light microscope. The other three families all show *polysiphonous* construction in which the central cell (derived from the apical cell) is surrounded by four or more *pericentral* cells. The Delesseiraceae have about 90 genera, some of which are highly attractive red algae and have been used as decorative specimens when pressed. The species tend to be foliose and thin, as seen in the small (0.5 to 1 cm wide) monostromatic blades of *Caloglossa lepieurii* (Fig. 6-35). The Dasyaceae show *sympodial* development in which the apical cell is constantly replaced by a lateral cell. *Dasya* has polysiphonous construction and sympodial growth (Fig. 6-36). The Rhodomelaceae include about 125 genera that are constructed of four or more pericentral cells. In many species, colorless hairs or *trichoblasts* are produced from the apical cell. *Laurencia* is a tropical genus that exhibits a typical triphasic life history (Fig. 6-37). Growth is by a group of apical cells that occur in branch pits (Fig. 6-37; arrow), with the polysiphonous construction being obscured by extensive cortication.

Figure 6-32. *Gracilaria cornea* (*G. debilis*) is a pale-yellow tropical agarophyte with irregular to secund branching axes (2 cm dia.).

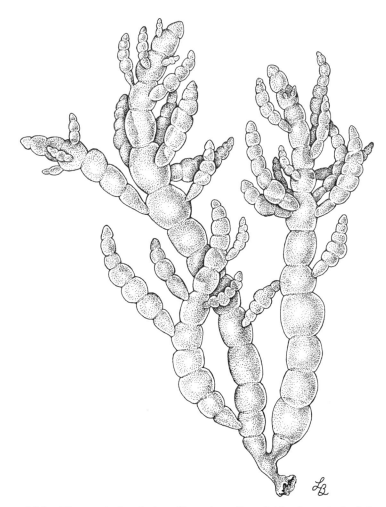

Figure 6-33. The tropical red alga *Champia salicornioides* is typical of the family Champiaceae, having hollow segments (to 4.0 mm dia.) and reaching 12 cm in length.

Figure 6-34. *Callithamnion cordatum* is a delicate, rose-pink filamentous plant that grows to 4 cm. It is abundantly branched forming dense tufts with the cells reaching 200 μm in diameter and attaches by basal rhizoids.

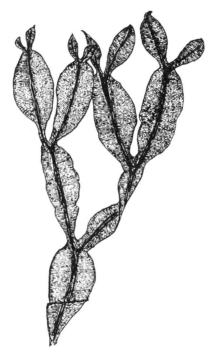

Figure 6-35.   The red alga *Caloglossa leprieurii* has small (1 to 3 cm) delicate (mono-stromatic) blades with polysiphonous "veins."

Figure 6-36. Habit and branch of *Dasya caraibica*. This bushy tropical red alga reaches 20 cm, is covered by pink to red monosiphonous branchlets, and has polysiphonous axes (arrow).

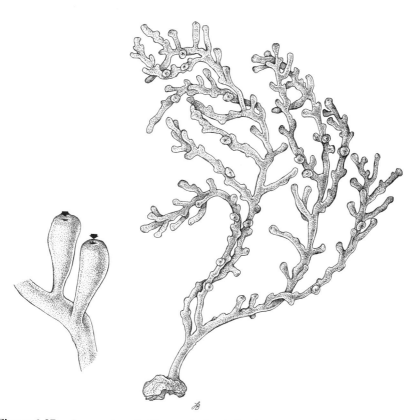

Figure 6-37. *Laurencia poitei* is a tropical red alga found in drift or attached. It reaches 1 dm and is covered by short, truncate ultimate branchlets with sunken apical meristems (arrows).

CHAPTER 7

Microalgae and Their Communities

Although the emphasis in this book is on marine macrophytes, a chapter on predominantly microalgal divisions and introduction to the types of microalgal communities is included. This is because microalgae have critical roles in primary production, nutrient cycling, and in food chains. A close examination of a rocky shore, a salt marsh, mangrove forest, or a seagrass community will reveal microalgae. These organisms can dominate coastal waters forming phytoplankton communities as free-floating or swimming cells and as benthic communities on plant (epiphytic) and sediment (epipelic and epipsammic) surfaces.

Two prokaryotic and four eukaryotic divisions of algae are included in this chapter. The divisions considered here are separated from macroalgae because of their small size, which usually requires the use of a compound light microscope. Species of marine microalgae occur in almost all the algal divisions with the exception of the Phaeophyta. The Rhodophyta are represented by only a few examples. The algal divisions presented here are dominated by microalgae, although they may contain some macroalgal species. Because the chlorophytes were previously treated, they will not be discussed again. Even so, over 90% of green algal species are microscopic coccoid or flagellated forms with the majority being freshwater species (Bold and Wynne, 1985; van den Hoek et al., 1995), although there are a number of marine monads (Butcher, 1959). The general features of each division and a brief review of their structural features are presented.

DIVISION CYANOPHYTA

The blue-green algae (Cyanobacteria) belong to the Eubacteria, which along with the Archaebacteria comprise the Prokaryota (Chap. 1). The number of species is open to question, although many treatments estimate the numbers at about 2000. Being prokaryotic organisms, one might question why these blue-green "bacteria" are considered "algae" (i.e., Bold and Wynne, 1985; van den Hoek et al., 1995). Blue-green algae have chlorophyll *a* and through photosystem II release oxygen in photosynthesis. In this sense, cyanophycean algae bridge the gap between prokaryotic and eukaryotic plants. Cyanophycean taxa have existed for at least 2200 million ybp, with fossil *stromatolites* or layered rocks formed by accretion around cells dating back to the Precambrian Period. Thus, blue-green algae are probably descendants of the oldest oxygen-producing organisms on Earth.

Blue-green algae are ubiquitous. They occur in seawater and freshwater, on unconsolidated substrata, as endolithic organisms in deserts and Antarctic habitats, and as epilithophytes on tropical intertidal limestone. Blue-greens are the photosynthetic component of some lichens; they cause toxic "red" tides (e.g., *Trichodesmium* off Sierra Leone, west Africa, see *Oscillatoria erythaea* in Hummand Wicks, 1980), and are important fixers of nitrogen in coastal communities. Some species approach macroscopic size and have been treated as seaweeds (Lobban and Harrison, 1994), but most are microscopic. In addition to phycological texts, there are extensive reviews of blue-green algal biology (Fritsch, 1945; Carr and Whitton, 1973) as well as taxonomic treatments (Tilden, 1910; Geitler, 1932; Desikachary, 1959; Humm and Wicks, 1980).

Some blue-green algae can fix nitrogen, which is of great significance in oligotrophic marine habitats, including coral reefs, seagrass beds, and estuaries. Nitrogen fixation is known to occur in *heterocyst*-bearing filaments (*Anabaena*), as well as in some unicellular species (*Gloeocapsa*) and nonheterocystic filamentous species (*Trichodesmium*). Heterocysts are specialized thick-walled cells in which nitrogen fixation can occur. As in bacteria, the nitrogen-fixing enzyme nitrogenase is inhibited by oxygen, so that nitrogen-fixing cells are isolated in thick-walled heterocysts, or by a gelatinous sheath if the species is filamentous.

Cytology

The peripheral *chromatoplasm* of a blue-green cell is usually more pigmented than its paler *centralplasm*, because of the greater concentration of thylakoids in the former than the latter. In addition to chlorophyll *a*, blue-green algae contain β carotene, a number of xanthophylls (myxoxanthin, zeaxanthin), and the phycoblins, c-phycoerythrin, c-phycocyanin, and allophycocyanin (Tab. 4-1). The phycobilins are organized into phycobilisomes (Fig. 7-1) on thylakoids, as seen in the eukaryotic red algae. The prokaryotic structure is similar to that of bacteria, lacking membraned organelles (plastids and mitochondria). The photosynthetic thylakoids are concentrated in the chromatoplasm of the cell, and the DNA is present as fine fibrils in the centralplasm (Fig. 7-1). The cytoplasm lacks tonoplast-bound vacuoles, is viscous, and difficult to plasmolyze. Gas vacuoles (protein-bound cylindrical vesicles), carboxysomes (RuBisCO containing polyhedral bodies), cyanophycin granules (amino acids), polyphosphate bodies, and polyglucan granules (cyanophycean starch) also occur in their cytoplasm.

Cell Wall and Reserve Food

Blue-green algae have a conspicuous cell wall and may have an outer gelatinous investment that can be divided into a sheath next to the cell wall and a defined

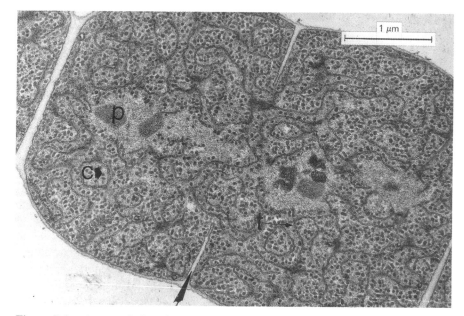

Figure 7-1. A transmission electron micrograph of the blue-green alga *Anabaena variabilis* showing a dividing cell (cross wall at arrow). Convoluted thylakoids are in the outer chromatoplasm along with cyanophycean granules, and polyhedral carboxysomes (p) occur in the centralplasm.

or indefinite mucilaginous film. The investment material appears to be made up of pectic and mucopolysaccharide materials. The cell wall is a complex multilayered structure (usually four layers) with an outer lipopolysaccharide coat similar to gram-negative bacteria. Beneath the outer layer are peptidoglycans consisting of muramic acid, glucosamine, alanine, glutamic acid, and α- and ε-diaminopimelic acids. If filamentous, the chain of cells with their cell wall is called a *trichome*, and the term *filament* includes the sheath materials as well.

The food reserves are polyglucan granules of cyanophycean starch and α-1,4,linked glucan. Cyanophycin granules are composed of co-polymers of aspartic acid and arginine and serve as a source of nitrogen. The polyphosphate bodies contain highly polymerized polyphosphate materials, which serve as reserves.

Motility and Reproduction

Although Cyanophyta have no flagella, some species that lack a distinctive sheath can glide. The oscillation of *Oscillatoria* appears to be due to contraction of protein microfibrils in the surface of the cell wall and the release of gelatinous substances through wall pores. Hader and Hoiczyk (1992) review the types and mechanisms of gliding motility in prokaryotes, including blue-green algae.

Cell division occurs by centripetal development of the two inner-wall layers (Fig. 7-1, arrow). In filamentous species, asexual reproduction occurs by fragmentation of filaments (i.e., *hormogonia*) either due to the presence of separation discs or by simple breakage. In some species, fragmentation occurs where resting cells (akinetes) occur. *Akinetes* are modified vegetative cells that have a thick cell wall and dense cytoplasm. Such resting cells may remain dormant for long periods and can germinate into new filaments when conditions change. *Endospores* develop by internal divisions of the cytoplasm, resulting in a number of thick-walled spores within a mother cell. There is no direct evidence of sexual reproduction as with bacterial conjugation. Even so, recombination experiments indicate that genetic exchange can occur, probably through *parasexual* processes of transformation and conjugation, as in bacteria. *Transformation* occurs when DNA that is released from a donor cell is incorporated into a recipient and replaces part of the host cell DNA. *Conjugation* is where two cells become connected and DNA material is exchanged; this type of parasexual process is poorly known for blue-green algae.

Taxonomy

The taxonomic delineation of the Cyanophyta has been questioned by bacteriologists (Stanier, 1977) and phycologists (Humm and Wicks, 1980; Drouet, 1981). The use of nucleic acid sequences (Komarek and Angnostidis, 1986) may help to clarify blue-green algal taxonomy, because there is much confusion regarding the separation of species. The classification used here is based on Geitler (1932) and Bold and Wynne (1985), in which the division contains a single class, the Cyanophyceae, and three orders: Chamaesiphonales, (pseudofilamentous, endospore production), Chroococcales (unicellular or colonial forms), and Oscillatoriales (filamentous forms). *Chrococcus dimidiatus* [Fig. 7-2(A)] and *Merismopedia elegans* are examples of unicellular and colonial Chroococcales, respectively. *Anabaena fertilissima* and *Calothrix crustacea* [Fig. 7-2(B)] are examples of the Oscillatoriaceae, both showing the presence of heterocysts.

DIVISION PROCHLOROPHYTA

The division Prochlorophyta contains three prokaryotic genera (*Prochloron*, *Prochlorothrix*, *Prochlorococcos*) that are characterized by the presence of chlorophylls *a* and *b*. Species of *Prochloron* occur as extracellular obligate symbionts and free-living organisms (Burger-Wiersma et al., 1986). Based on the nucleotide sequences of 16S rRNA, none of the genera appear related to one another, but do relate closely with the Cyanophyta (Urbach et al., 1992). At this time, it is difficult to know how this small group of prokaryotic organisms is related to one another and to the blue-green algae.

Prochloron didemni is symbiotic within colonial ascidians (sea-squirts) found in tropical and subtropical waters. It was described by Lewin (1976),

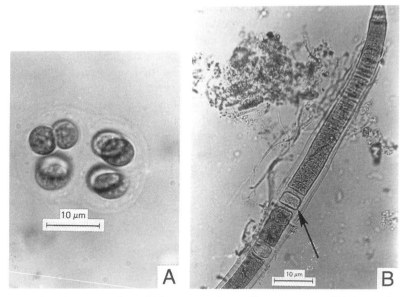

Figure 7-2. Two examples of blue-green algae. *Chrococcus* sp. (A) is a spherical colony of two to eight cells that retain their original sheaths. *Calothrix crustacea* (B) has tapering trichomes with a distinct sheath, and basal and intercalary heterocysts (arrow).

who created a new division, the Prochlorophyta, and co-edited a text on this unique prokaryote (Lewin and Cheng, 1989). The symbiont, whose cells are 10 to 30 μm in diameter, contains chlorophylls *a* and *b* and lacks phycobilins. Lewin suggested that *Prochloron* is a possible link between prokaryotes and green algae. Its cytology resembles a coccoid blue-green alga, but has paired thylakoids that extend around the cell periphery.

Prochlorothrix is a newly described filamentous prochlorophyte that produces blooms in shallow eutrophic lakes in the Netherlands. A coccoid picoplanktonic species, *Prochlorococcus marinus* has cells that are 0.6 to 0.8 μm in diameter and contains a divinyl form of chlorophyll *a*. It grows in subtropical and tropical oceans, being concentrated at 50 to 100 m (Kiyosawa and Ishizaka, 1995).

DIVISION EUGLENOPHYTA

Members of the Euglenophyta are grass-green motile unicells (*monads*) that are free-living and contain chlorophylls *a* and *b*. Further, they have a distinctive cytology and physiology when compared to unicellular Chlorophyta. Extensive reviews of this division are given by Leedale (1967), Walne (1980), and Walne and Kivic (1990). Although euglenoids are more common in freshwater environments, there are a number of estuarine and intertidal species. All species can absorb extracellular organic material, and thus exhibit heterotrophy. Ingestment of bacteria has also been seen in colorless forms. Euglenoids may represent protists that contain chloroplasts evolved from a symbiotic green alga (Gibbs, 1978), which is evidenced by these features:

1. A third, outer chloroplast membrane, which is not derived from the endoplasmic reticulum, may represent the plasmalemma of the original green symbiont.
2. The reserve food is paramylon, not starch, as in the Chlorophyta.
3. Protist organelles in euglenoids include an ampulla (gullet), a photoreceptor free of the chloroplast, mucilage bodies, and an outer proteinaceous pellicle; none of these occurs in the Chlorophyta.
4. Karyokinesis and cytokinesis are distinct.
5. A large number of nonphotosynthetic forms exist and chloroplast loss can be induced.

Cytology

Euglenoids have the same chlorophylls as the Chlorophyta (i.e., *a* and *b*), as well as β-carotene and, to a lesser degree, similar xanthophylls (neoxanthin; Table 4-1). A specialized xanthophyll, astaxanthin, can be so abundant as to give the cell a blood-red color. The anterior portion of the cell contains an *ampulla* or gullet, where the two flagella are attached. The cells are uninucleate with one to many chloroplasts that may contain pyrenoids (Fig. 7-3). A photoreceptor (*eyespot, stigma*) is found in the cytoplasm at the anterior end of the cell and it is not associated with a chloroplast as in other divisions. The eyespot is typical of other algae, having lipid droplets with red carotenoids.

Cell Wall and Reserve Food

Euglenoids are "naked" cells, being covered by a proteinaceous layer called the *pellicle*. The latter is composed of flat strips that overlap and are arranged helically around the cell. The pellicle can be rigid or flexible and may contain ferric hydroxide particles. The species with flexible pellicles may show *euglenoid movement* or a contraction-expansion motion. Reserve food is a refractive, crystalline starchlike *paramylon* (β-1-3 glucose polymer), which is similar to chrysolaminarin and found in the cytoplasm, not the chloroplast.

Motility and Reproduction

Except when encysted or in a nonmotile stage, euglenoids have two to several flagella that originate from basal bodies in the ampulla. In *Euglena*, one flagellum is emergent and covered with one or more rows of fine hairs (pleuronematic). A swelling on the emergent flagellum (*flagellar swelling*) is associated

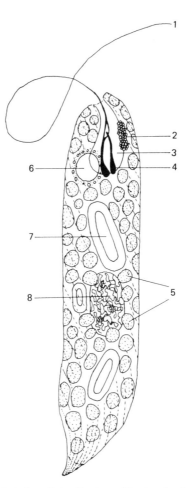

Figure 7-3. The green phytoflagellate, *Euglena*. Plant-animal cellular features include an emergent flagellum (1), an eyespot (2), reservoir (ampulla) (3), and nonemergent flagellum (4), discoid chloroplasts (5), contractile vacuole (6), paramylon starch graules (7), and a nucleus (8).

with the photoreceptor. The other flagella, if nonemergent, is appressed against the emergent one. Asexual reproduction is by longitudinal cytokinesis begin-ning in the ampulla. Karyokinesis is a form of nuclear fragmentation. Sexual reproduction is unknown.

Taxonomy

The five orders are placed in the single class *Euglenophyceae* and demon-strate the variety of trophic natures of these flagellated species, ranging from *phageotrophic* (ingestion of particles and organisms), *osmotrophic* (absorption of organic matter) to photosynthetic (Dawsn and Walne, 1994). The *Eutreptiales* are photosynthetic and have two emergent flagella of equal length and exhibit active euglenoid movements. The *Euglenales*, which are represented by *Euglena*, are also photosynthetic and have one emergent flagella. Other genera in the Euglenales are sessile and attached to substrata by a mucilaginous stalk (*Colacium*), and other free-swimming species produce a *lorica* or rigid wall consisting of pectic compounds that is impregnated with ferric and manganese salts (*Trachelomonas*). The last three orders are nonphotosynthetic, being phageotrophic (*Heteronematales*: colorless and lacking both photoreceptor and flagellar swellings), osmotrophic (*Rhabdomonadales*), or both (*Sphenomonadales*). Evolution in the Euglenophyta probably progressed from phageotrophic, to osmotrophic, to photosynthetic after the capture of photosynthetic cells (Dawsn and Walne, 1994).

DIVISION PYRRHOPHYTA

The division contains the *dinoflagellates*, most of which are biflagellated unicellular organisms that form a major component of oceanic and estuarine phytoplankton communities. Van den Hoek et al. (1995) refer to these organisms as the Dinophyta. There are about 2000 extant and almost an equal number of fossil species. Most species are unicellular and free-living cells, although some are coccoid and a few are filamentous. Different species show many unique modifications of eukaryotic cell structure, including cell shape, cell wall, flagellar attachment, and types and structure of various organelles. Because of these unique features, Tomas and Cox (1973) and Dodge (1983) proposed that dinoflagellates were protists that have captured photosynthetic "brown" algal symbionts. Dodge (1983) suggested an evolutionary sequence for dinoflagellates, with colorless species being the original protists and those with symbiotic chrysophycean algae the next stage. The present photosynthetic forms were the result of losing the nucleus of the endophytic "plastid" and the replacement of fucoxanthin by peridinin. Steidinger and Cox (1980), Loeblich (1982), and Taylor (1987) have reviewed this group.

Dinoflagellates show a wide range of trophic forms, ranging from autotrophs (photosynthetic, free-living), to auxotrophs (requiring vitamins), and heterotrophs (over 50% of the species). The latter include phageotrophic (e.g., *Gymnodinium* spp.) and parasitic (e.g., *Dissodinum pseudolunula*) species (Lee, 1989). A few of "colorless" dinoflagellates are photosynthetic due to the presence of symbiotic algae such as the bioluminescent *Noctiluca* that contains flagellated green endophytes. The ecology of dinoflagellates is very important to human affairs with some marine species forming blooms (*red tides*) and liberating toxins that can result in sickness or death. Other benthic species appear to be the source of fish poisoning (*ciguatera*). A number of dinoflagellates form mutualistic and parasitic relationships. Of special interest are the intercellular symbionts or *zooxanthellae* (Chap. 12) that occur in a wide variety of invertebrates, including hermatypic corals, sponges, jellyfish, sea anemones, gastropods, and various protistans such as foramainiferans, ciliates, and radiolarians.

Cytology

Chlorophylls *a* and *c*, β-carotene, peridinin, and diadinoxanthin are found in the dinoflagellates without "brown algal" symbionts (Table 4-1). Fucoxanthin and diatoxanthin are found in chloroplasts of the brown algal symbionts. All plastids exhibit the "brown algal" organization with thylakoids arranged in bands of 3 (Fig. 7-4). Chloroplasts are discoid or lobed and have a triple membrane; none is connected to the endoplasmic membrane. Some species have aberrant plastids that represent brown alga symbionts; many species lack plastids and are heterotrophic.

The interphase nucleus is large, with chromosomes remaining condensed (Fig. 7-4). The nuclear membrane and nucleolus remain intact during closed

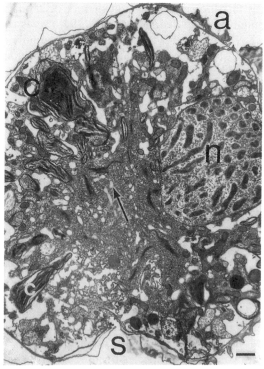

Figure 7-4. An ultrathin section of the armored dinoflagellate *Gonyaulax polyedra*. One lobe of the nucleus (n) and associated Golgi bodies (arrow) plus peripheral discoid chloroplasts (c) and the sulcus (s) are visible. The amphesima membrane and cellulose plates (a) are also present. Unit mark equals 1 μm.

mitosis. Spindle microtubules extend though the nucleus in cytoplasmic chan-nels, attaching to the nuclear membrane adjacent to sites of chromosomal attachment. The low level of chromosomal proteins (histones), the presence of hydroxymethyluracil, and the attachment of the DNA to membranes are similar to bacteria, suggesting that the Pyrrhophyta is an ancient group (Dodge, 1983).

In addition to typical eukaryotic organelles (Golgi bodies, mitochondria, chloroplasts), many dinoflagellates have specialized organelles. Hundreds of *trichocysts* or rods of protein material that extend up to 200 μm on ejec-tion occur in some species. *Nematocysts* (cnidocysts), or elaborate ejectile organelles, occur in two genera. A saclike *pusule*, occurs near the flagellar inser-tion of most species; it functions like a contractile organelle, probably regulat-ing floatation and osmoregulation. *Muciferous bodies* are vesicles along the cell membrane that contain mucilaginous material, which is used in attachment to sand grains by epipsammic species. *Eyespots*, uncommon in dinoflagellates, are found in several forms: (1) a mass of lipid globules that is independent of the chloroplast, (2) a single layer of globules in the chloroplast, (3) two layers of globules that are membrane-bound and separated from the chloroplast, or (4) a complex *ocellus* consisting of a lens with a retinoid and associated pigments.

Cell Wall and Reserve Food

The reserve food in Pyrrophyta is starch, which is not synthesized in the chloro-plast. The cell is covered by a membrane system called an *amphiesma*, which is divided into *thecal vesicles* or membrane sacs that can contain cellulose plates in *armored* (thecate) species (Fig. 7-5) or remain empty in the *unarmored* taxa (Fig. 7-6). The inner membrane of the amphiesma is considered to be the plas-malemma. The cell is divided into upper (*epitheca*) and lower (*hypotheca*) por-tions, which are separated by the *cingulum* (girdle) where the transverse flagella is positioned. In armored dinoflagellates, the cell wall consists of *thecal plates* that are arranged in specific patterns; the latter are used in species delineations.

Motility and Reproduction

The two flagella are located in groves. The cingulum contains the transverse, flattened flagellum. It is pleuronematic with two rows of hairs (mastigonemes) and vibrates in a helical fashion in the cingulum. The posteriorly directed (longi-tudinal) flagellum has one row of hairs and lies in the *sulcus*, a longitudinally oriented grove in the hypotheca (Fig. 7-5). The combination of the beating, trailing flagellum and the hemihelical beat of the transverse flagellum results in a forward, corkscrew type of motion, hence the name dinoflagellates. The forward movement has been measured at 1 to 2 m h^{-1}.

Karyokinesis is closed [Fig. 6-3(A)] and occurs by constriction of the nuclear envelope after anaphase, whereas cytokinesis occurs by an oblique cleavage of the cell in the naked forms. Armored species may either shed their theca (called *ecdysis*) or split the theca in cytokinesis; in both cases, the missing theca

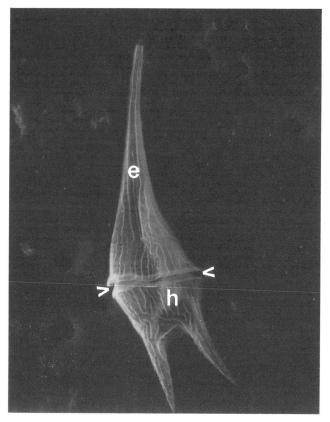

Figure 7-5. *Ceratium hircus.* This armored dinoflagellate is divided at the girdle (arrow) into an upper epitheca (e) and lower (h) hypotheca and is about 130 μm long (courtesy of K. Steidinger, Florida Marine Research Institute).

is reformed. Both isogamous and anisogamous sexual fusions are known in dinoflagellates. With the exception of the diplont taxa of *Noctiluca*, all known life histories are haplontic, with meiosis occurring during or after germination of the zygote.

Taxonomy

Five classes are recognized in here (Bold and Wynne, 1985) in contrast to van den Hoek et al. (1995) who list a single class and 12 orders. Three contain heterotrophic species, that is, the *Ebriophyceae* contain colorless naked species, whereas the *Ellobiophyceae* and *Syndiniophyceae* have parasitic naked cells. The first two classes are small and poorly known. The species of *Dinophyceae* range from parasitic (Blastodiniales) to ameboid (Dinamobeales), filamentous (Dinocloniales), colonial (Gleodiniales), coccoid (Pyrocystales), and

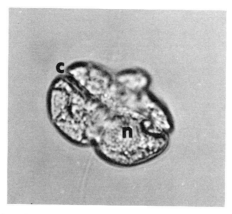

Figure 7-6. A light micrography of the naked dinoflagellate *Gymnodinium brevis* showing its nucleus (n) and cingulum (c) in the living (25 μm wide) motile cell (courtesy of K. Steidinger, Florida Marine Research Institute).

motile, photosynthetic cells (seven orders). The motile cell of all dinophycean species is biflagellate and has the "typical" dinoflagellate organization (Fig. 7-5). *Ceratium* (Fig. 7-5) and *Gonyaulax* are members of the order Peridiniales, which contains armored species. *Gymnodinium brevis* (formerly *Ptychodiscus*; Fig. 7-6) is a red-tide-forming species in the Gulf of Mexico; it is an example of a naked dinoflagellate of the order Gymnodiniales. The final class, the *Desmophyceae*, contains species with motile cells that have two apical or subapical flagella (not latterly inserted as in the Dinophyceae); one of these is directed forward, whereas the other beats in a plane perpendicular to the first and encircles the cell (*Prorocentrum*).

DIVISION CRYPTOPHYTA

The "cryptomonads" are a small division with about 24 genera of which about half are marine. Butcher (1967) illustrated many of the marine species, and Gantt (1980) and Gillott (1990) reviewed the division. Various species occur in salt marshes as well as in open coastal and pelagic waters. Sexual reproduction is unknown except for one species, where isogametes and subsequent karyogamy (fusion of the gametic nuclei) were observed (Krugens and Lee, 1988). The division can be characterized using the following features outlined by Gantt (1980) and as seen in *Chroomonas* (Fig. 7-7):

1. Cells are asymmetric with a dorsiventral flattening (2 to 20 μm long by 3 to 16 μm wide) and an anterior gullet where flagella arise.

2. The two pleuronematic flagella are unequal in length, they emerge from a shallow reservoir, and bear small (150-nm) organic scales.

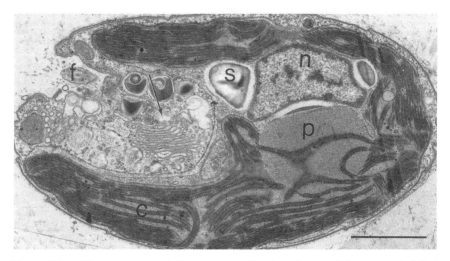

Figure 7-7. The cryptomonad *Chroomonas*. The flagellated cell has a single-lobed chloroplast (c) in which thylakoids are loosely paired and a large pyrenoid (p). The nucleus (n) and starch grains (s) are found in the central portion of the cell along with a Golgi body (arrow), and an oblique section of a flagellum (f) is visible in the shallow reservoir. Unit mark equals 1 μm (Gantt, 1980; photo with permission of the *Journal of Cell Biology*).

3. The proteinaceous periplast consists of thin plates arranged in rows.

4. *Ejectosomes* occur along the cell periphery with smaller ones in the reservoir; they are similar to trichocysts, but consist of membranous rather than protein material.

5. The single chloroplast is surrounded by four membranes, the inner two being the chloroplast envelope and the outer two the endoplasmic reticulum, which also encases the nucleus.

6. One to many pyrenoids may be present in the chloroplast and the thylakoids occur singly or in bands, an unusual organization compared to the other divisions. Starch is present as granules around the pyrenoid.

7. Species may be colorless or a variety of colors (red, blue, olive-yellow, brown, green). Chlorophylls *a* and *c*, and phycobiliproteins (as dense material in thylakoids) are present in the plastid (Table 4-1).

8. A *nucleomorph*, a double-membrane, DNA-containing body, is encased in the outer, double-chloroplast ER membranes. It is thought to be a vestigial nucleus of a primitive endosymbiont.

9. Mitosis includes a partial breakdown of the nuclear envelope and cell division is longitudinal.

DIVISION CHRYSOPHYTA

Six classes are recognized here. By contrast, other treatments delineate these as separate divisions (Lee, 1989) or as classes of a larger division, the Heterokontae (van den Hoek et al., 1995). The features common to all classes include the predominance of cartenoids over chlorophylls (brown to gold to yellow-green color) and the occurrence of a β-linked glucan reserve food (*chrysolaminarin, leucosin*) that resembles laminarin in the Phaeophyta (Bold and Wynne, 1985). Five of the six classes have chlorophylls *a* and *c*, and members of all six classes have chloroplast-endoplasmic reticulum (CER), as described for the Phaeophyta. All classes show a variation in flagellar number (1 to many) and type (acronematic to pleuronematic). The most common are two acronematic flagella.

Class *Chrysophyceae*

The "golden algae" are predominantly freshwater, but include some marine plankton, benthic and intertidal species. The species are mostly unicellar and flagellated, but amoeboid and coccoid forms also exist. Other morphologies include colonial, filamentous, and parenchymatous forms. Typically, fucoxanthin and other carotenoid pigments mask the chlorophylls (*a* and *c*). Even so, a number of species are colorless heterotrophic taxa. Other xanthophylls in pigmented taxa include zeaxanthin, antheraxanthin, violaxanthin, diatoxanthin, and diadinoxanthin (Table 4-1). If present, the cell wall, or *lorica* (investment of cell separated from protoplast), may contain cellulose or consist of scales of organic (e.g., cellulose) or inorganic (silicon dioxide, calcium carbonate) components.

Reproduction is usually by cell division. Mitosis is open (nuclear envelope breaks down) and the poles of the spindles are formed by flagellar basal bodies (*rhizoplasts*). Cytokinesis occurs by cleavage, which develops from the cell anterior. *Statospores*, or resting cells (cysts), are characteristic of the class; they are covered by distinctive siliceous walls. The wall of a statospore consists of two parts, a main, flask-shaped container and a small plug that fills the opening after the protoplast migrates into it. Sexual reproduction is only known for a few genera; sexual fusion is isogamous and the life history is haplontic (e.g., *Dinobryon*). In some species, a statospore results from sexual reproduction and is the zygote.

Two subclasses are recognized by Kristiansen (1982), the *Chrysophycidae*, which contain 99% of the 800 species, with the remaining species in the *Dictyochophycidae* (silicoflagellates). The order Ochromonadales of the Chrysophycidae contains biflagellated, unicellular species of which the marine genus *Ochromonas* is an example (Fig. 7-8). The two flagella are unequal in length and the CER (chloroplast-endoplasmic reticulum) envelopes the nucleus and the one or two chloroplasts. The chloroplasts are of the "brown algal type" with thylakoids in bands of 3. The order Sarcinochrysidales contains marine genera with motile cells and laterally inserted flagella (similar to Phaeophyta); they range

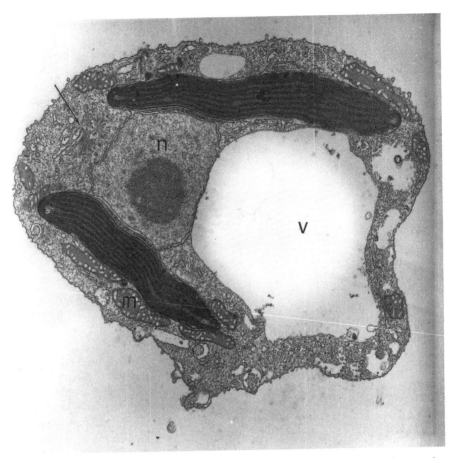

Figure 7-8. *Ochromonas danica*, a chrysophycean monad. A near median section showing the large central vacuole (v), nucleus (n), paired chloroplasts (c), mitochondria (m), and Golgi bodies (arrow). The chloroplasts are connected to the nucleus by its outer membrane (Slankis and Gibbs, 1972; with permission of the *Journal of Phycology*).

from gelatinous colonies (*palmelloid*) to filamentous or thalloid (parenchymatous) morphologies. *Olisthodiscus luteus*, a member of the Sarcinochrysidales, is a flagellated unicell found over a salinity range of 2 to 50% (Tomas, 1978). The species has been extensively studied (Cattolico et al., 1976) and was originally placed in the class Xanthophyceae. It is cultured as food for oyster and shellfish.

The Dictyochales, the only order of the subclass Dictyochophycidae, contain marine silicoflagellates with siliceous skeletons. Its cytoplasm, as seen in *Dictyocha fibula* (van Valkenburg, 1980), is differentiated into a dense *endoplasm* and vacuolated (frothy) *ectoplasm* [Fig. 7-9(A)] that is "draped" over the silica skeleton [Fig. 7-9(B)]. Because of its skeleton, fossil silicoflagellates are known since the early Cretaceous.

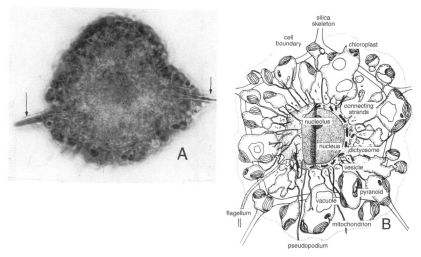

Figure 7-9. *Dictyocha fibula*, a silicoflagellate. Its internal silica skeleton (arrow) projects from the living cell (A). The cell structure is explained in (B), which shows the central nucleus and surrounding globular cytoplasm (van Valkenberg, 1980; part B with permission of the *Journal of Phycology*).

Class *Prymnesiophyceae*

Although described here as a class of the Chrysophyta, others include it as a division (Prymnesiophyta: Lee, 1989; Green et al., 1990; Haptophyta: van den Hoek et al., 1995). Most species have motile cells, but some are coccoid, colonial, and filamentous in morphology. Three features distinguish this class from other members of the Chrysophyta (Hibberd, 1980a):

1. Motile cells possess two acronematic flagella (unlike other chrysophytes) that may be equal or unequal in length.

2. Motile cells have a *haptonema*, or a specialized projection, adjacent to the flagella. In cross-section, the haptonemal projection is constructed of six to seven tubules arranged in an arc along with folds of the plasmalemma. The haptonema is flexible and can coil, but its function is not understood.

3. Most species have a covering of scales or a cellulose wall. The scales are either organic (cellulose) or inorganic, calcified *coccoliths*. The latter structures usually cover the cell and are a characteristic feature in the marine order Coccosphaerales. For example, *Chrysochromulina pring-shemii* bears four types of scales, including long-spined coccoliths and small convex organic ones [Lee, 1989; Fig. 7-10(A)].

Chloroplasts are yellow-brown to gold in appearance due to the presence of chlorophylls *a* and *c* plus fucoxanthin (Table 4-1). The plastid envelope is

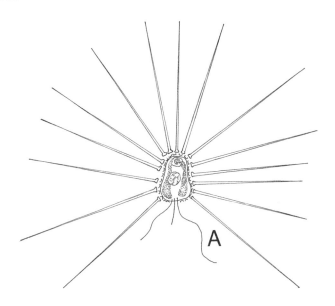

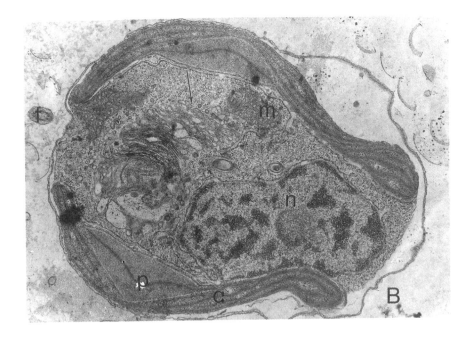

connected with the endoplasmic reticulum and the plastids contain thylakoids grouped into bands of 3 [Fig. 7-10(B)]. Mitosis is open [Fig. 6-3(A)] but unique in that the chromosomes congregate at a metaphase plate with openings for pole-to-pole spindle fibers. Asexual reproduction is by binary cell division or pinching of the plasmalemma. There appear to be haplontic as well as isomorphic and heteromorphic haplodiplontic life histories. However, little is known about the sexual cycle of most species.

Four orders are recognized: the Isochrysidales (haptonema reduced or absent; flagella equal or subequal), Pyrmnesiales (haptonema present; equal or subequal flagella), Coccolithophorales (haptonema reduced or absent; coccoliths present), and Pavlovales (haptonema and eyespot present; flagella unequal). Of the 500 species in the entire class, the majority belong to the marine order Coccolithophorales, which along with dinoflagellates form an important component of phytoplankton communities in terms of productivity in tropical and subtropical oceans (Green et al., 1990). The majority of *cocolithophorid* species occur as small monads of the nanoplankton size class (2-to-20-μm diameter cells). The fossil record of coccoliths indicates that ancestors of this order were abundant in the Jurassic (180 million ybp) and peaked in the Cretaceous (90 million ybp).

A number of taxa in this class generate reduced sulfur compounds such as DMS (dimethylsulfide), an osmoregulatory compound that causes production of sulfuric acid when released in air. Other species form blooms in the North Atlantic, including the noncalcareous member of the Prymnesiales, *Phaeocystis pouchetii* (Veldhuis et al., 1986), and the coccolithophorids, *Emiliania huxleyi* and *Chrysochromulina polylepis*. The latter species formed extensive blooms around the coast of Norway in 1988, causing death to fish, invertebrates, and even seaweeds (Nielsen et al., 1990). The dominant coccolithophorid in present-day oceans is *E. huxleyi*, which is considered to be the largest producer of calcite on Earth and a major component for long-term transfer of carbon to deep sediments (Heimdal et al., 1994). The monad is found in all oceans, from polar to tropical seas and in oceanic and coastal waters. Its blooms can turn the sea a milky-green. The importance of *E. huxleyi* is evident by the devotion of an entire issue of Sarsia (Heimdal et al., 1994) to its ecology, physiology, and productivity.

Class *Xanthophyceae*

The class, which is also called the Tribophyceae, shows a parallel morphological evolution to the Chrysophyceae and the Chlorophyceae, ranging from

---

Figure 7-10. *Chrysochromulina pringshemii*, a chrysophycean monad, shows (A) the long coccolith spines and small organic convex scales that cover the cell and the two unequal flagella. The electron micrograph (B) shows the central nucleus (n), Golgi bodies (arrow), mitochondria (m), and two peripheral chloroplasts (c) with pyrenoids (p). One of the two flagella (f) can be seen in cross-section (courtesy of R. Pienaar, University of Natal).

unicellular to colonial, filamentous, thalloid, and siphonaceous forms (Hibberd, 1990a). Only a few of the 600 species are ameboid or flagellated (paired, unequal-length flagella); most taxa occur in freshwater or edaphic environments. The yellow-green color is due to chlorophyll a, β carotene, and a variety of xanthophylls (vaucheriaxanthin, diatoxanthin, diadinoxanthin; Table 4-1). Chlorophyll c occurs in low concentrations, and fucoxanthin is lacking. In addition to chrysolaminarin, some of the species store oil, fats, free mannitol (which may be an osmolite), and oligopolysaccharides. The cell wall, if present, is cellulose and usually impregnated with silica.

Asexual reproduction is by cell division, *aplanospores* (nonmotile spores), or zoospores. Statospores (resting cells) are produced in some species, but, unlike the Chrysophyceae, they consist of two halves of equal size. The six orders range from naked, bilflagellated, or amoeboid cells (Heterochloridales) to colonial (Heterogloeales), coccoid (Mischococcales), filamentous (Tribonematales), and coenocytic (Vaucheriales) forms. *Vaucheria* (Fig. 7-11) contains coenocytic species, some of which occur on mud flats of salt marshes. The life history of *Vaucheria* is diplontic, with meiosis preceding gamete production and sexual reproduction via oogamy (Hibberd, 1990a).

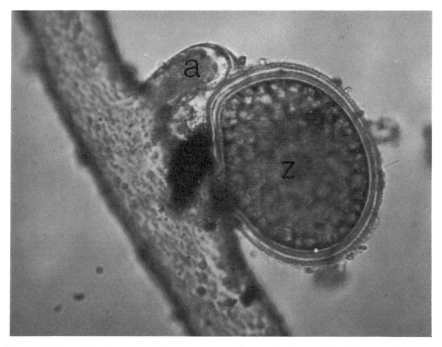

Figure 7-11. *Vaucheria sessilis*, a coenocytic xanthophyte. The filament bears an oogonium with a zygote (z) and an antheridial branch (a) and is approximately 30 μm in diameter.

Class *Raphidophyceae*

The class contains about nine genera of mostly freshwater unicellular monads, which lack a cell wall and have chlorophylls *a* and *c* (Heywood, 1990). Other pigments include β carotene, diadinoxanthin, vaucheriaxanthin, and heteroxanthin, with fucoxanthin being dominant in marine species (Table 4-1). The cells range from 50 to 100 μm in diameter and are flattened dorsiventrally with two apically inserted flagella. The anteriorly directed flagellum is pleuronematic, whereas the posteriorly directed one is acronematic. A *rhizostyle*, or microtubular net, connects the basal bodies of the two unequal flagella to the nucleus. The cytoplasm has a central dense area (endoplasm) where a large nucleus resides. Chloroplasts (thylakoids in bands of 3), vacuoles, trichocysts, and mucocysts (mucilaginous ejectile organelles) occur in the peripheral cytoplasm, or *ectoplasm*. Karyokinesis is closed and the flagellar basal bodies may act as centrioles.

A single order, the Raphidmonadales, contains two families, one being saprophytic and phageophytic (Thaumatomastigaceae) and the other having photosynthetic species (Vacuolariaceae). Species of two marine genera, *Chattonella* (*Hornellia*) and *Fibrocapsa*, can cause intense blooms, or "red tides," in enclosed bays of Japan. Species of the former genus are responsible for extensive fish kills off the Malbar coast of India (Heywood, 1990). Commonly called "chloromonads," these species are usually found in neritic environments (Loeblich, 1977).

Class *Eustigmatophyceae*

Members in this class were removed from the Xanthophyceae because of their unique eyespot, as well as the lack of a chloroplast-endoplasmic reticulum (CER) envelope and Golgi bodies. The class contains seven genera and twelve species (Hibberd, 1980b, 1990b). Most members are coccoid, freshwater, or edaphic algae, although *Nannochloropsis* is a marine picoplanktonic species having cells smaller than 0.2 μm in diameter. The coccoid cells have one or more bowl-shaped to lobed chloroplasts, whose thylakoids are organized into bands of 3. The motile cells have a unique eyespot that consists of a cluster of globules located outside the chloroplast and adjacent to a flagellar swelling. Zooids are elongated with one or two apically inserted pleuronematic flagella. Zoospores lack Golgi bodies and the CER is not continuous with the nucleus as it is in the coccoid cells and other members of the Chrysophyceae.

Class *Bacillariophyceae*

Members are called *diatoms* and are well represented in marine and freshwater environments. There are about 250 extant genera and 100,000 species (Round et al., 1990). In addition to being important components of the phytoplankton, diatom communities include *periphyton* (*aufwuchs*), benthic plants that grow on other aquatic plants (*epiphytic*), rocks (*epilithic*), or unconsolidated sediments either attached to sand grains (*epipelic*) or not (*epipsammic*). Planktonic diatoms have been called the "grass of the sea" because they supply up to 20% to 25% of the world's primary production, which is estimated to be 1.4×10^{14} kg dry mass y^{-1} (Werner, 1977; Raymont, 1980). Diatoms are well represented in the fossil record due to their silicious cell wall. Thus, present-day marine genera have ancestral species dating back to the Cretaceous Period. Fossil deposits of *diatomaceous earth* (diatomite) are mined in various parts of the world, and are used for insulation, filtration, and a variety of other commercial purposes.

Diatoms are unicellular (some forming colonies) and uninucleate, appearing brown to golden brown to slightly green in color. The cytological features of the diatoms are those of the Chrysophyta, including the reserve food, which is chrysolaminarin and oil. The primary photosynthetic pigments are chlorophylls *a* and *c* and β-carotene, which are masked by fucoxanthin and other xanthophylls (Table 4-1). Chloroplasts are connected to the nucleus by a chloroplast endoplasmic reticulum, the thylakoids are grouped into bands of 3, and pyrenoids may be present (Fig. 7-12).

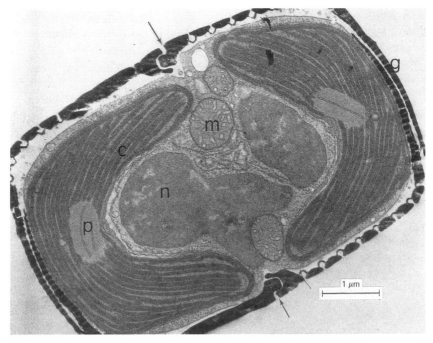

Figure 7-12. *Navicula tripunctata*, a pennate diatom. Raphes (arrows) occur on the epitheca and hypotheca, which overlap extensively at the girdle (g). Two laminate chloroplasts (c) with pyrenoids (p), large mitochondria (m), and a central, lobed nucleus (n) are visible in the electron micrograph (courtesy of E. F. Stoermer, Great Lakes Research Station).

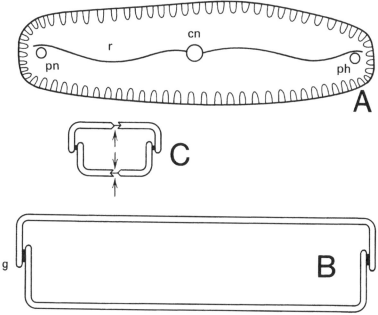

Figure 7-13. A pennate diatom frustule. The valve view (A) shows the raphe (r), plus the central (cn) and polar (pn) nodules, and the girdle view (B) shows the overlapping valves. A cross-section (C) shows how the raphe is constructed (arrows, see also Fig. 7-12).

The cell wall, or *frustule*, is unique to this class and is divided into an upper *epitheca* and a lower *hypotheca* that fits together similar to a box with its cover (Fig. 7-13). The frustule can be viewed from the top (*epivalve*), bottom (*hypovalve*), or from the side (*girdle*). Each valve has its own girdle region, the epitheca having the *epicingulum* and the hypotheca having the *hypocingulum*. Frustules primarily consist of polymerized silicic acid along with protein, pectic acid, and lipid. The frustule shape will have bilateral, radial, or irregular symmetry (Figs. 7-14 and 7-15). The structure and ornamentation of the frustule form the basis for their taxonomy (Barber and Hayworth, 1981; Round et al., 1990). Frustule formation is the result of valve initiation (i.e., cell division) within a *silica deposition vesicle* (silicalemma), which controls the exact morphology and ornamentation. Silica is an absolute requirement for mitosis and frustule formation in diatoms.

Cell division is the most common mode of asexual reproduction. Karyokinesis is open [Fig. 6-3(A)], the nuclear envelope breaks down, and a spindle is formed. Cytokinesis occurs in the girdle region by a cleavage furrow of the protoplast after the two valves have separated (via swelling of the protoplast). Both parental thecas (hypotheca and epitheca) act as the new epithecas. In many species, this results in a progressive reduction in size of one of the two daughter

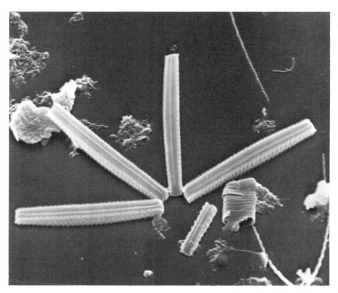

Figure 7-14. The pennate diatom *Thalassionema nitzschioides*. The diatom forms stellate chains. Unit mark equals 1 μm (courtesy of G. Frexyll, Texas A&M University).

Figure 7-15. The centric diatom *Actinoptychus senarius* seen in valve view. Unit mark equals 1 μm (courtesy of G. Frexyll, Texas A&M University).

cells with each cell division. Some species do not show size reduction, although they follow the pattern of producing new hypothecas. The lack of reduction may be due to the plasticity of the frustule or expansion of the girdle region. At some stage after the progressive reduction in cell size, the cell may return to its original size via protoplast extrusion from the cell wall, swelling, and regeneration of the frustule.

More commonly, the smaller-sized frustules become sexually active, producing gametes after meiosis. The life history of diatoms is diplontic. Thus, on fusion of the gametes, the zygote protoplast usually swells and generates an *auxospore* that is characterized by a distinctive cell wall and a surrounding fertilization membrane, or *perizonium*. An auxospore can be either a resting cell or can produce a new, enlarged frustule characteristic of the species. Auxospores also can be produced via *autogamous* (two haploid nuclei from a *single* cell fuse) or via *apogamous* (parental cell undergoes mitotic divisions, each of which develops into a new spore) reproduction.

Two orders are recognized (Werner, 1977; Bold and Wynne, 1985; Round et al., 1990). The *Centrales* have valves that are centric (radial; Fig. 7-15), multipolar, or irregular. Centric diatoms usually contain many discoid plastids, and gametogenesis is oogamous with sperms being uniflagellated. The flagella is pleuronematic and anteriorly placed in the sperm. The *Pennales* show bilateral symmetry (Fig. 7-14) in which the valve is arranged in reference to a central or other hypothetical line. Some families of pennate diatoms are characterized by a frustule having a *raphe*, which is an opening or fissure running along the apical axis (Fig. 7-13), on one or both valves. The raphe connects the *polar nodule* on each end of the valve to the single *central nodule*. Pennate species with raphes can glide via cytoplasmic flow that occurs along the exterior part of the fissure. The cytoplasm returns via the nodules to the inner fissure to flow inside the frustule. Thus, the cytoplasm acts similar to a belt or tread pulling the cell forward (Hader and Hoiczyk, 1992). The polar and central nodules are pores that allow the cytoplasm from the cell to flow into the outer raphe fissure. In contrast to centric diatoms, sexual reproduction in pennate diatoms is via isogamous amoeboid gametes and vegetative cells contain one or two large, laminate chloroplasts.

PHYTOPLANKTON COMMUNITIES

Studies of microalgal communities are as varied as those with macroalgae (Chap. 8). Hence, they have utilized a variety of techniques to measure population dynamics, patterns of distribution, and adaptation with these microscopic photosynthetic organisms. Although this text focuses on macrophyte communities, the importance of the phytoplankton is considerable because they occur in all marine plant communities. *Phytoplankton* are autotrophic free-floating microscopic plants or microalgae that are mostly unicellular, although clonal and filamentous species also occur. Phytoplankton, which occur in rivers, lakes, and seas, are the most widespread group of microalgae, and, because of their significant primary production, they are have been extensively studied. Texts dealing with phytoplankton include ecological assessments (Bougis, 1976; Raymont, 1980; Round, 1981; Fogg and Thake, 1987; Harris, 1986; Sommer, 1989), physiological studies (Morris, 1980; Platt, 1981), and technique manuals (Sournia, 1978).

Classification and Diversity

Phytoplankters are a diverse group of microalgae and include species from all algal divisions, except the Rhodophyta and Phaeophyta (but see *Sargassum* spp. of the Sargasso Sea). Most classes of marine phytoplankton are motile, with the most thoroughly studied motile species being dinoflagellates, with speeds ranging from 50 to 500 μm s^{-1} (Sournia, 1982). There are about 5000 species of phytoplankton in the sea, and the basis for this limited number in such a vast environment has not been easily explained (see Tett and Barton, 1995).

Phytoplankton are usually grouped according to *cell size*, which is important with regard to nutrient uptake; a cell size larger than 20 μm is needed before swimming or sinking can increase the flux of nutrients, while cells 60 μm or more in diameter will increase turbulence and thus reduce the boundary layer effect (Karp-Boss et al., 1996). The authors also found that large chains of diatoms or filaments of blue-green algae are best able to take advantage of turbulence that will decrease the boundary layer, resulting in more rapid uptake of nutrients (see Chap. 3). Regarding size classification, the larger size, commonly called *net phytoplankton*, includes species larger than 64 μm (as retained by a net of that mesh size); they primarily include diatoms and dinoflagellates. Other classifications consider cell sizes of 20 to 200 μm as net or microplankton (Morris, 1980; Fogg and Thake, 1987). Until the end of the 1970s, the other phytoplankton size class was the *nanoplankton*, with cells being 2 to 20 μm. Nanoplankton are dominated by cyanophytes, prymnesiophytes, and coccolithophorids and can account for 75% of the total productivity. In the 1980s, it became apparent that yet a smaller size class existed, the *picoplankton* (0.2 to 2 μm) or submicroscopic organisms that can account for 50% of the primary production in oceanic waters. The class is dominated by prokaryotes, including bacterial plankton, blue-green and prochlorphytic species, as well as a variety of eukaryotic algae (Sieburth et al., 1978). As noted before, a member of the Prochlorophyta, *Prochlorococcus marinus*, is a coccoid species that occurs in the picoplankton of subtropical and tropical oceans, extending to 250 m. The newly discovered coccoid picoplanktonic species can account for more than 50% of the total chlorophyll *a* in the central Pacific Ocean (Kiyosawa and Ishizaka, 1995).

Responses to Submarine Light

Three points regarding submarine illumination should be noted in terms of phytoplankton photosynthesis. First, less than 5% of all light that enters oceanic water will escape (Yentsch, 1980). Second, there is a relationship between the "clear window" of water to visible light and absorption by the photosynthetic pigments. Chlorophyll absorbs in the 425-to-450 nm and 665-to-680 nm range, cartenoids in the 525-to-575 nm range, and phycobilins in the 525-to-650 nm range (cf. Chap. 3). Third, euphotic zones can exceed 100 m in Type I oceanic water (Fig. 2-8), resulting in changes to the *spectrum* and *intensity* of submarine irradiation (Chap. 2).

Phytoplankton that encounter changes in underwater irradiance may modify pigment concentrations, plastid size, density of thylakoids, or number and size of the photosynthetic antennae [Figs. 4-5(B) and (C)]. For example, Sournia (1986) listed 20 species in 5 algal classes that are permanent *shade phytoplankton*, or those only occurring in deep (100-m) oceanic water. Shade species modify their photosynthetic apparatus via increases in pigment concentration and shifts in the types of pigments present (Richardson et al., 1983). Some shade genera such as *Phaeodactylum* can survive in irradiances below 1 μM cm^{-2} s^{-1}, yet increase their photosynthetic rates under higher intensities (Geider et al., 1986). Such responses indicate broad acclimation ability via modifications of the photosynthetic apparatus.

The use of *P-I* curves is the cornerstone of phytoplankton ecology, because such curves are fundamental to modeling photosynthesis over depth (Yentsch, 1980). Further studies of phytoplankton responses to low submarine irradiance suggest that an increase in total pigment ("black-box" concept) does not occur as frequently as shifts in types of pigments (chromatic acclimation), which is in contrast with macroalgae (cf. Chap. 3). Chromatic acclimation has been reported through shifts in pigments to match the available spectra for blue-green and green phytoplankton (Prezelin and Boczarn, 1986).

Nutrients

As with macroalgae, phytoplankton require both inorganic (phosphorus, nitrogen, trace metals, silicon) and organic (vitamins) nutrients. In addition, many microalgae will take up organic carbon, as shown in the light-stimulated uptake of acetate (*photoheterotophy*). *Vitamins* are organic nutritional cofactors that are used in very low concentrations (10^{-13} to 10^{-10} mol L^{-1}) and required for growth by a variety of phytoplankton (Swift, 1980). Thus, even though microalgae show *autotrophic* growth in terms of carbon fixation, many are *auxotrophic* being dependent on one or more vitamins. Of 400 clones of phytoplankton, 44% required vitamin B$_{12}$ (cyanocobalamin), 21% required thiamine, and 4% needed biotin (Swift, 1980). Similar vitamin requirements have been shown for macroalgae (cf. Stein, 1973). Vitamin requirements also vary according to species; most centric diatoms require B$_{12}$, and 50% of pennate species are auxotrophic. Few diatoms require thiamine and none need biotin, whereas most dinoflagellates (80% to 90%) studied require one or more of these same vitamins (Swift, 1980). Vitamin concentrations are very low in oceanic and coastal waters (e.g., B$_{12}$: Sargasso Sea, 0.01 to 0.1 ng L^{-1}; Long Island Sound, 16 ng L^{-1}; Swift, 1980). However, the high turnover rate and low requirements of vitamins result in them not being limiting for growth. Sources of vitamins include bacterialplankton, other microalgae, and zooplankton.

Of the four inorganic nutrients, *trace metals*, or micronutrients, are usually found in nanomolar concentration (<0.01 μg g^{-1}) required by organisms. Four elements are usually listed: zinc, iron, copper, and manganese (DeBoer, 1981). Of these, iron and manganese are thought to be limiting, and trace levels of copper can be toxic (Huntsman and Sunda, 1980). Of the other three inorganic nutrients, *nitrogenous* compounds usually have the lowest concentrations in oceanic waters, with nitrate-N being highest and ammonium-N and nitrite-N in lower amounts (Chap. 2; McCarthy, 1980)). However, the rapid turnover and preferential utilization of reduced forms of nitrogenous nutrients suggest that the effect of low concentrations may not be as limiting as thought. *Phosphate* (PO_4-P) is the preferred form of phosphorus for most microalgae, although polyphosphates and organic phosphorus compounds can also be used (Nalewajko and Lean, 1980). As with nitrogen, soluble reactive phosphate (SRP) is usually low in open oceanic water (Chap. 2), particularly in regions where summer stratification occurs in the euphotic zone (0 to 0.06 μg atoms L^{-1}). Again, turnover rates can be high in areas of higher concentrations of phytoplankton. *Silicon* is changed from dissolved silicic acid to amorphous silica by organisms. Because they utilize silicon, diatoms and silicoflagellates play a major role in marine phytoplankton communities in its biological turnover (Paasche, 1980). The latter substance is present as monomeric orthosilicic acid [$Si(OH)_4$] in oceanic water with concentrations ranging from 0 (Naragansett Bay at minimum) to 100 μmol L^{-1} (Antarctic Ocean; Paasche, 1980). Although silicon occurs in low concentrations in the sea (Chap. 2), it is usually present in higher levels than nitrogen or phosphorus compounds. Further, although often reported as undetectable in coastal waters, silicon may be actually tied up in diatom frustules during blooms. For example, in euphotic regions, silicon may occur in lower concentrations than nitrogen or phosphorus during periods of diatom blooms (Paasche, 1980).

The growth of microalgae in oligotrophic oceanic waters in part may be due to the rapid recycling of nutrients among unicellular algae. In an unlimited nutrient environment, phytoplankton that are growing rapidly exhibit uptake ratio of 106 C: 16 N: 1 P (atomic ratios), which is called the *Redfield* ratio (Harris, 1986). Uptake ratios of Si are about the same as N in the North Atlantic, so that the ratio is 16 Si: 15 N: 1 P (Harris, 1986), reflecting the element's importance in diatom growth. The Redfield ratio is used as evidence of rapid growth of natural populations of phytoplankton in oceanic waters. The exact value of this ratio depends on species composition, with large departures from this ratio suggesting nutrient limitations. For example, nitrogen limitation causes decreased cellular development, and phosphorus limitation delays regeneration.

After rapid growth of phytoplankton in temperate waters during the spring, phosphorus or nitrogen depletion can result in reduced summer growth, followed by a rise in species and biomass in the fall, as nutrients become avail-

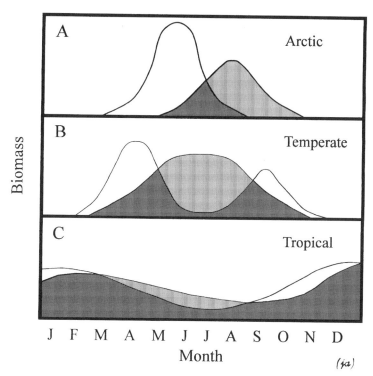

Figure 7-16. Seasonal fluctuations of phytoplankton (clear areas) and herbivorous zooplankton (solid, hatched areas) in the Arctic (A) where there is a strong seasonal growth pattern, temperate (B), and tropical (C) oceans where the seasonal patterns are modest (after Cushing, 1959).

able (Fig. 7-16). In contrast to temperate waters, seasonal maxima and minima of nutrient levels are much reduced in tropical areas, resulting in a more stable phytoplankton biomass, but a variable species composition. Upwellings and other changes in oceanic water can result in nutrient replenishment, which in the tropics is probably the most critical factor controlling phytoplankton growth (Steemann-Nielson, 1959, 1974).

Primary production by phytoplankton can be divided into new and regenerated production based on the source of nitrogen compounds (Dugdale and Goering, 1967; Eppley and Peterson, 1979; Harrison et al., 1987). The main source for new production in the open ocean is via upwelling and diffusion of particulate organic matter (POC) from deeper water into the entrophic zone (Eppley and Peterson, 1979). Secondary sources for new production are from terrestial (runoff, rivers), atmospheric (e.g., nitrate from rain), and nitrogen fixation. Regenerated production includes the remainder of primary production and is based on recycled nutrients present in the oceanic food web (Eppley and Peterson, 1979). The ratio (f) between new and total production indicates the probability of assimilation of nitrogen via new production (Eppley and Peterson, 1979). Thus, the f ratio gives qualitative data on amount of POC that is lost from the euphotic zone via sinking as well as the importance of new production. The total primary production for all oceans is 23.7×10^9 tons C y^{-1}, and new production is 4.7×10^9 y^{-1}, or about 20% of the total, with the remainder coming from regenerated nitrogen compounds showing a recycling efficiency of 80% (Eppley and Peterson, 1979).

Buoyancy

The floatability, or buoyancy, of phytoplankton is critical if the cells are to remain in the euphotic zone to carry out photosynthesis. However, sinking allows cells to encounter deeper, nutrient-rich water and avoid predation or damaging levels of irradiance (Round, 1981; Fogg and Thake, 1987). All algae, except the cyanophytes with their gas vacuoles, tend to sink (Smayda, 1970). Flagellated species can swim toward light, and nonmotile forms rely on the viscosity of the water, convection cells, or wind-induced rotations in surface waters, as well as modifications to their cells. Morphology and microscopic size along with mechanisms that reduce cell density (specific gravity) are the most common adaptations to reduced sinking rates (Smayda, 1980).

Smayda (1970) suggests that *morphological features* like bladder- and needle-shaped cells, plus branching frustules and colonial forms, can increase diatom surface area and slow sinking. Reduction of the cell's specific gravity also will slow the rate of sinking, which can be caused by mucilage production, selective accumulation of ions, gas vacuoles, and reduction in heavy components (Fogg and Thake, 1987). Production of *mucilage* is effective because its density is less than water. The density of vacuole fluid can be reduced though selective accumulation of *lighter ions* (K^+, Na^+(K^+, Na^+) and reduction of heavier ones (SO_4^{-2}, Ca^{+2}) in diatoms and dinoflagellates, respectively. The filamentous cyanophyte *Trichodesmium* forms extensive blooms in tropical seas; it produces *gas vacuoles* relatively rapidly (hours) to provide buoyancy, as seen in other blue-green algae (Fogg and Thake, 1987). In some species of *Trichodesmium*, gas vacuoles can withstand hydrostatic pressures of 20 bar, which is equivalent to 200 m depth. A reduction in *heavy components* of cells such as fats, carbohydrates, and coccoliths (Coccosphaeriales) will increase buoyancy. Thus, in contrast to earlier information, accumulation of carbohydrates or fats does not reduce the sinking rate in the diatoms *Thalassiosira weissflogii* or *Ditylum brightwellii* (Fogg and Thake, 1987; Fisher and Harrison, 1996).

Productivity and Biomass

The *primary production* of marine phytoplankton has been estimated to be 23.7 to 31×10^9 metric tons C y^{-1} (Platt and Rao, 1975; Eppley and Peterson, 1979). The Pacific Ocean has the lowest productivity due to the limited upwelling of nutrient-rich water as compared to the Indian and Atlantic oceans (Table 7-1).

TABLE 7-1. Primary Production (PP; Carbon-Fixed) by Ocean, Latitude, and Region

By Ocean	Ocean Area (10^6 km^2)	PP/Year $(10^6 \text{ tons C y}^{-1})$	$PP y^{-1} ocean^{-1}$
Indian	73.82	6,600	89.4
Atlantic	92.57	7,760	105.0
Pacific	117.56	11,400	64.2
By Latitude		PP $(m^{-2} d^{-1})$	
Arctic/Antarctic		0.3–1.0	
Temperate		0.1–0.4	
Equatorial			
Pacific Ocean		0.5	
Indian Ocean		0.2–0.3	
Tropical (general)		0.1	
Rice field		4.0	
Pine forest		2.0–3.0	

By Region	Surface Area (10^6 km^2)	PP/Year $(g \text{ C m}^{-2} y^{-1})$	Total Net PP (10^9 C y^{-1})
Oceanic	50	326	16.3
Neretic	100	36	3.6
Upwelling	300	0.4	0.1
Benthos	600	2	1.2
Total marine	58	364	21.2
Total continents	358	149	53.4
Total world	145	513	74.6

Sources: From Ryther (1969); Platt and Rao (1975); Dring (1982).

Phytoplankton primary productivity in atoll lagoons is low (1 to 5 mg C m^{-3} d^{-1}) when compared to fringing and barrier reef waters (50 to 300 mg C m^{-3} d^{-1}). Thus, atoll lagoon production is an order of magnitude higher than surrounding oceanic water (Sorokin, 1990a). When productivity is compared on a kg C m^{-2} y^{-1} basis, the highest producers are corals and large seaweeds (0.5 to 2.5), followed by benthic microalgae (0.2 to 2.0), salt marsh grasses and seagrasses (0.4 to 1.5), mangroves (0.5 to 1.0), and, finally, coastal (0.1 to 0.5) and oceanic (0.2) phytoplankton (Valiela, 1984). In comparison, the productivity of tropical rain forests is higher than oceanic phytoplankton communities (0.5 to 2.0 kg C m^{-2} y^{-1}), whereas productivity of deserts and tundra is about equal (0.05 to 0.2 kg C m^{-2} y^{-1}). Hence, the importance of oceanic primary production is not due to high rates, but because of oceanic coverage (ca. 70% of Earth). Steeman-Nielson's (1959) calculations suggest there is a close relationship between the overall productivity of land (2 $\times 10^{10}$ tons C y^{-1}) and marine (1.5 $\times 10^{10}$ tons C y^{-1}) habitats, while other estimates are higher for marine habitats (2.97 $\times 10^{10}$ tons C y^{-1}; Walsh, 1988).

The *biomass* of marine phytoplankton varies with latitude and season (Fig. 7-16). Cold water (boreal) areas [Fig. 7-16(A)] tend to show a single summer phytoplankton peak, whereas temperate areas [Fig. 7-16(b)] have spring and late summer peaks, and tropical waters [Fig. 7-16(c)] show no strong seasonal peak (Cushing, 1959). There is also a difference in species dominance according to latitude. Nutrient-rich temperate waters are dominated by diatoms, whereas coccolithophorids are more abundant in warm oligotrophic waters (Smayda, 1980). In coral reef habitats, diatoms tend to dominate, with this possibly reflecting net size and collecting procedure (Sorokin, 1990a). *Bacterioplankton* can account for a major portion of plankton production, being 10 to 60 times the biomass (200 to 900 vs. 10 to 40 mg wet wt m^{-2}) of coral reef phytoplankton. When tropical water samples are analyzed, more than 50% of the total phytoplankton are naked dinoflagellates, small pennate diatoms, nanoplankters such as *Platymonas*, and picoplankton. Neretic phytoplankton, which are primarily influenced by runoff and benthic nutrients, usually differ in biomass and species from oceanic floras.

Grazing and Succession

Changes in phytoplankton biomass during summer can be due to shifts in water temperature, photoperiod, or nutrients, as well as changes in *grazing* pressures. The importance of even moderate grazing is evident if one considers that a population of 100 cells would produce 6400 cells after six cell divisions, but only 1692 cells at a grazing rate of 20%. The effect of grazers varies seasonally and geographically and it may be due to a variety of predators such as protozoa (ciliates), copepods, amphipods, and tintinnids (Fig. 7-17). Mid- and high-latitude phytoplankton communities usually show strong seasonal variations in biomass because grazers decline during winter and there is a lag in the spring before grazing is effective. In stable tropical regions, dissolved nutrients and plankton biomass may not fluctuate as greatly, but grazing continues to play an important role (Frost, 1980). Hence, phytoplankton communities are dynamic because of extensive grazing and shifts in nutrient availability. Jackson (1980) found that even the oligotrophic Sargasso Sea and North Pacific Gyre supported highly productive communities with rapid turnover in species.

Cell size has been cited as a means of preventing grazing with the larger dinoflagellates surviving and smaller species showing a decrease in number. Grazing on a given phytoplankton species is influenced by the amount of biomass and size-dependent grazing rates (Frost, 1980). Microzooplankton, especially protozoans and invertebrate larvae, primarily graze on nanoplankton, whereas macrozooplankton (e.g., copepods) will remove cell sizes of 4 to 100 μm. Grazing can be beneficial to a species if it can pass through the gut unharmed and pick up nutrients from the animal along the way (Frost, 1980).

There are conspicuous shifts in species composition and relative phytoplankton dominance in both temperate and tropical waters (Fig. 7-18), pointing to a "nonsteady" state in phytoplankton assemblages. The nonequilibrium status of

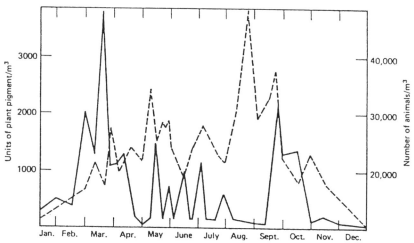

Figure 7-17. Relationship between phytoplankton density (solid line, units of plant pigment m^{-3}) and zooplankton (dashed line). When zooplankton counts rise, the phytoplankton counts decline, reflecting grazing (after Harvey et al., 1935).

phytoplankton ecosystems reflects the effect of different time scales for physical, chemical, and biological processes (Harris, 1986). Thus, studies of productivity will need to match both narrow- (e.g., storm) and broad-banded (long-term environmental changes) events (see Walsh, 1988). Comparison of biological seasons by latitude (Bogorov, 1958) show major differences among tropi-

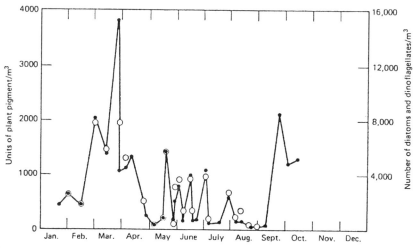

Figure 7-18. Yearly changes in phytoplankton density. The solid circles represent total chlorophyll *a*, and the open circles represent the number of diatom and dinoflagellate cells m^{-3} (after Harvey et al., 1935).

cal, subtropical, moderate and Arctic climates. Although a winter season is lacking in tropical oceans, a summer season is lacking in the Arctic basin. Thus, the phytoplankton communities exhibit "bottom–up" control of oceanic food chains (Harris, 1986). Most phytoplankton communities show seasonal progressions from diatom to coccolithophorids or dinoflagellates (Smayda, 1980). For example, the diatom *Skeletonema costatum* will occur year-round, whereas *Rhizosolenia alata* peaks in the summer in the eastern Gulf of Mexico. Sverdrup et al. (1964) found seasonal shifts in diatoms in North Atlantic waters from April (*Thalassiosira nordenskioldii*) and *Chaetoceros diadema*) to May (*C. debilis*), to June (*C. compressus*), to August (*C. constrictus* and *Skeletonema costatum*). Sorokin (1990a) described seasonal and diurnal fluctuations of phytoplankton in coral reefs, with the latter being caused by grazing at night by migrating zooplankton.

The process of regular, continuous changes in species composition of phytoplankton communities, irrespective of the biogeographic region or annual cycles, is a form of *succession*. Successional sequences are known for temperate seas (e.g., Iceland; Gulf of Maine), subtropical waters (e.g., Sargasso Sea), shallow coastal waters (Long Island Sound), and brackish systems such as the Baltic Sea (Smayda, 1980). Regulatory factors for phytoplankton succession can be separated into *allogenic* (salinity, temperature, light, turbulence, anthropogenic), *autogenic* (life cycle, nutrients, water quality, predation), and *sequential* (Smayda, 1980; Sommer, 1989). Allogenic factors are not under the control of the organism, whereas autogenic factors can be regulated to some degree by the species. Sequential factors include hydrographic disturbances (e.g., storms) and the resultant environmental modifications.

A regular sequence in species succession is fundamental in phytoplankton communities (Harris, 1986). Thus, phytoplankton succession can be identified using the following six autogenic and allogenic factors (Margalef, 1963): (1) stable water column that shows minimal mixing; (2) progressive depletion of resources that will drive succession in a fixed direction; (3) shifts in production/biomass (P/B) ratios away from initial high energy per unit biomass; (4) changes in plant pigments of individual species and in species composition: (5) increases in grazing efficiency along with grazing selection; and (6) increases in concentrations of dissolved organic carbon and detritus.

Patchiness, Blooms, and Toxic Tides

Patchiness refers to the spatial distribution of phytoplankton communities in four dimensions, consisting of three physical (one horizontal, two vertical) and a temporal aspect. A further complication to patchiness is the grazing pressure that can modify the distribution of a community (horizontal and vertical). Phytoplankton patchiness is important because of the vast areas of the ocean and the variety of communities (Mann and Lazier, 1991). Thus, a nonuniformity in available food (type of species present) will affect the trophic dynamics of the entire area. For example, the most productive fisheries in the world are restricted to areas where high densities of phytoplankton occur (e.g., Georges Bank, Gulf of Maine), which are usually sites with nutrient-rich water (e.g., upwellings).

In some cases, patch dynamics of a phytoplankton community can develop into *blooms* (also macroalgal blooms; Chap. 5) or highly concentrated populations of a single or group of species. The best known blooms are the *red tides*, so called because of the rust to red colored water caused by high concentrations of dinoflagellates (Steidinger and Vargo, 1988). The two genera, *Chattonella* and *Fibrocapsa*, of the class Raphidophyceae are also responsible for red tides in the coastal waters of Japan. Other phytoplankton cause blooms, including the blue-green alga *Trichodesmium* (Red Sea; coral reef lagoons) and the pyrmnesiophyte *Phaeocystis* (gelatinous blooms in the North Sea).

The earliest record of a red tide is found in Exodus 7:20–21, where the Nile is described as turning to blood and fish are killed during the plagues of Egypt (Steidinger and Vargo, 1988). Red tides primarily occur in tropical or subtropical coastal waters, although they can also develop in temperate areas during late summer. Surface waters with monotypic red tides often contain 1 to 20 million cells L^{-1} and may be caused by species of *Gonyaulax*, *Gymnodinium*, *Prorocentrum*, or *Ceratium*. Three factors seem to be common to all red tides: (1) an increase in population size; (2) support for the bloom; and (3) the maintenance and movement of the blooms by hydrological and meteorological factors. A red tide bloom is thought to begin with the germination of diploid resting cells (hypozygote), which can reach 6×10^3 cysts m^{-2} in the sediment. Thus, most red tides occur in shallow, coastal waters. A combination of physical and chemical forces allow a dinoflagellate population to develop into a bloom, including weather fronts, isolation of water masses, availability of limiting nutrients (e.g., iron), and light. Red tides may disappear within a few days of forming, suggesting that controlling factors have changed or grazing is occurring.

Toxic red tides can be grouped into three primary categories: (1) those that primarily kill fish and a few invertebrates (e.g., *Gymnodinium*); (2) those that primarily kill invertebrates (e.g., *Gonyaulax*); (3) and those that do not kill, but whose toxins are concentrated by filter feeding bivalves (e.g., *Protogonyaulax*). Toxins of the last type will cause paralytic shellfish poisoning (PSP), which can cause paralysis and death (by suffocation) when contaminated molluscs are eaten (Steidinger and Vargo, 1988). The neurotoxin PSP is produced by the dinoflagellate and accumulated by filter-feeding shellfish, which if eaten will result in poisoning.

Another toxin produced by dinoflagellates (not in red tides) appears to be the origin of *ciguatera*, or fish poisoning (Steele, 1993). Originally, fish poisoning was thought to occur from the accumulation of blue-green algal toxins by herbivorous tropical fish (e.g., Parrot fish), which were then transferred up the food chain to top predators such as Barracuda. Evidence now suggests that benthic dinoflagellates, including species of *Coolia*, *Ostreopsis*, and *Gambierdiscus toxicus*, living on seaweeds are the primary source of ciguatera (Bagnis et al., 1979; Besada et al., 1982; Holmes and Lewis, 1993). It appears that gambiertoxin-4b is produced by *G. toxicus*, which is the precursor to ciguatoxins 1, 2, and 3; with ciguatoxin 1 the most toxic form (Holmes and Lewis, 1993). Thus, when the macroalgae and their epiphytic dinoflagellates are eaten, the toxin is stored by herbiverous fish that are then consumed by predatory ones. Upon eating the fish, which store the toxins, severe illness follows and includes vomiting, diarrhea, and neurologic abnormalities (Steele, 1993).

Sampling and Measurement

Detailed procedures for the collection and processing of phytoplankton are available (Sournia, 1978; Raymont, 1980). Phytoplankton collection can be done with nets, water bottles, and plankton pumps. Nets (20-to-60-μm mesh size) permit qualitative studies because their mesh size will determine the species collected. Water bottles such as the reversing Nansen can sample in situ without damaging the algae (Fig. 7-19). Similarly, the Van Dorn sampler

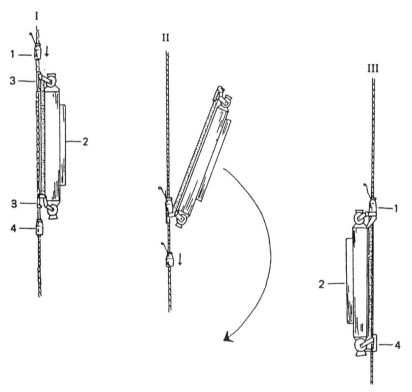

Figure 7-19. Diagram of a Nansen bottle in action. I. A weight (1) slides down the line (arrow) to the upright bottle and thermometer (2) positioned at the desired depth. II. The weight hits the release pin (3), causing the bottle to reverse and release the lower weight (4), so that both ends open at the same time. III. Once reversed, both caps close and the thermometer (2) is fixed when the bottle is hoisted up the line.

(two spring-loaded covers that close when the messenger hits the spring) and the Niskin sampler (with reversing thermometer and O-ring seals) can be employed. Pumps are used in integrating samplers, which provide a continuous stream of water, with the phytoplankton being concentrated through continuous filtration. Concentration of cells can be done with centrifugation or filtration. Samples can be preserved using Lugol's solution (10g I_2, 20g KI in 200 mL water with 20 mL glacial acetic acid) and kept in the dark (Sournia, 1978; Raymont, 1980).

Biomass can be determined in terms of volume, and carbon or chlorophyll content (Fogg and Thake, 1987). Chlorophyll is most commonly measured by absorption spectrophotometry (sensitivity: to 1 mg Chl m^3) and fluorometry (sensitivity: to 0.05 mg Chl m^3). Cell counts using electronic counters (e.g., Coulter counters) and inverted microscopes (i.e., counting slides) can enumerate the number and types (size) of cells of each species, from which biomass can be calculated (Sourina, 1978).

Measurement of in situ phytoplankton photosynthesis can be done by following changes in dissolved inorganic carbon (DIC) content, dissolved oxygen (DO), and the uptake of radioactive (^{14}C) carbon (Strickland and Parsons, 1968; Sournia, 1978; Morris, 1980; Parsons et al., 1989). The DIC procedure follows changes in seawater pH, as this reflects uptake (or release) of CO_2 (Chap. 2; Eqs. 2-13 and 2-14). Determination of ^{14}C uptake has been a popular procedure, although there are a number of errors associated with it, including standardization of the original ^{14}C ampoles, simple absorption by cells, and filtration errors (Morris, 1980). The first two techniques may not be sensitive enough if the phytoplankton biomass is very low. Levels of DO can be measured using an oxygen meter or a Winkler titration procedure (Chap. 2).

ATTACHED MICROALGAE

Periphyton microalgae are found growing on other plants (epiphytic) or animals (epizoic). As treated here, epiphytism does not include hemiparasitism (e.g., *Polysiphonia lanosa* on *Ascophyllum nodosum*), true parasitism, or commensalism (Chap. 3). *Benthic* microalgae grow on consolidated (e.g., rocks, *epilithic*) or on unconsolidated substrata (e.g., sand) as unattached (*epipelic*) or attached (*epipsammic*) organisms.

Epiphytic Communities

Epiphytic microalgal communities are common on seaweeds and seagrass blades (Fig. 7-20), mangrove roots [Fig. 10-13(B)], and seaweeds (Fig. 7-21). The overwhelming density of epiphytes demonstrates how microalgae, like macroalgae (Chap. 8), have attempted to solve the "space problem" or lack of suitable substrata (Chap. 3). The electron micrograph (Fig. 7-20) demonstrates the "sandwich" effect of a crustose green algal epiphyte growing on the seagrass *Ruppia maritima*, which is also epiphytized by microalgae that include

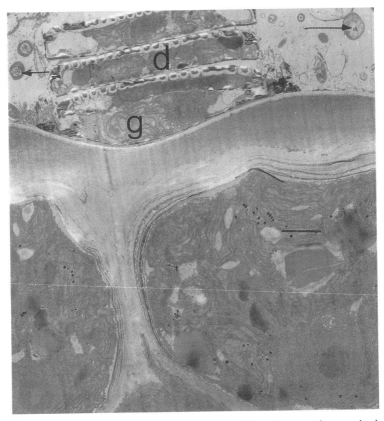

Figure 7-20.   Microalgal epiphytes of seagrasses. The electron micrograph shows a green alga (g), diatoms (d), and bacteria (arrows) attached to the epidermis of *Ruppia maritima*. Unit mark equals 1 μm.

pennate diatoms, bacteria, and blue-green algae [see also Figs. 11-7(B) and 11-12]. Many microalgae occur as epiphytes on the roots of mangroves and salt marsh plants and play an important role in primary production and as direct or detrital food for snails and small grazers (Chaps. 9 and 10).

Benthic Microalgal Communities

Some diatom and blue-green algae are *epilithic*, as they grow attached to hard substrata, and wood pilings, shells, and so on. Other species are *epipelic*, as they occur on the surface of unconsolidated sediments of mudflats and tidal marshes. They are characterized by not being attached to the sediment; they are often dominated by pennate diatoms and cyanophytes [Fig. 7-22(A)]. Diatoms appear to serve an important role in shaping sedimentary structure and influencing the rates of erosion in mudflats and intertidal marshes (Paterson, 1990,

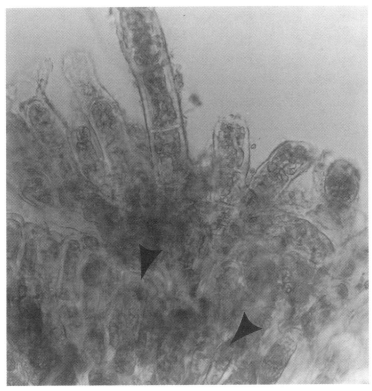

Figure 7-21.   The filamentous brown alga *Elachista fucicola* epiphytizing the fucoid *Ascophyllum nodosum.* The host's epidermis is penetrated by filaments (40 μm dia.) of *Elachista* (arrows) as seen under the light microscope.

1995). That is, diatoms secrete mucilage, which consolidates and stabilizes sediments, reducing erosion, and stabilizing their habitats [Fig. 7-22(B)]. The intimate association of the microalgae and easily disturbed sediments has been difficult to study until the perfection of frozen sampling techniques and use of low-temperature, scanning electron microscopes (Paterson, 1995).

Endogenous vertical migrations of pennate diatoms can occur with the cells moving to the surface at daybreak (positive phototaxis or negative geotaxis) and downward (a few millimeters) into the sediment at night (Round, 1981). Diatoms, along with blue-green algae, can produce a highly complex biogenic matrix [Fig. 7-22(B)]. Quantitative data of diatom vertical migration can be obtained by placing lens paper on the sediment surface at night, an hour or so before sunrise. The papers are removed a few hours after daybreak and examined under a dissecting microscope, where the diatom species are identified and individuals counted (Darley et al., 1981).

Epipsammic microalgae grow attached to sediment particles where disturbance is minimal. The flora is usually dominated by diatoms, some of which

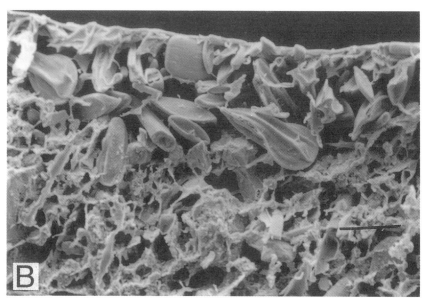

Figure 7-22. Epipelic diatoms. The surface of unconsolidated tidal marsh sediment is colonized by pennate diatoms like *Pleurosigma angulatum* (arrow, A). View of the frozen-fractured sediment shows diatoms binding (B) clay and silt particles. Unit marks equal 3 and 1 μm, respectively (courtesy of D. Paterson, St. Andrews University).

have mucilage stalks. Other microalgae include coccoid chlorophytes and cyanophytes. Epipsammic microalgal floras can account for a large portion of the primary productivity on sands of protected bays or tidal marshes. For example, high rates of carbon fixation was noted for a mixture of dinoflagellates, euglenoids, blue-greens, and diatoms in a sand beach on Long Island Sound (Burkholder et al., 1965). Subtidal communities of diatoms can be exceedingly rich, as shown in the bottom silts off Barrow, Alaska (Matheke and Horner, 1974).

CHAPTER 8

Macroalgal Communities

Lithophytic seaweed communities are some of the most productive and extensive marine plant communities in the world. Because of their availability for study and rapid response to manipulation, seaweed communities have been used in experimental studies to test ecological hypotheses. The zones of the intertidal and subtidal regions, introduced in Chap. 1, occupy a narrow coastal area and account for less than 1% of Earth's surface (10^6 km^2), as seen in a hypsographic curve (Chap. 2, Fig. 2-2). Yet the productivity of this region can equal or excel that of most terrestrial communities. For example, production of intertidal and subtidal lithophytic seaweeds is estimated to be 0.5 to 1.0 kg C m^{-2} y^{-1} and 0.5 to 2.5 kg C m^{-2} y^{-1}, respectively (Mann and Chapman, 1975; Valiela, 1984). By comparison, tropical rain forests have an annual productivity of 0.4 to 2.0 kg C m^{-2} y^{-1}.

Structural and ecological aspects of lithophytic, psammophytic, and drift seaweed communities and epiphytic macroalgae are the topic of this chapter. Other texts dealing with these subjects include Lüning (1990), Lobban and Harrison (1994), and Little and Kitching (1996). Field survey methods are described in Chap. 14, as well as procedures for determining productivity, pigment, and chemical constituent levels. Detailed field and laboratory procedures for the study of macroalgae can be found in Littler and Littler (1985) and Lobban et al. (1988).

ROCKY COASTS

A rocky shore is one of the most complex and compact habitats, with few other types of biospace being as tightly organized as lithophytic communities (Morton, 1991). The most impressive feature of rocky shores is how all the available space is so intensively zoned, causing sessile plants and animals to be organized into distinct vertical bands. The vertical zonation of seaweeds on rocky shores has been described in numerous texts (Lewis, 1964; Stephenson and Stephenson, 1972; South and Whittick, 1987; Mathieson, 1989; Mathieson and Nienhuis, 1991; Lobban and Harrison, 1994). Lüning (1990) describes the seaweed floras throughout the world, and Little and Kitching (1996) outline animal and plant interactions in many rocky habitats.

Abiotic Factors of Rocky Coasts

The two primary abiotic factors for benthic seaweeds are substrata and light. Probably, seaweeds evolved as plants adapted to grow on firm substrata, as evidenced by the high productivity and diversity of lithophytic seaweeds compared to psammophytic and drift populations. In addition to a need for firm substrata, transmittance of light plays a major role in the depth distribution of marine plants. As noted in Chap. 2, the lower depth limit of seaweeds is about 200 m in oceanic waters, where 0.01% of surface light is present. Depth distribution is more limited in coastal waters, where light is absorbed and reflected so that seaweed communities "thin out" as one approaches the lower edge. Thus, benthic communities exhibit competition for space and light.

Geologically, rocky coasts can be divided into three landforms: sea cliffs, shore platforms, and limestone coasts (Trenhale, 1987; Chap. 2). Sea cliffs (Fig. 8-1) occur on all of the world's oceanic coasts, with temperate areas showing the largest gradients in terms of wave energy. Shore platforms (Fig. 2-4) are horizontal or gently sloping surfaces due to wave erosion of substrates such as shale. Sea cliffs and shore platforms are most common in temperate latitudes.

Figure 8-1.   Sea cliffs of the Apostle Island, South Australia (Dawes, 1995; with permission of Academic Press).

TABLE 8-1. Functional Forms of Benthic Algae and Grazing Difficulty

Functional Group	Examples	Morphology	Size (average)	Grazing (0–6)[a]
Microalgae	Diatoms	Cells, chains	50 μm	0
Filaments	*Ceramium*	Filamentous	300 μm	1
Foliose	*Ulva*	Thin blades	3-5 cm	2
Thalloid	*Hypnea*	Corticated	5–10 cm	3
Leathery	*Fucus*	Tough, thick	10–50 cm	4
Branching, calcareous	*Halimeda*	Segmented, calcified	1–20 cm	5
Crustose, calcified	*Petrocelis, Clathromorphum*	Crusts	1–3 cm	6

Source: Modified from Steneck and Watling (1982).
[a]0 to 6: increasing level of difficulty.

Limestone coasts (Fig. 2-5) consist of geologically young calcareous substrata and are most common in lower latitudes and warm climates. In contrast to the first two types, calcareous coasts usually encounter lower wave energy and are shaped in part by chemical and biological activity. As might be expected, each coastal type influences seaweed diversity and morphology.

Biotic Factors on Rocky Coasts

As noted in Chap. 3, competition and grazing are two fundamental biological factors in marine plant distribution. A *functional-form model* (*F-F model*) for seaweeds was developed by Steneck and Watling (1982) to consider the ease of grazing benthic algae (Table 8-1). Thus, filamentous or thin-bladed macroalgae offer little resistance to grazers, whereas leathery, calcified, or crustose species are difficult to eat. The developmental stage of a seaweed also must be considered in the F-F model; mature seaweeds may be difficult to graze (e.g., corticated macrophytes), whereas the sporeling stage would not.

Littler et al. (1983) presented another type of F-F model to explain how morphological (form, edibility) and functional (productivity, nutrient uptake) are related (Table 3-1). Seaweed productivity was related with respect to availability of photosynthetic tissue. For example, *Ulva*, being a distromatic blade, showed the highest productivity (g dwt d^{-1}), whereas a crustose seaweed such as the coralline alga *Lithothamnion* had much lower levels. The F-F model predicts that with increased complexity in plant structure (morphology and tissues), there will be a decline in photosynthetic tissue. Thus, increased nonphotosynthetic tissue will result in reduced productivity (in terms of g C fixed gdwt^{-1} h^{-1}). Another outcome of an increase in plant complexity is a lower surface area to volume ratio (SA : V), resulting in decreased nutrient uptake (expressed as g dwt^{-1} of tissue). The combination of increased complexity, lower photosynthesis potential, and reduced nutrient uptake results in lower growth rates. On the other hand, complex seaweeds are usually perennial and make structural investments that prevent predation or reduce the effects of disturbance. These latter groups resemble "K" strategists, and sheet and filamentous groups resemble "r" strategists (cf. Chap. 3).

The "cost" and allocation of biomass for reproduction by seaweeds forms another aspect of lithophytic communities (Vernet and Harper, 1980). It is surprising that reproductive ecology is poorly understood when compared to higher plants, considering the variety of life histories (Chap. 6) and multiplicity of sexual and vegetative reproductive modes (Mathieson and Guo, 1992). Perhaps, quantification of seaweed reproductive structures is difficult because they often are embedded within vegetative tissue (e.g., cryptic) that is photosynthetic and not distinct from the vegetative branches (Santelices, 1990). By contrast, flowering plants produce easily identified, nonphotosynthetic organs, including flowers, fruits, and seeds (Lovett Doust and Lovett Doust, 1988). Thus, studies of reproductive effort have focused on fucoids and kelps, which produce distinctive structures. Eight species of fucoids (*Fucus* and *Ascophyllum*) in the Great Bay Estuary of New Hampshire allocated 52 to 62% of their biomass for reproduction (Mathieson and Guo, 1992), whereas *A. nodosum* showed a wider range in Sweden (31 to 72%; Åberg, 1990). Kelps also showed wide variation in total reproductive effort, ranging from *Ecklonia radiata* (10 to 20%; Novaczek, 1984) to species of *Laminaria* (1 to 37%; Klinger, 1985) and *Macrocystis pyrifera* (4%; Neushul, 1963). Reproductive effort has also been determined for *Ulva lactuca* (20 to 60%; Niesembaum, 1988) and *Chondrus crispus* (0.6 to 2.4%; Pringle and Mathiesan, 1987).

The wide range in reproductive effort may reflect the low "cost" for macroalgae (Klinger, 1985) because their structures are often photosynthetic and the energetic trade-offs (i.e., reproduction versus growth and increased mortality) are not large (Santelices, 1990). Further, some fucoids utilize both vegetative (e.g., see ecads in "Drift Seaweeds and Blooms," this chapter) and sexual means to propagate (Russell, 1986). For example, *A. nodosum* shows a high degree of reproductive effort, is a perennial, and harbors a dense understore of suppressed shoots, called "meristem banks" by Mathieson and Guo (1992).

INTERTIDAL COMMUNITIES

Biological *zonation* of intertidal organisms is not random and can be viewed as a series of vertical zones with horizontal continuity (Figs. 1-6 and 8-2). Lewis (1964) pointed out that "plants and animals are probably distributed in accordance with their differing abilities to endure exposure to areal conditions," and this has been demonstrated through measurements of seaweed tolerances (e.g., Einav et al., 1995). Within these zones, random distributions can occur due to recruitment (Menge et al., 1993). Thus, in a study of turf-forming algae on wave-exposed coasts of Oregon, they found that recruitment contributed to the

Figure 8-2. The intertidal zone of a granite bridge piling at Dover Point, New Hampshire, within the Great Bay estuarine system. The front face of the piling (arrow) experiences strong tidal currents and shows severe erosion. The side of the piling experiences little wave activity and is also protected from the current (white line); thus, zonation follows tidal levels closely (compare with the exposed cliff shown in Fig. 1-6). Here, an upper band of green algae (*Blidingia, Urospora, Enteromorpha* = g) occurs at the upper tidal mark, followed by the barnacle *Semibalanus* (b), the red alga *Porphyra* (p), and fucoids (f) *Ascophyllum* and *Fucus* (courtesy of Don Cheney, Northeastern University; see Mathieson et al., 1977).

maintenance of a diverse mosaic of species. The key factors controlling distribution of intertidal algae are tides, exposure, substrata, and biotic factors (Chap. 2), and they are considered in what follows. Classical publications on marine algal communities of rocky intertidal shores include Doty (1957), Lewis (1964), Stephenson and Stephenson (1972), and Chapman (1974), with more recent reviews by Lüning (1990), Morton (1991), and Lobban and Harrison (1994).

Tides

Of all abiotic factors, *tides* are the primary controlling feature of intertidal zonation (Lewis, 1964). Tidal amplitude (vertical range) and frequency (period of cycle) are both critical (Chap. 2). If the spring tides coincide with hot, dry weather, the severe exposure of intertidal organisms can result in death. For example, bleaching and death of intertidal *Laurencia papillosa* and *Padina santae-crucis* populations occur in May and June in the Florida Keys when

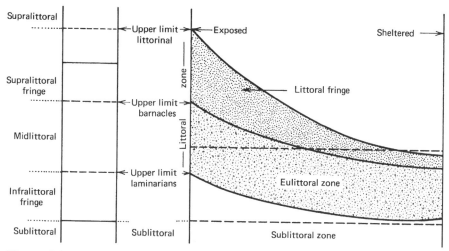

Figure 8-3. Intertidal schemes of Stephenson and Stephenson (1972; left), Lewis (1964; middle), as well as a comparison of exposed and protected zones (right).

unusually low spring tides occur during midday. A similar event occurs in the fall during low spring tides in southern California coasts, causing extreme damage to *Corallina officinalis* and *Laurencia pacifica*.

The effectiveness of tides is seen in the *critical tidal-level* (*CTL*) hypothesis that correlates specific tidal marks with the range of algal species (Doty, 1946). CTLs occur where there are major shifts in intertidal exposure associated with the daily and monthly tidal periods (Fig. 4-11). CTLs are most discernible in sheltered areas in contrast to exposed shores with extreme wave activity. Specific plants and animals can be used to identify *biological exposure indices* (*BEIs*), which correlate with CTLs (Little and Kitching, 1996), as shown in Figure 8-3. The number of BEI zones can range from 5 for rocky British shores (Lewis, 1964) and the Bay of Fundy (Morton, 1991) to 8 on subtropical to Arctic shores of the western North Atlantic (Stephenson and Stephenson, 1972). Use of BEI zones can demonstrate levels of exposure by comparing positions of key organisms on exposed and protected sites (Fig. 8-3). The schemes of Stephenson and Stephenson and Lewis utilize similar dominant BEI organisms to identify zones of temperate rocky coasts, namely, fucoids, *Porphyra*, periwinkle, barnacles, and kelps. The problem with BEI zones is that they do not quantify the effects of exposure, so that a circular argument results between the species present and the level of exposure (Little and Kitching, 1996), nor are they universal.

Exposure

The effect of *wave action*, *shore topography*, and *desiccation* can be used to define *exposure* in intertidal seaweed communities. Persistent wave activity will cause an *upward expansion* (dilation) of zones on high-energy coasts versus sheltered coasts (Fig. 8-3; also compare Figs. 6-1 and 8-2). Topography will influence the effect of wave activity. A jutting headland (Fig. 8-1) will encounter large waves compared with protected, irregular, or gently slopping shores, which dissipate the energy. On high-energy coasts, the dominant seaweeds are tough, flexible, and structurally complex, including the brown algae *Eisenia* and *Postelsia* in California, *Durvillaea* in Chile, and heavily calcified crusts (i.e., corallines) on windward Pacific coral reefs. Intertidal communities on wave-exposed seastacks in California exhibited an upward expansion and increased diversity versus a protected boulder beach (Seapy and Littler, 1979). The importance of wave disturbance on intertidal coasts is critical for the recruitment of the sea palm *Postelsia palmaeformis*, which occurs on high-energy coasts [Fig. 8-4(A)]. Winter storms result in removal of mussel beds, beneath which young sporophytes of the sea palm are protected and can develop [Fig. 8-4(B)].

Desiccation is controlled by exposure to climatic factors, including air temperature and sunlight. The effect of *climate* can be seen by comparing intertidal seaweeds of tropical (Fig. 2-5) and temperate coasts (Fig. 1-6). The large, robust fucoids (*Ascophyllum nodosum*, *Fucus* spp.) and turf-forming red seaweeds (e.g., *Chondrus crispus*, *Mastocarpus papillatus*) in temperate intertidal communities have no parallels in tropical coasts. Kanwisher (1957) found that the brown alga *Fucus vesiculosus* could lose 91% of its water during a normal intertidal exposure. Even so, desiccation need not be detrimental, as brown algae like the high intertidal fucoid *Pelvetia canaliculata* can tolerate long periods of desiccation and increase its photosynthesis within minutes of rehydration. Thus, it inhabits the highest zone in Great Britain because of its tolerance to desiccation and inability to outcompete seaweeds below it (Rugg and Norton, 1987). Brinkhuis et al. (1976) showed that the rockweed *Ascophyllum* had enhanced photosynthetic rates after being desiccated. Similarly high rates of photosynthesis were measured in the air for three partially desiccated red algae (*Caloglossa*, *Bostrychia*, and *Catenella*) that grow on the bases of *Juncus roemerianus* (Dawes et al., 1978) and the algal epiphytes of *Spartina alterniflora* (Jones, 1980) in salt marshes.

Morphological adaptations to desiccation by intertidal seaweeds include hollow (saccate) or thickened branches. For example, saccate seaweeds such as *Colponemia* (Oats, 1988) and *Halosaccion* (Oats, 1986) can avoid some desiccation by using the water stored in their central cavities; further, this "stored" seawater also supplies some carbon for photosynthesis (Chap. 4). The red alga *Mastocarpus papillatus* also exhibits thicker thalli in upper than lower intertidal habitats, with the enhanced thickness reducing water loss (Bell, 1995). Plants in the lower intertidal zone are often more branched, which increases their surface area for nutrient absorption and also increases their sensitivity to desiccation and temperature. An added advantage for seaweeds with thick thalli is that they retain heat while exposed to low temperatures in temperate intertidal shores. In contrast, the effect of high body temperatures probably is one

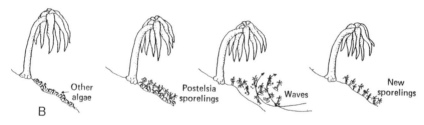

Figure 8-4.   The sea palm *Postelsia palmaeformis*. It is restricted to the low intertidal zone of high-energy shorelines of the Pacific northwest (A). The plants are outcompeted by mussels and other algae for space (B). However, they produce numerous sporelings that attach to their competitors, acting as drags, and are removed in storms. The newly cleared space is then controlled by the remaining sporelings (Dayton, 1975a).

of the factors preventing larger or thicker macroalgae from dominating tropical shores.

Type of Surface

The composition and texture of the *substrata* will influence the type and diversity of seaweed communities. Limestone, because of its light color, reflects light; it is also more soluble and will hold water in its porous structure. Macroalgae can penetrate limestone more easily or grow in its solution holes and crevices. For example, many tropical seaweeds like the green algae *Valonia ocellata* and *Dictyosphaeria cavernosa* grow under projecting limestone ridges

Figure 8-5. A tropical intertidal turf community at Quintin Roo (Yucatan), Mexico. Dense, interwoven branches of brown (*Turbinaria turbinata, Sargassum platycarpum, Dictyota dichotoma*) and red (*Gracilaria crassissima, Gelidium pusillum*) macroalgae form dense turfs on limestone reefs in the low intertidal zone.

within the intertidal zone. In general, the more mosaic the surface topography, the more diverse the intertidal seaweed communities (Seapy and Littler, 1979). One major exception to the lack of intertidal seaweeds within tropical areas is the occurrence of turf-forming macrophytes (Fig. 8-5). Tropical turf algae are tough, highly productive (see Adey and Goertemiller, 1987), densely branched plants that form a thick 1-to-3-cm-high carpet, within which seawater is retained while exposed at low tide. Thus, the functional morphology of *Turbinaria, Sargassum, Gelidium*, and *Pterocladia* allows them to persist within intertidal habitats, which are stressed due to wave activity, desiccation, and grazing (Hay, 1981a).

Biological Interactions

Succession, competition, and grazing play significant and interactive roles within intertidal seaweed communities (Little and Kitching, 1996). The upper limit of intertidal seaweeds is usually limited by abiotic factors, including desiccation and temperature (Bell, 1995), and lower limits are typically controlled by biotic factors such as grazing and competition (Chap. 3). *Grazing* within the intertidal zone is primarily done by a variety of molluscs, including gastropods (limpets, winkles, neritids) and chitins (Little and Kitching, 1996). Sea urchins also can extend into the lowest intertidal area, but they are more common subtidally. On rocky intertidal boulder fields, barnacles provide a refuge for algal mats that are otherwise grazed by limpets. By manipulating limpet densities, van Tamelen (1987) found that barnacles had no direct effects on algae, whereas the removal of the algal turf allowed barnacles to establish. In a similar study, Farrell (1991) observed that two brown and one red macroalgae (*Pelvetiopsis, Fucus*, and *Endocladia*) colonized cleared rocks only after the barnacle *Semibalanus* was established. The limpet *Acmaea*, which is a major herbivore, can slow succession of cleared plots by initially delaying the establishment of barnacles and then grazing algal turfs. However, as long as limpet predators (e.g., the starfish *Pisaster ochaceus*) keep the herbivore at low densities, the equilibrium between macroalgae and grazer can be maintained (Dayton, 1975b). One example of the effect of grazing on competition is demonstrated by the interaction between the periwinkle *Littorina littorea* and the green alga *Enteromorpha* sp. (Lubchenco, 1978). That is, the snail removes *Enteromorpha* from tide pools, resulting in colonization by the red alga *Chondrus crispus*. If the snail is removed, then *Chondrus* will be overgrown by the opportunistic green seaweed. Further, if the shore crab *Carcinus maenas* is added, it will eat the snails, allowing the green alga to become reestablished. Because sea gulls prey on the crab, it primarily occurs under the green alga in pools where the small snails are and where it is not apparent to the birds.

An elegant example of interspecific *competition* was shown for the red alga *Mastocarpus papillatus*, where its lower extension was controlled by competition and its upper level by desiccation (Hodgson, 1980). Another red alga *Gastroclonium coulteri* forms a turf at about 0.3 m above MLW. If *M. papillatus* and the co-dominant red alga *Rhodoglossum affines* are removed from the high intertidal, *G. coulteri* does not expand upward into the cleared areas. However, *M. papillata*, a slow-growing seaweed, can extend down into the area previously occupied by *G. coulteri*. In a similar fashion, *G. coulteri* expands its lower distribution if the surfgrass *Phyllospadix torreyi* is removed. However, the surfgrass cannot extend upward into the *G. coulteri* zone. Hodgson found that *G. coulteri* could not tolerate more than a 50% water loss and thus was prevented from occupying the upper intertidal zone. Thus, the lower boundary of *G. coulteri* is set by surfgrass, which is a stronger competitor, and its upward limit is determined by sensitivity to desiccation. Stewart (1989) found a similar relationship between the surfgrass and the red calcified *Corallina* (Chap. 3). Not all expansions upward in the intertidal zone are controlled by abiotic factors. For example, Hawkins and Hartnoll (1985) found that two fucoids (*Fucus vesiculosus, F. serratus*) expanded their range upshore when the species zoned above were removed. In the same study, the highest fucoid, *Pelvetia canaliculata*, was able expand downward if lower shore species were removed.

Succession in intertidal seaweed communities has been studied on newly formed volcanic rock in Hawaii (Doty, 1957) and after removal of the kelp forest in Alaska (Duggins, 1980; Chap. 3). In both cases, an intertidal *disturbance* was followed by a sequential recovery of the original community. Disturbance, whether due to biotic (grazing) or abiotic factors (e.g., ice or log scouring, wave activity), can control the succession of intertidal seaweed communities. The erect brown sea palm *Postelsia palmaeformis* competes for space on exposed rocky shores along the North Pacific coast of the United States [Fig. 8-4(B)]. Barnacles (*Chthamalus dalli, Balanus glandula, B. cariosus*), mussels, and algae can out-complete the sea palm in this high-energy community. Dayton (1973) suggested that the zoopores of the sea palm can settle and germinate on other algae, mussels, and barnacles. After producing eggs and sperm, young sporophytes mature while attached to the gametophytes, causing drag and removal of the other organisms. Hence, space is opened up for the remaining young sea palm sporophytes [Fig. 8-4(B)]. Paine (1979) also suggested natural disturbances from waves and log abrasion can clear regions for the remaining sporophytes to develop. Patch size in intertidal communities can influence the type of seaweeds that recruit when a disturbance gap occurs (Kim and DeWreede, 1996). The authors removed the three dominant perennial algae (*Mazzaella cornucopiae*, formerly *Iridaea*; *Fucus distichus*; and *Pelvetiopsis limitata*) along with some ephemeral algae to create three gap sizes (25, 100, 400 cm^2). Within one year, the intermediate gap size was filled with a well-mixed combination of all species, then the smallest was filled, and, finally, the largest.

EXAMPLES OF INTERTIDAL ZONATION

Three sites are used to demonstrate intertidal zonation of seaweeds, a subtropical site in the Florida Keys and two cold temperate sites in Maine. The temperate sites at Otter Point, Maine (44°11′ N, 68°20′ W), demonstrate the importance of wave energy on intertidal seaweed communities, and the limestone reef of the Content Keys (24°40′ N, 81°28′ W) in the Florida Keys shows the stresses of a subtropical shore (see also Chap. 12).

The Exposed Shore of Otter Cliffs

Otter Point, containing Otter Cliffs, is an exposed granite ridge of Mount Desert Island that projects (south) into the cold temperate waters of the Gulf of Maine. It is 50 km from the Bay of Fundy and has a spring tidal amplitude of 4.6 m. The site (Fig. 8-6) experiences extreme northeastern storms with waves often being 3 to 6 m. An earlier study of zonation at this site (Johnson and Skutch, 1928) determined elevations, and the same site was reexamined in 1996 using the original benchmark (Fig. 8-11, arrow).

The lower *maritime* zone of Otter Cliffs (9.2 to 15.3 m above MLW) grades from bare rock to gravely soil. The former supports terrestrial (*Xanthoria* and *Caloplaca*) and marine lichens (*Physia, Leconia, Ramilina, Verrucaria* spp.) plus blue-green algae (*Calothrix*). Halophytic flowering plants dominate where adequate soil is present and include *Plantago oliganthos* and *Empetrium nigrum*. Both of the latter species can extend into the upper spray zone, where they often grow in crevices. The upper part of this zone contains prostrate forms

Figure 8-6. The exposed intertidal community at Otter Cliff, Mount Desert Island, Maine. Algae occur in horizontal zones that are expanded due to high wave activity and can be divided into a spray (1), upper (2), and lower (3) intertidal, and a subtidal fringe (4).

of Juniper (*Juniperus horizantalis, J. communis depressa*) and black spruce (*Picea canadensis*).

The *spray* zone (Fig. 8-6, zone 1), influenced by maximum spring tides (3.9 m) and intense waves, is extensive (3.5 to 9.2 m above MLW) and the rocks are blackened by the lichen *Verrucaria* (3.5 to 6.3 m). Blue-green algae (*Calothrix, Rivularia, Entophysalis*) grow intermixed with the lichen, forming dark green to black crusts within the lower region of the spray zone. *Fucus spiralis* forms a distinct yellow-brown band at 4.3 m, and the periwinkle *Littorina saxatilis* occurs within diverse crevices. Bands of lithophytic green algae occur at the lower edge of this zone and include *Blidingia minima* (3.9 m), *Codiolum pusillum, Urospora penicilliformis,* and *Ulothrix flacca* (3.6 m).

The *upper intertidal* zone (Fig. 8-6, zone 2; 2.0 to 3.5 m) is primarily delineated by the barnacle *Semibalanus balanoides*, as reported for other Northeast Atlantic sites (Stephenson and Stephenson, 1972; Morton, 1991). The red alga *Porphyra umbilicalis* extends to 3.0 m and *Enteromorpha intestinalis* to 2.9 m. The fucoid brown algae *Ascophyllum nodosum* and *F. vesiculosus* (to 2.8 m) dominate the cliffs and form dense patches in areas of reduced energy, being absent on more exposed rocky slopes. Periwinkles (*Littorinia obtusata, L. saxatilis*) are abundant in areas of reduced exposure, while *L. littorea* is also present. Only stunted blades of *Porphyra umbilicalis* (to 1.4 m) occurred in the lower, highly exposed regions of this zone.

The *lower intertidal* zone (0.6 to 2.0 m) marked the end of the two fucoids (Fig. 8-6, zone 3) and the upper extension of *Mastocarpus stellatus* and *Ulva lactuca* (to 1.8 m). Two bands of red algae are evident on the cliffs, a dense turf of *M. stellatus* and *Chondrus crispus* (to 0.9 m) and patches of foliose *Palmaria palmata* (to 1.5 m). The blue mussel *Mytilus edulis* (to 0.8 m) marks the lower edge of this zone.

The *subtidal fringe* (−0.7 to 0.6 m) is continually washed by waves (Fig. 8-6, zone 4) during low tides. *Fucus distichus* ssp. *edentatus* marks the upper extension of the fringe, and *Alaria esculenta* dominates the lower portion with *Laminaria digitata* and *L. saccharina* (to 0.4 m) occurring in pools and coves. Several seaweeds are abundant in coves, including green (*Ulva lactuca, Spongomorpha spinescens*) and red algae (*Polysiphonia harveyi, Devaleraea ramentacea*). Crustose (*Clathromorphum circumscriptum; Petrocelis cruenta*) and turflike red algae (*Ceramium nodulosum*) grow near the lower (to MLW = 0.0 m) portion of the fringing community of the cliffs. In highly exposed areas, *Alaria esculenta, Polysiphonia urceolata, D. ramentacea,* and *Ulvaria obscura* extend to 0.7 m. Also severely eroded tufts of *Ahnfeltia plicata, C. crispus,* and *Rhodomela confervoides* occur at MLW on the outer rocks.

A Protected Temperate Shore

A lack of expansion and reduced diversity of intertidal seaweeds are evident at the more sheltered Otter Cove, which is located 0.5 km from Otter Cliffs, that is, on the opposite side of Otter Point (Fig. 8-7). Both sites have the same tidal amplitude, but Otter Cove experiences much lower wave activity.

The lower *maritime* zone extends to 4.6 m above MLW with tufts of *Panicum* (to 5.4 m) on ledges, and *Rosa* and *Ulnus* grow to the edge of a boulder shore (to 4.9 m). Even so, *Verrucaria* may extend upward to 6.3 m on bare rock. Trees (*Abies, Picea*) grow to the lower edge of the maritime zone and show little wind or salt damage, which was characteristic at Otter Cliffs. The *spray* zone (Fig. 8-7, zone 1; 3.0 to 4.5 m) is mostly barren above, and green algae form a narrow band at 3.0 to 3.2 m (*Blidingia, Urospora, Enteromorpha*).

The *upper* (1.2 to 2.8 m) and *lower* intertidal zones (0.1 to 1.2 m) are dominated by fucoid algae (Fig. 8-7, zones 2 and 3) except on outer rocks that project into the cove covered with barnacles. The upper intertidal zone is marked by *Semibalanus*

Figure 8-7. The protected intertidal community of Otter Cove, Mount Desert Island, Maine. The lower maritime almost extends to the upper intertidal zone due to lack of wave activity and a restricted spray zone (1). The upper (2) and lower (3) intertidal zones are covered by fucoid algae except at the exposed outer rocks, and the subtidal fringe (4) is narrow.

balanoides, and *Polysiphonia lanosa* epiphytic on *Ascophyllum nodosum* marks the lower end. *Fucus spiralis* only forms a narrow upper band (2.0 to 2.2 m); *A. nodosum* (0.2 to 2.0 m) and *Fucus vesiculosus* (0.0 to 2.0 m) extend from the mid- to lower intertidal zones as a dense carpet. The lower intertidal zone is delineated by the upper extension of epiphytic *Pyliella littoralis* (1.0 m) and *Chondrus crispus* (0.9), and the mussel, *Mytilis edulis* (0.8 m).

The *subtidal fringe* (Fig. 8-7, zone 4; −0.6 to 0.1 m) marks the upper extension of *Mastocarpus stellatus* and various crustose algae (*Clathromorphum circumscriptum, Pseudolithoderma extensum*). Most seaweeds were below the MLW mark (*Ulva, Laminaria, Petrocelis*). An "urchin-barren," caused by the green sea urchin *Strongylocentrotus droebachiensis*, occurs within the subtidal fringe, and only the herbivore-resistant kelp *Agarum cribrosum* and crustose corallines (*C. circumscriptum*) grow abundantly there.

An Exposed Subtropical Shore

The subtropical seaweed community of the intertidal limestone reef on the north-facing coast of the Content Key in the Florida Keys at first appears to be

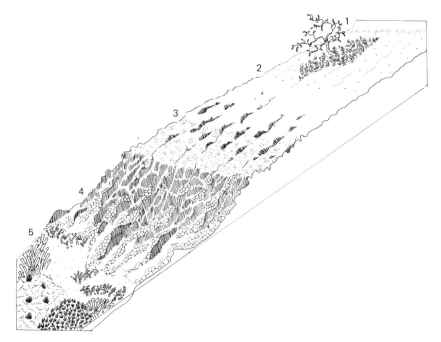

Figure 8-8. Subtropical intertidal zonation at Content Key, Florida. The intertidal zones on this limestone reef (see also Fig. 2-5) can be separated based on color. Four zones are evident: (1) the maritime to spray zone is gray and may have halophytes (e.g., *Batis*, *Distichlis*) and the black mangrove (*Avicennia*); (2) the spray zone is black and dominated by blue-green algae; (3) the upper intertidal zone is yellow and has barnacles (*Chthamalus*), blue-green algae, and macroalgae (*Valonia, Halimeda, Laurencia*) that grow under rocks and in pools; (4) the lowest reef platform grades into the subtidal fringe (5) and contains a variety of macroalgae (*Halimeda, Laurencia, Cystoseira*), sea urchins (*Echinometra* and *Diadema*), and seagrasses.

nonexistent (Figs. 2-5 and 8-8). However, 258 taxa of macroalgae were identified (Croley and Dawes, 1970), a majority growing within crevices and under limestone fragments of the intertidal zone. The tidal range is small (ca. 1 m) and wave energy moderate, with winter storms producing 1-to-2-m waves, similar to the Caribbean (Dawes et al., 1991).

The *maritime* zone (1 m above MLW) at Content Key is a 3-m-high sand berm vegetated by coastal plants, including *Ipomoea stolonifera* (morning glory vine), *Opuntia* spp., *Sesuvium portulacastrum*, *Batis maritima*, *Panicum* spp., and the exotic *Schinus terebinthifolius* (Brazilean Pepper). The berm separates the limestone intertidal zone from a large overwash mangal. The orange lichen *Pyrenula cerina* epiphytizes mangrove trunks, and blue-green algae (*Oscillatoria*, *Merismopedia*) grow on shells and coral rubble.

A narrow *spray* zone (45 to 60 cm) is sparsely vegetated by stunted black mangroves growing in fragmented limestone along the base of the maritime berm and extending into the upper intertidal zone. The limestone color changes from white, to grey, to black (Stephenson and Stephenson, 1972; van den Hoek et al., 1972) primarily due to the increasing abundance of blue-green algae. Storm waves and wind damage the trees and spring tides wet the epiphytic (*Catenella, Bostrychia, Monostroma* on black mangrove pneumatophores) and lithophytic blue-green (*Calothrix, Lyngbya, Rivularia*), green (*Caulerpa, Cladophora*), and red (*Amphoria, Lithothamnion, Laurencia*) algae. Whereas the blue-green species grow on the surface, macroalgae are found in solution holes or under rubble. Small snails are common, including *Nerita, Littorina*, and *Tectarius*.

The *upper* intertidal zone (30 to 45 cm) is yellow in color and contains barnacles (*Chthamalus*) and a variety of blue-green and macroalgae. Stunted forms of coenocytic green algae (*Penicillus, Udotea, Caulerpa, Halimeda*) are partially buried in sand-filled solution holes, and *Dasycladus* and *Neomeris* grow with the brown alga *Padina vickersiae* under rock ledges. *Laurencia papillosa* produces a 2-cm-tall, sand-impregnated turf in shallow depressions along with members of the Gelidiales (*Gelidium, Pterocladia*, and *Gelidlella*).

The *lower* intertidal zone (10 to 30 cm) is yellow to gold in color and is primarily bare limestone due to wave action. Mosslike growths of red (*Centroceras, Herposiphonia, Hypnea, Laurencia*), tufts of green (*Cladophoropsis, Avrainvillea, Halimeda*), and dense growths of brown (*Cystoseira*) algae are common in depressions. Blue-green algae continue to dominate the surface, and *Mytilis* is attached in crevices.

Depressions in the *subtidal fringe* (−10 to 20 cm) contain seagrasses (primarily *Thalassia testudinum*) and coenocytic green algae (*Halimeda, Penicillus, Udotea*). A few macroalgae (e.g., *Laurencia papillosa, Padina vickersiae*) grow on the surface. Dense tufts of *Gelidium, Laurencia*, and *H. opuntia* trap sand in the shallow depressions.

SUBTIDAL COMMUNITIES

The *subtidal* zone, or region below the lowest tide, also shows zonation of seaweeds that is evident in their functional-form characteristics (Vadas and Steneck, 1988). In the upper subtidal zone, thick, leathery morphologies of seaweeds relate to their exposure to intense water movement. The mid-subtidal zone usually contains a variety of morphologies, including coarsely branched, sheetlike, and filamentous algae. Due to the low submarine illumination, seaweeds in the lower subtidal zone are more sparsely distributed and are dominated by morphologies that have maximum surface area available for light, namely, peltate, broadly bladed, prostrate, and crustose species. Variations in species diversity that occur between locations with similar hydrographic conditions usually reflect differences in substrata and light. Based on studies of New Hampshire subtidal seaweed communities, Mathieson (1979) concluded that (1) few species are restricted to the deepest region; (2) the majority of deep-water algae are limited by light and substrate;

(3) most deep subtidal species are crustose or fleshy forms; and (4) most deep subtidal species are perennial red algae. As noted in Chap. 1, subtidal seaweed communities now can be studied directly by marine botanists using SCUBA and submersible vehicles. For example, based on 170 submersible dives, Hanisak and Blair (1988) demonstrated a complex, deep-water macroalgal community on the east coast of Florida with 208 taxa and maximum number of taxa from 30 to 40 m, but some extending to 98 m.

Abiotic Factors

Typically, submarine illumination and availability of substrate are the major controlling abiotic factors affecting subtidal seaweed communities. Such communities may be particularly limited in estuaries or nearshore open coastal regions with turbid water. Thus, increased sediment and plankton can decrease underwater light and mask available hard substrata. Usually, the number of macroalgae and depth limits of subtidal seaweeds decrease from open coastal to estuarine habitats. For example, the depth distribution of macroalgae is greater than 30 m at the offshore Isle of Shoals in the Gulf of Maine, about 27 m on the open coast of New Hampshire 10 km away, and less than 1 m at Chapman's landing in Great Bay 20 km inland (Mathieson and Hehre, 1986). By contrast, subtidal seaweed communities can extend below 200 m (Littler and Littler, 1984) in oligotrophic waters on continental shelves or offshore islands that lack sediment and plankton.

In addition to substratum and irradiance, another important abiotic factor is *water movement*. Mathieson et al. (1977) described the seaweed flora of an inland tidal rapids, the water movement supporting a community that was more characteristic of the exposed coast. Neushul (1972) found that the giant kelp *Macrocystis pyrifera* will grow through three levels of water movement. Water currents near the ocean surface were about 1 m s^{-1}, a surge effect occurred in intermediate depths, and essentially no water movement occurred near the boundary layer over the substratum (Fig. 8-9). Thus, zoospores released by the reproductive sporophylls of *M. pyrifera* must pass through a surge zone, become established within the boundary layer, and then grow into filamentous gametophytes. The removal and recovery of a two-tiered subtidal seaweed community growing at 20 m on the west coast of Florida showed the significance of storm-caused disturbances (Dawes and Lawrence, 1990). The 5-m storm waves carried sand that eroded the plants and covered the limestone outcroppings with 0.5-m sand ripples. Within two years, the upper layer of *Sargassum filipendula* and understory algal turf had recovered. A combination of waves, *ice*, and water clarity control subtidal zonation of macroalgae in the tideless Baltic Sea (Kiirkki, 1996). Macroalgae in the upper subtidal zone (+40 to −10 cm: *Ulothrix* and *Urospora*) were most influenced by wave energy, those in the midzone (−10 to −100 cm: *Dictyosiphon*, *Cladophora*, and *Pilayella*) by ice scraping, and those in the deepest zone (−25 to −200 cm: *Fucus*, *Furcellaria*) by ice and water clarity.

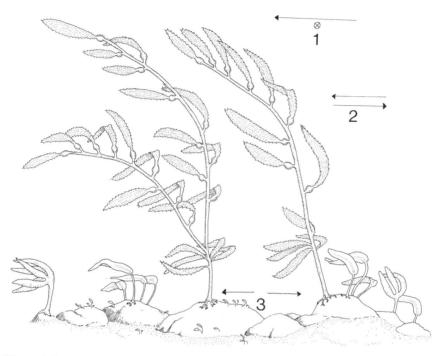

Figure 8-9. Water movement and the giant kelp *Macrocystis pyrifera*. Surface waves and currents (1) distribute spores and create drag on the giant kelp, and subsurface currents and wave surges (2) pull and yank the stipes. At the substrata level (3), the boundary layer will limit water movement and diffusion of gases and nutrients (after Neushul, 1972).

Biotic Influences

As noted by Dayton et al. (1984), both competition and grazing influence the distribution and composition of subtidal communities in California. Hence, the stability of kelp communities is related to four types of adaptations to biotic (grazing, competition) and abiotic (wave surge) factors: (1) Early stage, opportunistic brown algae (*Nereocystis*, *Desmarestia*) develop in disturbed areas; (2) *Macrocystis* outcompetes (exploitative competition) other seaweed taxa for light and nutrients in areas of reduced wave action and grazing; (3) kelps like *Eisenia* and *Dictyoneurum* dominate in areas of high wave action and reduced grazing; (4) other algae (*Agarum*, corallines) grow in patches where grazing is heavy. Grazing by sea urchins (Lawrence, 1975) can destroy seaweed communities, causing "barren grounds" in Nova Scotia, Newfoundland, and other locales (Elner and Vadas, 1990; Little and Kitching, 1996). Sea urchins are prominent subtidal grazers in both tropical and temperate latitudes, whereas herbivorous fish are more apparent in tropical waters. Reed (1990) studied a three-layered kelp community in California and showed that competition for light influenced population density of the kelp *Pterygophora californica* by controlling growth and reproduction. New recruits of this kelp and other macroalgae only established in areas of low density. The sea urchin *Lytechinus anemesus* and two species of the sea hare *Aplysia* removed large amounts of biomass in both high- and low-density areas, but did not appear to affect their density.

Succession has been studied in subtidal communities, including artificial reefs that had been established as fish havens. Ohno (1993a) found that *Sargassum* communities on concrete blocks positioned on a mixture of sand and rock substrata increased from 229 to 5137 g wet wt m^{-2} after one and six years, respectively. Artificial reefs established on rocky substrata did not show increases in biomass because they were heavily grazed by local sea urchins. Disturbance of subtidal seaweed communities can be due to storms (Dawes and Lawrence, 1990) and grazing (Reed, 1990). Coexistence of kelp and mussels may result in the Gulf of Maine, where storms and grazing occur (Witman, 1987). The dominant kelps (*Laminaria digitata* and *L. saccharina*) will overgrow the subtidal mussel *Modiolus modiolus*, which will then be dislodged by storm waves, causing cleared areas and allowing colonization by kelps. In turn, the kelps were prevented from growing below 10 to 12 m by the sea urchin *Strongylocentrotus droebachiensis*.

Structure in Subtidal Communities

The complexity of subtidal seaweed communities will range from *single-layered* (i.e., crustose corallines and algal turfs; Fig. 8-10) to *multilayered* habitats

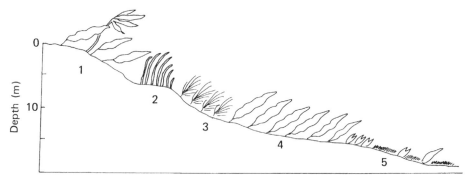

Figure 8-10. Subtidal zonation of seaweeds at San Juan Island in Puget Sound, Washington. The subtidal slope includes five zones: (1) an upper kelp zone dominated by *Laminaria* and *Nereocystis*; (2) beds of the seagrass *Zostera* in soft sedimented areas; (3) a number of fleshy red algae including *Agardiella* and *Iridaea*; (4) a dense *Agarum* community; and (5) a deep (ca. 5 m) red algal turf of *Rhodoptilum* and *Callophyllis*.

(i.e., kelp canopies). Single-layered communities may occur where urchin grazing is intense, as in Newfoundland, where most fleshy algae are removed and only crustose corallines remain. Analogous communities can occur on coral reef ridges of atolls due to a combination of intense wave activity and grazing by fish and sea urchins. Two-layered communities consisting of crustose and turf algae are common in high-energy or moderately grazed regions (Hay, 1981b). Turf-forming algal communities, which are common in the tropics, can be complex floristically, having more than 15 genera of red, green, and brown seaweeds (Hay, 1981a; Fig. 8-5). Multilayered seaweed communities are usually dominated by large brown algae, which are kelps and fucoids in temperate waters (Fig. 8-2) and species of *Sargassum* in warmer waters. Dayton (1975a) divided the subtidal kelp communities of Amchitka Island, Alaska, into four vertical zones (Fig. 3-2). The large kelp *Alaria fistulosa* (up to 22 m tall) formed the uppermost canopy below which a second layer of *Laminaria* spp. developed. A third layer, dominated by the kelp *Agarum clathratum*, covered a fourth layer consisting of a turf of small red and green algae. In a later study, Dayton et al. (1984) found that plants at each level of a multilayered community have specific adaptations. Plants forming the upper canopy showed morphological adaptations to exploit light and nutrients (*Macrocystis*), whereas subcanopy browns were better adapted to tolerate water surge (*Eisenia, Dictyoneurum*). At the substrata level, corallines and *A. clathratum* were able to resist intense grazing. Dayton's study found an interaction between biotic and abiotic factors; the relative stability of seaweed communities was determined by their biological relationships, whereas abiotic factors were critical between patches.

Tide Pools

Although a component of the intertidal zone, tide pools can be considered a subset of subtidal communities, where abiotic factors of temperature, pH, and salinity are usually more important than light or wave activity. In general, abiotic factors seem to play a larger role in algal diversity than biotic ones, although grazing can cause variations in pool communities (van Tamelen, 1996). Even to the casual observer, it is evident that tide pools vary greatly with regard to diversity and abundance of seaweeds (Fig. 8-11). The classic studies of Klugh (1924) in New Brunswick, Canada, demonstrated the importance of several abiotic factors, particularly temperature on the diversity of tide-pool organisms (Table 8-2). The larger the range in temperature, characteristic of high intertidal pools, the lower the diversity and abundance of macroalgae. Femino and Mathieson (1980) also found that climatic (rainfall, sunlight, air temperature) and biotic (photosynthetic, respiratory rates) factors determined the presence and abundance of tide pool in Maine. In Rhode Island tide pools, species were most similar when pool zonation, snail density (*Littorina*), and pool volume were analogous (Wolfe and Harlin, 1988).

Tide-pool communities have been characterized by vegetational types (Gross-Custard et al., 1979), abiotic features (Huggett and Griffiths, 1986), and

Figure 8-11. A tide pool at Otter Cliff, Mount Desert Island, Maine 2.0 m above MLW. The large (40 m^2) pool (1) is wetted by storm waves at low tide and is separated from fresh seawater for less than 4 hours each tidal cycle. The pool contains many subtidal algae (e.g., *Laminaria, Fucus, Devaleraea*). A small tide pool (2) is visible at 3.5 m above MLW and the arrow points to the benchmark of Johnson and Skutch (1928).

the ecology of dominant plants (Lubchenco, 1982). Using multivariate analyses, Kooistra et al. (1989) found macroalgal diversity increased progressively downward in the intertidal belt. The high-tide pools on an exposed rocky shore near Roscoff, France, were subjected to high temperatures, extreme salinities, and desiccation due to evaporation. Hence, they only supported monospecific

TABLE 8-2. The Effect of Elevation on Tide Pools in New Brunswick, Canada

Height Above MLW (m)	Monthly Exposure (h)	Exposure (at end)		Species Number
		Temp. (°C)	pH	
1.0	13	14.0	8.1	High
1.6	88	16.5	8.0	High
2.1	175	17.6	8.4	Moderate
3.6	285	20.3	8.1	Moderate
4.2	420	22.7	7.9	Low
5.1	560	24.3	7.9	Low

Source: After Klugh (1924).

stands of stress-tolerant algae such as *Chaetomorpha aerea* and *Hildenbrandia rubra*. In the same study, midlevel tidal pools showed four vertical zones and pronounced seasonal shifts in species. The upper zone was heavily grazed and dominated by a crustose coralline, and the lowest zone contained shade-tolerant algae, including *Gelidium latifolium* and *Laminaria hyperborea*. The low-tide-pool flora were essentially intertidal fringe seaweeds due to wave wash and spillover during low tide.

OTHER MACROALGAL COMMUNITIES

Psammophytic and drift macroalgal communities have not been studied to the extent of epiphytic ones. The *psammophytic* (rhizobenthic; Round, 1981) macroalgae grow attached to unconsolidated sediments and may form extensive communities. Yet, only limited information regarding their productivity, biomass, and ecological importance is known. *Epiphytic* algae, which grow extensively on other plants (Chap. 3), demonstrate how seaweeds and microalgae (Chap. 7) utilize any firm substrata. Because they are abundant, epiphytic algae have been studied more frequently. *Drift* algae are common throughout the world, particularly in estuaries. Even so, few studies have characterized their ecology except where they produce nuisance blooms (Chap. 5).

Psammophytes

Seaweeds that are adapted to grow and attach in sand and mud are called psammophytes and are most abundant in subtropical and tropical waters, particularly within seagrass communities (Fig. 8-12; Chap. 10) and lagoons of atolls (Chap. 12). Psammophytic seaweeds are usually dominated by members of the Caulerpales (Fig. 8-12, insert) and, to a lesser extent, the Dasycladales (Chlorophyta), which have the highest rhizoidal development (Chap. 6). In addition, a variety of smaller forms, including tubular (*Enteromorpha*) and filamentous (*Cladophora, Chaetomorpha*) greens and blue-green algae, can form mats that bind the sand particles of protected shores (Scoffin, 1970; van den Hoek et al., 1972) or mangrove forests (Dawes, 1996). Diversity in psammophytic algal communities is limited compared to lithophytic habitats. Further, macroalgal diversity in temperate eelgrass (*Zostera marina*) and tropical seagrass meadows increases when hard substrata is available (Heijs, 1985; Wheland and Cullinane, 1985).

Species of the calcified green psammophyte *Halimeda* can dominate tropical sandy communities showing high (2.3 g C m^{-2} d^{-1}) productivity (Colinvaux, 1974). Common psammophytic species include *Halimeda*, *Udotea*, and *Caulerpa* that may grow intermixed with algae growing on coral fragments

Figure 8-12. A tropical psammophytic algal community. The unconsolidated sediments of Glover's Reef lagoon (Belize) is populated by sponges (arrow) and a mix of seagrasses (*Thalassia* and *Syringodium*) and psammophytic green algae (e.g., *Penicillus* and *Udotea*) of which species of *Caulerpa* (inset: *C. sertularioides*) are common.

(*Agardhiella, Laurencia, Dasya*). Little is known about the importance of psammophytic seaweeds in terms of productivity; even so, their biomass and that of associated drift algae can reach 85% in a seagrass community (Dawes et al., 1985) and equal or surpass the total organic mass of the seagrasses (Dawes et al., 1979).

Psammophytes compete for space and light with seagrasses and can play a role in development of the substratum, allowing succession and stabilization of sediment for seagrasses. Scoffin (1970) found that *Batophora* (Dasycladales) will remain attached to buried shells in currents up to 50 cm s^{-1} (0.1 kn h^{-1}) whereas larger siphonous algae like *Halimeda*, *Udotea*, and *Rhipochephalus* will not be affected until currents reach 80 cm s^{-1} (0.16 kn h^{-1}). The same large green algae can act as an early successional stage, improving sediment nutrients for the ultimate succession of seagrass communities. Williams (1990) found that rhizobenthic green algae (*Caulerpa*, *Udotea*, *Halimeda*) invaded cleared plots in a coral reef lagoon in St. Croix (Caribbean) and were subsequently replaced by seagrasses. As noted in Chap. 5, the psammophytic green alga *Caulerpa taxifolia*, which was introduced into the Mediterranean, can invade and outcompete the seagrass *Posidonia oceanica* (de Villele and Verlaque, 1995). Seaweeds associated with salt marshes and mangals can be psammophytes, including the filamentous algal mats that form on sediments in mangals (Chap. 10), as well as being epiphytic on the stalks and roots of the tidal marsh plants (Chap. 9).

Epiphytes

The "space race" by seaweeds is evident by the number of smaller macroalgae (and microalgae; Chap. 7) that occurs as *epiphytes* on other seaweeds, the *basiphytes*. Algal epiphytes are ubiquitous and, in many cases, are opportunistic species, which form an important component of marine communities (Linskens, 1963; Arrontes, 1990). There is a large body of literature on algal epiphytes due to their frequency, high production, and importance as a food source for herbivorous invertebrates or mesograzers (Lubchenco and Gaines, 1981). Algal epiphytes have been studied in terms of species diversity (e.g., Tokida, 1960), cellular and metabolic activities such as nutrient uptake (e.g., Ducker and Knox, 1984), and ecological functions (e.g., Novak, 1984; Orth and van Montfreans, 1984) as reviewed by Russell (1988b). In addition, epiphytes are involved in human affairs, as when they increase the cost for mariculture of economically important seaweeds (Friedlander, 1992). This is because epiphytes (1) remove oxygen, carbon dioxide, and nutrients from the media; (2) increase load and drag on the basiphyte, causing breakage; shade the basiphyte and reduce photosynthesis; (3) and may even penetrate host tissues (Fletcher, 1995). Removal of epiphytes can be carried out by physical (mechanical brushing, rapid water movement), chemical (rinsing in chlorine, copper, or changes in pH), and biological means (use of epiphyte herbivores).

Typically, epiphytes are most common on older or damaged portions of seaweeds (Ballantine, 1979; Arrontes, 1990). Competition between seaweed basiphytes and epiphytes has been studied under natural (Arrontes, 1990) and culture (Friedlander and Ben-Amotz, 1991) conditions, but only limited information is available on epiphyte-basiphyte interactions. Epiphytes attach via single cells, filamentous bases, or massive rhizoidal structures (Chap. 7; Fig. 7-21). Algal basiphytes can exhibit a variety of defenses against attachment (Ducker and Knox, 1984), including production of slime (*Ascophyllum*), rapid growth (annuals), changes in pH at the plant surface, sloughing of outer cell walls (*Enteromorpha*, coralline algae), release of toxic chemicals (*allelopathy*, fucoids), and by having ephemeral life histories. *Enteromorpha* will shed its outer layers of cell-wall glycoprotein, whereas *Ascophyllum* may lose an entire epidermal layer. *Eucheuma*, a carrageenophyte, has a slimy surface during growth and is free of epiphytes; but when it ceases growth, the plant surface loses that slime and becomes epiphytized (Dawes et al., 1974). Allelopathy occurs through the exudation of antifouling compounds such as phenolics by the two species of *Sargassum* in the Sargasso Sea (Sieburth and Conover, 1965) and antibiotics by species of *Laminaria*. One of the most extensively studied relationships is the obligate partnership of the red alga *Polysiphonia lanosa*, which is a hemiparasite of the fucoid *Ascophyllum nodosum* (Garbary et al., 1991; Levin and Mathieson, 1991). Perhaps, hemiparastism is an intermediate step between simple epiphytism and parasitism (Chap. 3).

Drift Seaweeds and Blooms

Most seaweeds are attached, which prevents them from being removed by water currents. However, seaweeds do not *need* to be rooted like a terrestrial plant because the seawater delivers the nutrients to them. Over 230 species of unattached, or *drift*, seaweeds are known throughout the world (Norton and Mathieson, 1983), occurring in bays, salt marshes, estuaries, and the open ocean (e.g., the Sargasso Sea). Most drift forms originated as attached plants (but see *Sargassum* in what follows), and because they are detached, they may show distinct changes in morphology and usually a loss of sexual reproduction. For example, salt marsh ecads typically lack sexual reproduction, are reduced in size, and have fine proliferations. Examples include ecads of *Ascophyllum nodosum* [Fig. 8-13(A1); *mackaii, minor, scorpioides*], *Fucus vesiculosus* [Fig. 8-13(A2); *limicola, muscoides, nanus*], and *Pelvetia canaliculata* (*coralloides, radicans*).

Unattached seaweeds may be *entangled* or intertwined with other plants (e.g., *Hypnea musciformis*); *loose-lying* on the sea floor (e.g., *Chorda filum*), *aegagropilus*, or loose-lying species that form balls (e.g., *Cladophora*); *embedded* amongst sand or mud (e.g., *Fucus* spp.); or *floating* (e.g., *Sargassum natans* and *S. fluitans*), as seen in the Sargasso Sea. Such plants originate from propagules (spore, zygote) and vegetative fragmentation of branches due to brittleness or abscission of the attached parent plant (Norton and Mathieson, 1983).

Examples of drift seaweed studies range from Florida to Japan. More than 63 species of drift algae occur in the Indian River Lagoon on the east coast of Florida, providing oxygen, food, and habitat for smaller animals [Benz et al., 1979; Fig. 8-13(B)]. During the spring, their biomass averaged 164 g dry wt m^{-2} over a 15-ha area (Virnstein and Carbonara, 1985). Floating mats of mixed species can serve as nurseries and habitats of commercial fish in Japan, including yellowtail, or *Seriola quinqueradiata* (Ohno, 1984), or as dispersal methods for various fauna. Crustose coralline red algae also can grow unattached, forming spherical structures, or *rhodoliths* (Fig. 8-14), that will cover the benthos. As described by Dawson (1960), their growth pattern can range from *fruticose* (cylindrical branches) to *foliose* (flattened branches) morphologies. Rhodoliths occur worldwide, extending from the low intertidal to over 100 m. They are also found as fossil populations and have been used to interpret paleoenvironmental conditions (Steller and Foster, 1995). Thallus shape and branch density also can be used as indicators of environmental factors, such as light, nutrients, and temperature (Bosence, 1991), and their shape can be used as an indicator of water motion (Prager and Ginsburg, 1989).

Drift macroalgae form *harmful blooms* (Chap. 5) similar to those produced by microalgae (e.g., red tides; Chap. 7). Exponential increases in macroal-

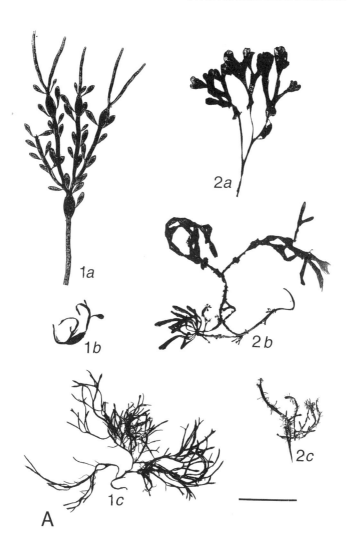

Figure 8-13. Drift macroalgae. A. Drift *Ascophyllum nodosum* (1) and *Fucus vesiculo-sus* (2) are common in salt marshes of the Great Bay Estuary, New Hampshire. Branches from attached plants (1*a*, 2*a*) can be compared with ones recently detached (1*b*, 2*b*) and the resulting ecads *scorpioides* (1*c*) and *limicola* (2*c*). Unit mark equals 5 cm (courtesy of A. Mathieson, University of New Hampshire). B. Subtropical and tropical seagrass beds of *Thalassia* may contain large mats of red algae (e.g., *Laurencia, Acanthophora, Hypnea*) that drift through the community (courtesy of M. Hall, Florida Marine Research Institute).

Figure 8-13.   (*Continued*)

gal biomass can foul beaches, impact commercial fisheries, produce noxious gases, cause anoxia, and alter native plant communities. Examples of macroalgal blooms caused by eutrophication given in Chap. 5 include *Cladophora vagabunda* and *Gracilaria tikvahiae* in Waguoit Bay, Massachusetts (Valiella et al., 1992), *Cladophora prolifera* in Bermuda (Bach and Josselyn, 1978; Lapointe and O'Connell, 1989), and the "green tide" by *Ulva lactuca* in Venice and the French coast (Merrill and Fletcher, 1991; Maze et al., 1993). The causes of blooms of the brown alga *Pilayella littoralis* in Nahant Bay, Massachusetts (Wilce et al., 1982), and the green alga *Codium isthmocladum* on coral reefs near West Palm Beach, Florida (Fig. 5-3; Lapointe, in press), may be a combination of natural and anthropogenic factors.

Figure 8-14. Rhodoliths. Crustose corallines can break free and grow into spherical structures as they are rolled around by currents. The fruticose and foliose branching rhodoliths were photographed at 8 m in Baja California (courtesy of M. Foster, California State College, Moss Landing).

CHAPTER 9

Salt Marshes

INTRODUCTION

Of the two types of vascular plant tidal communities in the world (Figs. 9-1 and 9-2), *salt marshes* are found where freezing air temperatures occur with some regularity, whereas *mangals* are limited to latitudes where air temperatures usually remain at 20°C or higher. Even so, the latter can tolerate short exposures below 5°C but will be killed back to their roots with prolonged freezing (Chap. 10). Both tidal communities are restricted to low-energy coastal regions, such as estuaries within drowned river basins or coastal habitats protected by barrier islands (Chapman, 1977). Other factors controlling tidal communities include amounts of fresh water input, sediment availability, and intertidal topography.

Because salt marshes occur from subtropical to arctic waters, they have been extensively studied in Europe (Long and Mason, 1983; Adam, 1990), North America (Teal, 1969; Pomeroy and Wiegert, 1981; Nixon, 1982; Frey and Basan, 1985; Mitsch and Gosselink, 1993), as well as worldwide (Chapman A. R. O., 1974; Chapman, V. J., 1977). Salt marshes have considerable economic value, as they serve as nursery grounds for invertebrates, fish, and birds. For example, Odum (1974) estimated that salt marshes in Georgia return $2600 acre^{-1} y^{-1} in direct commercial value to the state. Salt marshes also have been extensively studied, because they often have been impacted by man due to their use as harbors. They are also considered to be common-use land that can be exploited by anyone. Thus, salt marshes have been seriously damaged worldwide, being "used, manipulated, and exploited for centuries, usually for some economic gain" (Daiber, 1986). Recently, the conservation, management, and restoration of salt marshes have also become important topics (Daiber, 1986; Allen and Pye, 1992).

ADAPTATIONS OF SALT MARSH FLOWERING PLANTS

Salt marshes are typically areas of natural salt-tolerant herbs, grasses, or low shrubs growing on unconsolidated (e.g., alluvial) sediments bordering saline water bodies whose water level fluctuates tidally or nontidally (Daiber, 1986).

Figure 9-1. Brave Boat Harbor salt marsh near Kittery, Maine. The tidal stream and lower intertidal mud are visible, fucoid ecads occur along the marsh edge. The lower border of *Spartina alterniflora* (Sa) is being eroded by the stream. The lower slope is dominated by *S. patens* (Sp), although succulent halophytes (*Suaeda*) grow there as well. The upper slope has a thin band of *Juncus gerardii* (Jg) that is contiguous with the upland forest.

The key features of the vegetation are adaptation to saline conditions, as well as exposure to waterlogged sediments, including peat deposits.

    Warming (1909) first proposed that flowering plants could be grouped into three classes based on their adaptations for retaining and obtaining water. *Mesophytes* or glycophytes (e.g., wheat, beet) grow in habitats where freshwater is available in the sediment and lack specialized adaptations to prevent water loss. *Hydrophytes* are plants that live in water, partially or wholly submerged (e.g., seagrasses; Chap. 11), and salt marsh plants and mangroves (Chap. 10) are *xerophytes*, as they have morphological, anatomical, and reproductive adaptations to aid in the retention and uptake of water (Fahn and Cutler, 1992). Further, *halophytes* can be contrasted with xerophytes in that they have adaptations to prevent water loss, but also are able to grow in saline habitats. However, all ecotypes occur in salt marshes because of the gradation of salinity between estuarine and upland communities. Most, if not all, halophytic salt marsh plants are *facultative halophytes*, as they do not require saline conditions for growth. However, even facultative forms usually show increased root development when the water potential decreases due to salt, whereas glycophytes do not (Adam, 1990). *Suaeda aegyptica* and perhaps a few other succulent species have been recognized as *obligate halophytes*, as they show a specific requirement for sodium and not potassium

Figure 9-2. Weeki Wachee River salt marsh in Florida. The lower intertidal zone, consisting of muddy sand and oysters, is evident at low tide. Occasional clumps of *Spartina alterniflora* (Sa) grow on the lower border. The extensive lower slope is dominated by *Juncus roemerianus* (Jr), with occasional clumps of the fern *Acrosticum* (arrows) growing where the elevation is slightly (10 cm) greater. The upper slope and border zones surround the distant palm hammocks.

(Eshel, 1985). Obligate halophytic marsh plants can grow in soil salinities greater than 100 M m^{-3} (Flowers et al., 1986) and require salt water to complete their life cycle. In this text, salt marsh plants are considered to be any species that grow in the tidal marsh community, including glycophytic plants.

Taxonomy

Presently, a listing of families and species of salt marsh plants is not possible because of the lack of worldwide data and different opinions regarding what constitutes a salt marsh plant. Of the 18 families of flowering plants with salt marsh species listed by Adam (1990), he considers 6 to contain the dominant plants worldwide, namely, the Gramineae, Chenopodiaceae, Juncaceae, Cyperaceae, Plumbaginaceae, and Frankeniaceae. The Compositae (Asteraceae) should also be added to this list. By contrast, Tiner (1987) lists 280 plants in 60 families as being found in coastal wetlands of the northeastern United States. Salt marsh floras contain over 400 species, which form 9 maritime formations throughout the world (Chapman, A. R. O., 1974; Chapman, V. J., 1977). The most prominent species belong to the genera *Spartina, Salicornia, Juncus, Arthrocnemum, Suaeda,* and *Plantago* (Chapman, 1977). Adam (1990) lists 325 species of vascular plants in British salt marshes, with 45 species and 12 families constituting the halophytic element. Duncan (1974) lists 347 species and 75 families for the eastern United States, and Eleutherius (1980) lists 200 species (3 ferns, 1 gymnosperm, 89 monocots, 107 dicots) and 63 families for Mississippi.

The salt marsh floral diversity is greatest in temperate and mediterranean climates and smaller in both high (arctic) and low (subtropical) latitudes (Adam, 1990). The salt marsh vegetation of the world is not highly diverse when compared with terrestrial grasslands, has a limited halophytic element, and has widespread species with related *vicariant* species on different continents. Chapman (1977) proposes that the wide generic distribution of the halophytic element is due to continental drift, whereas Adam (1990) has considered dispersal by birds or vegetative propagation as primary factors. The halophytic element is more diverse than in mangles, with many plant families showing independent parallel evolution of salt adaptations on both a local and continental scale (Adam, 1990). This is especially true of high-marsh (upper-zone) plants, where glycophytic species can occur.

Taxonomic and reproductive details of the highly diverse group of flowering plants adapted to salt marshes can be found in local floras. The flowering plants found in salt marshes do not show any clear pattern for flower and seed adaptations, which are similar to their glycophytic counterparts. Thus, a section on biological reproduction is not included, in contrast to the mangroves (Chap. 10). Three salt marsh flowering plants that are widely distributed are presented here. The examples are used to briefly introduce their morphological, structural, and reproductive adaptations to coastal saline habitats.

Spartina alterniflora Loisel (smooth cordgrass; Fig. 9-3) is a monocot placed in the grass family or Poaceae (Gramineae). The specie is found on both coasts of North and South America and Europe and the genus has a worldwide distribution (Mobberley, 1956). The erect stems (culms) grow to 3 m (tall form) along tidal creeks or 0.2 to 0.8 m (short form) in the upper marsh. The two forms apparently retain some differences in their culm sizes, even in common garden experiments suggesting genetic differences (Gallagher et al., 1988). However, much of the difference in culm heights may also be a phenotypic expression based on edaphic factors, as seen for *S. anglica* (Thompson et al., 1991) and other species (Howes et al., 1986). The species are clonal, with a horizontal subterranean rhizome, which branches sympodially and produces the culms and roots. The roots are of two types, unbranched anchoring roots, which are covered by a corky covering, and ephemeral, fine, much-branched, matted absorbing roots. The culms and their leaves account for about one-third to one-tenth of the plant biomass (Gross et al., 1991). The culms have air spaces, including a hollow center and a ring of lacunae that alternates with the vascular bundles in the periphery of the stem and is continuous into the rhizome and roots (Anderson, 1974). The leaves, produced by a basal intercalary meristem, are smooth, flat blades with longitudinal furrows. The leaves have a dry, thin anatomy of a xeric grass and show halophytic adaptations with two-cell, epidermal salt glands

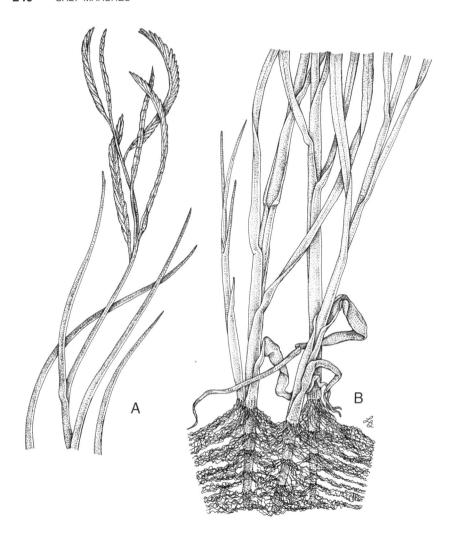

Figure 9-3. *Spartina alterniflora* or cord grass. The upper (A) portion of the culm contains flowers typical of the grass family. The lower (B) portion of the culm has adventitious roots; and the rhizome is not shown.

(Fahn and Cutler, 1992). The blade also shows the Krantz-type construction that is related to its C_4 photosynthesis (see what follows). Flowering and seed production are typical of the grass family.

Juncus roemerianus Scheele (black rush; Fig. 9-4) is a monocot and member of the family Juncaceae, which contains eight genera and is best represented in the Southern Hemisphere (Seibert, 1969). The genus *Juncus* contains both freshwater and salt marsh species. Like *Spartina*, it is a clonal plant with a subterranean, branching rhizome, erect culms, and a fibrous root system. The

Figure 9-4. *Juncus roemerianus*, or black rush. The upper portion of the stem (A) bears typical rush-type flowers and shows terete, needlelike leaves that reach 2 m in height. The lower portion (B) has thick, densely branched ramets and roots arising from a rhizome with scale leaves.

rhizome is covered by suberized scale leaves, has air canals (lacunae) in its cortex, and an endodermal layer limiting the cortex from the pericycle. The culms contain lacunae and produce long needlelike (to 2 m tall) leaves that are oval in cross-section. The blades develop from a basal intercalary meristem. The epidermis is lignified at maturity, has a thick cuticle, and covers a dense chlorenchyma tissue. The central portion of the blades consists of a spongy parenchymatous ground mesophyll with numerous vascular bundles (Anderson, 1974). Lacunae occur centrally as well as between vascular bundles and are separated by plates of parenchyma. The plant's flowers occur in dense cymes, typical of the family.

Salicornia virginica Linneaus (pickle weed, perennial glasswort; Fig. 9-5) is a dicot genus in the goosefoot family Chenopodiaceae, which contains more halophytes than any other family. The stem and leaves are *succulent* and appear swollen because of the abundance of water-containing cells (DeFraine, 1912).

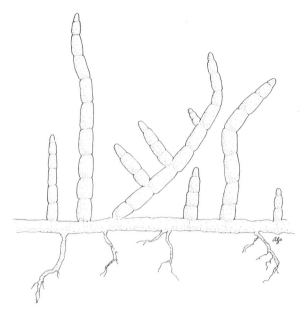

Figure 9-5. *Salicornia virginica* or perennial glasswort. The succulent stem can reach 20 cm in height but it is usually prostrate. The stem appears segmented due to the modified fleshy leaf petioles.

The plant's common name indicates the perennial and glossy nature of its succulent stem, which also has a thick, waxy cuticle. The stem tends to be procumbent, producing short, erect branches. A typical dicot tap root is lost and the stem produces adventitious roots as it extends over the sediment. The blades are reduced to scales, and the succulent petioles are wrapped around the stem, producing a segmented appearance (Fig. 9-4).

Geographic Distribution

The worldwide distribution of salt marsh communities can be divided into nine geographical seres, subclimax groups (Chapman, 1974), or simple groups (Chapman, 1977) based on floristic and vegetational criteria (Fig. 9-6). The *Arctic Group* includes the circumpolar North American and Russian Arctic as well as Greenland and Iceland. The group is not subdivided and grows where extreme winter air temperatures result in fragmentary salt marshes being dominated by the grass *Puccinella phryganodes*, with species of *Carex* occurring with the higher intertidal zones. The *Northern European Group* has five subgroups based on soil or salinity differences; it includes the Iberian Peninsula northward, including the North Sea coasts, Baltic Sea, and coasts of Ireland and England. Annuals (*Salicornia* spp., *Puccinella maritima*, and *Juncus gerardii*) dominate. The *Mediterranean Group* includes three subgroups having affini-

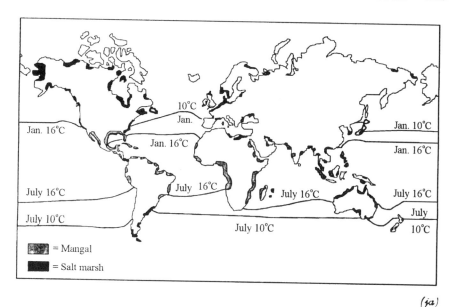

(ja)

Figure 9-6. World distribution of salt marshes and mangals. The northern and southern 16°C winter isotherms generally enclose the distribution of mangals, and the 10°C winter isotherms usually mark the lowest latitudes where salt marshes are found (after Chapman, 1977).

ties to North Europe and is dominated by *Juncus acutus*, along with species of *Arthrocnemum* and *Limonium*. The *Western Atlantic Group* extends from the St. Lawrence River in Canada to the Gulf coast and forms a transition to mangroves on the southern coast of North America. It also contains three subgroups, ranging from the Bay of Fundy, New England, and the southeastern Coastal Plain. Besides, *Spartina alterniflora*, the dominant plants include *Puccinellia maritima* and *Juncus balticus* (Bay of Fundy), *S. alterniflora* and *S. patens* (New England), and *Juncus roemerianus* and *Sesuvium portulacastrum* (Coastal Plain). The *Pacific American Group* has three subgroups, including Alaska (*Puccinellia phryganodes*), which has unstable marshes due to ice damage (see what follows, Fig. 9-14). The central north Pacific subgroup (*S. foliosa*, *Suaeda* spp.) grades into the southern subgroup, where salt marshes are complex (e.g., San Francisco Bay). The latter has *Salicornia virginica* and *Scripus maritimum* in the low marsh and *Distichlis spicata* and *Juncus balticus* in the high marsh. The *Sino-Japanese Group* includes *Carex ramenskii* in the northern region and *Triglochin maritima* and *Salicornia brachystachya* further south. The *Australasian Group* contains the Australian and New Zealand subgroups and includes Southern Hemisphere species such as *Salicornia australis*, *Suaeda novae-zelandiae*, and *Triglochin striata*. The *South American Group* includes *Spartina brasiliensis* and *S. montevidensis*, as well as species of *Distichlis* and *Heterostachys*. The final, *Tropical Group* is limited to higher elevations, where tidal flooding is infrequent and inland temperatures damage the mangroves. The group can be found in Florida (*Juncus roemerianus, Sesuvium portulacastrum*) and arid Baja California (*Salicornia virginica, Spartina foliosa*).

Morphological Adaptations

Species with clonal growth that are *hemicryptophytes* or perennials with buds at the soil surface (e.g., *Spartina, Juncus*) dominate salt marshes (Chapman, 1977; Adam, 1990). Other common morphotypes include *chamaephytenanophanerophytes* or dwarf shrubs (e.g., *Baccharis*) and *therophytes* or annuals (*Puccinellia*). The last group varies according to latitudinal position and salt marsh group, being more conspicuous in Mediterranean and semiarid climates (Chapman, 1974). As noted in the previous species descriptions, the clonal nature is common and characterized by their *rhizomes* for most perennial salt marsh plants. Further, clonal integration (see also Chap. 11) of *S. patens* culms has been shown for nitrogen and salinity gradients (Hester et al., 1994). The horizontal stems occur a few centimeters to 1 m below the surface (Reimold, 1972) and serve as storage organs during periods of dormancy. They are also responsible for vegetative expansion. Rhizomes on salt marsh plants have thick anchoring and delicate absorbing roots, which bind unconsolidated sediments and serve to reduce erosion. The secondary roots of dicots and fibrous roots of monocots often can release oxygen into the surrounding substrata, modifying the reducing anaerobic *rhizosphere* (see "Edaphic Features," Chap. 9). The extensive development and binding power of the below-ground biomass are highlighted by the difficulty of taking sediment cores within salt marshes.

Anatomical Adaptations

The xerophytic adaptations of salt marsh plants are evident by their anatomical complexity, which aids in water retention. Thus, their stems, leaves, and roots show (1) increased lignification, (2) complex epidermal development, (3) and well-developed bundle sheaths having an endodermal layer with an extensive Casparian strip (Anderson, 1974). In addition, many salt marsh plants show the development of *aerenchyma* and *lacunae*, which facilitate diffusion of oxygen between the stem and roots (also found in mangroves, Chap. 10; and seagrasses, Chap. 11). Aerenchyma is formed via *schizogenous* separation of parenchyma cell walls to form irregular air passages. In contrast, the regular air channels called lacunae are formed either schizogenously or *lysigenously* (e.g., via digestion of cells), resulting in columnar air passages through the plant (Fahn and Cutler, 1992). The anatomical adaptations of leaves in salt marsh plants are particularly evident when compared to mesophytic forms. All three types of xeric leaves can be found in different salt marsh plants, namely, succulent, thick, and dry-type (thin) morphologies. Each xeric morphology shows distinct adaptations for water retention. Dry-type leaves show enhanced cuticular resistance to water loss, produce epidermal hairs, and may curl. In contrast, succulent leaves store water and thus dilute internal salt concentrations, whereas thick leaves increase vascular, water-storage, and photosynthetic tissue (Fahn and Cutler, 1992). The two dominant leaf types found in salt marsh plants are dry-type and succulent morphologies and are described for *Spartina* and *Sesuvium*, respectively.

Being a grass, the *dry-type leaves* of *Spartina alterniflora* are linear, with parallel venation (Anderson, 1974; Fahn and Cutler, 1992). A cross-section of a mature leaf (Fig. 9-7) shows deep grooves in the upper (adaxial) side of the leaf that runs parallel to the leaf length, each with a vein. The lower (abaxial) epidermal cells of dry-type leaves tend to have thicker, lignified walls and cuticle than the upper epidermis. Further, the stomates are concentrated in the upper epidermis (unlike mesophytic leaves) lining the groves. Two-cell salt glads occur in both epidermal layers, being most common in the adaxial grooves. Many dry-type grass leaves (i.e., *S. alterniflora*) can curl by the collapse of thin-wall cells, called *hydrocytes*, found at the base of the grooves on the upper epidermis. Thus, the leaf becomes a tube with its lower epidermis exposed, limiting water loss. The ground mesophyll is sparse and does not have a spongy mesophyll with air spaces. The species is a C_4 plant with *Krantz* anatomy that consists of two bundle sheaths around each of the parallel veins. The outer sheath consists of large cells with starch-bearing chloroplasts, and an inner sheath of smaller endodermal cells with suberized Casparian strips in their cell wall (Kuramoto and Brest, 1979).

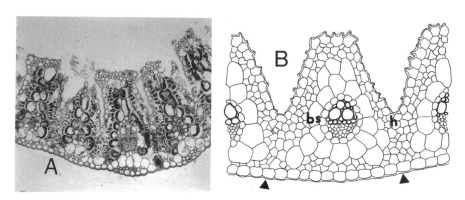

Figure 9-7. A cross-section of a cord grass (*Spartina alterniflora*) leaf. The light micrograph (A) and diagram (B) show deep parallel groves on the upper (adaxial) side of the leaf where the stomates and thin-walled hydrocytes (h) occur. The top of each grove consists of lignified epidermal and sclerenchyma cells. The abaxial epidermis (arrows) has a thick cuticle. The C_4 Krantz anatomy is shown by two concentrically arranged bundle sheaths (bs) around each vein. The outer sheath contains large cells with starch-bearing chloroplasts where C_3 carbon fixation occurs, and the inner sheath consists of small cells without plastids. A ground mesophyll contains chloroplasts where C_4 carbon fixation occurs.

The petioles of *Salicornia virginica* are *succulent*, wrapped around the stem, and replace the leaf blade (reduced to a scale) as the photosynthetic organ (Fig. 9-8; DeFraine, 1912; Anderson, 1974). Succulence is an adaptation for water storage, dilution of inorganic salts, and reduction in surface area to reduce water loss. The epidermal cells are papillate and have a thick outer wall and cuticle; the latter is so thick that it can be separated from the wall. The stomates occur around the stem. The outer portion of the modified petiole has one to three rows of columnar palisade cells containing most of the plant's chloroplasts. Interspersed with the palisade cells in the petiole are tracheoid idioblasts or storage tracheids that function in water storage, similar to that found in cactus and orchids. Up to two-thirds of the modified petiole consists of an inner spongy mesophyll of irregularly shaped water-containing cells, or *hydrocytes*. The ground mesophyll is an area where high salt concentrations can occur

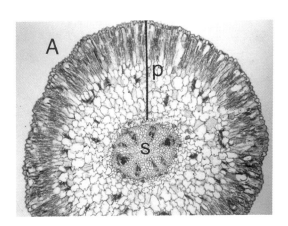

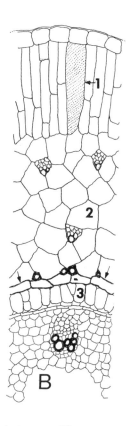

Figure 9-8. Cross-sections of a glasswort (*Salicornia virginica*) stem. The cross-section (A) shows the stem(s) encased by the modified petiole (p), which is fleshy and has papillate epidermal cells. Diagram (B) shows a storage tracheid among the palisade cells (1), an inner ground mesophyll with small veins (2) that border on what may be the modified abaxial leaf epidermis (arrows), and the suberized stem pericycle (3).

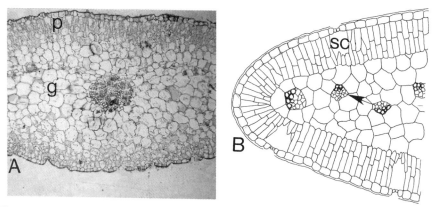

Figure 9-9.   Sections of a leaf of *Sesuvium portulacastrum*. The succulent leaf has an epidermis with a thick cuticle (arrow), irregular outer palisade (p) and an inner ground mesophyll (g) that functions in water storage and dilution of salt (A). Diagram (B) shows the stomatal chambers (sc) and secondary veins (arrow).

in the hydrocyte vacuoles. Fiber bundles and small veins occur in ground meso-phyll. The "lower" epidermal layer of the petiole and "outer" epidermis and cor-tex of the stem are not evident, so that the vascular stele begins with the outer large-cell endodermis in which the Casparian strip is difficult to discern. At the end of the growing season, abscission of the modified petiole reveals a layer of suberized cork cells, which is formed from subepidermal cells of the stem.

Sesuvium portulacastrum is another example of a salt marsh succulent dicot. Its elongated leaves contain a central ground mesophyll and vascular bundles (Fig. 9-9). The veins are surrounded by well-developed bundle sheath cells with a suberized Casparian strip in the cell wall that function as an endodermis. The ground mesophyll functions in water storage and salt storage, and the outer, loosely organized palisade mesophyll contains most of the chloroplasts and air spaces. The epidermal cells are cuboidal and stomates occur on all sides of the leaf. Surprisingly, neither *S. virginica* nor *S. portulacastrum* have C4 type carbon fixation (see "Photosynthesis"), although they are succulent xerophytes.

ECOPHYSIOLOGY OF SALT MARSH PLANTS

Salt marsh plants cope with a less than optimum environment compared to ter-restrial habitats. In the soil, sodium and chloride concentrations are high, essen-tial nutrient ions may be limited due to ionic composition and ionic effects of seawater, and anaerobic conditions occur. Further, tidal immersion can result in temperature shock, changes in photoperiod, mechanical effects of tidal cur-rents, and siltation of leaves by sediments that block the stomata. Thus, it is interesting to consider physiological features of salt marsh plants in relation to water conservation and salt tolerance.

Salinity and Plants

The effects of salinity on plants has been the subject of extensive research, particularly in relation to selection of agricultural species and varieties for growth on saline soils. Thus, there has been a strong interest in the physiological responses of salt marsh plants to their environment (e.g., Flowers et al., 1986; Adam, 1990). Soil salinities vary across a salt marsh, with the highest levels usually occurring in the middle to upper regions, although the level of rain and freshwater input can modify this (e.g., arid Baja California marshes). Generally, high elevation within marshes limits tidal flushing and if evaporation is high, as in marshes of lower latitudes (e.g., southern United States), the level of ions may increase and become toxic (hypersalinity) to many halophytes.

Vascular plants show distinct tolerances to soil salinity, ranging from low (soy bean: 5 to 10 M m^{-3} Cl) to high (*Suaeda* sp.: 340+ M m^{-3} Cl) concentrations. High concentrations of Na$^+$Na$^+$ and Cl$^-$ in the rhizosphere of salt marsh plants will influence the ability of plants to take up nutrients as well as water. Halophytes have higher uptake rates of most macro- and micronutrients than glycophytes, even where soil salinity is low. This appears to be a distinct feature of such plants (Adam, 1990).

Salt Regulation

A glycophyte has an internal ionic concentration of 100 to 200 M m^{-3} (mostly potassium; Storey and Wyn Jones, 1979), which results in a cell osmotic potential of -0.5 to -1 MPa (see "Salinity and Plants," Chap. 4). In order for water to migrate to the shoot, there must be a gradient in water potential. Seawater at 35 ppt salinity will have a water potential of -2.5 mPa and thus the roots of a halophyte must have an osmotic potential *below* that to take up water. Although the distinction is not sharp, most halophytes can be considered either *osmoconformers* or *osmoregulators* (Storey and Wyn Jones, 1979), with the former showing a gradient in osmotic potential between the soil and plant and the latter lacking a gradient and exhibiting sharp changes in ion content of tissues when subjected to changes in external salinity. Depth of rooting in saline soils is also an adaptation, as shallow-rooted plants are subjected to irregular shifts in soil salinity (due to evaporation, rain, tidal flooding), whereas more deeply rooted species are not. It would be interesting to correlate root morphology and depth with zonation and the type of salt regulation in marsh plants. Adam (1990) suggests three ways halophytes may respond to increased external salinity: (1) by synthesizing organic solutes, (2) by taking up inorganic salts, and (3) by dehydration.

Production of organic solutes is an effective method to prevent uptake of salts. For example, cells with a concentration of 20% to 40% sugar will acclimate to 100 M m^{-3} NaCl. The enzymes of halophytic salt marsh plants show some tolerance to higher levels of Na$^+$ and Cl$^-$, but increases in salt in the cytoplasm is toxic, so that cellular osmotic potentials must be adjusted by organic solutes. In a study of 16 salt marsh plants, sucrose was abundant in 3 mono-cots, maltose in 2 dicots, polyols in 4 dicots and monocots, and the remaining species showing a mixture of soluble carbohydrates, polyols, betaines and free proline (Briens and Larher, 1982). Nitrogen-containing compounds, such as the amino acid proline and quaternary ammonium compounds (e.g., glycinebetaine) and carbohydrates (sorbitol, polyhydric alcohols, and sugars), are examples of *osmolites* found in salt marsh plants. Proline and quaternary ammonium com-pounds are osmolites that are common to halophytes, occurring in the cytoplasm and stored in cell vacuoles (Adam, 1990), but can also become toxic at high levels. Thus, there is usually a mixture of osmolites in the cells, with sugars dominant in the roots and ammonium compounds relatively more important in leaves of most salt marsh plants (Briens and Larher, 1982; Burdick, 1989). Although the production of osmolites requires energy, the synthesis of organic solutes also can be a sink for photosynthates and is often reversible (Chap. 4).

A second response to soil salinity is via accumulation of salts, as shown by the high ash content (<50% dry wt) in the leaves and shoots of halophytes. The ions most commonly accumulated are Na^+ and Cl^-, although SO_4^{-2} and NO_3^- can be high. In contrast to glycophytes, halophytes will tolerate levels of Na^+ and Cl^- equal to 10 to 13 ppt, whereas higher concentrations will cause ribosomal breakdown (Storey Wyn Jones, 1979). The third, short-term, response of increasing solute concentration is via dehydration, as shown by the grasses *Leptochloa fusca* and *Spartina anglica* and other monocots (Adam, 1990). Loss of water increases cellular ionic strength (through concentration), which is effective. However, the loss in turgor pressure results in photosyn-thetic and metabolic slowdown, curling, and wilting, which can be lethal when cellular plasmolysis and permanent wilting of leaves occurs.

In addition to osmoregulatory responses of cells to changes in salinity, *reg-ulation* of shoot salt content can be accomplished in several ways (Chap. 4; Adam, 1990):

1. *Ion Exclusion in the Roots.* Salt exclusion by absorbing roots is carried out primarily by the endodermal layer. Ions are transported from the root hairs through the outer cortex via the cell wall (apoplastic movement) until reaching the suberized Casparian strip of the endodermal cell walls. There, all water and ions must pass through the endodermal cell membranes into the proto-plast (symplastic moment). The endodermal cell membrane is extremely impor-tant in controlling which ions will pass into the root interior (Flowers et al., 1986).

2. *Growth and Succulence.* Dilution of ions carried into the plant can occur through the growth of new plant material and by an increase in water containing tissue or succulence. The latter feature is less common in monocot salt marsh plants versus dicots (but see *Triglochin* spp.). Different ions will affect succu-lence; in some species, Cl^- increases succulence more than SO_4^{-2} (Fahn and Cutler, 1992).

3. *Shedding.* The concentration of ions and then shedding of salt-containing tissues and organs is a common strategy employed by halophytes, both perennials and annuals. Leaves die and are shed when salt concentrations become toxic in the cell; in other cases, salt is stored in the cell walls of the leaf that is shed at the end of each season.

4. *Secretion.* A variety of mangrove and salt marsh plants have glands (specialized trichomes) that secrete salt from their leaves (Haberlandt, 1914; Fahn, 1988; Fig. 10-13). Excretion is an active metabolic process in salt glands and requires energy.

5. *Root Discharge.* Salt removal also can be carried out through translocation from growing regions in the stem and rhizome to the roots, which then discharge excess amounts into the rhizosphere. Species with limited salt tolerance often discharge salt from their roots (Flowers and Yeo, 1986).

6. *Controlling Water Loss.* Lastly, reducing transpiration will lower the amount of water needed, so that salt uptake in the roots will be limited. It is interesting that most salt marsh plants show xeric adaptations to prevent water loss, yet few have the photosynthetic CAM (Crassulacean Acid Metabolism) adaptations that are found in desert plants (see what follows).

Photosynthesis

The Calvin cycle (C_3) is the type of carbon fixation found in the great majority of cold temperate to arctic salt marsh species (Chap. 4; Adam, 1990). In contrast, most grasses (*Spartina alterniflora, Distichlis spicata*) of subtropical salt marshes show Hatch-Slack carbon fixation (C_4), while most rushes and sedges are C_3 plants. The advantage of C_4 carbon fixation is the high affinity of PEP carboxylase for CO_2 and the more efficient use of nitrogen for growth and osmolites. Perhaps a reason for so few C_4 cold temperate salt marsh plants is the reduction in photosynthetic efficiency at lower temperatures.

A more puzzling fact is that none of the succulent salt marsh plants has a CAM C_4 carbon fixation (Adam, 1990). In arid habitats, CAM plants conserve water by opening their stomates at night, when humidity is high and air temperature is cool; hence, they reduce leaf transpiration. The CO_2 that diffuses into the leaf at night is fixed via C_4 photosynthesis and then it is released in the day and converted to sugars via C_3 photosynthesis in the day while the stomates are closed. The advantage for succulent salt marsh plants is CAM carbon fixation would reduce transpiration, as stomates would be closed during the day and reduce salt loading because water uptake would decrease due to conservation.

Tidal Effects

Salt marsh plants have adapted to tidal flooding and waterlogged soils. At high tide, salt marsh sediments are submerged, resulting in edaphic changes such as lowered soil aeration and redox potential. While the plants are flooded, photosynthesis is also limited. In addition, mechanical damage to stems or uprooting of plants may occur from water movement. Waterlogged salt marsh soils are anaerobic; hence, unless oxygen is introduced from above-ground shoots, aerobic respiration ceases in the roots where toxic compounds can accumulate. Oxidation of the rhizosphere through direct release of oxygen and enzyme activities of salt marsh plants lowers the effect of potential soil toxins such as sulfides and reduced metal ions. In addition, the roots of salt marsh species show higher tolerances than glycophytes to potentially toxic compounds containing reduced manganese, iron, and sulfur.

The air channels (lacunae, aerenchyma) in salt marsh plants can account for up to 60% of the total plant body (Adam, 1990) and 50% of its root volume (Burdick and Mendelssohn, 1990), thus allowing the diffusion of oxygen into the roots to support growth and metabolism. A similar organization can be seen in other marine plants growing in waterlogged substrata, including mangrove roots (Chap. 10) and seagrasses (Chap. 11). The level of aerenchyma or lacunar development can be related to the degree of waterlogging, as shown for *Suaeda maritima* (Flowers et al., 1986) and *S. patens* (Burdick and Mendelssohn, 1990). *Spartina patens* responds to the onset of waterlogging by an increase in root aerenchyma (Burdick, 1989) and the activity of enzymes that can support anaerobic metabolism such as alcohol dehydrogenase (ADH) during the first three days. However, after root aeration increases, the level of ADH activity declines (Burdick and Mendelssohn, 1990). Thus, tolerance to anaerobiosis via changes in metabolic activity, which includes the production of nontoxic intermediate metabolites, is a feature of many salt marsh plants that complements aerenchyma development, which allows roots to avoid anaerobiosis.

Costs

The expense of being a halophyte is high in terms of anatomical structure and energy expenditure (Adam, 1990). However, the advantage to adapting to an intertidal habitat is a reduction in competition for light and nutrients, resulting in wider niches. Adaptations to saline soils causes lower growth rates (Schat, 1984), because plants must expend energy to obtain water against an ionic gradient; uptake and transport of nutrients is energetically more costly, the sediments are often anaerobic to toxic; and photosynthesis may be reduced to limit water loss via transpiration. Thus, the surprising point may be not how few halophytes are able to vegetate salt marshes, but rather the diversity and number of species worldwide.

OTHER SALT MARSH PLANTS

In addition to monocots and dicots that dominate salt marshes, a variety of ferns, fern allies, bryophytes, algae (micro- and macro-), and seagrasses occur in salt marshes. They are found growing on the sediment beneath the flowering plants, as epiphytes on their stems or as benthic forms in tidal creeks and pools, either drift or attached. Although overlooked in many studies, their role as primary producers can be substantial.

Ferns and Fern Allies

Acrosticum is the most common genus found in mangals and in subtropical and warm temperate salt marshes of both hemispheres. A variety of ferns can occur as a fringing understory in mangals and species of *Acrosticum* may overtop the salt marsh plants, as seen on the west coast of Florida (Dawes, 1974) and African marshes (Chapman, 1977). Some of the fossil fern allies such as the lycopod *Pleuromeia longicaluis* were probably the dominant salt marsh vegetation during the Carboniferous Period (Adam, 1990).

Bryophytes

Few species of mosses or liverworts have been described in salt marshes within lower latitudes. However, over 50 moss species and one liverwort are known for British marshes (Adam, 1990) and their presence in arctic and Scandinavian marshes is well documented (de Molenaar, 1974). In general, they form an understory in the upper elevations of tidal marshes.

Algae

The algal flora of salt marshes can be significant in terms of productivity (Pomeroy, 1959; Blum, 1968; Zedler, 1980). For example, Pomeroy (1959) reported gross algal productivity of 200 g C m^{-2} y^{-1} in a Georgia salt marsh. The algal flora can be divided into four communities: *sediment microalgae*, *macroalgal turf* (drift and mat) on the substrata, intertidal *epiphytic macroalgae* on the marsh plants, and *subtidal* drift or attached macroalgae.

Sediment microalgae form colonies on lower slopes below the salt marsh, as well as on the sediments beneath salt marsh plants [Fig. 9-10(A)]. Microalgae usually form a continuous cover [Fig. 7-22(A)] and often consist of epipelic and epipsammic blue-green (*Lyngbya*, *Rivularia*) and diatoms (*Cylindrotheca*, *Navicula*). Microalgae can account for 10% of the primary production in Mississippi salt marshes (Sullivan, 1978), with a number of species also occurring in New Jersey salt marshes (Sullivan, 1977). Microalgal mats composed of filamentous blue-green and green algae and diatoms found beneath the salt marsh plants exhibit high production (185 to 341 g C m^{-2} y^{-1}) in southern California marshes (Zedler, 1980). Microalgal species composition is dynamic, seasonal, and localized, with nutrient limitations being a common factor, as shown in Delaware (Sullivan and Daiber, 1975) and Georgia (Darley et al., 1981) salt marshes.

The macroalgal turf or mat community includes attached filamentous species such as green (*Ulothrix*, *Rhizoclonium*), brown (*Ectocarpus*, *Pylaiella*), and red (*Catenata*, *Griffithsia*, *Polysiphonia*) algae. Seaweed mats also can include drift

Figure 9-10. Algae of salt marshes. The sediment in the lower *Spartina alternifora* border of a Florida salt marsh will have a covering of diatoms, blue-green, and green algae (A). Epiphytic macroalgae (e.g., *Bostrychia, Catenata*) grow on the culm bases (B).

forms like *Gracilaria, Ulva,* and *Enteromorpha* in warm temperate marshes and fucoid ecads in cold temperate marshes (Chap. 8, "Drift Seaweeds and Blooms"). Of special interest are the ecads of *Fucus, Pelvetia, Hormosira,* and *Ascophyllum* that exhibit an unattached life-style, with populations occurring in both hemispheres (Norton and Mathieson, 1983). Such ecads, which lack sexual reproduction and primarily reproduce via vegetative fragmentation, have been extensively studied (Brinkhuis, 1976; Chock and Mathieson, 1976). Interestingly, the same fucoid species when attached to rock outcrops within the marsh will be sexually active (Chock and Mathieson, 1976).

Although algal epiphytes of salt marsh plants (e.g., *Catenella, Bostrychia, Caloglossa*), have been studied less than seagrass epiphytes [Fig. 7-22(B)], they are known as important primary producers (Pomeroy, 1959), even showing high rates of photosynthesis during periods of desiccation (Dawes et al., 1978). In addition to cyanobacteria, epiphytic macroalgae form dense growths on the bases of old culms [Fig. 9-10(b)] and stems. Some genera are worldwide in occurrence, including several taxa of green (*Chaetomorpha, Rhizoclonium, Cladophora*), brown (*Ecotocarpus,* fucoids), and red (*Bostrychia, Catenella, Centroceras, Polysiphonia*) algae. The broad tolerances of these algae to temperature, salinity, light and desiccation have been studied in Florida (Dawes et al., 1978) and elsewhere (Chap. 4).

The number of subtidal drift and attached macroalgae found in tidal creeks and bays of salt marshes varies widely due to differences in abiotic factors (climate, freshwater input, sedimentation, substrata). Drift species include the fucoid ecads (Norton and Mathieson, 1983; Chap. 8), as well as other macroalgae (Adams, 1990). Attached subtidal algae occur in subtropical to warm temperate (*Gracilaria, Spyridia, Batophora*) as well as cold temperate (*Chondrus, Fucus, Ascophyllum*) marshes. Seagrasses may dominate shallow marsh basins

(*Ruppia maritima, Zostera noltii, Z. marina*). In Texas tidal marsh channels, over 70 taxa of benthic marine plants occur, averaging 4.0 kg wt m^{-2} during the growing season of July and August and declining to 1.5 kg in January (Conover, 1958). A comparison of estuarine seaweeds in the Goleta Slough (California) and estuaries of New Hampshire (Mathieson and Hehre, 1994) showed strong differences between the two types of floras that were attributed to climate. The arid California salt marsh contained 26 species (16 ephemeral) and the larger New England estuaries contained up to 150 taxa. Even so, many of the green algae were common to both marshes.

ECOLOGY OF SALT MARSHES

Tidal marsh communities include large regions of natural populations, having spatially structured (zoned) vegetation and low diversity. Thus, they offer the plant ecologist a unique opportunity to test concepts of niches in zonation studies, the importance of competition and facilitation (Shat, 1984; Bertness, 1991; Bertness and Shumway, 1993; Chap. 3), and the influence of distinct abiotic factors within different zones (e.g., Burdick and Mendelssohn, 1990). The physiography of tidal marshes is influenced by the degree of protection from waves, tidal regime, rate of sea-level rise, topography of the coast, availability of sediments, and type of substrata (Chapman, 1977). The most extensive salt marshes in the world occur on transgressive coasts with gentle slopes that are protected from offshore waves, as found in drowned river basins, river deltas, and behind barrier beaches. Overall, tides and elevation control the degree of flooding and therefore soil salinity and waterlogging (Adam, 1990). Other abiotic factors interrelated with tides include topographical features (steep or gradual slopes), textural structure of substrata (fine to coarse particles; low to high organic matter), climatic events (rainfall, wind patterns), and soil chemistry (chlorinity, soil water, sulfide levels, pH) (Long and Mason, 1983; Allen and Pye, 1992).

Edaphic Features

Physiochemical factors of salt marsh soils limit the type of plants to those having the appropriate physiological adaptations, as seen in a low-salinity northern Florida marsh (Kurtz and Wagner, 1957) and higher-salinity Louisiana marsh (Burdick et al., 1989; Table 9-1). Usually, the lower edge of salt marshes, whether along a tidal creek or bay, are drained more completely than inland zones. Compared to the lower edge, soil of the inland zone (upper slope) has low redox potentials or Eh, that is a higher reducing potential with increasing depth, high pH, free sulfides and high pore-water salinity. The data from both studies show that the effects of increased waterlogging and salt content are associated with a change in species dominance. The physiological features of *Spartina patens* in the Louisiana salt marsh included the leaf AEC (adenylate energy-charge) ratio, a measure of cellular energy, and root ADH (alcohol

TABLE 9-1. Edaphic Factors in Louisiana and North Florida Salt Marshes and Responses of *Spartina* spp. to Waterlogging in Louisiana[a]

Soil Factors	Marsh Zones			
	Edge (Lower Border)	Berm (Lower Slope)	Inland (Upper Slope)	Upland Forest
Bayou Faleau, Louisiana (15-cm core)				
Moisture (%)	62.5	65.8	74.0	—
Salinity (ppt)	11.2	12.0	13.4	—
Eh (mV)[b]	−9.0	−150	−192	—
pH	6.86	7.08	7.16	—
NH_4^{+2} (mg L^{-1})	1.69	3.9	2.33	—
Free sulfides	0.6	5.4	24.8	—
Species Responses	*S. alt.*	*S. patens*	*S. alt.*	—
AEC ratio	0.705	0.751	0.673	—
ADH (μM gdwt^{-1})	91	96	210	—
Apalachecola, Florida (7.5-cm core)				
Dominant Species	*S. alt.*	*J. romer.*	*S. patens*	Pine, palms
Moisture (%)	66.8	46.8	18.0	10.2
Salinity (ppt)	2.8	2.7	6.8	0.1
pH	5.6	5.3	8.6	4.3

Sources: For Louisiana, Burdick et al. (1989); for North Florida, Kurtz and Wagner (1957).
[a]*S. alt.* = *Spartina alterniflora*, *S. patens* = *Spartina patens*, *J. romer* = *Juncus roemerianus*, AEC = adenylate energy-charge ratio, and ADH = alcohol dehydrogenase activity.
[b]Samples taken at 15 cm.

dehydrogenase) activity, a measure of potential anaerobic activity. The AEC ratio was higher at the berm, where stress was lowest, and, by contrast, root ADH activity was high in the inland zone, where reducing potential and sulfide levels were also high. Thus, *S. patens* showed physiological responses to differing edaphic environments.

Elevation is critical in controlling the period of submersion at high tide, as well as rates of drainage and length of exposure to air temperature (Zedler, 1977). Soil drainage was found to be critical for *Spartina alterniflora* distribution in North Carolina (Mendelssohn and Seneca, 1980), with poor drainage leading to waterlogging and reduced oxygen availability. The diffusion rate of oxygen into waterlogged soils is limited, so that both inorganic and organic substances occur in reduced states, as seen by the low redox potential of salt marsh substrata (Burdick et al., 1989; Table 9-1). The health and vigor of salt marsh plants reflects, in part, the level of oxygen diffusion from the erect stems through the rhizome and into the roots in waterlogged, anaerobic sediments. For example, an increase in subsurface drainage resulted in a significant increase in culm height of *S. alterniflora* (Wiegert et al., 1983). Planting the same species into waterlogged areas caused a decline in growth with sulfide toxicity and extended periods of anaerobic metabolism in the roots being major factors (Mendelssohn and McKee, 1988). Similarly, the distribution of the border shrub *Iva frutescens* is controlled by elevation and degree of waterlogging (Bertness et al., 1992); marsh elders occur only where they are not subjected to prolonged flooding. In another study, sedimentation, soil salinity, and period of tidal flooding were the prime factors determining the zonation of *Spartina* and *Salicornia* populations in a salt marsh in San Francisco Bay (Mahall and Park, 1976).

Salt marsh root systems have active phosphorus pumps, transferring phosphates produced through the phosphorus cycle in the substrata to leaves and stems (Reimold, 1972). The availability of nutrients is also an important factor in tidal marsh development, as shown in a Louisiana study of *S. patens* (Burdick et al., 1989) as well as New England salt marshes (Levine and Bertness, 1996). The latter study used tiller number and biomass to demonstrate changes in dominance in pairwise manipulations of *S. alterniflora*, *S. patens*, *Juncus gerardii*, and *Distichlis spicata*. The study showed that under "typical" nutrient levels, competitive dominance results from competition for nutrients. However, when nutrients are in excess, competition for light controls the dominant taxon. Thus, the competitive hierarchy will change between the preceding four species depending on nutrient level and shading effects.

Zonation

The striking zonation of plants in coastal tidelands is generated by different, but frequently overlapping, vertical ranges of individuals (Gray, 1992). Zonation is most visible with regard to the marsh vegetation, but also exists for the microalgal, seaweed, and faunal components (Adam, 1990). In most zonation studies of tidal marshes, research has focused on physical rather than biotic factors (Bertness and Ellison, 1987). However, the actual mechanisms controlling zonation patterns in salt marshes are not well understood. Usually, there are more discrete clustering patterns of species in the upper marsh regions with species richness increasing as one approaches the terrestrial communities, whereas the lower regions, which have fewer tolerant species, show weaker boundaries. This distribution supports the concept that competition is highest in the "mildest" habitats and lowest in harsher ones (Bertness, 1991).

The number of recognized plant zones within salt marshes varies according to latitude, physical factors, and the criteria used for identifying zones. For example, four vegetational zones were described in North Carolina (Wells, 1928), eight in northwest Florida (Kurtz and Wagner, 1957), two on the east coast of Florida (Provost, 1973), and three for all of the Gulf of Mexico (Stout, 1984). Detailed examples of plant zonation have been presented for salt marshes throughout the world (Chapman, 1977).

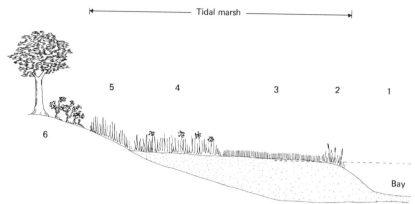

Figure 9-11. Profile of a New England salt marsh. The zones are as follows: (1) exposed intertidal mud (blue-green algae, green filamentous algae); (2) lower border of *Spartina alterniflora*; (3) lower slope of *S. patens*; (4) upper border of succulent halophytes (*Salicornia, Aster, Limonium*), *Distichlis spicata*, and *S. patens*; (5) upper slope of *Juncus gerardii* and *Iva fructescens*; and (6) upland scrub and forest (after Miller and Egler, 1950).

Two examples of plant distribution in salt marshes are presented here to demonstrate latitudinal differences, one from New England on the east coast of North America (Fig. 9-11; Niering and Warren, 1980) and the other from northern Florida in the Gulf of Mexico (Fig. 9-12; Kurtz and Wagner, 1957). Both intertidal marshes have similar zones, the lowest area barren of vegetation, a low border of *Spartina alterniflora*, distinct ecotones in the lower and upper slopes, and an upper border adjacent to the upland schrub and forest. Both marshes have their lower intertidal bands free of angiosperms, being composed of mud or sand and colonized by edaphic algae (blue-green, green, diatoms). *Spartina alterniflora* usually forms a lower border, and *S. patens* will dominate the lower slope in the higher-latitude marshes. The large tidal range (2 to 3 m) is thought to be a controlling factor (Provost, 1976) that supports an extensive lower border and slope (Nixon and Oviatt, 1973) and lower barren zone. By contrast, the limited tidal range (0.5 to 1 m) in the northern Gulf of Mexico (Kurtz and Wagner, 1957) can result in a narrow lower band of *S. alterniflora*. Further, the black rush *Juncus roemerianus* is the dominant meadow plant in the lower slope (Dawes, 1974), in contrast to that described for Louisiana (Burdick, 1989). In New England, the upper slope is usually dominated by *J. gerardii* or *S. patens*, whereas in northern Florida, a combination of *Distichlis spicata* (spike grass) and *S. patens* occurs. The upper borders of both salt marshes usually will be vegetated by *Salicornia* sp. as well as halophytic herbs or woody perennial bushes (*Iva imbricata* and *Fimbristylis castanea* in Florida; *I. fructescens* and *Solidago sempervirens* in New England).

A further feature in many salt marshes where elevation is gradual, evaporation is high, and rain is limited is the occurrence of *salt barrens*, or *salterns*

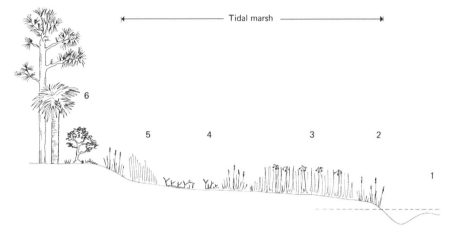

Figure 9-12. Profile of a black rush marsh in Florida. The zones are as follows: (1) and (2) as in Fig. 9-11; (3) lower slope of *Juncus roemerianus*; (4) upper slope of *Spartina patens*; (5) upper border of a mix of shrubs and grasses (*Iva frutescens, Fimbristylus castana, Distichlis spicata*); and (6) upland forest of pine, palm, and palmetto. The regions between zones 4 and 5 may also contain halophytes (*Salicornia, Batis, Sesuvium*) and salterns (see Fig. 9-13; after Kurtz and Wagner, 1957).

[Fig. 9-13(A)], between upland communities and salt marshes. Salterns are flat expanses of firm sediment with fringing halophytes, including species of *Batis, Sesuvium,* and *Salicornia* [Fig. 9-13(B)]. The central exposed sediment contains blue-green algae and has a pink hue due to sulfur bacteria. Sediment salinities in mangrove and salt marsh salterns of Florida are high during periods without rainfall, ranging from 40 to 170 ppt (Hoffman and Dawes, 1996). Hypersalin-

Figure 9-13. Salterns. The aerial photo (A) of a north Florida salt marsh shows the fringing salterns (arrows) that lie between the upper slope and upper border that fringe the upland forest (Hoffman and Dawes, 1996). Salterns barrens are fringed by *Salicornia* and *Juncus roemerianus* (B).

ity occurs as seawater evaporates, after spring tides flood the upper regions of the marsh. Elevational surveys show the salterns of Florida do not result from salt-water trapped by depressions in the high intertidal region but are due to accelerated rates of evaporation (Hoffman and Dawes, 1996). The effect of rain is obvious, with sediment salinities falling to 20 ppt after extensive precipitation. Three ecological roles have been suggested for salterns: storage and release of salts and orthophosphates (Ridd et al., 1988), transitional sites for plant succession (Adams, 1963), and as habitat for fish and wildlife due to the abundance of brine shrimp and open areas for wading birds (State of Florida, 1986).

Salt Marsh Dynamics

Earlier studies of salt marshes proposed that elevation controlled the abiotic factors, which influenced zonation of ecotones (e.g., Yapp et al., 1917). However, the zones occupied by species will widen when diversity decreases, as seen in natural and manipulated (e.g., removal) communities (Gray, 1992). It appears that the ecological niche of many marsh species reflects both abiotic (e.g., flooding) as well as biotic (*facilitation* and *competition*) factors. For example, *Spartina alterniflora* dominates the low marshes of New England and is excluded from the high marsh by the dense turf of *S. patens* and *Juncus gerardii* (Bertness and Ellison, 1987). In a study in the Netherlands, Scholten and Rozema (1990) removed either *Spartina* or *Puccinellia* from parts of two adjacent plots that differed in elevation by 4 cm and a 15-to-20-min difference in tidal submergence. Removal of *Spartina* in the plot at the lower elevation allowed *Puccinellia* to significantly increase its biomass, and the same happened to the former when the latter species was removed. The authors concluded that there is competitive interaction between these two species over an elevation difference of 4 cm. A three-species manipulative experiment using *Puccinellia maritima*, *Festuca rubra*, and *Agrostis stolonifera* found that competitive interaction between the first two was strongly affected by waterlogging (Gray and Scott, 1977). In contrast, *A. stolonifera* was strongly competitive, but could not tolerate higher salinities; hence, it was restricted to higher elevations. These studies illustrate the importance of abiotic factors, including waterlogging and salinity, as well as the biotic factor of competition in zonation salt marsh plants.

The distribution of salt marsh plants also can be due to positive interactions or facilitation (Bertness and Shumway, 1993). Artifical and natural bare patches in the *Juncus gerardii* zone become hypersaline due to evaporation and prevent recolonization of that species. However, *Spartina patens* and *Distichlis spicata* will grow into these bare areas and reduce evaporation via shading, which results in a decline in soil salinity. The reduction in salinity facilitates *J. gerardii* to recover the patches by shading out the other species (Bertness, 1991).

The idea of *succession* in tidal marshes has invoked considerable discussion (Chapman, 1977; Adam, 1990; Gray, 1992). For example, Niering (1990) prefers the terms "biotic change" to succession and "relative stability" to climax. A. R. O. Chapman (1974), V. J. Chapman (1977), as well as others, view salt marshes as serial or subclimax stages that followed zonation based on elevation in which ecotones replace those found in the zone immediately below them in a succession toward a terrestrial climax community. Such scenarios are evident in areas with high rates of sedimentation or where sea level is dropping. Salt marshes then can be viewed as a continuum from submerged aquatic to terrestrial communities. In this respect, Gray (1992) proposes that abiotic factors act along elevation gradients, while competitive interactions are involved at zone boundaries. Others consider tidal marshes as relatively stable, complex zones of vegetation that show a predictable sequence in their recovery after disturbances (Dawes, 1981). The view of salt marshes as communities in equilibrium is supported by their ability to maintain stable zonation patterns during periods of rising sea levels (Redfield, 1965; Niering, 1990) provided rates of sedimentation are sufficient. A projected rise in sea level by the year 2100 is estimated to be 31 to 110 cm versus a slower rate (ca 100 cm) worldwide over the past 1000 years (Allen and Pye, 1992; Chap. 5). Changes in rates of sea-level rise will influence salt marsh structure. For example, salt marshes along the New England coast were established about 3000 to 4000 ybp when the postglacial sea-level rise slowed to 1 mm y^{-1} (Redfield, 1965). In the past 50 years, eustatic sealevel has increased to 2.5 mm y^{-1} and documented changes in the vegetation of New England salt marsh has been toward a wetter, more open type dominated by stunted *S. alterniflora* and *Triglochin maritima* (Warren and Niering, 1993).

Studies of *disturbances* (e.g., storms, fires, ice rafting) in salt marshes indicate that there is a sequence of species replacements during recovery. A common disturbance is the deposition of drift vegetation (wrack mats) that often remain for three or more months on salt marshes (Valiela and Rietsma, 1995). The level of dieback due to the wrack does not affect species richness, but may require facilitated recovery (Bertness, 1991). Another disturbance, ice rafting, is well known in northern latitudes (Hardwick-Witman, 1984; Chap. 4). That is, large pieces of marsh vegetation, including rhizome and sediment, can be removed and floated to new estuarine sites during spring thaws (Fig. 9-14). Patches of salt marsh then settle in the midst of mudflats, sometimes surviving and spreading, but more commonly degrading over time. Such island-type patches could prove interesting to study in terms of size-faunal relationships. In contrast, the pools produced in the salt marsh from ice rafting may last for decades due to lack of sediment, but gradually return to the surrounding community structure.

Ecological Roles

The consideration of salt marshes as ecosystems is recent, and few multifaceted studies are available similar to those conducted at Sapelo Island, Georgia (Pomeroy and Wiegert, 1981). The importance of salt marsh ecosystems to the coastal environment can be presented as five ecological roles, including primary production, food sources, habitats, stabilization of sediments, and filtration.

Figure 9-14. The effect of ice rafting on salt marshes. Clumps of frozen salt marsh near Bath, Maine, were ripped out and floated to new sites (arrows) during winter thaws. This is a common disturbance to northern latitude intertidal communities.

1. Perhaps the most significant contribution of salt marshes is their *primary production*, which includes the production of salt marsh biomass, the release of dissolved organic carbon (Gallagher et al., 1976), the production of peat and detritus (Squires and Good, 1974), and algal productivity (Zedler, 1980; Pomeroy et al., 1981). Most productivity estimates consider the above-ground component, particularly that of *Spartina* marshes (Squires and Good, 1974; Smith et al., 1979; Howes et al., 1986). Long and Mason (1983) found that below-ground biomass exceeded the above-ground portion, as did Smith et al. (1979), who presented a flow sheet of organic production and constituents (Fig. 9-15). In Smith's study, the highest level of biomass had a below-ground contribution of 89% (2.8 kg m^{-2}). In another study of *Spartina* marshes, Squires and Good (1974) estimated that total production was 1.5 kg C m^{-2} y^{-1} in New Jersey, and in Oregon, Eilers (1979) reported above-ground production from 0.2 (winter) to 2.8 (summer) kg C m^{-2} y^{-1}. There appears to be a strong relationship between latitude and macrophytic production, with the higher productivity occurring in lower-latitude salt marshes (Turner, 1976).

2. Salt marshes serve in a limited capacity as a *direct food source* for a wide variety of animals (cf. Chapman, 1977; Adam, 1990, for detailed reviews). The dominant herbivores in salt marshes are sap-sucking and leaf-chewing insects (Pfeiffer and Weigert, 1981). There is a low level of herbivory, presumably due to the poor nutritional status of such vegetation when compared to terrestrial meadows. Other important grazers include wildfowl (geese, ducks), crabs,

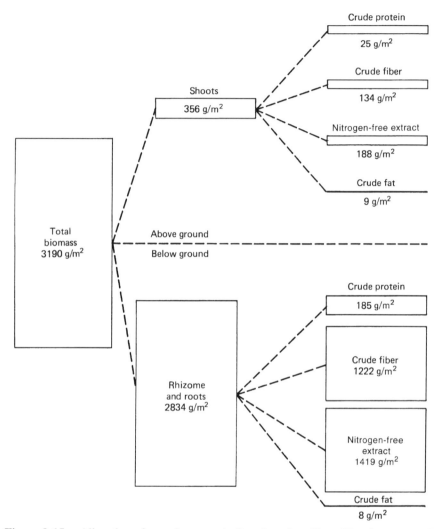

Figure 9-15. Allocation of organic matter in *Spartina alterniflora*. The above-ground shoots contain about 11% of the total biomass as compared to the below-ground rhizome and roots, which contain the remaining 89% (after Smith et al., 1979).

and occasional mammalian herbivores (e.g., muskrats, rabbits, and deer; Daiber, 1977; Nixon, 1982).

More common than consumption as direct food, salt marsh plants become part of the *detrital food chain* as they decompose. Bacteria and fungi colonize fragmented plant material and these packets of organic material become available for detrital feeders (protozoa, snails, worms, clams, fish, insect larvae, and an assortment of crustaceans). Cell-wall components (i.e., cellulose, lignin) provide little nutrition to the animals unless these compounds are digested by microbes (Pomeroy and Weigert, 1981: Long and Mason, 1983). Initial rates of decomposition by bacteria and fungi are rapid and sensitive to temperature and moisture, with the more easily metabolized cellular constituents being initially metabolized (Buth and de Wolf, 1985). Often overlooked, seaweeds can account for a large component of salt marsh detrital food chains, because algae decompose more quickly than *Spartina* (Tenor and Hanson, 1980).

3. Salt marshes serve as *habitats* for animals with marine, freshwater, and terrestrial affinities (Adam, 1990). Detailed faunal studies have been published for the Wadden Sea (Dijkema, 1984), northern European coasts (Smit et al., 1981), and North America (Daiber, 1982, 1986). Marsh fauna in the United States are also reviewed in a series of community profiles for San Francisco Bay (Josselyn, 1983), southern California (Zedler, 1982), New England (Nixon, 1982; Teal, 1986), southeastern United States (Wiegert and Freeman, 1990), and the Gulf of Mexico (Gosselink, 1984; Stout, 1984). The linkages between salt marshes and fisheries or rookeries are complex (Nixon, 1982; Herke et al., 1992); a number of studies indicate an essential link between the two (Kutkuhn, 1966; Turner, 1982; Boesch and Turner, 1984). Further, the role of salt marshes as nurseries and support for offshore communities need further study similar to that carried out in Georgia (cf. Wiegert and Pomeroy, 1981) and Louisiana (Herke et al., 1992).

4. *Stabilization* of coastal sediments is an important role of salt marsh plants via their rhizomatous growth and shallow root systems (van Eerdt, 1985). Thus, salt marsh vegetation traps sediment, resulting in vertical and horizontal accretion if changes in sea level are moderate and sediment is available. Marsh expansions of 50 m y^{-1} have been recorded in Great Britain. If increases in sea-level flooding are too rapid, rates of sedimentation too slow, or water movement too high, then salt marshes will retreat. One dramatic example of marsh loss is occurring in Louisiana at a rate of 65.6 km^{-2} y^{-1} due to subsidence and reduced input of sediment. The channelization of the Mississippi River has limited its flood waters from supplying alluvial sediment to the marshes, so that erosion and subsidence is more rapid then sedimentation (Webb et al., 1995).

5. The *filtration* of coastal runoff and removal of organic waste by marshes (Teal and Valiela, 1973) lowers the sediment and nutrient loading to adjacent coastal waters (van Eerdt, 1985). Examples of this important role can be seen in estuaries where salt marshes have been removed through dredging and development (e.g., Lewis and Estevez, 1988). The adjacent coastal waters typically show higher levels of fine sediment and increased plankton, which reduce water transparency and adversely impact water quality and the survival of benthic communities (e.g., seagrass beds).

Field Methods

Appendix A describes the use of quadrats and line transects for field studies of salt marsh macrophytes. Above-ground stem densities can be determined used standard quadrat measurements. Studies in plant zonation can use line transects and leveling devices to determine horizontal and vertical distribution. Above- and below-ground biomass can be determined using a coring device made out of a 15.2-cm PVC pipe that has a beveled edge with teeth in order to cut through the dense mat of roots and rhizomes.

ANTHROPOGENIC CONSIDERATIONS

Uses and Impacts

Historically, salt marshes have been manipulated and used for a variety of purposes (Daiber, 1986), including farming, strip or open coast mining (sediment removal), salt and chemical (potash, minerals) production, land reclamation, recreation, industrial and urban sites, scientific studies, insect control, wildlife management, waste disposal, and marsh rehabilitation (Daiber, 1986; Adam, 1990). Because salt marshes occur in estuaries that are usually populated, pollution (industrial, agricultural, domestic) is common. Nutrification and the introduction of heavy metals and toxic compounds usually results in their being sequestered by plants and sediments (Chap. 5). Most metals appear to be precipitated as sulfides or organically complexed in the pore water (Giblin et al., 1986). However, the compounds may be reintroduced into the environment through plant herbivory and decomposition, as well as oxidation of sediments.

The most common use of salt marshes is agricultural, including grazing by livestock (sheep, cattle), harvesting (for hay, mulch, and thatch), turf cutting (for building, peat), and farmland after diking and draining. Cattle and sheep have been grazed on salt marshes for centuries in Great Britain (Adam, 1990), so that the lower marsh shows a loss in floral diversity and direct damage by trampling. In contrast, diversity may increase in the mid- to upper-marsh zones due to grazing through removal of highly competitive dominant species. When grazing is reduced in the upper marsh, litter increases, but dominance is usually limited to a few species, which results in a "coarse grained vegetation mosaic" (Bakker, 1985). It has been the custom to burn off the old vegetation at the end of the growing season to improve muskrat habitat (Louisiana), eliminate nonedible forage such as the succulent chenopod *Sclerostegia arbuscula* (Tasmania), and to promote new growth for cattle (Queensland) and waterfowl (Texas).

Loss of freshwater and sediment via channelization, dams, and agriculture causes increased salinities and sediment starvation in marshes downstream. For example, the building of levees along the Mississippi River for flood control has contributed to a decline in Louisiana salt marshes (Stevenson et al., 1988), resulting in a severe reduction in sediment load (380×10^6 tons in the 1900s vs. 240×10^6 tons in the 1950s). The reduction in sediment in the face of a rising sea level ultimately can cause the loss of a marsh (Chap. 5). For example, a study of Louisiana salt marshes indicated that higher salinities and longer periods of submergence would decrease recruitment of seed bank plants and thus reduce their abundance (Baldwin et al., 1996). Removal of tidal marshes and use of storm water outfalls also reduce the filtration of surface runoff and result in an increase in fine sediments and introduction of pollutants (agricultural, industrial, road wastes) to estuarine waters. Industrial uses of estuarine water also can result in thermal pollution if the water temperature exceeds the tolerance of salt marsh vegetation (Chap. 5). All of these occurrences will impact a "sensitive" system such as salt marshes, where environmental stress is always present due to wide fluctuations of temperature, salinity, and oxygen levels.

Probably the most significant present-day destruction of salt marsh estuaries is urban expansion (construction, dredging, filling) due to their use as harbors and ports. In the past, the primary impact on marshes has been the reclamation (diking, draining) of estuaries for agriculture and human development as well as disposal areas for waste. The percent of salt marsh loss is unknown and perhaps has reached 60% worldwide (Daiber, 1986). The loss of marsh habitat impacts not only the immediate estuary, but also other marine and coastal ecosystems because of the significant role estuaries have as nurseries and sources of primary production for far-ranging fish and birds.

The invasion of salt marshes by *exotic* species is well known and is becoming a significant problem (Chap. 5). One example is *Spartina anglica*, which has altered salt marsh ecology in Europe (Gray et al., 1991). The species appears to be a polyploid of *S. townsendii*, which is a sterile hybrid between *S. maritima* and *S. alterniflora* (Guenegou et al., 1988). It is thought that *S. townsendii* gave rise to *S. anglica* through doubling of chromosomes (Gray et al., 1991). Being an aggressive plant, *S. anglica* was planted to reclaim and stabilize lower inter-tidal mudflats, particularly in France. However, it has outcompeted the native lower marsh *Salicornia* spp. as well as replaced the seagrass *Zostera noltii* on mudflats. Unfortunately both native species are important sources of food for a variety of migratory birds (Gray et al., 1991). Another example of a salt marsh exotic is *S. alterniflora*, which was introduced to the Pacific Coast of North America and now dominates some of the salt marshes (Callaway and Josselyn, 1992).

Management and Restoration

Until the early 1950s, tidal wetlands were treated as a "commons" in Great Britain, where each person could pursue his or her own best interest (Daiber, 1986). The evolution of a *Public Trust Doctrine* (*PTD*) in Britain and legislative protection in the United States have come about because it became evident that management and protection of tidal wetlands were needed. This is due to continued expansion of human population and awareness of the importance of estuarine habitats. Some problems with enforcement of a PTD are the rights and liabilities involved in alteration of marshes, boundaries of private ownership, and the jurisdiction of regulators. The outcome in the 1980s has been legal conflicts between regulatory agencies and private owners regarding use and alteration of tidal wetlands (Daiber, 1986). For example, the questions of where the intertidal (beach) begins and whether the public should have access through private land to public estuaries has resulted in legislation and lawsuits. In the United States, the result is that much of the eastern coast from Maine to Florida is inaccessible due to lack of public rights-of-way.

Conservation and preservation of estuaries have become international concerns, resulting in detailed *management* plans to restrict inflow of pollutants, stop dredge and filling operations, and restore coastal wetlands. Establishment of salt marsh preserves must take into account their dominant physical processes, primary functions, and cultural values. Primary functions of marshes include their ecological roles and hydrologic functions, and cultural values include socioeconomics, commercial and recreational fisheries, recreation, and aesthetics (Daiber, 1986). It is evident that not only the tidal marsh itself but also the entire upland drainage basin must be monitored and protected if the estuarine ecosystems can avoid pollutants or serious changes in freshwater input that may lead to instability. Detailed reviews have been performed in the United States for most of the larger estuaries and their drainage basins, including Tampa Bay (Lewis and Estevez, 1988), Chesapeake Bay (U.S. Army Corps Engineers, 1977), Great Bay (Short, 1992), and San Francisco Bay (Kockelman et al., 1982; Josselyn, 1983). The management of marshes through the establishment of wetland preserves, restoration, mitigation, and creation of marsh land (Chap. 5) is becoming a major effort throughout the world (Lewis, 1982). Although it is costly to restore many salt marshes, restoration activities and increased public awareness are resulting in a return to higher productivity of adjacent coastal waters (cf. Daiber, 1986).

Tampa Bay, a 1030-km^2 open-water estuary on the west coast of Florida, can be used as an example (Lewis and Estevez, 1988). By 1985, 44% (3972 ha) of the coastal wetlands (mangrove and salt marsh) had been removed through dredge and fill and 80% of the submerged seagrass community had been lost (Chap. 11). The loss of habitat coincided with the collapse of the shrimp and fin fish industry in the 1960s. The dumping of primary and secondary treated domestic waste and direct runoff from urban and agricultural areas led to massive macroalgal blooms (*Ulva, Gracilaria, Hypnea*). The algae would decay, resulting in decreased levels of dissolved oxygen (BOD = biological oxygen demand), which resulted in summer fish kills in the upper parts of Tampa Bay. Enforcement of new water-quality guidelines for Tampa Bay began to take effect by the mid-1980s after tertiary treatment of sewage was begun by the city of Tampa (Johansson, 1991). Existing coastal wetlands were placed under protection and restoration of tidal marshes began by the late 1970s. By 1995, there was a 10% increase in salt marsh and mangrove communities over the 1980 estimate, and water quality (chlorophyll *a* level, turbidity, nutrient load) continued to improve (Johansson, 1991). Fish populations have begun to increase, although shellfish (scallops, shrimp) have only made limited gains. Through the involvement of the Tampa Bay National Estuarine Program, management of the bay has become a partnership of regulatory agencies, conservation groups, municipalities, and private owners.

CHAPTER 10

Mangals

INTRODUCTION

Mangrove forests form impressive intertidal communities on protected, tropical to subtropical coasts and atolls. Being composed of diverse trees and shrubs (MacNae, 1968), some mangroves grow to 30 m (Fig. 10-1). Mangroves were described in the 325 B.C. Chronicle of Nearchus and reported by Du Terte in 1967 as dangerous due to "wild boars and savage beasts that live in them" (see Lugo and Snedaker, 1974). The term *mangrove* is derived from a combination of the Portuguese word for tree (*mangue*) and the English word for a stand of trees (grove). This chapter describes both the mangroves and associated tidal marsh communities, namely, *mangales* (Walsh, 1974), or *mangals* (Tomlinson, 1986). Reviews of Caribbean mangals are given by Lugo and Snedaker (1974) and Walsh (1974), and the more diverse Asian-Pacific mangals are reviewed in several texts (Clough, 1982; Hutchings and Saenger, 1987; Asia-Pacific Symposium on Mangrove Ecosystems, 1995). Global summaries include a bibliography (Rollet, 1981), an introduction to mangroves (Rutzler and Feller, 1996), edited texts on mangrove ecology (Chapman, 1976; Robertson and Alongi, 1992), and a botanical review (Tomlinson, 1986). Research methods for studying mangrove ecosystems are also available (Snedaker and Snedaker, 1984).

Mangrove forests have been classified (Lugo and Snedaker, 1974; Cintròn et al., 1985) as coastal fringes, overwash islands, riverine, basin, hammock, and dwarf communities (Fig. 10-2) with a modified version by Twilley (1995). *Fringing mangals* occur along protected shoreline berms [Fig. 10-2(A)], while *overwash mangals* [Fig. 10-2(B)] are low, intertidal islands. *Riverine mangals* [Fig. 10-2(C)] occur along rivers and streams, often extending several miles inland. The trees are usually largest in riverine forests due to the availability of freshwater, nutrients, and sediments. *Basin mangals* [Fig. 10-2(D)] occur in depressions or basins behind a berm or fringing mangals; they are connected to freshwater streams and coastal waters via tidal creeks. Being inland, basin mangals tend to be smaller and have a more limited flora due to fluctuations

Figure 10-1. A Caribbean mangal. The pneumatophores of the black mangrove, or *Avicennia germinans*, form a dense covering inside the fringing prop roots of the red mangroves, or *Rhizophora mangle*.

in salinity and prolonged periods of flooding. *Dwarf*, or *scrub*, *mangrove* communities [Fig. 10-2(E)] occur where abiotic conditions are severe due to limited exchange of water. Typically, poor water exchange will result in low nutrients, increased soil salinities (due to evaporation), or waterlogging. The stunted physiognomy and leaf selerophylly of dwarf mongroves appears to reflect nutrient, especially phosphorus, limitations (see "Anatomical Adaptations"; Feller, 1995). *Hammock* forests [Fig. 10-2(F)] occur in inland tropical wetlands and are isolated by freshwater. In subtropical areas, hammock mangals may be replaced by salt marshes due to low temperatures or frost (Lugo and Snedaker, 1974).

The relationship between the six functional types of mangrove forests is demonstrated in Figure 10-2(G), with the three extremes being *river-dominated*, *tide-dominated*, and *interior mangals* (Woodroffe, 1992). River-dominated habitats are forests with an outwelling of nutrients, sediment, and organic matter; tide-dominated communities have bidirectional fluxes; and interior habitats are sinks where organic and inorganic materials are deposited (Woodroffe, 1992). The triangle demonstrates a close continuum among riverine, fringing, and overwash mangals, the more "central" relationship of basin mangals, the less related scrub, and isolated hammock habitats with regard to fluxes of inorganic and inorganic materials. In addition, oceanic mangals growing on carbonate platforms can be considered a variety of overwash mangles (Woodroffe, 1987; Twilley, 1995).

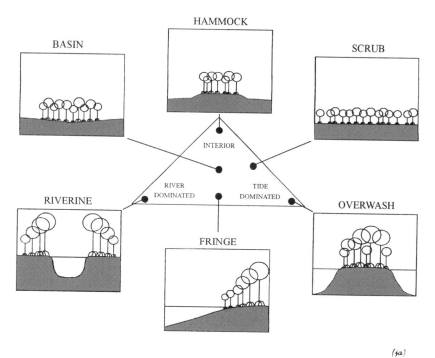

Figure 10-2. Six types of mangals and their relationship to one another based on hydrodynamic conditions (modified from Lugo and Snedaker, 1974; Clintròn et al., 1985; Woodroffe, 1992).

ADAPTATIONS OF MANGROVES

Mangrove forests occur in sheltered tropical depositional and saline environments, with the plants being exposed to salt water, causing an expenditure of energy to conserve water. Hence, mangroves and associated vascular plants have *xeric* and *halophytic* adaptations similar to salt marsh plants, including growth in aerobic, saline, and frequently waterlogged sediment. Warming (1883) listed five adaptive characteristics of mangroves:

1. The trees show mechanical adaptations for attachment in soft or loose substrata.
2. Aerial roots are common and show specialization for diffusion of gases to subterranean portions.
3. Many species have evolved *vivipary* or germination of seedlings while the fruit remains attached to the tree.
4. The seeds and seedlings can survive in salt water and utilize seawater as a means of dispersal.
5. The trees exhibit xerophytic and halophytic modifications.

Taxonomy

The diversity of taxa and families that contain mangroves indicates they are an ecological rather than natural group (Duke, 1992). The taxonomic diversity also suggests that there has been frequent and independent adaptation of flowering plants to tropical tidal habitats. The total number of trees and shrubs that authors consider as mangroves varies according to the criteria employed. Waisel (1972) lists 12 dicot genera, whereas Walsh (1974) includes 15 plus the palm *Nypa fruticans*. Dawes (1981) lists 23 genera, Tomlinson (1986) includes 9 major and 11 minor genera, and Duke (1992) lists 26 genera. According to Tomlinson, the major genera have all or most of the following features:

1. Species are restricted to mangals.
2. Trees exhibit a major role in community structure.
3. Plants show morphological specializations, including aerial roots and vivipary.
4. Plants exhibit salt-exclusion physiology.
5. Taxonomic isolation from terrestrial relatives occurs at least at the generic level.

The orders Myrtales and Rhizophorales contain about 25% of the families and 50% of the species (Duke, 1992). Table 10-1 lists the species composition and distribution of these tidal forests (Duke, 1992). Diagnostic vegetative keys for major New and Old World mangroves are given in Table 10-2 according to Tomlinson (1986). Two Caribbean mangrove species are briefly described in what follows; descriptions of all major and minor mangrove genera can be found in Tomlinson (1986).

Rhizophora mangle L. (red mangrove; Fig. 10-3) is one of eight species (including three putative hybrids) in this pantropical genus of evergreen trees that can reach 30 m tall. The family Rhizophoraceae contains 16 genera of which four (*Rhizophora, Kandelia, Ceriops, Bruguiera*) contain exclusively mangrove species that have stilt ("drop" and "prop" roots; Fig. 10-1), paired and interlocked stipules, and opposite, leathery leaves [Fig. 10-3(A)]. The tap root is short-lived, with the seedling producing adventitious "stilt" roots that develop from the stem (prop) and branches (drop). The common name is associated with the red hue of the bark of young roots and stems and in the wood. A young branch will show an annular scar from paired stipules, with the petiolar scar just below it [Fig. 10-3(B)]. It is possible to age red mangrove seedlings using these leaf scar nodes (Duke and Pinzon, 1992). The apical bud is elongated and recurved. The leaves (10 to 15 cm by 6 cm) are simple and elliptical to obvate with an entire, slightly recurved margin. One or more perfect flowers are formed in the axis of the leaves; the four fleshy, green sepals are fused into a calyx [Fig. 10-3(C)] that persists on the fruit after dehiscence of the four delicate, white petals [Fig. 10-3(D)], usually, flowers will have eight sessile

TABLE 10-1. A Classification of Mangrove Genera by Order, Family, and Distribution

Order/Family	Genera	No. of Species	New World	Old World
Plumbaginales				
Plumbaginaceae	*Aegialitis*	2		×
Theales				
Pellicieraceae	*Pelliceria*	2		×
Malvales				
Bombacaceae	*Camptostemon*	2		×
Sterculiaceae	*Heritiera*	3		×
Ebenales				
Ebenaceae	*Diospyros*	1		×
Primulales				
Myrsinaceae	*Aegiceras*	2		×
Fabales				
Caesalpiniaceae	*Cynometra*	1		×
	Mora	1		×
Myrtales				
Combretaceae	*Conocarpus*	1	×	
	Laguncularia	1	×	
	Lumnitzera	3		×
Lythraceae	*Pemphis*	1		×
Mytraceae	*Osbornia*	1		×
Sommeratiaceae	*Sonneratia*	9[a]		×
Rhizophorales				
Rhizophoraceae	*Bruguiera*	5		×
	Ceriops	3		×
	Kandelia	1		×
	Rhizophora	9[a]	× (3)	× (6)
Euphorbiales				
Euphorbiaceae	*Excoecaria*	2		×
Sapindales				
Meliaceae	*Aglaia*	1		×
	Xylocarpus	2		×
Lamiales				
Avicenniaceae	*Avicennia*	8	× (4)[b]	× (5)
Scrophulariales				
Acanthaceae	*Acanthus*	2		×
Bignoniaceae	*Dolichandrone*	1		×
Rubiales				
Rubiaceae	*Scyphiphora*	1		×
Acrecales				
Palmae	*Nypa*	1	×[b]	×
Totals	26 genera	66 species	12 species	58 species

Sources: After Dawes (1981), Tomlinson (1986); and Duke (1992).

[a]Three species of *Sonneratia* and *Rhizophora* are possible hybrids.

[b]*Avicennia marina* and *Nypa fruticans* were introduced to the New World; the latter was present during the Tertiary Period (Graham, 1995).

TABLE 10-2. A Dichotomous Key to the Major Mangrove Genera

1 Species found in both the Old World and New World.................	2
2 Species restricted to either the Old World or New World...............	3
2 Stipules overwrap and leave an annular scar above petiole insertion, stilt roots may be present, no pneumatophors, viviparous...........	*Rhizophora*
2 Stipules absent, no stilt roots, cable roots produce pneumatophores, not conspicuously viviparous......................................	*Avicennia*
3 Species restricted to New World......................................	4
3 Species restricted to Old World......................................	5
4 Leaves alternate, aerial roots absent, trunk base highly ribbed.......	*Pelliceria*
4 Leaves opposite, small knee-like (pointed) pneumatophores develop from cable roots...	*Laguncularia*
5 Leaves compound, two to three pairs of leaflets......................	*Xylocarpus*
5 Leaves simple...	6
6 Leaves opposite...	7
6 Leaves alternate..	12
7 Stipules present	8
7 Stipules absent...	11
8 Stipules overwrapping, producing an annual, nodal scar (Rhizophoraceae)..	9
8 Stipules not overwrapping, persistent..............................	*Scyphiphora*
9 Trees usually lacking pneumatophores, but may have small, inconspicous, erect surface roots....................................	*Kandelia*
9 Trees with horizontal, looping roots becoming rounded or knobby with age	10
10 Leaves 10+ cm long with acute apex, terminal buds are not flattened..	*Bruguiera*
10 Leaves shorter than 10 cm with rounded tip, terminal buds flattened..	*Ceriops*
11 Leaves lack basal grove and not aromatic, a pair of nodal glands at base of petiole on stem..	*Sonneratia*
11 Leaves have basal grove and aromatic (clear dots), no nodal glands	*Osbornia*
12 Plants shrubs with swollen trunks, leaf bases encircle stem, leaving an annular scar	*Aegialitis*
12 Trees, no swollen base (but may have buttresses), leaves do not encircle the stem and leave a narrow scar	13
13 Stipules present, leaves 15 to 30 cm long, with white underside........	*Heritiera*
13 Stipules absent, leaves less than 15 cm long, underside not white.......	14
14 White latex from cuts, leaf margin notched........................	*Excoecaria*
14 No latex, leaf margin entire	15
15 Minute scales on lower leaf surface, petiole not showing gradual widening to blade...	*Camptostemon*
15 No scales on lower epidermis of leaf, normal widening of petiole at base of blade...	16
16 Leaves with hairs on both surfaces, less than 3 cm in size..........	*Pemphis*
16 Leaves waxy, lacking hairs, larger than 3 cm	17
17 Leaves ovate with glandular dots, fruits elongated, one-seeded	*Aegiceras*
17 Leaves obovate, fleshy, lacking glandular dots, fruits not elongated, drupelike...	*Lumnitzera*

Source: After Tomlinson (1986).

Figure 10-3. *Rhizophora mangle*. The red mangrove branch (A) bears leaves in an opposite arrangement that shows leaf and stipule scars (B). The flower (C) has four thick sepals and hairy petals, branching stamens, and a central ovary. The pencil-like seedlings (D) have germinated from the fruit while attached to the tree, thus exhibiting vivipary.

stamens (lacking a filament), an inferior bilocular ovary, and a bilobed sessile (lacking a style) stigma. The fruit is green and viviparous, remaining on the tree during germination. The hypocotyl (15 to 30 cm long) extends from the attached fruit and the paired cotyledons are fused into a collar [Figs. 10-3(D) and 10-3(E)].

Figure 10-4. *Avicennia germinans*. Black mangrove leaves are oppositely arranged on a branch (a) that bears young flowers at its tip. The floral cluster (b) shows one open, zygomorphic flower with its tubular corolla. The fruit (c) is a leathery capsule that is flattened and is part of a cluster.

Avicennia germinans (L.) L. (black mangrove; Fig. 10-4) is in a pantropical genus of eight mangrove species within the monotypic family Avicenniaceae. Trees can reach 30 m, although in the Caribbean, they rarely exceed 20 m. The main stems have a flaky, black bark, which, along with the grayish leaves, gives the species its common name. Horizontal cable roots grow from the stem just below the surface and produce (negatively geotropic) pneumatophores that extend 10 to 30 cm above the substrata and function in gas exchange. A short petiole supports leathery obvate leaves, which have a gray color due to the occurrence of dense, club-shaped hairs on the lower epidermis [Fig. 10-4(A)]. Leaf size varies according to the type of mangal. In overwash communities, leaves will reach 9 cm in length and 4 cm in width, whereas in hypersaline scrub forests, the leaves are small. Four to eight flowers are found on a reduced spike, with each flower being partially enclosed by three bracts [Fig. 10-4(B)]. The flowers have five free ovate sepals and a tubular (at base) corolla that develops into four lobes; the latter is zygomorphic in the Caribbean black mangrove. The four free stamens surround a hairy ovary, which is superior, flask-shaped, and unilocular. The ovary produces a short style and bilobed stigma. The fruit is a single-seeded, laterally beaked capsule that is flattened to ovoid [Fig. 10-4(C)] with the embryo showing cryptovivipary (see what follows).

Evolution and Biogeographic Distribution

The two main centers of mangrove diversity are in the Old World and New World (Chapman, 1977). There are about five times more "true mangroves" in Eastern than Western Hemispheres (Tomlinson, 1986). However, the inclusion of mangrove associates reduces this ratio (Table 10-1). The latitudinal distribution of mangroves is usually limited to the 24°C isotherm (Fig. 9-6). The presence of mangals in subtropical latitudes reflects patterns of warm ocean currents and historical tolerance of mangroves. For example, the Gulf of Mexico waters support mangrove forests up to 27°N on the west coast of Florida, while *Kandelia candel* extends into southern Japan. Subtropical mangals are more common at higher latitudes in the Northern Hemisphere, presumably because of their greater tolerance to cold (i.e., frequent glacial periods) than within the Southern Hemisphere (Tomlinson, 1986). Thus, *Avicennia germinans* is found as a shrub on the Texas coast of the Gulf of Mexico. Overall, the highest latitude where mangrove forests (stunted, scrub) occur is in Westonport Bay, Victoria, Australia (38° 45′ S).

The *center of origin* (Chapman, 1977) and *vicariance* (McCoy and Heck, 1976) hypotheses have been used to understand mangrove distribution and evolution (see Chap. 3). The high mangrove diversity in the Indo-Pacific may indicate either an evolutionary center or retention of more species in that region, particularly with respect to the Western Hemisphere with its reduced diversity. However, based on fossil pollen deposits, Gentry (1982) suggests that the present Western Hemisphere floras are relics of historically diverse mangals. The evolution and present distribution of mangroves, in part, may be due to separation of Gondwanaland during the late Cretaceous (Chapman, 1977).

Using the vicariance hypothesis, McCoy and Heck (1976) suggest that mangroves and seagrasses evolved in the early Cretaceous Period when early Angiosperms were migrating into shallow seas between separating continents. Based on the history of plate tectonics, the separation of Gondwanaland occurred in the late Cretaceous Period. The formation of the continents that resulted in land and oceanic barriers then allowed the speciation of mangroves and seagrasses. Thus, the vicariance hypothesis can be used to explain the presence of *sister* species (*Rhizophora*, *Avicennia*) having similar features. Further, the hypothesis has been used to propose that these genera are the oldest mangroves, with other genera evolving after separation of the continents. Unfortunately, the majority of mangrove species are poorly understood with regard to their adaptations, affinities, and fossil records (Duke, 1992).

The possibility of global warming due to the Greenhouse Effect and a subsequent rise in sea level (Chaps. 5 and 9) places long-term survival of mangals into question (Ellison and Stoddart, 1991). Calculations made for low island mangals in the Pacific Ocean with limited sedimentation (8 to 9 cm 100 y^{-1}) indicate that they would be drowned if eustatic increases in sea level are greater than 12 cm 100 y^{-1} (Ellison and Stoddart, 1991), which is the present rate (Gronitz et al., 1982) and predicted to increase from 30 to 100 cm 100 y^{-1} (Davis, 1986; Milne, 1995). Other physical factors that are coupled with a rise in sea level, such as decreased rain fall and runoff and increased salinity, could cause sediment loss (via decomposition of organic sediment) and decreased mangrove production (Snedaker, 1995). However, fluctuations in sea level have been rapid on a historical basis (see Chap. 5) and mangrove forests have persisted through the Quaternary despite substantial changes in sea level (Woodroffe, 1992).

Morphological Adaptations

As trees, mangroves have architectural forms that allow for efficient absorption of light and stability on soft substrates. Of 23 models described for tropical trees (Halle et al., 1978), the unique, repeating crowns produced from lateral or adventitious buds of *Rhizophora mangle* show an adaptation for above-ground vegetative expansion (Fig. 10-5). The crown structure of mangroves results in a dense canopy, thus there is little or no understory growth, except for suppressed, unbranched seedlings (Tomlinson, 1986). The evergreen habit, leaf shape, and

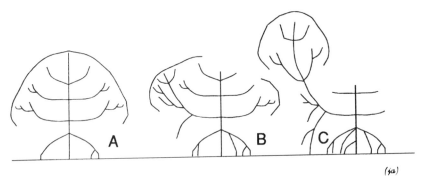

Figure 10-5. The reiteration model of *Rhizophora* architecture. Three examples of vegetative expansion are shown from left to right: (A) prop root development, (B) drop roots from branches, and (C) a second tree developing from a drop root (modified from Hallè et al., 1978; with permission of Springer-Verlag, Berlin).

thickness are characteristic of mangroves. Leaves are generally ovate to elliptical with either a pointed (*R. mangle*) or blunt to indented (*Laguncularia racemosa*) tip. Leaf texture is usually firm to leathery and flexible. Secretory structures are found on the leaves of *R. mangle* and petioles of *L. racemosa* and
in some cases function as salt glands (*Avicennia germinans*; Balsamo et al.,
1995).

Vegetative propagation by mangroves is limited (but see the reiteration
growth of *R. mangle* and the coppicing abilities of *Avicennia*, *Laguncularia*
and *Conocarpus*), as shown when a forest is severely damaged by a hurricane.
For example, Hurricane Andrew passed through southern Florida, snapping off
black and red mangroves at about 5 m, which was the level of the tidal surge
(Dawes et al., 1995a). Canopy loss resulted in exposure of seedlings and benthic macroalgae to intense sunlight. Most understory plants died after three
months and recovery is predicted to take 10 to 20 years. The slow recovery is
because *Rhizophora* is dependent on new seedlings and cannot resprout from
damaged stems unlike *Avicennia*, *Laguncularia*, and *Sonneratia*, which have
reserve meristems (Tomlinson, 1986).

Aerial roots are common on highly specialized mangroves; they are formed
adventitiously from stems and branches in waterlogged soils (Fig. 10-6). For
example, *Rhizophora mangle* and *Avicennia germinans* occur in the lower intertidal zones and produce extensive aerial roots when compared with *Laguncularia racemosa*, which grows in the upper region (see "Zonation"). Aerial roots
include stilt roots [drop and prop roots of *Rhizophora*; Fig. 10-6(a)], pneumatophores [erect roots of *Avicennia*; Fig. 10-6(b)], root knees (aerial loops in
horizontal roots of *Ceriops*), and plank roots (vertical extensions of horizontal
roots of *Xylocarpus*). In addition to their conspicuous aerial roots, mangroves
have *absorbing*, *anchoring*, and *cable roots*. The delicate absorbing (feeding)
roots are fibrous and develop from aerial roots at the point of attachment to
the substrate, or from fleshy anchoring roots, and from horizontal cable roots
arching above (*Rhizophora*) or growing below (*Avicennia*) the substrate. The
development of a tap root is limited to seedling germination and is surpassed
by lateral and adventitious roots in all species.

Anatomical Adaptations

The xerophytic adaptations of mangroves (Fahn and Cutler, 1992) is demonstrated in the increased complexity in leaf anatomy (Roth, 1992). As noted
for salt marsh plants, the *leaf anatomy* of mangroves demonstrates conspicuous xerophytic features that are easily seen using freehand sections. Detailed
descriptions of Caribbean mangrove leaves are given by Roth (1992), and Tomlinson (1986) presents an overview of all mangroves. The epidermis of both
Avicennia germinans and *Rhizophora mangle* is single-layered, with thick outer
walls and cuticle (Figs. 10-7 and 10-8; Roth, 1992). Specialized epidermal cells
include glandular hairs (*A. germinans*; Fig. 10-11) that function in salt secretion
and lenticels or "cork warts" (*R. mangle*) that may secrete water and chloride

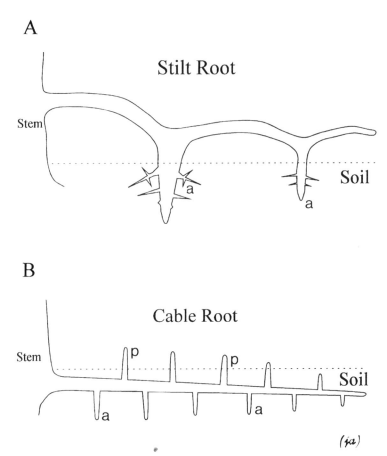

Figure 10-6. Aerial roots of *Rhizophora* (stilt) and *Avicennia* (pneumatophores). The adventitious prop roots of *Rhizophora* (A) reattach via anchoring roots (a) and are capable of producing new stems (reiteration, Fig. 10-5). Horizontal cable roots of *Avicennia* produce both negatively geotrophic pneumatophores for air exchange (p) and positively geotrophic anchoring roots (a) (modified from Tomlinson, 1986; with permission of Cambridge University Press).

(Roth, 1992). The hypodermis in *R. mangle* consists of an upper layer (1 to 2 cells thick) of tannin-containing cells and a lower (2 to 3 cells thick) layer of large *hydrocytes* or water-containing cells. In *A. germinans*, the hypodermis only consists of 4 to 5 layers of hydrocytes. In both species, *mesophyll* tissues occur below the hypodermis, consisting of palisade and spongy parenchyma. The mesophyll tissue of both red and black mangrove leaves is multilayered, containing sclerids, idioblasts, or *cystoliths* that contain calcium oxalate druses (*Rhizophora*; Fig. 10-8) and laticifers (*Exoecaria agallocha*). Conspicuous bundle sheaths have endodermal-like suberin wall thickenings (Casparian strips) but lack extensive bundles of fibers, allowing the leaves to be flexible. The lower

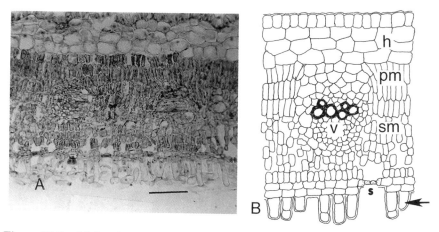

Figure 10-7. Light micrograph (A) and diagram (B) of a leaf cross-section in the black mangrove *Avicennia germinans*. The hypodermis consists of hydrocytes (h) that cover the palisade mesophyll (pm) and spongy mesophyll (sm) tissues. Veins (v) occur in the central part of the leaf. The lower epidermis is covered by "hammer hairs" (arrow) and contains stomata (s). Unit mark equals 100 μm.

epidermis is the usual site of stomates that may (*A. germinans*) or may not (*R. mangle*) be sunken. The lower side of the black mangrove leaf appears gray due to "hammer hairs" that consist of a basal cell, a unicellular stalk, and an asymmetric head cell (Fig. 10-7). Similar to terrestrial plants, sun and shade

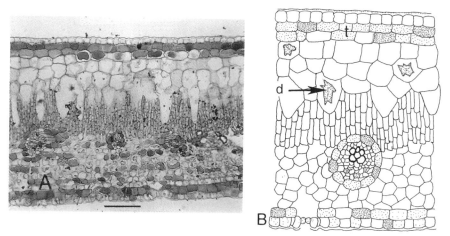

Figure 10-8. Light micrograph (A) and diagram (B) of a leaf cross-section in the red mangrove *Rhizophora mangle*. Compare with *Avicennia germinans*. The red mangrove leaf lacks epidermal hairs, and its hypodermis contains one or more layers of tannin cells (t) and calcium oxalate druses (arrow). Unit mark equals 100 μm.

leaves of mangroves will differ in terms of leaf thickness, layers of tissue, and stomatal density, as shown for *R. mangle* (Farnsworth and Ellisar, 1996a).

Schlerophylly, or thickening, of mangrove leaves through tissue complexity (production of multiple hypodermal layers) is due to limited nutrients (Feller, 1995, 1996) and not because of water stress. In a multiple-year study, leaves of dwarf *Rhizophora mangle* trees fertilized with phosphorus or a combination of nitrogen, phosphorus and potassium (NPK) were only 40% to 80% as thick as leaves of control trees. However, the leaves of fertilized trees, although thinner than the controls, were still xerophytic in construction, reflecting adaptations to water stress.

The *root anatomy* of mangroves, especially their aerial roots, has been extensively studied (Chapman, 1976; Tomlinson, 1986). The anatomy of primary roots of mangrove dicots includes exarch protoxylem, alternating with primary phloem strands, an endogenous origin of lateral roots, and the production of a root cap. Unlike mesophytic dicot roots, the cortex usually has a well-developed lacunar system (*Rhizophora mangle*) or aerenchyma tissue (*Avicennia* spp.) that is continuous between aerial and subterranean portions (see Chap. 9). The path of air from the atmosphere through the lenticel into the aerenchyma (cortex) and then into the root has been vizualized using the scanning electron microscope (Ish-Shalom-Gordon and Dubinsky, 1992). Aerial roots, such as pneumatophores (Fig. 10-9), show early periderm development, which includes an outer cork tissue, cork cambium, and phelloderm. Localized cell divisions of the phelloderm and cork cambial cells will produce dense masses of cells that result in breaks in the bark called *lenticles* (Fig. 10-10). The pattern and shape of lenticles is species-specific; they function in gas exchange and are critical for root survival.

Reproductive Biology

Most mangroves have perfect flowers with sepals, petals, stamens, and carpels [e.g., *Rhizophora*, Fig. 10-3(b)] and are hermaphroditic (both sexes present). Dioecy (male and female flowers on different trees) and monoecy (both sex flowers on same tree) account for only 15% of the species (Tomlinson, 1986). Mangroves appear to be self-fertile, which may be an adaptation for colonization of isolated areas. Pollination is almost completely done by animals, except for wind pollination of *Rhizophora* flowers. Animal pollinators range from bats (*Sonneratia* spp.), to moths (*Ceriops tagal*), birds (*Bruguiera gymnorrhiza*), bees (*Avicennia* spp.), butterflies (*B. parviflora*), and small insects (*Kandelia candel*). Because mangrove species occur in a variety of families, pollen structure varies greatly and it is used to identify different living (Tomlinson, 1986) and fossil genera (Gentry, 1982).

Seedling development is critical for mangroves because of the salt water, water movement, stressful intertidal environment, and need for long-term dispersal by ocean currents. As with terrestrial plants, various types of germination occur, including *hypogeal* (cotyledons not developed or exposed), *epigeal*

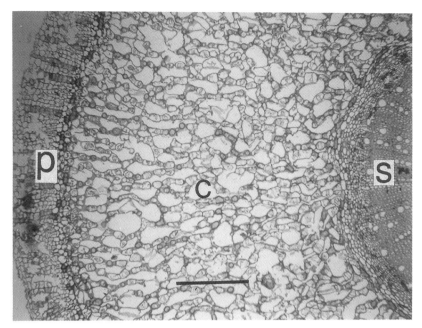

Figure 10-9. A light micrograph of a cross-section of a black mangrove (*Avicennia germinans*) pneumatophore. The outer periderm (p) consists of cork cells, and the cortex (c) is honeycombed with lacunae and surrounds the vascular stele (s). Unit mark equals 300 μm.

(cotyledons developed and exposed), and a unique *Rhizophora* type in which the cotyledons remain in the fruit and the radical (root tip) and hypocotyl are exposed. The latter type of germination is common among mangrove species (Tomlinson, 1986). The fruits and seeds of all mangrove species are buoyant and are dispersed via floatation. About one-third of all mangrove species exhibit *vivipary*, which ranges from zero to limited dormancy of their embryo before germination. *True vivipary* occurs where the embryo penetrates through the fruit pericarp before dispersal (Elmqvist and Cox, 1996). The embryo shows no dormancy and the *propagule* (seedling) germinates from the seed and fruit (Rhizophora type) while attached to the tree [Figs. 10-3(D) and 10-3(E)]. The three Caribbean species range from true vivipary (*R. mangle*), to *cryptovivipary* (*Avivennia germinans*) where the embryo grows out of the seed but not the fruit before dropping from the tree, to no vivipary (*Lacuncularia racemosa*). The types of germination correlate with the zonation habit of each species, with the red mangrove extending into the subtidal and the white mangrove dominating the upper intertidal zone. Presumably, vivipary allows a rapid establishment of seedlings in unstable regions. Further, the large red mangrove propagules will be trapped on the land margin rather than being carried into the upper intertidal zone as described by Rabinowitz (1978) but contested by Smith (1988).

Figure 10-10. A lenticle of black mangrove (*Avicennia germinans*) pneumatophore as seen in cross-section. The outer cork tissue (c) is torn apart due to increased cell division by cork cambial and phelloderm cells (arrows), producing a lenticle that allows gas exchange. Unit mark equals 300 μm.

ECOPHYSIOLOGY OF MANGROVES

The distribution of mangroves and salt marsh species can be linked to seven coastal features: air temperature, coastline protection, shallow shores, currents, salt water, tidal range, and substrata (Chapman, 1977). Unlike salt marsh plants, mangroves also occur on oceanic islands such as atolls and limestone reefs where freshwater and sediment may be limited.

Sediment Ecology

The zonation of mangroves is directly related to elevation and degree of tidal flooding, which in turn influences the ability for nutrient and water uptake (McKee et al., 1988). Thus, consideration of sediment ecology for mangroves has a number of ecophysiological aspects, including pore-water chemistry, gas exchange, and nutrient cycles. Although usually thought of as being *anoxic*, the sediments in mangals vary greatly throughout the world. Deltaic forests have well-sorted silts and clay sediments and fine fibrous root materials, whereas riverine systems have fine sands with a mud veneer (Alongi and Sasekumar, 1992). Two features of mangrove sediments that are also seen in other tropical estuarine communities are low having micromolar concentrations of pore-water nutrients (ammonium, nitrate, phosphate) as well as soluble and condensed tannins from the leachates of roots and leaves.

Soil and litter bacteria, which abound in mangrove communities, play significant roles recycling *nutrient* and organic matter, which is primarily derived from the trees (Lacerda *et al.*, 1995). Mangrove muds contain 0.8 to 1.0 $\times$ 10^9 cells g dwt^{-1}, whereas litter bag studies have documented over 10^9 cells g dwt^{-1} (Alongi and Sasekumar, 1992). The micromolar concentrations of dissolved inorganic nitrogen and phosphorus are low (NH_4^+: 0.1 to 24.7; NO_3^-: 0 to 36.6; PO_4^{-3}: 0 to 20) and sediments (NH_4^+: 3 to 760; NO_3^-: <0.1 to 21; PO_4^-: 0 to 40) within the tidal streams of mangals (Alongi et al., 1992). All components of the nitrogen cycle occur in or on mangrove sediments (Fig. 2-16). Nitrogen fixation is higher on roots and litter than on the sediments, being related to the abundance of blue-green algae. Although ammonification may be high, denitrification and nitrification are low, presumably because of waterlogging, which restricts the activities of soil-nitrifying bacteria. Denitrification is also limited by the lack of nitrate in estuarine and coastal marine sediments that are anoxic. Mangal sediments remove both NO_3^- and NH_4^+ from the tidal waters. Further, most of the nitrogen utilized is not reduced to N_2 gas but is immobilized (Rivera-Monroy et al., 1995). It appears that the uptake of inorganic nitrogen from tidal waters does not result in a nitrogen sink via denitrification (ammonia or nitrate to N_2); instead, the majority of nitrogen (<90% NO_3^-) is immobilized in the sediments (Rivera-Monroy and Twilley, 1996). Mangroves do not use nitrate in situ, but assimilate ammonia through the glutamate synthase cycle that is also involved in recycling of ammonia released in photorespiration (Stewart and Popp, 1987). However, they exhibit nitrate reductase activity if given the nutrient (Stewart and Orebamjo, 1984). Thus, mangroves take advantage of the ammonia-enriched anoxic substrata where they grow.

Pore-water chemistry has been used to understand nutrient fluxes in mangals. Coastal and basin mangals tend to be net exporters of particulate nitrogen (PN) and dissolved organic nitrogen (DON), although overwash islands [Fig. 10-2(B)] may act as sinks (Rivera-Monroy et al., 1995). The availability of iron, sulfides, phosphate, and nitrates is coupled to seasonal and short-term variations in tidal flooding. Thus, regular tidal changes maintain most of these nutrients (Carlson et al., 1983), whereas infrequent flooding causes nutrient limitations, as found within scrub mangrove forests. In a comparison between fringing and scrub red mangroves, Feller (1995) found that phosphorus availability limited growth in two dwarf red mangrove communities. In order to conserve nutrients, dwarf red mangroves develop sclerophylly, their leaves showing enhanced hypodermal tissue. Thus, sclerophylly is not due to herbivory, but caused by phosphorus limitation (Feller, 1996).

The fact that *gas exchange* occurs between the aerial and subterranean roots of mangroves is shown in a number of studies. Thus, the redox potentials and sulfide concentrations in pore water of mangrove sediments are significantly lower near prop roots of *Rhizophora mangle* and cable roots of *Avicennia germinans*, indicating that leakage of oxygen occurs from mangrove roots (McKee et al., 1988). Further, there is a direct relationship between pneumatophore development in black mangroves and the degree of flooding (Stewart and Popp, 1987). Black mangroves create larger oxidized rhizospheres than red mangroves, which help *Avicennia* to establish in the mid-intertidal zone where anaerobic conditions are highest (Thibodeau and Nickerson, 1986). The average volume of mangrove roots devoted to transport of gas is about 40% of their total size and their continuity can be demonstrated by blowing air through a cut root (Tomlinson, 1986). The importance of pneumatophores and stilt roots acting as "chimneys" for oxygen was demonstrated by Scholander et al. (1955) and expanded on by Skelton and Allaway (1996). In both root systems, flooding at high tide lowered oxygen levels, with these being replenished again at low tide. The driving force for oxygen to move into the roots is not simple diffusion, but oxygen removal through respiration *and* the loss of the respired carbon dioxide as it dissolves in the flooding seawater. Thus, a lowering of gas pressure during high tide results from the removal of respiratory carbon dioxide from the gas space (Skelton and Allaway, 1996). Because of the utilization of oxygen and loss of carbon dioxide, a negative pressure results, causing oxygen from the aerial roots to flow into the below-ground roots at low tide (Scholander et al., 1955). However, four species of mangrove seedlings did not show the same responses compared to roots of trees (Youssef and Saenger, 1996). The authors found a constriction in the air-flow path at the hypocotyl junction in seedlings resulting in a low level of oxygen loss by the roots. Thus, waterlogging tolerance of mangrove seedlings is probably through conservation of oxygen for aerobic metabolism in the roots. In contrast, roots of mature plants utilized oxygen also for rhizosphere oxidation and nutrient uptake.

Mangroves can exploit reduced sediments as long as gas exchange occurs via root and stem lenticles (Fig. 10-10). The importance of lenticles can be demonstrated in two ways. First, they are lacking in submerged or subterranean portions of roots where gas exchange is impossible. Second, an obstruction of lenticles on mangrove roots results in a rapid depletion of oxygen in the belowground portions (Stewart and Popp, 1987). Hence, blockage of lenticles after oil spills (Chap. 5) results in the death of mangroves due to an oxygen deficit in the roots and a subsequent increase in toxic sulfides in the sediment.

Water Relations and Osmoregulation

Based on their salt-balance physiology, mangroves are usually thought of as being *facultative* rather than *obligate* halophytes (Chap. 9; Scholander, 1968; Stewart and Popp, 1987). The trees can grow in freshwater, but are probably limited to saline habitats because of competitive exclusion by more rapidly growing glycophytes. All evidence to date indicates that the trees will use freshwater if available, but they can take up water from seawater. For example, isotopic ratios of stem water in mangroves along a riverine salinity gradient demonstrated that the trees were able to obtain water from seawater and freshwater (Sternberg and Swart, 1987). In another study, *Rhizophora mangle* was found to utilize the surface water rather than sediment pore water of higher salinity, which suggests a selective uptake during rainfall or freshwater runoff (Lin and Sternberg, 1994). Finally, the levels of sulfate and chloride in sediment pore water increase significantly when leaves have high rates of transpiration (Carlson et al., 1983). Hence, this is cited as evidence for the filtration and rapid uptake of water by roots, which results in a concentration of ions in saline pore waters.

Mangroves have been divided into *salt secreters* (*Aegiceras*, *Aegialitis*, *Avicennia*) and *nonsecreters* (*Sonneratia*, *Rhizophora*, *Lumnitzera*), yet have transpiration rates (1.5 to 7.5 mg dm^{-2} min^{-1}) similar to terrestrial plants (Ball, 1988). This division between secreters and excluders is not complete, because both functions actually exclude the salt (Ball, 1996). Salt secreters have multicellular salt glands that terminate in a modified trichome (Fig. 10-11). The glands of *Aegiceras corniculatum* and *Avicennia marina* produce 47% and 33%, respectively, of the total sodium and chloride absorbed in a 50% seawater solution, thus demonstrating their importance in salt removal (Ball, 1988). Salt transport is carried out *apoplastically*, or via the cell wall, and *symplastically*, or through the cytoplasm in the gland, with the latter process requiring metabolic energy. Salt is carried into the modified hair through a decreasing ionic gradient to the secretory cell. In a study of salt glands on *A. germinans*, Balsmao et al. (1995) found that salt secretion utilizes cation channels and permeases and an electrochemical proton gradient generated by the cell membrane H$^+$/ATPase. Highly vacuolated hypodermal cells accumulate the salt; at higher cation concentrations, the salt gland stalk and collecting cells transport the ions to the secretory cells (Fig. 10-11). The function of the secretory cells is to rapidly transport the cations to the collection chamber, where the salt is released at the leaf surface via cuticular pores. Exudation from the secretory cells to the exterior is primarily via formation and rupture of vesicles (merocrine) and, to a lesser extent, via rupture of the cuticle (holocrine) in *A. marina* (Ish-Shalom-Gordon and Dubinsky, 1990).

Nonsecreting species can exclude 90 to 95% of the NaCl found in pore water at the level of their roots. Scholander (1968) proposed that the exclusion mechanism in the roots involves a nonmetabolic ultrafiltration that is based on a strong salt gradient. He showed that there were significantly higher levels of salt in the sap of salt secretors than nonsecretors. However, the nonmetabolic process has not been demonstrated and the rates of transpiration and level of salt content in the leaves indicate that high levels of salt are still being delivered even to the leaves of salt excluders. Salt may be removed from the xylem during conduction (mangrove wood has high levels of salt) or translocated via the phloem from young to old leaves (Stewart and Popp, 1987). All mangroves accumulate large amounts of salt (<500 mM) in their leaves regardless of whether they

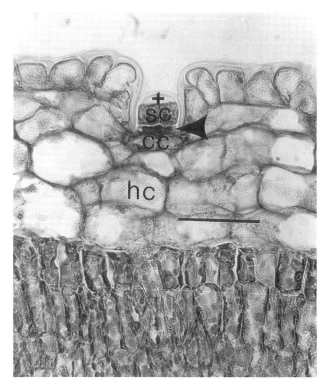

Figure 10-11.   A light micrograph of a salt gland on the upper epidermis of *Avicennia germinans*, the black mangrove. The highly vacuolated hypodermal cells (hc) accumulate salt and carry out osmoregulation. The salt is moved to collecting (cc) and stalk cells (arrow) via apoplastic and symplastic transport. The salt is then moved to the secretory cells (sc), where it is pumped into the collecting chamber (+). Unit mark equals 80 μm.

are "salt secretors, salt excluders, or salt accumulators" (Chaps. 2 and 9). Like other halophytes, mangroves adjust to high external salt concentrations, at least in part, by accumulating sodium and chloride ions. The accumulation of salt results in an increase in leaf succulence with the age of the leaf. Because salt buildup continues through the life of a leaf, cell volume must increase, causing enhanced succulence. When the old leaves are shed, the salt is also lost and thus accumulation and shedding are part of the same process. Hence, mangroves also can be considered as shedders like some salt marsh plants.

Although mangroves can tolerate low levels of NaCl in their cytoplasm, a variety of enzymes in their leaves are inhibited at higher levels. To avoid enzyme inhibition, Na^+ and Cl^- are isolated in the vacuole of leaf mesophyll cells (Stewart and Popp, 1987). Similar to salt marsh halophytes, osmoregulation in the cytoplasm is accomplished with low-molecular-weight organic solutes (Chap. 4), including the amino acid proline (*Xylocarpus mekongensis*), the methylated ammonium glycene betaine (*Avicennia marina*), and the sugar-alcohols mannitol

(*Lumnitzera littorea*) and pinitol (*Rhizophora apiculata*). It is also interesting that glycine betaine and other solutes enhance heat stability of enzymes as well as protect against oxalate destabilization of membranes (Stewart and Popp, 1987), both features being important for tropical halophytic plants.

In addition to taking up water, mangroves must obtain their nutrients from the sediment pore water. Thus, any mechanism for salt exclusion must be selective. There is some evidence in mangroves for preferential absorption of K^+ over Na^+, and NH_4^+ over NO_3^-. The nitrogen source as well as level of salinity show relationships to mangrove leaf biomass (Naidoo, 1990). The leaf biomass in *Bruguiera gymnorrhiza* increases with increasing levels of inorganic nitrogen (either nitrate or ammonium), whereas biomass decreases when exposed to salinities above 50 mol m^{-3}. Fertilization of dwarf red mangroves with nitrogen, phosphorus, and potassium resulted in a reduction of leaf sclerophylly and increased growth (Feller, 1995). Thus, the plant's nutrient uptake is similar to that of glycophytes, being partially independent of salinity.

Photosynthesis

Photosynthesis-irradiance (*P-I*) curves of mangrove leaves show saturation (I_k) and compensation (I_c) levels of 700 to 800 and 30 to 50 μmol photons m^{-2} s^{-1}, respectively (Stewart and Popp, 1987). Although mangroves are tropical trees, they exhibit moderate saturation and compensation levels, being analogous to those of understory plants. Perhaps such levels are be due to the long periods (> 10 y) young mangroves survive as subcanopy plants were percent transmittance is reduced by 80% to 90% (Dawes, 1996). Photosynthetic rates are temperature-insensitive over the range of 17 to 30°C, but it declines rapidly thereafter (Stewart and Popp, 1987). It might appear that C_4 carbon fixation would dominate in mangrove photosynthesis because of the high irradiances and temperatures that occur at the mangrove canopy. However, isotopic ^{13}C values of thirteen North Australian mangrove species show only C_3 carbon fixation (Andrews and Muller, 1985). Further, gas exchange by *Rhizophora stylosa* is closely coordinated with stomatal conductance and CO_2 assimilation that allows efficient water use, which is an important difference between C_3 and CAM (C_4) plants (Chap. 4). In a study by Joshi et al. (1984), evidence was given for a modified C_4 pathway in *R. mucronata*. These latter studies also suggest that halophytic C_3 trees are able to able to obtain water and prevent desiccation without the involvement of a CAM pathway.

Costs

The cost of xerophytic and halophytic adaptations is paid in part through slower growth when compared to glycophytes (Chap. 9). The ecophysiological features of mangroves suggest that they are facultative halophytes (Chap. 9). Hence, they grow slowly in freshwater using the same nutrient-uptake mechanisms as glycophytes, while they are able to restrict, store, and secrete salt.

The lack of mangroves in temperate tidal marshes is apparent, as they are sensitive to frost. However, the question of why there are no frost-tolerant mangroves was raised by Stewart and Popp (1987). A number of families containing mangroves have subtropical and temperate species, yet there are no temperate salt-tolerant trees. Salt marshes are dominated by grasses, rushes, and sedges (Chap. 9) with low woody shrubs having a more limited role and all of the species being dormant in the winter. It may be that halophytic trees have a particularly high energy demand to regulate water uptake so an evergreen habit is required. It is true that the leaves are essential for salt regulation by mangroves, producing the energy needed via photosynthesis for metabolic restriction and excretion of salt. As evergreens, mangroves would experience the effects of frost and low temperatures in temperate areas that limit metabolic activities, including water uptake by the roots. Also, thick or succulent leaves would be disadvantageous in terms of freezing. A deciduous tree probably cannot survive in a temperate tidal habitat because the seasonal loss of leaves would prevent year-round desalinization and salt removal.

OTHER PLANTS OF MANGALS

In addition to the trees themselves, mangals contain vascular plants, bryophytes, macroalgae and microalgae, as well as fungi and various animals. With the exception of faunal and macroalgal studies, only limited, regional information is available regarding the other components (Chapman, 1977). The "Community Profile" of South Florida mangals by Odum et al. (1982) appears to be one of the most comprehensive of these regional reviews.

Other Vascular Plants

In comparison to salt marshes, understory vascular plants are more limited and this may reflect the low irradiance of mangrove forests. The limited understory in mangals has been a topic of discussion (Janzen, 1985; Corlett, 1986; Lugo, 1986). If woody associates are included within the mangrove classification, as presented in this chapter (but see Tomlinson, 1986), then other vascular plants are either hanophytic forms found in salt marshes or glycophytic forms growing at the freshwater or terrestrial interface. Seagrass communities will be found along fringing and riverine mangals with strong energy linkages between the two communities (Zieman and Zieman, 1989). Both air temperature and tidal flooding control mangrove distribution on protected coasts. An example of temperature control can be seen with subtropical fringing mangals on the west coast of Florida that grade into salt marsh communities (*Spartina* spp., *Juncus* spp.) due to lower winter temperatures inland. The warm coastal waters of south Florida moderate the air temperatures in the winter and prevent frosts along the immediate coastal area but not just a few hundred meters inland. The effect of tidal flooding and freshwater is seen where mangals grade into either freshwater or terrestrial communities, depending on land elevation. In Indo-Pacific mangals with freshwater runoff, the mangrove fern *Acrostichum speciosum* and palm *Nypa fruticans* may grade into the upland coconut palm *Cocos nucifera*. In the Caribbean, an upland transition in the Caribbean might include a variety of salt-tolerant species that fringe the mangals, but cannot grow under dense canopies. Species of the terrestrial maritime Caribbean would include the "fringing mangrove" buttonwood (*Conocarpus erecta*), seagrape (*Coccaloba uvifera*), and cactus (*Opuntia* sp.). In areas where the sediment is saline or where sand berms occur, species of the fern *Acrostichum* (*A. danaefolium*, *A. aureum*), and the xeric monocots (*Yucca*) and dicots (*Suadea, Opuntia, Ipomoea*) may dominate.

In the Florida Everglades, riverine mangals diminish in diameter and stature, forming isolated, stunted hammocks; they are ultimately replaced by cypress hammocks (*Taxodium* spp.) and sawgrass (*Cladium jamaicense*) prairies in freshwater wetlands. In terrestrial interfaces, the fringing mangals are replaced in the maritime zone by cabbage palm (*Sabal palmetto*) and slash pine (*Pinus elliotii*) forests. In tropical Caribbean islands, hardwood forests occur in the maritime zone, including buttonwood, (*Conocarpus erectus*), manchineel (*Hippomane mancinella*), and mahogany (*Swietenia mahoagni*). Aerial (epiphytic) vascular plants can be prolific in tropical mangals; they are usually dominated by bromeliads and orchids (Odum et al., 1982; Murren and Ellison, 1996).

Microalgae

Cyanobacteria are abundant in mangal habitats, forming mats in shallow lagoons and tidal streams and growing on various substrata. By contrast, diatoms and other microalgae have low standing stocks, as demonstrated by reduced levels of chlorophyll *a* (2.6 to 6.1 μg gdwt sediment) in the sediment (Alongi and Sasekumar, 1992). Microalgae in mangals include phytoplankton, benthic diatoms, and blue-green algae; all these forms are known to contribute to primary production, but limited specific data are available. For example, blue-green algae form extensive mats in Red Sea mangals; ninety species (14 bearing heterocysts) are recorded with *Scytonema* being the dominant taxon (Hussain and Kohja, 1993). Zonation is also evident in the Red Sea mats, with these prokaryotic species tolerating "normal" salinities of 39 to 45 ppt, and hypersaline conditions reaching 59 ppt. Microalgae are light-limited within the mangals of northeastern Australia, causing low standing stocks and productivity (Alongi, 1994). Similar findings are reported for southeast Asia (Kristensen et al., 1988), where primary production (^{14}C, O_2 production) is 4% to 20% that of *Rhizophora apiculata*. The estimated oxygen demand of benthic decomposers was 73% of the total oxygen sediment uptake. Gross primary production in the Australian mangal (Alongi, 1994) ranged from -281 to 1413 μmol O_2 m^{-2} h^{-1} with highest levels during tidal exposure.

TABLE 10-3. Macroalgae of Mangals[a]

Region	Algae				Habitat		Reference
	O	C	P	R	Epi	Turf	
Australia							
Victoria	1	6	3	13	+	–	Davey and Woelkerling, 1985
South Wales	–	11	6	15	+	–	King and Wheeler, 1984
South Aust.	2	10	9	28	+	–	Beanland and Woelkerling, 1982
Malaysia	–	3	2	5	+	+	Aikanathan and Sasekumar, 1994
New Guinea	–	5	5	15	+	–	King, 1990
Philippines	16	19	4	33	+	–	Fortes, 1987
Florida	3	7	–	4	+	+	Dawes, 1996
Mexico	–	8	1	10	+	–	Rivandeneyra, 1989
Puerto Rico							
Guayacan	–	7	–	11	+	–	Almodorar and Pagan, 1971
La Paguera	–	11	3	8	+	–	Burkholder and Almodovar, 1973
Laguna Joy	4	6	–	2	+	–	Rodrriguez and Stoner, 1990

[a]Abbreviations: O = other; C = Chlorophyta; P = Phaeophyta; R = Rhodophyta. Habitats reported in study include epiphytes (Epi) and turf.

Macroalgae

Although the data are limited, the diversity and importance of mangal macroalgae indicate that they are important primary producers worldwide. For example, the biomass of epiphytic microalgae on red mangrove prop roots can equal the annual leaf litter of the red mangrove fringe in Puerto Rico (Rodriguez and Stoner, 1990). The flora of a single red mangrove root produced as much organic matter per day as phytoplankton found in a cubic meter of a tropical lagoon (Burkholder and Almodovar, 1973). Macroalgal epiphytes show vertical zonation on mangrove roots that is similar worldwide, and demonstrated through studies in South Africa (Phillips et al., 1996) and Florida (Dawes, 1996). In both studies, macroalgal zonation includes upper (green algae: *Rhizoclonium*, *Chaetomorpha*), mid (*Bostrychia*), and lower (*Caloglossa*) zones on the pneumatophores of *Avicennia* spp. Blue-green algae occur along the entire aerial roots.

The macroalgal flora of mangals can be diverse (Table 10-3), as seen in mangals of the Old World, including Australia (23 to 47 spp.; Beanland and Woelkerling, 1982; King and Wheeler, 1984; Davey and Woelkerling, 1985) Malaysia (10 spp.; Aikanathan and Sasekumar, 1994), New Guinea (25 spp.; King, 1990), and the Philippines (56 spp.; Fortes, 1987). Mangal macroalgae in the New World (Table 10-3) ranged from 14 to 22 species (Almodovar and Pagan, 1971; Burkholder and Almodovar, 1973; Rivandeneyra, 1989; Rodriguez and Stoner, 1990; Dawes, 1996).

Most studies of mangal macroalgae have concentrated on epiphytic populations of pneumatophores and stilt roots [Fig. 10-12(A)], except for a few deal-

Figure 10-12. Mangal algae. Macroalgal epiphytes (A) of pneumatophores can form dense growths (e). Turf algae (B) form green feltlike coverings (t) on the sediment.

ing with macroalgal turfs [Fig. 10-12(B); Table 10-3]. The importance of the epiphytic and turf communities can be seen in their biomass and productivity. Algal turfs vary from 49 to 795 g dwt m^{-2} in subtropical mangals (Dawes, 1996) to 0.4 to 12.9 g dwt m^{-2} in tropical mangals (Aikanathan and Sasekumar, 1994). In Florida, epiphytic biomass values vary from 0.07 to 22.0 g dwt pneumatophore^{-1} (Hoffman and Dawes, 1980; Dawes, 1996), and in Australia, values of 0.6 to 20.4 mg dwt m^{-2} are known (Beanland and Woelkerling, 1982). Production values for epiphytic populations in Florida mangals vary from 0.14 g C m^{-2} d^{-1} (Hoffman and Dawes, 1980) to 0.16 to 0.67 g C pneum.$^{-1}$ d^{-1} (Dawes, 1996) for epiphytes versus 0.12 to 0.75 g C m^{-2} d^{-1} for turf algae (Dawes, 1996).

Physiological studies of epiphytic, turf, and drift mangal seaweeds have demonstrated broad tolerances to light, temperature, salinity, and desiccation (Dawes et al., 1978), particularly for species of *Bostrychia* (Hoffman and Dawes, 1980; Davis and Dawes, 1981). Epiphytic macroalgae show high rates of photosynthesis after three hours of desiccation, whether monitored in the air or submerged. By using photosynthesis-irradiance curves, mangal macroalgae were found to be shade plants in terms of saturation (I_k) and compensation (I_c) irradiances and photosynthetic efficiency (α), with the added ability to tolerate high (1500 μmol photons m^{-2} s^{-1}) irradiances. The last feature probably reflects their ability to grow under a dense but uneven canopy (Dawes, 1996).

ECOLOGY OF MANGALS

Three aspects of mangals are described in what follows, including zonation, succession, and ecological roles (see also Lugo and Snedaker, 1974). Mangroves, like salt marshes, carry out a number of ecological functions, including land formation (Warming, 1925; Davis, 1940), primary production, sediment stabilization, filtration of land runoff, and serve as habitats and nurseries. Mangroves are greatly influenced not only by tidal and topographic features, but also by climatic factors, in particular rainfall (Smith, 1992). Thus, their diversity and forestry must be considered in terms of their ecology.

Zonation

Vertical *zonation* is evident moving landward from the fringing species to highest intertidal regions and is controlled primarily by abiotic factors at the lower boundaries and biotic factors at the upper ones (see Chap. 3). Although there are zonation patterns in mangals throughout the world (Chapman, 1977; Smith, 1992), local topography and storm events will modify the general pattern. In the Caribbean, *Rhizophora mangle* occupies the shallow subtidal and lower intertidal zone; *Avicennia germinans* dominates the mid-intertidal region, and *Laguncularia racemosa* occurs in the highest intertidal region (Fig. 10-13). This zonation pattern will be modified with increased water movement (i.e., waves, boat wakes, tidal currents) usually eroding the fringing red mangroves. Monotypic forests of red mangroves occur in overwash islands or in basins where there is no high intertidal zone. If drainage is limited (i.e., few tidal channels, blockage by storm berms), the basin sediment can become highly saline through evaporation, so that only a scrub forest of black mangroves can survive.

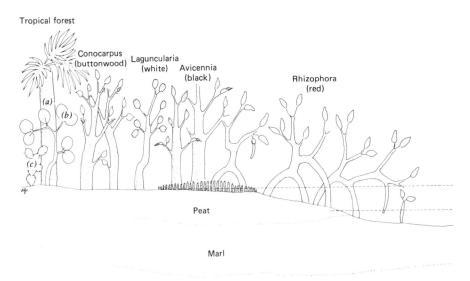

Figure 10-13. Zonation of a south Florida fringing mangal (after Davis, 1940). The red mangrove grows in the shallow subtidal zone and the buttonwood marks the upper edge of the mangal. The fringing tropical forest contains sabal palm (*a*), sea grape (*b*), and cactus (*c*). Tidal range is about 1 m.

The physiological adaptations of mangroves to *salinity* and *flooding* are reflected in species zonation. Because mangroves are facultative halophytes (Thom, 1967), salinity can be viewed as an abiotic eliminator of competition between mangroves and glycophytes. Topography, tidal flooding, and water movement are the determining factors in mangrove zonation, as they influence the degree of salt in the sediment. Primary production for each species is highest in specific parts of the intertidal zone (Lugo and Snedaker, 1974) that reflect their differential abilities to obtain freshwater as well as their competitive ability (Ball, 1980; Ellison and Farnsworth, 1993). Species of *Avicennia* are efficient in obtaining freshwater from highly saline sediments (Stewart and Popp, 1987). One of the problems in determining the major factors controlling zonation is the interaction of salinity, pore-water sulfide concentration, and the frequency of tidal inundation (Smith, 1992). Inferences regarding physiological adaptation of adult mangrove populations must be considered with care, as seedling tolerances are often narrower than with trees. In addition, changes in environmental conditions over time and the complex interactions of multiple physiochemical parameters should be emphasized (Smith, 1992). Mangroves can be divided into two groups, including those with broad and those with restricted salinity tolerances, with the former growing in a variety of soil salinities and the latter being found at less than 40 ppt salinity in upstream riverine ecosystems.

Propagule size has also been used to explain zonation. The hypothesis is that larger propagules are limited to the lower intertidal edge, where they are initially trapped, and due to their ability to survive longer periods of submergence (Rabinowitz, 1978). Caribbean species seem to support this idea, with red mangrove propagules surviving in the lower intertidal region, whereas the black and white species show their highest rates of survival in the upper intertidal zone. However, evidence for tidal sorting as a cause in mangrove zonation is not strong; rather, postdispersal factors (shade and salinity tolerances, predation on propagules) probably play a larger role (Smith, 1988, 1992).

Although not as common as found in salt marshes due to higher levels of rainfall, *salterns* (salt barrens; Chap. 9) can also form an upper intertidal zone of a mangal habitat. Thus, they form a continuum with the upland or terrestrial community [Hoffman and Dawes, 1996; Fig. 10-14(A)]. As the saltern is approached, Caribbean mangals show dwarfing and sclerophylly; they also grade into monotypic stands of black mangroves. The black mangroves decrease in size (scrub morphology) and are replaced by succulent halophytes that fringe the saltern [Fig. 10-14(B)]. Pore-water salinities of salt barrens during the dry season can exceed 100 ppt, resulting in a band of bare sand fringed on both sides by *Sessuvium* spp. and *Salicornia* spp.

Succession and Land Building

Ecological views of *succession* in mangrove communities are similar to those described for salt marshes (Chap. 9). Succession in terms of predictable shifts in community species (Chap. 3) has been demonstrated in mangals disturbed by

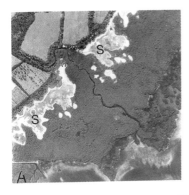

Figure 10-14. Mangal salterns. The aerial view (A) of the saltern at Cockroach Bay, Florida (see S), shows its fringing nature and a nearby mangal stream. The saltern (B) consist of salt-encrusted sediment surrounded by the halophyte *Salicornia* and then stunted black mangroves (Hoffman and Dawes, 1996).

hurricanes (Davis, 1940; Smith, 1992; Dawes et al., 1995a). Such disturbances demonstrate the ability of mangroves to recover albeit rather slowly. For example, Hurricane Andrew in 1992 destroyed or severely damaged almost 150 km^2 of mangroves in south Florida; the same area was previously damaged by Hurricane Donna in 1960. After 32 years, tree heights suggested that recovery was still in progress; hence, it appears recovery from the 1992 hurricane will last at least 40 years (Dawes et al., 1995a). Mangrove succession can be viewed as a sequence from pioneer colonizers to mature climax terrestrial forests. Low intertidal species trap sediments, resulting in changes in elevation, and the invasion of competitive high intertidal species. Thus, mangroves, according to this view, are serial stages progressing toward an upland forest. Davis (1940) proposed a serial-stage succession of mangals in Florida, stressing their importance in land building through the trapping of sediments. Chapman (1977) has presented a similar view and included specific conditions that differ for mangals worldwide. The alternative view of succession is that the forests are not building land or evolving toward a terrestrial community; rather, they exist within the intertidal zone as a climax community (Tomlinson, 1986; Smith, 1992). Support for this view comes from geological records that show historical dominance of key species in steady-state, 2000-year-old *Rhizophora* forests on Grand Cayman Island in the Caribbean (Tomlinson, 1986).

Whereas biologists view mangrove ecosystems as sources of organic matter, geologists consider them as sinks involved in sediment stabilization. The views of Davis (1940) and Chapman (1977) emphasize what Curtiss (1888) proposed, namely, that mangroves are *land builders*. However, Smith (1992) and Tomlinson (1986) point out the lack of evidence for the connection between land building and succession in mangals. Land building occurs when there is a *prograding* environment, that is, where sediment is being imported and changes in sea level

TABLE 10-4. Mangrove Productivity and Litter Fall (after Day *et al.*, 1989, Mitsch and Gosselink, 1993)

Feature	Type of Mangal in South Florida			
	Riverine	Basin	Fringe	Dwarf
Height (m)	81	9	8	1
Basal area (m^{-2} ha^{-1})	41	19	18	1
Net production				
(g dwt m^{-2} d^{-1})	13	6	—	0
(g C m^{-2} y^{-1})	—	1099	1764	—
(kcal m^{-2} d^{-1})	57	25	—	0
Litter fall				
(g dwt m^{-2} y^{-1})	1170	730	906	120
(kcal m^{-2} d^{-1})	14	9	—	2

Sources: After Day et al. (1989) and Mitsch and Gosselink (1993).

are gradual. When the environment is *erosive*, mangals will be removed because they cannot override the abiotic processes of a rapid increase in sea level, limited sedimentation, and increase in water motion. The forest will respond to such changes through reduction of coastal habitat (Smith, 1992). In fact, during historical periods of rapid sea-level rise, mangals decreased worldwide. In the recent past, gradual changes in sea level and moderate rates of sedimentation have supported mangrove expansion and land formation through trapping of sediment. Thus, Davis's (1940) cores in *Rhizophora mangle* forests demonstrated sediment accretion. Young and Harvey (1996) report a strong positive correlation between the density of *Avicennia marina* pneumatophores and sediment deposition over distances less than 70 m. It appears that pneumatophores are produced at a density that will balance the need for aeration of the subterranean root system, which results in sediment accretion.

Forestry and Standing Stock

Biomass, or standing stock, data, as well as forestry and architecture information concerning mangals are limited (Cintròn and Novelli, 1984; Smith, 1992). The lack of data is partially due to the difficulties of working in mangals and the diversity of habitats that range from scrub to large riverine forests (Table 10-4). Forestry measurements, detailed by Cintròn and Novelli (1984), require measurement of tree diameters taken 1.3 m above the ground (diameter at breast height = dbh) with a dbh tape that calculates diameter from circumference ($c = \pi \times d$). Tree basal area (g), or the space covered by a tree stem, is expressed in m^2 of tree base per hectare and is determined by the following formula, where the average *dbh* is in cm, g is expressed in m^2, and n is the number of trees per hectare in Eq. 10-1:

$$g = (dbh)^2 (0.00007854)(n) \qquad\qquad (10\text{-}1)$$

Stem dbh, basal area (g), forest density, tree height, and canopy coverage measurements have been done in a few mangals, mostly in Australia and the Caribbean. Mature trees of fringing and riverine mangals of south Florida and Puerto Rico have tree heights of 5 to 11 m, whereas those in a northern mangal of Florida show the effect of dieback due to frost, with heights of 2 to 5 m (Dawes, 1996). The area covered by tree bases (g) is an important expression of the size of the trees and thus a major feature of forests. Basal areas of south Florida mangals range from 20 to 38 m^2 ha^{-1}, whereas those of central west Florida are smaller, 2 to 31 m^2 ha^{-1} (Dawes, 1996).

Ecological Roles

Mangals exhibit many significant roles similar to salt marshes, including filtering land runoff, stabilization of sediments, and trapping sediments. The latter two roles are dependent on sea level fluctuations (Ellisan and Stoddart, 1991; Ellisan and Farnsworth, 1996b). Primary production as well as habitat and nursery functions are discussed in what follows.

1. *Primary production* of mangroves has been determined by measurements of photosynthesis, demographic growth analysis, and litter production. *Photosynthetic* studies using mangrove leaves includes following changes in CO_2 concentrations (gas analytical methods), CO_2 partial pressure (physical-chemical pCO_2), and changes in the ^{14}C-fixation rate (assimilation). Net primary production of mangrove forests (g C m^{-2} y^{-1}) range from 572 in a fringing mangal in Puerto Rico to 2442 for a basin mangal in South Florida (Mitsch and Gosselink, 1993). Total organic production shows the same trend, as seen in photosynthetic rates of different forest types (Table 10-4), with riverine communities showing the highest daily rates. For example, organic production was similar for mangals from different parts of the world based on carbon fixation (g dwt m^{-2} y^{-1}, where 1 g C = 2 g dwt tissue). Thus, a fringing *Rhizophora mangle* forest in Mexico produced 1607 to 2458 g dwt m^{-2} y^{-1}, similar to a *Kandelia candel* mangal in Hong Kong (1950 to 2440 g dwt m^{-2} y^{-1}).

Productivity of mangrove trees also can be measured indirectly through *litter* traps (Fig. 10-15), where an enumeration of leaves, twigs, fruits, and seedlings is made. Litter data, whether in terms of dry weight (g dwt m^{-2} d^{-1}) or energy (kcal m^{-2} d^{-1}), support the photosynthetic and organic production data. Further, the decomposing litter plays a significant role in the carbon budget of adjacent estuaries as well as sustaining the microbial food chain and nutrient regeneration (Wafar et al., 1997). Fringing and riverine mangals grow more rapidly and have higher production of biomass than other types of mangals (Table 10-4). The larger the tidal exchange of water, the greater the litter production (riverine > fringe > basin > scrub). Also, the amount litter fall varies according to lat-

Figure 10-15.   A litter trap (1.0 m^2) in a mangal. Leaves, twigs, flowers, and fruits are separated and weighed periodically (1 to 2 wk) to determine litter production.

itude, with high-latitude mangals producing more than low communities, and this may be due to winter damage and subsequent loss of tree biomass. However, latitudinal differences in litter production may be subtle, as seen in a northern mangal in Tampa Bay (27° 59′ N) with a litter fall of $3.1 \text{ g dwt m}^{-2} \text{ d}^{-1}$ (Dawes, unpublished), while that of a riverine forest at Rookery Bay (27° 34′ N) was almost identical ($3.2 \text{ g dwt m}^2 \text{ d}^1$; calculated from Table 10-1).

Mangal litter production is similar in different parts of the world, with lower levels reported for south Florida ($1.2 \text{ kg dwt m}^{-2} \text{ y}^{-1}$; Table 10-4) and for Hong Kong and also Pacific Mexico ($1.2 \text{ kg dwt m}^3 \text{ y}^1$; Mitsch and Gosselink, 1993). Annual litter production for continental Australia is either comparable or higher; ranging from 1.6, 2.4, to $1.3 \text{ kg dwt m}^{-2} \text{ y}^{-1}$ for *Avicennia marina*, *Rhizophora stylosa*, and *Ceriops tagal*, respectively (Bunt, 1995). Similar, high levels of litter production have been reported in mangals consisting of *A. officinalis* ($1.0 \text{ kg dwt m}^{-2} \text{ y}^{-1}$), *R. apiculata* and *R. muronata* ($1.2 \text{ kg dwt m}^2 \text{ y}^{-1}$), and *Sonneratia alba* ($1.7 \text{ kg dwt m}^3 \text{ y}^1$) in India (Wafar et al., 1997). In lush mangals of Queensland, Australia, seasonal changes are evident in littler fall of *A. marina* (Mackey and Smail, 1995). Over a two-year period, leaf litter accounted for 47% versus 30% for reproductive and 20% for woody litter. Leaf litter peaked during the wet summers and wood litter only correlated with high winds. Reproductive litter (flowers, fruits, seeds) showed a biannual pattern, with fruit fall occurring the dry winter season. The highest litter occurred in the midtidal range. Similar findings in south Florida showed that distinctive litter falls could be correlated with climatic zones throughout the distribution of each species.

2. The role of mangals as *nurseries* and *habitats* and the diversity of their fauna are documented in regional (MacNae, 1968; Odum et al., 1982) and worldwide reviews (Chapman, 1976, 1977; Clough, 1982; Hutchings and Saenger, 1987, Robertson and Alongi, 1992). As seen in salt marshes, there is a broad range of inhabitants and "users" of mangals, including zooplankton, benthic infauna and epifauna, nekton, insects, and birds, as well as terrestrial wildlife. The epizoic community of mangrove roots, just as the epiphytic algae, shows complex interactions (Ellison and Farnsworth, 1992; Farnsworth and Ellison, 1996b), including short time-scale variations in larval recruitment and long time-scale effects of physical factors. A diverse set of insects, particularly herbivorous species, are recorded for mangals (Murphy, 1990; Lee, 1991; Farnsworth and Ellisar, 1993) and defenses against herbivory (de Lacerda, 1986). In his studies of animal distribution in Indo-West Pacific mangals, MacNae (1968) reported that larger animals showed preferences for specific habitats but not tidal zones. Animal distribution was dependent on resistance to water loss and the need for protection from the Sun; further, the level of the water table, consolidation of the sediment, and availability of food sources were also important. Walsh (1967) found that faunal distribution in Hawaiian mangals was linked to salinity and oxygen gradients as well as substrata. Mangals, such as those in Laguna de Terminos, Mexico, serve as nurseries and habitats for over 100 species of fish, many of which are economically important (Day et al., 1989). Higher vertebrates, such as birds, mammals, reptiles, and amphibians, frequent mangals and are cited in preservation and protection arguments for these ecosystems. Not only do wildlife need mangal habitats, but the mangals need the wildlife in order to balance flora (need for pollinators) and faunal (control of grazers) processes (Day et al., 1989).

Crabs are frequently the most abundant group of benthic macrofauna in mangals, that is, in terms of numbers and biomass. Because of their control in litter production, soil oxidation, and ammonia levels, crabs have been considered as *keystone* species (Smith et al., 1991). In Queensland, Australia, mangals, removal of crabs resulted in a rise in soil sulfide levels, a decrease in soil ammonia levels, and a subsequent loss in tree productivity. Mangroves support complex, highly diverse root-fouling communities, consisting of hydrozoans, ascidians, sponges, anemones, hard corals, and isopod crustaceans (Ellison and Farnsworth, 1992; Farnsworth and Ellison, 1996). Further, species richness in Belizean mangals increases with distance from land or toward the fringing edges of mangals. While wood-boring isopods can cause severe root damage in mainland mangals, they have a more limited effect on offshore islands, where epibiontic sponges and ascidians inhibit isopods (Ellison and Farnsworth, 1992).

Modeling

Qualitative and quantative *ecosystem models* have been developed for south Florida mangals. The conceptual models (Fig. 10-16) consider major stresses (siltation, harvesting, hurricanes, herbicides, drainage, thermal additions), and simulation models include productivity and nutrient cycling (Mitsch and Gosselink, 1993). Information obtained from these models has helped to explain mangal survival. For example, model simulations showed that attainment of a steady-state mangal required about 24 years, the same average period between hurricanes (Sell, 1977). This suggests that recovery of mangroves is an adaptation to the hurricane cycle. Simulation models also can differentiate between the importance of tidal exchange and terrestrial sediments as sources of mangal nutrients.

The importance of mangrove production to the coastal zone is evident if a flux model is produced in which import of dissolved organic carbon (DOC) as well as export of particulate organic carbon (POC) are examined (Robertson et

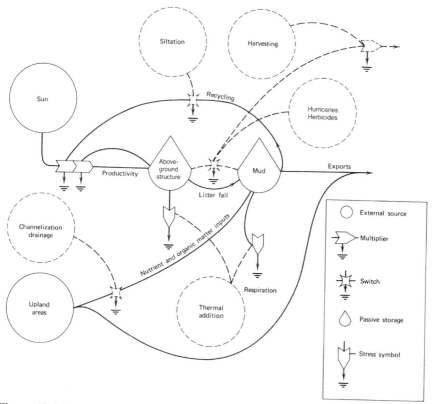

Figure 10-16. Conceptual model of a mangal showing the major stresses of siltation, harvesting, storms, drainage, and thermal additions (after Lugo and Snedaker, 1974).

al., 1992). The major fluxes (kg C ha^{-1} y^{-1}) of a mangrove forest in Missionary Bay, Australia, included 3322 kg C of POC from the forest to the creeks, of which 8 kg C was processed in nearshore embayments and 520 kg C in the benthic coastal zone. The low level of DOC (73 kg C) delivered to the mangrove forest from nearshore waters indicate their high productivity.

Methods

Research methods for the study of mangroves, including forestry techniques (basal area, height, litterfall, photosynthesis), have been described (Snedaker and Snedaker, 1984). Techniques are also available to determine biomass and productivity of the turf and root epiphytic macroalgae (Dawes, 1966) and microalgae (Darley, 1982). Above-ground densities can be determined using standard quadrat measurements, and zonation can be determined with line transects and leveling devices to determine horizontal and vertical distribution (App. A).

MANAGEMENT OF MANGALS

Tomlinson (1986) notes that "mangroves are an entrepreneur's dream" in that they produce lignocellulose using sunlight. Further, they do not require expensive fertilizers, as they obtain nutrients from land runoff and tidal flushings of seawater. As with the management of terrestrial forests, there must be a balance between conversion to capital (harvesting) and sustaining a renewable resource (Fortes, 1988). There is a further caveat for mangals, namely, that they require more time to recover after harvesting due to their slow growth. Mangal ecosystems are complex, highly structured, and naturally engineered forest and drainage systems driven by water fluxes (tidal, riverine), sedimentation, evaporation, and geomorphic factors. Thus, any impacts on the input or outflow of water will directly impact the community. Dams, channelization, or other water diversions upstream and coastal habitat removal will rapidly damage mangrove communities. In summary, mangals are often overexploited and in many cases destroyed (Linden and Jernelov, 1980; Fortes, 1988), as well as being sensitive to upland modifications.

Human Impacts

Although mangrove forests have survived natural destruction, they are rapidly disappearing due to human impacts (Chaps. 5 and 9); in particular, extraction, pollution, and reclamation (Ellison and Farnsworth, 1996a). In the past, mangals were considered to be wastelands; hence, they were destroyed for fish and shrimp ponds, as well as rice farming in Malaysia and Ecuador. In the Philippines and Southeast Asia, mangals have been overharvested for wood. In south Florida, mangals have been destroyed via dredge-fill operations to create real estate, as well as port and industrial facilities. Mangal destruction results in a chain of reactions that affect offshore and commercial activities. The example of wetland loss (44%) in Tampa Bay, Florida (Chap. 9), is known to be linked to declines in fin fish and commercial shrimping. In Malaysia, destruction of mangals for fish ponds is correlated with the decline of shrimp harvesting in adjacent grass beds (Fortes, 1988). The connection between coastal fin fish and shrimp with mangroves is via detrital and benthic microalgal production. For example, the juvenile shrimp (*Penaeus merguiensis*) live in the mangals and are dependent on them for their nutrition (Newell et al., 1995). In Ecuador, mangrove deforestation to create shrimp mariculture ponds has caused a decline in postlarval shrimp production. Parks and Bonifaz (1994) found that deforestation directly impacted shrimp farming through the loss of the juveniles; a case of "biting the hand that feeds you."

The impact of domestic and industrial pollution is also seen in mangals (Chap. 5). Long-term pollution occurs from oil spills due to residual hydrocarbons that continue to leak and cause mutations in mangroves (Klekowski et al., 1994). Such effects are in addition to the immediate damage to trees by oil coating on the roots, which seal the lenticles and prevent gas exchange. Five major oils spills have happened within the Straits of Malacca (near Singapore) between 1976 and 1979. In one spill alone (1975), 7000 tons of crude oil from the tanker Showa Maru resulted in death of 300+ hectares of mangals in Sumatra (Fortes, 1988). In addition to point source spills, about 4,705,000 tons of oily waste are discharged into the South China Sea yearly, some of which finds it way into mangals (Fortes, 1988).

Uses of Mangroves

Mangroves are used as a source of wood (boards, poles, fuel), tannins, and dyes (Walsh, 1977). Although mangals comprise a small part of forests in most tropical countries (>2%), many countries (Malaysia, Kenya, Thailand) have developed management schemes. Unfortunately, most of these programs do not consider the importance of the habitat or attempt to limit the size of the areas harvested (Fortes, 1988). The diameter of harvestable trees has been decreased a number of times over the past years in the Malaysian sustained yield management program, reflecting the lack of sufficient time allotted for tree regrowth. Thus, it takes 20 to 30 years for *Rhizophora apiculata* to grow to 15 m. *Nypa fruticans*, a common mangrove of Asia and Oceania, has been introduced into West Africa because almost all of the tree can be used by humans (Hamilton and Murphy, 1988). The sap is used to produce "beer," sugar, vinegar, and an alcoholic fuel. The leaves are woven together to produce walls and palm thatch for homes. It is not uncommon in the Sulu Sea of the Philippines to see homes of *Eucheuma* farmers mounted on stilts made from *R. apiculata* stems and roots and walls of woven *Nypa* leaves.

The tannin content in the dried bark of *Rhizophora*, *Bruguiera* and other mangroves ranges from 23% to 39%, and it was used in the production of leather up to the 1960s (Walsh, 1977). With the advent of synthetic tanning compounds, this use of mangroves declined, reducing forest removal. However, many mangals in Malaysa and the Philippines are being destroyed for fuel and lumber for construction. Coupled with destruction of mangals for shrimp and fish farms throughout the tropics (Ecuador to Thailand), habitat loss has also become a serious problem worldwide.

Management and Restoration

The legal protection of mangals and other wetlands in the United States is reviewed by Mitsch and Gosselink (1993), and restoration is discussed in Chap. 5. The 1971 Ramsar Convention provides an international framework for wetland protection, as 74 countries currently have endorsed this document. Unfortunately, differences of opinion exist even among the co-signers regarding the need for mangal protection. Economic demands are often being placed before ecological evaluation of mangrove ecosystems. On a positive note, mangals can be created, mitigated, and restored through use of the seedlings (Carlton and Moffler, 1978; Barnett and Crewz, 1990), with the costs being reasonable (Hoffman and Rodgers, 1981) and little follow-up maintenance being required. The benefits of restoration and protection of mangals are enormous. For example, conversion of shrimp ponds in Ecuador back to mangals would restore a habitat that not only would support the production of larval shrimp needed for the farms, but serve as a nursery for a large variety of coastal fish and birds.

CHAPTER 11

Seagrass Communities

Only 0.01% (58 species) of the 300,000+ flowering plants have adapted to the submerged marine habitat. Two questions come to mind: Why are there so few submerged marine Angiosperms? Why are they all monocots? Although few in number these monocots form extensive meadows in tropical and temperate waters (Fig. 11-1). The literature on seagrasses is extensive, including texts dealing with research methods (Phillips and McRoy, 1990), taxonomy (den Hartog, 1970), biology (McRoy and Helfferich, 1977; Phillips and McRoy, 1980; Larkum et al., 1989a), and ecosystems (Kuo et al., 1996). Regional publications include Australian seagrasses (*Austral. J. Mar. Freshwat. Res.* **44(1),** 1993), as well as documentations of temperate eelgrass communities from the Pacific (Phillips, 1984) and Atlantic (Thayer and Fonseca, 1984), and mixed tropical communities of the Gulf of Mexico and Florida (Zieman, 1982; Durako et al., 1987; Zieman and Zieman, 1989).

GENERAL CHARACTERISTICS

Seagrasses are most closely related to plants in the class Helobiae and are not true grasses of the family Poaceae (Tomlinson, 1982). Seagrasses, having adapted to a submerged aquatic environment, show morphological and anatomical features that are similar to freshwater *hydrophytes*. However, there is sufficient variation in reproduction and structure to indicate the independent evolution of species.

Taxonomy

The low diversity of seagrasses (0.001% of Angiosperms) is surprising when the high numbers of terrestrial and freshwater flowering plants and the vastness (18% of Earth's surface) of the coastal environment are considered. Den Hartog (1970) recorded 49 species and 12 genera in two families, namely, the Hydrocharitaceae and Potomogetonaceae of the class Helobiae (Monocotyledonae). In a later treatment (Kuo and McComb, 1989), 58 species and 12 genera were delineated, with these placed in four families, two orders (Hydrocharitales, Potomogetonales), and one class (Liliopsida). Three genera were placed in the family Hydro-

charitaceae (*Enhalus, Thalassia, Halophila*), one in the Posidoniaceae (*Posidonia*), five in the Cymodoceaceae (*Syringodium, Halodule, Cymodocea, Amphibolis, Thalassodendron*), and three in the Zosteraceae (*Zostera, Heterozostera, Phyllospadix*). *Ruppia* (Ruppiaceae) also was included as a seagrass by many authors, although some species occur in lakes where Na_2CO_3 content is high. Regardless, *R. maritima* will form dense beds in both oceanic and estuarine waters (Koch and Dawes, 1991; Lazar and Dawes, 1991; Short, 1992). Keys to the genera have used vegetative features (den Hartog, 1970) or a combination of vegetative and reproductive characteristics (Table 11-1). Three species representing different families are briefly described in what follows.

Zostera marina L., or eelgrass (Zosteraceae), is one of four species in the genus (Fig. 11-2). The species is widely distributed in temperate North Atlantic and eastern Pacific habitats; it also extends to the Arctic Circle, as well as in the Mediterranean and Black seas. Eelgrass primarily occurs subtidally, but occasionally extends into the lower intertidal region in tide pools and exposed on the substrata. It is considered to be a "k," or climax, strategist, being a perennial and using the carbohydrate reserve in its rhizome for maintenance if its leaves are removed (Dawes and Guiry, 1992).

Eelgrass shows monopodial growth, producing one or more thick, unbranched roots and leaf sheaths at each node on the 2-to-3-mm diameter rhizome (Fig. 11-2). Internodes are 10 to 35 mm long, with short shoots (determinant branches) arising at regular intervals to produce flat, ribbon-shaped leaves having 5 to 11 parallel veins. Leaves develop from a basal meristem and are distichously arranged. The lower part of the leaf forms a sheath that is compressed and persistent. Two to three spathes can occur on a single, short shoot, each bearing up to 10 male (Fig. 11-2; m) and female flowers (Fig. 11-2; f). The spadix sheath is sessile on a short shoot, with the male and female flowers being alternately arranged. The ovary develops before the stamens (proterogyny) and a single plant can produce more than 200 seeds in a season. The seeds are cylindrical, 1×2 mm in size, and have up to 20 ribs.

Thalassia testudinum Banks ex König, or turtle grass (Hydrocharitaceae), occurs throughout the Caribbean (Fig. 11-3), and its "sister" species, *T. hemprichii*, is found in the Indian Ocean and western Pacific. Turtle grass occurs throughout the Gulf of Mexico and on the east coast of Florida to Cape Kennedy, and its southern distribution is limited to the coasts of Columbia and Venezuela. The plant primarily occurs subtidally and it is a "k" strategist (Dawes and Lawrence, 1980); it forms extensive meadows to 10 m, but can extend to 15 m in highly transparent waters, such as near the barrier reef of Belize. *Thalassia testudinum* is usually found in salinities above 25 ppt,

---

Figure 11-1. Beds of turtle grass, or *Thalassia testudinum*. A partially exposed grass bed found among overwash mangal islands (A) contains a variety of psammophytic green algae (B) such as *Penicillus* (p) and *Halimeda* (h).

TABLE 11-1. A Key to the Genera of Seagrasses Primarily Using Vegetative Features

1 All stems produce only foliage leaves[a]	**2**
1 Stems of two types, either bearing only scale or foliage leaves	**9**
2 Flowers and fruits spicate and pedicellate, leaves narrow (to 0.5 mm)	*Ruppia*
2 Flowers and fruits solitary in axial clusters, leaves wider than 0.5 mm	**3**
3 No specialized, erect short shoots (ramets[b]), stems all of one type, flowers on specialized lateral shoots	**4**
3 Stems differentiated into primary horizontal rhizomes and erect short shows, both bearing foliage leaves	**5**
4 Flowers in alternate leaf axis, roots unbranched	*Enhalus*
4 Flowers irregularly produced, roots branched	*Posidonia*
5 Short shoots annual and distinct from rhizome, roots unbranched, inflorescence obvious, spikelike	**6**
5 Short shoots not distinct from rhizome, flowers inconspicuous and in terminal pairs	**8**
6 Rhizomes congested, monopodia	*Phyllospadix*
6 Rhizomes elongated, not congested	**7**
7 Rhizomes elongated, branching monopodial	*Zostera*
7 Rhizomes elongated, branching sympodial	*Heterozostera*
8 Rhizomes and short shoots distinct, internodes easily seen, roots branched, flowers not petaloid	*Cymodocea*
8 Rhizomes and short shoots not distinct, internodes not clearly visible, roots unbranched, flowers petaloid	*Halodule*
9 Scale-bearing rhizome with sympodial branching, short shoots long (to 20+ cm) and distinct from rhizome	**10**
9 Scale-bearing rhizome with monopodial branching, short shoots long or short	**11**
10 Scale-bearing rhizome having diffuse branching with two to seven to many branch-free nodes	*Amphilbolis*
10 Scale-bearing rhizome having regular, periodic branching every fourth node	*Thalassodendron*
11 Branching continuous, flowers not petaloid	**12**
11 Branching periodic, short shoots produced at regular intervals and separated by 9 to 13 nodes, flowers in lateral clusters and petaloid	*Thalassia*
12 Rhizome internodes uniformly long, short shoots variable and usually determinate, leaves cylindrical	*Syringodium*
12 Rhizome internodes alternate between long and short intervals, short shoots determinate, leaves flat	*Halophila*

Sources: Modified from Hutchinson (1958) and Tomlinson (1982).
[a]The first leaves produced are always scale-like but all others are vegetative blades.
[b]Ramets or short shoots are all erect determinant stems that produce foliage leaves.

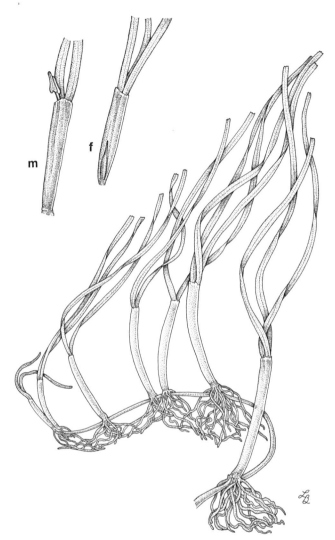

Figure 11-2.   Eelgrass, or *Zostera marina*. The clonal plant consists of a rhizome and short shoots (rametes) with clusters of roots. The male (m) and female (f) flowers are produced in clusters on short shoots.

replaced by *Halodule wrightii* and *Ruppia maritima* in lower salinities. Vegetative growth is usually the principal method of expansion, although seed production is high in tropical waters.

   Turtle grass produces a terete (3 to 6 mm dia.) rhizome with 4-to-7-mm-long internodes and scale leaves (Fig. 11-3). Single, unbranched, anchoring roots

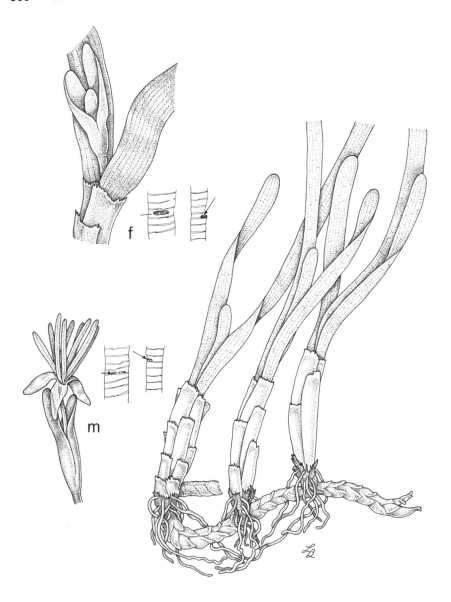

Figure 11-3. *Thalassia testudinum.* Each short shoot of turtle grass produces three to seven flat (to 15 mm wide) leaves and thick anchoring roots from a rhizome covered by scale leaves. The single female (f) flower produces a large scar (arrow) on the short shoot; by contrast two to three small scars (arrows) are left by the male (m) flowers (Witz and Dawes, 1995).

are produced at each node and contain internal air channels (lacunae). Delicate absorbing root hairs are produced near the root apices. Rhizomaceous nodes also have erect short shoots (ramets) with three to seven photosynthetic leaves, which grow by a basal leaf meristem. Leaf and flower scars (Fig. 11-3; f and m) are visible on the short shoot, so that historical analysis of reproduction, sex ratios, and leaf turnover rates is possible (van Tussenbroek, 1996). Leaves are up to 2 cm wide and differentiated into a persistent colorless basal sheath and a chlorophyll-bearing blade with 9 to 17 parallel veins. Leaf turnover rates vary from less than 10 to more than 20 days and growth rates can be calculated using leaf-marking techniques (Zieman, 1974; Tomasko and Dawes, 1989). The species is dioecious, with separate female (Fig. 11-3; f) and male flowers (Fig. 11-3; m) borne on short (subsessile) peduncles; both flowers have a single whorl of white tepals. Male flowers have 3 to 12 subsessile stamens, and female flowers have a single ovary with 6 to 8 carpels; the style is divided into two filiform stigmata. Pollen grains are spherical, embedded in mucilage, and arranged into moniliform chains. The fruit is globose and has a pointed tip; it contains one to few seeds that show cryptovivipary (Chap. 10).

   Syringodium filiforme Kutzing, or manatee grass (Cymodoceaceae), has a "sister species" (*S. isoetifolium*) in the Indo-Pacific. Manatee grass occurs throughout the Caribbean and it extends into the Gulf of Mexico; it also occurs northward to Cape Kennedy (Indian River Lagoon) on the eastern Atlantic coast of North America. Typically, *S. filiforme* occurs in areas of oceanic salinities, and it can grow down to depths of 20 m in highly transparent waters, such as found in Belize or St. Croix.

   The rhizome of manatee grass (2 to 4 mm dia.) is a creeping and herbaceous structure. It produces two to four thick roots with abundant root hairs, and scale leaves and a short shoot at each node. The short shoots produce two to three leaves that grow via intercalary meristems. The leaf base forms an extensive sheath around the short shoot, and the upper portion of the blade is terete. The plant is dioecious, with both flowers occurring in raceme. Flower production is controlled by temperature (McMillan, 1980). Thus, plants from the northern Gulf of Mexico flower at 20 to 24°C, whereas those from St. Croix require temperatures above 25°C. Female flowers have two free ovaries, are sessile, and are found within a reduced leaf (sheath) on the short shoot. In contrast, male flowers have two anthers and are pedunculate. The fruit is ovoid to oblique and about 6 mm long and 3 to 5 mm wide.

Evolution and Biogeographic Distribution

Seagrasses may have originated from freshwater and estuarine hydrophytic relatives (Arbor, 1920) or from xerophytic salt marshlike plants (den Hartog, 1970). The proposal of a gradual transition of hydrophytic species into saline habitats was the prevailing view until den Hartog (1970) suggested that fossils from Cretaceous deposits in Japan (*Archeozostera*) and the Netherlands (*Thalassocharis*) represent primitive seagrasses. If seagrasses evolved from xerophytic plants that tolerated salt, they then would have to become tolerant of a hydrophytic habitat. The idea of a gradual evolution from freshwater relatives is attractive, but to date, the only "hard" evidence consists of the xerophytic fossils of presumed seagrass ancestors. Whatever the origin of seagrasses, they probably arose in the mid- to late Cretaceous (65 to 40 m ybp) after Angiosperms began to evolve and spread on land in the earlier portion of this period (ca. 120 m ybp).

Theories regarding the diversification (see Chap. 3) of seagrasses have used the vicariance hypothesis (McCoy and Heck, 1976), the center-of-origin concept (den Hartog, 1970), and foraminiferan fossil data (Larkum and den Hartog, 1989). It is believed that seagrasses initially evolved in the tropical Tethys and Paratethys seas. They then extended westward to the Neotropics (present Caribbean) by the middle to late Eocene as suggested by the vicariance hypothesis (McCoy and Heck, 1976). Eocene fossils include species of *Posidonia*, *Cymodocea*, *Thalassodendron*, and *Thalassia*. Because of their limited ability for dispersal (fruits are not buoyant for long periods), the center-of-origin theory (Chap. 3) is useful in explaining the diversification of seagrasses in Indonesia, Borneo, and New Guinea (Larkum and den Hartog, 1989).

Seagrasses are found on all continents, except Antarctica. Seven genera occur in tropical seas and five in temperate waters (den Hartog, 1970). Species of all seven tropical genera occur in the Indo-West Pacific area, a region of highest diversity, whereas the neotropical Caribbean has four genera. Only two species are common to both parts of the world, that is, *Halophila decipiens* and *Halodule wrightii*. This disjunct distribution of most species of seagrasses suggests a long period of separation and diversification from centers of origin, and the vicariance hypothesis may explain the existence of the following sister species:

Indo-West Pacific	Caribbean
Thalassia hemprichii	*T. testudinium*
Syringodium isoetifolium	*S. filiforme*
Halodule uninveris	*H. wrightii*

Morphological Adaptations

In adapting to a submerged habitat, seagrasses have evolved similar hydrophytic features (Dawes, 1981; Kuo and McComb, 1989). At least five abiotic properties (Chap. 2) should be considered when studying the adaptations of seagrasses (Arbor, 1920; den Hartog, 1970). They include (1) the osmotic effects of salt water; (2) availability of dissolved CO_2 and nutrients; (3) intensity and quality of submarine illumination; (4) higher density (than air) and mechanical drag of an aqueous medium; and (5) the effects of an aquatic medium on the dispersal of pollen and seeds.

The morphological similarities between different species of seagrasses are evident and include a creeping rhizome and apical meristem, which produce one or more roots and erect short shoots at each node. The *roots* are adventitious, they range from thick to fibrous in form, have a root cap, and produce root hairs.

The *rhizome* is an underground primary stem, which is usually herbaceous (but see *Amphibolis* and *Thalassodendron*); branching is either sympodial or monopodial, and its morphology is cylindrical to compressed. The clonal nature of a seagrass is demonstrated by its persistent rhizome and production of short shoots. Thus, the individual *genet* (genetic individual) consists of a series of interconnected *ramets*, or short shoots. Tomlinson (1974) stresses the importance of the rhizome in vegetative expansion, which, in the case of *Cymodocea nodosa*, is more effective than seedling establishment in nutrient-poor areas (Durate and Sand-Jensen, 1996).

The erect stems or *short shoots* of most species produce the flowers and foliage leaves, which are *determinate* and not involved in vegetative propagation. With the exception of *Halophila*, all seagrasses have basal *leaf sheaths* that are differentiated from the upper leaf *blades*. The former structure is colorless and persists after the blade portion has eroded away. In addition, the sheath covers the short shoot, protecting its apical meristem. The leaf sheaths of *Posidonia australis* also provide a gradient in osmotic pressure with higher concentrations of sucrose and amino acids occurring at their bases, which protects the meristematic leaf cells from exposure to the surrounding seawater (Tyerman, 1989). *Blades* develop from a basal (turtle grass) or intercalary (manatee grass) meristem, so that the oldest and most heavily epiphytized portion of the leaf is at its tip. In most species, the leaves are flexible yet strong due to internal fiber bundles. Thus, they are adapted to a denser medium with strong movement (drag). Leaf shapes vary from wide and flat (eelgrass, turtle grass), narrow and thin (*Halodule*), to terete (manatee grass) morphologies.

Anatomical Adaptations

The internal hydrophytic adaptations of seagrass organs show *reduction* and are more uniform between species than the diverse and complex xerophytic anatomy of salt marsh plants and mangroves. A characteristic feature of all seagrasses is the occurrence of air spaces, or *lacunae*, which are formed by schizogenous breakdown of cells, and *aerenchyma*, that is, a specialized tissue that occurs in the cortex or subepidermal tissues (Haberlandt, 1914). Air chambers are continuous in larger seagrasses; thus, they extend through the blades to short shoots, rhizomes, and the roots and provide gas exchange throughout the plant (Larkum et al., 1989b). The lacunae do have diaphrams or septae that have pores, which allow air exchange but block water to prevent flooding. Detailed summaries of seagrass anatomy can be found in den Hartog (1970), Tomlinson (1982), and Kuo and McComb (1989); Ancibar (1979) compared the functional anatomy of three genera in the Hydrocharitaceae (*Halophila, Thalassia, Enhalus*).

The *root* is covered by an epidermis below which usually an exodermis consists of one or more cell layers with lignified or suberized walls (Fig. 11-4). Root hairs develop from epidermal cells adjacent to transfer cells of the underlying exodermis, which probably facilitate lateral conduction of nutrients and water into the cortex. The cortex, which is usually parenchyma, contains air channels

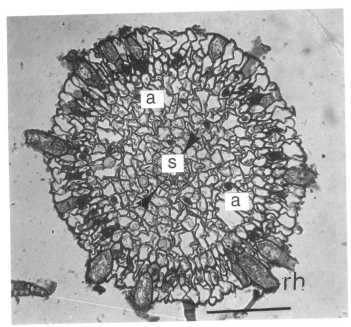

Figure 11-4.   A light micrograph of a young root of shoal grass, or *Halodule wrightii.* The small central stele (s) is surrounded by a single layer of endodermal cells (arrows). The cortex contains regularly spaced lacunae (a) and is covered by a hypodermal and epidermal layers, the latter producing root hairs (rh). Unit mark equals 1 mm.

(aerenchyma, lacunae). The root stele has an outer endodermis with a distinct, suberized Casparian strip in its cell wall. The stele contains a central xylary strand, with clusters of peripheral phloem (sieve elements) cells.

The anatomy of *rhizomes* and *short shoots* is similar, with a central stele of primary xylem, clusters of phloem (sieve) cells, and an outer endodermis (Fig. 11-5). Small vascular and weakly lignified fiber bundles may occur in the cortical region. In most species, lignification is limited (even in the central stele), making xylem identification difficult. The cortex may be divided into an outer, lignified, thick-wall exodermis and inner, thin-wall, starch-containing parenchyma with air spaces. Cell walls of the exodermis may have a suberized middle lamella. In most seagrasses, the epidermis is distinct in young stems and is covered by a cuticle.

The anatomy of *leaf sheaths* is similar to the *blade*, but the epidermal cells lack chloroplasts, making the sheath transparent. In contrast to the blade, the ground mesophyll of sheaths contains more air channels and lignified fiber bundles. After loss of the leaf, the sheaths usually break down, except for the persistent fiber bundles that appear as "hairs" at the base of older, short shoots. The anatomy of seagrass *blades* varies and it can be used for taxonomic delineations as well as demonstrating hydrophytic adaptations. The anatomy of a

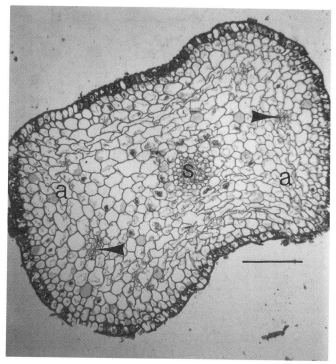

Figure 11-5. A light micrograph of the rhizome of *Halodule wrightii*. The central vascular stele (s) is surrounded by a single layer of endodermal cells. Cortical vascular bundles (arrows) are small as compared to the large cortex with lacunae (a). The young stem is covered by a single layer of epidermal cells containing tannin. Unit mark equals 200 μm.

mesophytic monocot leaf such as *Triticum vulgare* [wheat; Fig. 11-6(A)] can be compared with *Thalassia testudinum* [Fig. 11-6(B)] to demonstrate a number of hydrophytic adaptations (Table 11-2).

Epidermal cells of seagrass blades lack stomata and associated guard cells, and contain most of the blades' chloroplasts. As seen in *Syringodium filiforme* [Fig. 11-7(A)] and *Ruppia maritima* [Figs. 11-7(B) and 11-10], the outer wall is thick and covered by a thin, porous cuticle. Based on uptake studies, it is apparent the cuticle does not prevent absorption (e.g., CO_2, cadmium, manganese) by the blade (Larkum et al., 1989b). With the lack of an "effective" cuticle, any desiccation tolerance of intertidal seagrasses blades at spring low tides is probably due to the overlapping of blades rather than epidermal protection (Gessner, 1971). Tannin cells are common in the epidermis and ground mesophyll of many genera such as *Syringodium* (Fig. 11-9(a) and *Thalassia*. Epidermal cells may have masses of gelatinous polysaccharides and wall ingrowths; they have been cited as means to bind ions and increase surface area for osmoregulation

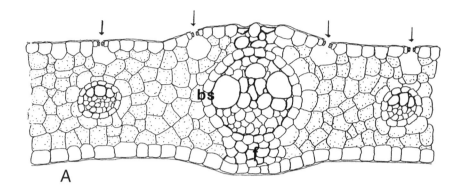

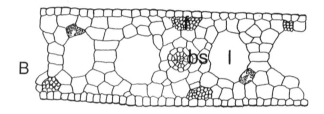

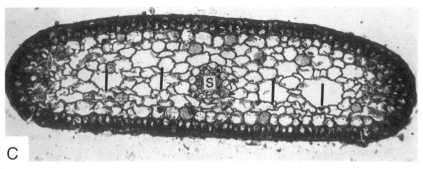

Figure 11-6.   Leaf diagrams of *Triticum vulgare* (wheat) and *Thalassia testudinum* (turtle grass) and a light micrograph of *Halodule wrightii* (shoal grass). The mesophytic (C_3) wheat leaf (A) has a main and two secondary veins, each surrounded by an endodermal bundle sheath (bs). Stomata (arrows) occur in the upper epidermis, and the leaf rolls upward, exposing the lower epidermis if water loss is high. Fiber bundles (f) are seen on the upper and lower side of the main vein. The diagram of a turtle grass leaf (B) shows the main vein with its bundle sheath (bs), fiber bundles (f), and lacunae (l). The light micrograph of the narrow leaf of shoal grass

(Fig. 11-12; Jagels, 1983; Kruzcynski, 1994). The apparent lack of plasmodes-mata between the epidermal and inner mesophyll cells (but between epidermal cells) suggests that photosynthates and nutrients diffuse through the cell wall via apoplastic movement.

In contrast to a wheat leaf [Fig. 11-6(A)], the mesophyll of a seagrass leaf [Fig. 11-6(B)] has few chloroplasts and a highly developed lacunar system. The lacunae have septa that prevent flooding if the leaf is damaged; they are spaced about every 0.5 cm in *Thalassia*. Septa consists of thin-walled parenchyma cells with small (up to 1 μm) intercellular air spaces that allow diffusion of gases along the lacunae from the leaf to the roots. Bundles containing unlignified fiber cells may be present (*Posidonia*) or absent (*Syringodium*). The vascular bundles (veins) of seagrass leaves are similar to those of mesophytic monocots, except that lignification is usually lacking or limited and the veins are reduced in size (Fig. 11-8). The veins are surrounded by a bundle sheath or endodermis with a suberized Casparian strip.

Reproductive Biology

An evaluation of seagrass flower, fruit, and seed structure supports the con-cept that the 13 genera show diverse evolutionary backgrounds (McConchie and Knox, 1989). Seagrasses show hydrophytic adaptations via hydrophyllous pol-lination, dioecious flowers, and modifications for fruit and seed dispersal. The absence of complete flowers demonstrates a reliance on cross pollination. Male and female flowers may occur on the same (monoecious, Posidoniaceae) or sep-arate plants (dioecious, Cymodoceaceae). The other two families have genera with both types of flowering, with 9 of the 13 genera being dioecious. The flow-ers may be stalked (pedunculate) or sessile, and the perianth may be lacking (*Posidonia*), reduced to a single whorl of tepals (*Thalassia*), or have three sepals and three petals (*Enhalus*). Usually, there are one to two stamens (but three in *Thalassia*); the carpels are either free (*Halodule*) or fused (*Enhalus*), and the style and stigma vary in length and size. Pollen grains range from spherical (*Thalassia*), to slightly elongate (*Halophila*), to moniliform (*Zostera*), and may be released in mucilagious strands (*Thalassia*) that aid in the attachment to the stigma. Fruits vary from achenes (*Zostera*) to capsules (*Enhalus*), and the fruits of some species (*Posidonia*) float. By contrast, no seagrasses have seeds that float. A few seagrasses show vivipary (*Thalassodendron*, Cymodoceaceae) where the embryo develops into a seedling before release of the fruit from the parent. Although sexual reproduction is known for all seagrasses, vegetative growth can be the primary means of expansion when sexual reproduction is curtailed (Witz and Dawes, 1995) or the seeds eaten (Fishman and Orth, 1996).

---

contains four lacunae (vertical bars), a central vein (S), and a dense epidermal layer in which the chloroplasts are confined. Both seagrass leaves have a modified mesophyll (aerenchyma), lack stomata, and have a thin, porous cuticle.

TABLE 11-2. A Comparison of the Leaves of Seagrass (*Thalassia testudinum*) and Wheat Plant (*Triticum*)

Structure	Mesophytic Grass (*Triticum*)	Seagrass (*T. testudinum*)
	Morphology	
Leaf shape	Linear, 1–2 mm	Ribbonlike, 5–10 mm
Venation	Parallel	Parallel
Meristem	Intercalary/basal	Intercalary/basal
	Anatomy:	
Cuticle	1–2 μm thick	Porous: 0.2–0.3 μm
Stomata	Lower epidermis, with trichomes and cork cells	Absent
Mesophyll (Palisade, spongy)	Mesophyll not distinct, surrounding veins	Ground mesophyll only, with lacunae
Xylem	Lignified	Lignin absent or limited
Bundle sheath	Obvious, layer of thin-wall endodermal cells	Present but not obvious in leaf
Fiber bundles	Lignified, at leaf margins	At margins, reduced, not lignified

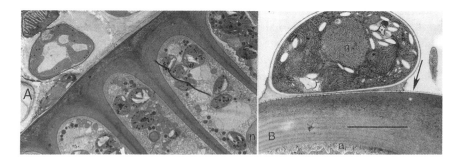

Figure 11-7. Electron micrographs of the leaf epidermis of manatee grass, or *Syringodium filiforme* (A), and widgeon grass, or *Ruppia maritima* (B). The calcified red algal epiphyte *Fosliella* sp. on *S. filiforme* has a pit connection between the basal and upper "star" cell. The seagrass epidermal cells have thick outer walls, numerous chloroplasts with plastoglobli (lipid deposits) and starch, and a nucleus (n) and mitochondria. A red algal spore is attached to the thin cuticle (arrow) covering the outer wall of *R. maritima*, which has a thin, convoluted layer of acid mucopolysaccharies (a) between the plasmalemma and cytoplasm. Unit marks equal 5 μm.

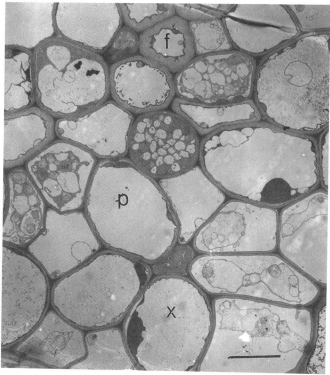

Figure 11-8. Electron micrograph of the main vein of the leaf of *Halodule wrightii*, or shoal grass. The phloem (p) has fiber cells (f) and the xylem (x) shows little differentiation when compared with a main vein of a mesophytic monocot leaf. Unit mark equals 10 μm.

PHYSIOLOGICAL ECOLOGY

Seagrass physiology also shows adaptations to a submerged marine habit. Thus, their rhizomaceous growth habit, the use of rhizomes as storage organs, and the occurrence of a lacunar system for gas exchange are all adaptive features. Kuo and McComb (1989) identify nine possibly important marine stresses that influence the distribution and abundance of seagrasses: salinity, temperature, water motion, anaerobiosis, nutrient limitation, epiphytes, irradiance, infection, and herbivory.

Clonal Habit

As noted previously, all seagrasses are clonal, with individual plants or genets being capable of expanding vegetatively via a rhizome that produces short shoots or ramets (Tomlinson, 1974). There are at least two possible advantages for being clonal: the ability to carry out vegetative expansion, or foraging (Cain, 1994), and sharing resources between ramets growing in areas of stress or low nutrients (Tomasko and Dawes, 1989; Wijesinghe and Handel, 1994). In a study of the clonal high marsh grass *Spartina patens*, Hester et al. (1994) found that severed ramets produced high levels of the osmoticum proline, whereas intact ramets did not. Thus, they concluded that there was physiological integration between ramets, with water being transported from intact ramets to those growing in high-salinity conditions. Similar findings were reported for the seagrass *Thalassia testudinum* with shaded ramets whose rhizome was severed on both sides, showing significantly lower blade growth than shaded ramets with intact connections (Tomasko and Dawes, 1989). Further, rhizomes and short shoots adjacent to shaded short shoots showed a depletion of soluble carbohydrate and protein, whereas those with severed rhizomes did not. Thus, the rhizome and short shoots function to mobilize sugars and amino acids to short shoots, whose productivity and growth are lowered due to shading.

Vegetative growth of seagrasses differs between species; some show aggressive growth (*Halodule wrightii*) and others do not (*Thalassia testudinum*). Species that exhibit aggressive growth are often chosen for revegetation of disturbed areas (Fonseca, 1994; Wyllie-Echeverria and Thom, 1994). For example, *Halodule wrightii* is preferred over *Zostera marina* and *T. testudinum* for transplanting. The types of growth (foraging) strategies of seagrasses parallel those of terrestrial clonal plants. Rhizomes of *H. wrightii* show a *guerilla strategy*, as they have the ability to reduce branching frequency and can spread rapidly into a heterogenous habitat, whereas *Z. marina* and *T. testudinum* rhizomes show a *phalanx strategy*, with the ability to branch more extensively and through slow, steady growth take advantage of available nutrients.

Organic Composition

Because of their conspicuous role in productivity and biomass (see "Ecology and Ecological Roles"), both the organic composition and caloric values of seagrasses have been studied (Dawes, 1981). As with other flowering plants, the proximate constituents of different seagrasses show pronounced seasonal and interspecific variations (Table 11-3). Rhizomes are the primary storage organ for soluble carbohydrates and proteins in *Thalassia testudinum, Syringodium filiforme, Halodule wrightii* (Dawes and Lawrence, 1980), *Halophila engelmannii* (Dawes et al., 1986), *Posidonia oceanica, Cymodocea nodosa* (Pirc, 1989), *Zostera marina* and *Z. noltii* (Dawes and Guiry, 1992). Further, rhizomes are the source of soluble carbohydrates and protein for regrowth of cropped blades and for initiation of spring blade growth (Dawes and Lawrence, 1979, 1980; Pirc, 1985, Dawes and Guiry, 1992). Sucrose and inositols are the two most important soluble carbohydrates mobilized for growth, and starch is the storage product in rhizomes of *Z. noltii* and *C. nodosa* (Pirc, 1989). Extractable carbohydrates from the rhizomes of *P. australis* come primarily from stelar rather than cortical tissues; further, they are more abundant in older regions rather than the

TABLE 11-3. Proximate Constituents in Seagrasses Including Kilojoules, Protein, Lipid, Soluble Carbohydrate, Insoluble Carbohydrate, and Ash[a]

Species	kJ	Protein (%)	Lipid (%)	SolCarb. (%)	InsolC. (%)	Ash (%)
Zostera marina						
Galway Bay, Ireland, 12-month data (Dawes and Guiry, 1992)						
Leaves	13–15	10–14	2.5–3.4	8–19	40–52	24–33
Short Sh.[b]	13–15	3–7	1.0–2.0	29–51	27–44	18–31
Rhizome[c]	13–15	4–8	0.6–2.8	31–59	18–38	18–28
Roots	8–11	4–6	0.6–2.8	13–20	27–40	38–62
Puget Sound, Washington, 14-month data (Dawes, unpub.)						
Leaves	13–16	4–16	1.7–5.7	9–17	32–56	19–34
Rhiz./Rt.	12–14	1–8	0.7–1.7	18–37	28–46	22-36
Izembeck Lagoon, Alaska, 12-month data (Dawes, unpub.)						
Leaves	10–16	4–13	1.7–3.9	12–28	20–52	14–47
Rhiz./Rt.	8–15	2–10	1.5–2.9	9–32	17–48	18–57
Posidonea oceanica						
Port-Cros, France (Lawrence et al., 1989)[d]						
Leaves	18	3.7–4.3	1.9–3.2	13–19	55–60	18–22
Rhizome	18	4.5, 4.8	1.3, 1.6	40, 47	42, 49	26
Roots	—	1.2, 1.5	3.2, 3.9	15, 21	58, 60	11, 16
Thalassia testudinum						
Gulf of Mexico, 12-month-data (Dawes and Lawrence, 1980)						
Leaves	10–13	8–22	0.9–4.0	6–9	34–41	29–44
Short Sh.	8–11	7–16	0.5–1.0	8–12	24–43	39–56
Rhizome	12–14	7–16	0.2–1.6	12–36	20–53	24–36
Syringodium filiforme (see *T. testudinum*)						
Leaves	10–13	8–13	1.7–6.2	16–22	32–42	28–33
Short Sh.	10–14	10–14	0.9–3.6	13–27	27–45	27–41
Rhizome	15–16	5–12	<.1–4.7	36–50	31–38	16–19
Halodule wrightii (see *T. testudinum*)						
Leaves	13–15	14–19	1.0–3.2	13–19	34–46	25–32
Short Sh.	12–13	5–9	0.8–3.5	16–31	25–53	25–36
Rhizome	14–16	7–9	<.1–1.6	40–54	20–34	14–22

[a]Abbreviations: kJ = kilojoules; SolCarb. = soluble carbohydrate; InsolC. = insoluble carbohydrate; Short Sh. = short shoots; Rhiz. = rhizomes; and Rt. = roots.
[b]First 4 cm of terminal rhizome.
[c]Second 4 cm of terminal rhizome.
[d]Data from analysis of plants from two sites.

TABLE 11-4. Cell Wall Constituents of Seagrass Leaves Including Lignin, Cellulose, and Soluble Carbohydrate[a]

Species.	Site (Ref.)	Lignin (%)	Cell. (%)	SolCarb. (%)
Thalassia t.	Gulf of Mexico (1)	0.2–2	18–32	1–9
Syringodium f.	Gulf of Mexico (1)	1–3	19–26	10–16
Halodule w.	Gulf of Mexico (1)	2–4	25–33	4–14
Halophila s.	Gulf of Acaba (2)	2	36	—
Halophila o.	Gulf of Acaba (2)	<1	33	—
Halodule uninervis	Gulf of Acaba (2)	<1	41	—
Heterozostera t.	Philip Bay (3)	5	20	—

Sources: (1) Dawes (1986), (2) Baydoun and Brett (1985); (3) Webster and Stone (1994).
[a]Abbreviations: Cell. = cellulose; SolCarb. = soluble carbohydrate.

growing tips (Ralph et al., 1992). Thus, carbohydrate storage is isolated inside the endodermis (i.e., stellar tissue) and is used by the young rhizome apex in growth.

Mean concentrations of carbon, nitrogen, and phosphorus in the leaves of 27 species of seagrasses were 34, 2, and 0.2%, respectively, of the dry weight giving a mean C:N:P ratio of 474:24:1 (Durate, 1990). The ratios are less than known for other marine macrophytes (550:30:1). In a review, Duarte found that the mean nitrogen and phosphorus contents were 1.8 and 0.2% of the dry weight, respectively, and suggested that changes in the N:P ratio might indicate changes in the water quality of seagrass habitats (e.g., entrophication).

Although the cell walls of seagrasses contain lignin, celluloses, and hemicelluloses, they have different levels than terrestrial monocots (Table 11-4). Soluble, nonstructural carbohydrates (hemicelluloses) can account for 1 to 16% of cell wall dry weight of seagrass leaves. In contrast, structural cellulose and lignin respectively account for 18 to 40% and 0.2 to 5% of the cell-wall dry weights of seagrass leaves. By comparison, tropical grasses such as tall fescue (*Festuca arundinaceae*) and bermuda grass (*Cynodon dactyon*) have 37 and 40% hemicellulose, 32 and 27% cellulose, and 10 and 6% lignin per g dwt^{-1}, respectively (Dawes, 1986). The adaptations of seagrasses to a hydrophytic habitat are seen in their flexible blades (i.e., few fiber bundles) and low amounts of lignin. The short life span of many seagrass leaves, ca. 15 days in *Thalassia testudinum* (Witz and Dawes, 1995; van Tussenbroek, 1996) and up to 100 days in other species, does not allow significant production of lignin but does make the blades edible for diverse grazers.

Photosynthesis

In general, seagrasses are shade plants (Hillman et al., 1989), with compensation values (I_c) of 25 to 50 μM photons m^{-2} s^{-1} and saturation irradiances (I_k) of less than 200 μmol photons m^{-2} s^{-1} (see Chap. 4), although submarine irradiances can be much higher. Studies comparing photosynthetic responses of seagrasses to different combinations of irradiance (i.e., I_c, I_k, P_{max} values) are dependent on temperature (Bulthuis, 1983) and salinity (Dawes et al., 1989). Typically, these physiological responses increase with increasing temperature within the genetic tolerance range of a species. The initial slope of the *P-I* curve α shows little relationship to thermal changes, as it is related to the light reactions of photosynthesis and is temperature-independent. Seagrasses photoacclimate to low irradiances via increased photosynthetic efficiency (α), decreased I_c and I_k, and increased chlorophyll content (Dunton and Tomasko, 1994). Photoinhibition occurs in *Halophila ovalis* (Ralph and Burchett, 1995) and in *Halophila decipiens* (Dawes et al., 1989), but not in *H. johnsonii*. However, the extent and type of photodamage in most seagrasses is not well understood (Hillman et al., 1989).

The management and protection of coastal and estuarine waters must consider limitations to light attenuation (via pollution) and its effect on the depth distribution of seagrass meadows (Dawes and Tomasko, 1988). Duarte (1991) predicted that seagrasses will colonize greater depths than freshwater angiosperms as long as the water transparencies are greater than 11% of surface irradiance. He concluded that differences in depth distribution (Z_e) in seagrass communities are due to differences in the light attenuation coefficient (K), as determined by a Secchi disc, and shown in Eq. 11-1. However, the proposed 11% average light requirement appears to be lower than that required by *Thalassia testudinum* (Fourqurean and Zieman, 1991) and other seagrasses (Onuf, 1991) when whole-plant photosynthetic quotients and daily irradiances were considered in Florida Bay and Texas, respectively.

$$Z_e = 0.26 - 1.07 \log\ K(\mathrm{m}^{-1}) \qquad (11\text{-}1)$$

Reduction of light via shading will limit photosynthesis in *Zostera marina* and decrease P_{max} leading to death (Goodman et al., 1995). Turbidity and resultant light attenuation are critical abiotic factors in the survival and distribution of seagrasses. For example, reduced water quality caused by a brown tide in the Laguna Madre, Texas, caused a decline in the biomass of *Halodule wrightii* within 2 years and the death of over 2.6 km^2 after 3.5 years (Onuf, 1996). The localized persistence of *H. wrightii* was probably due to utilization of stored reserves and reclamation of nutrients. Site selection for transplantation of seagrasses must take into account both short- and long-term fluctuations in turbidity, as seen for *Zostera marina* in San Francisco Bay (Zimmerman et al., 1991). Shading experiments showed that with reduced submarine irradiances, leaf growth, short shoot density, biomass, and primary production were reduced in *Posidonia sinuosa* (Gordon et al., 1994), *P. australis* (Fitzpatrick and Kirkman, 1995), and *Thalassia testudinum* and *Halodule wrightii* (Czerny and Dunton, 1995).

To date, carbon-fixation studies in all species of seagrasses, except possibly *Cymodocea nodosa* (Beer et al., 1980), have demonstrated the (C_3) Calvin cycle and ribulose-bis-phosphate carboxylase-oxygenase, or Rubisco (Abel and Drew, 1989). However, high levels of fixed carbon are incorporated into malate, aspartate, and other C_4 acids suggesting that there are some CAM-like variations of the traditional C_3 fixation scheme in seagrasses. Because the lacunae store respired CO_2, it is difficult to determine the level of photorespiration in seagrass blades, although it appears to be lower than those of terrestrial plants (Abel and Drew, 1989).

Internal Gas Transport

In contrast to terrestrial plants, seagrass blades have a thin cuticle and no stomata. The lacunar system, which is formed through schizogenous development of the mesophyll, is continuous throughout the plant, allowing for storage of CO_2 and diffusion of O_2 into the subterranean rhizome and roots. Seagrasses, being hydrophytes, experience slow diffusion rates of CO_2 and O_2 across the "unstirred" boundary layer of water covering the blades (Chap. 4; Fig. 11-9). During photosynthesis, the epidermal cells rapidly accumulate O_2, which diffuses through the cell wall and accumulates in the lacunae, reaching concentrations of more than 35% of the gases (Larkum et al., 1989b). Once in the lacunae, the gases diffuse from high to low concentrations and the lacunar diaphragm plates (septae) that occur every few millimeters do not affect this diffusion. The importance of gas diffusing from seagrass blades to their rhizome and roots is evident when the rhizosphere is studied. For example, there is a relationship between lacunar development and degree of anoxia of the rhizosphere (Penhale and Wetzel, 1983), with higher levels of O_2 around rhizomes of species having an extensive lacunar system. The relationships between O_2 and CO_2 levels in seagrass lacunae (i.e., the diffusion rates between blade and root), and their importance in the rhizosphere are still poorly known.

The boundary layer also limits CO_2 diffusion rates, but it can be overcome by increased water movement. For example, blade friction velocities (μ^*) of at least 0.25 cm s^{-1} will saturate photosynthesis in *Thalassia testudinum*, whereas 0.64 cm s^{-1} is required for *Cymodocea nodosa* (Koch, 1994). In the carbon dioxide equilibrium equation (Eqs. 2-12 and 2-13; Chap. 2), CO_2 accounts for 13 μM of the total inorganic carbon, and HCO_3^- accounts for 2.2 mM in seawater (Beer and Koch, 1996). The carboxylating enzyme Rubisco only uses CO_2. Even so, the more abundant HCO_3^- dissolved in seawater can be taken up by a variety of seagrasses (Beer and Waisel, 1979) due to the enzyme carbonic anhydrase that controls CO_2 movement, although not at the level of macroalgae (Beer and Koch, 1996; Beer 1996). The importance of carbonic anhydrase is shown in Figure 11-9 for the gas exchange of seagrass blades (Larkum et al., 1989b). Controlling factors for uptake of HCO_3^- and CO_2 by seagrass epider-

---

*The slow diffusion rates in water, relatively low CO_2 levels, and low affinity of Rubisco for CO_2 all must be overcome by submersed macrophytes (Beer, 1996).

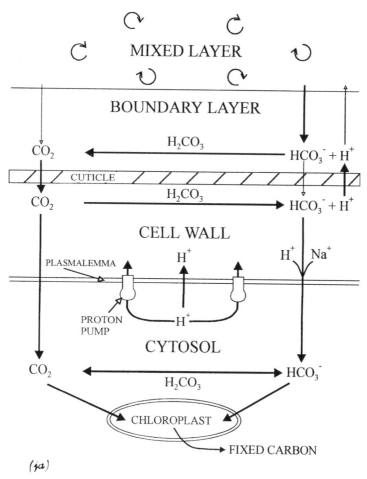

Figure 11-9. The movement of CO_2 and HCO_3^- into the epidermal cells of a seagrass (modified from Larkum et al., 1989, with permission of Elsevier Science Publishers). Carbon dioxide diffuses from seawater through the boundary layer, cuticle, cell wall, and plasmalemma, following a high to low concentration gradient. By contrast, after diffusion through the boundary layer, transport of $H_2CO_3^-$ requires energy and the use of proton pumps (changes in pH) to cross the cell membrane.

mal cells include low pH, carbonic anhydrase, and availability of the hydrogen ion.

Salinity and Water Relations

Photosynthetic responses can be used to measure tolerances of seagrasses to salinity (Dawes et al., 1986, 1989). For example, estuarine populations of *Halophila engelmannii* and *H. johnsonii* show broader tolerances than oceanic populations of *H. engelmannii* and *H. decipiens* after being grown in various combinations of temperature and salinity. The ability of seagrasses to tolerate salinity can be seen by their anatomical and physiological traits; in some cases, they parallel halophytic and xerophytic plants (Tyerman, 1989).

Light and electron microscopic studies of epidermal cells of *Thalassia testudinum*, *Zostera marina*, and *Ruppia maritima* from high salinities show an invaginated plasmalemma with associated mitochondria; also, they lack plasmodesmata, have a reduced vacuole, and increased amounts of acidic mucopolysaccharides (Jagels, 1983; Jagels and Barnabas, 1989; Iyer and Barnabas, 1993). The increase in cell membrane surface area and associated mitochondria is characteristic of cells active in osmoregulation, and acidic carbohydrates may be involved in ion transport and binding. In a study of populations of a single genetic strain of *R. maritima*, Kruzcynski (1994) found that plants grown in high salinities exhibited these features (Fig. 11-7B), whereas those grown in lower salinities did not. The leaf sheaths of *Posidonia australis* and *Zostera capricorni* have an osmotic gradient (i.e., high to low salt), which protects the meristematic leaf tissue from direct contact with seawater (Tyerman, 1989). As the growing leaf elongates, the maturing leaf is gradually exposed to ambient salinities, whereas the dividing cells below are protected by the outer sheaths.

Osmotic roles of proline, alanine, and glutamate have been identified in seagrasses, including *Ruppia* (Brock, 1981; Adams and Bate, 1994) and *Zostera* (van Diggelin et al., 1987). The amino acid proline increases with increasing salinities in a variety of seagrasses (Pulich, 1986; van Diggelen et al., 1987; Adams and Bate, 1994). The ability of seagrasses to regulate internal NH_4 levels was positively correlated with their ability to osmoregulate (Pulich, 1986). Although osmoregulation data are limited, seagrass blades appear to respond to increased salinities by initial use of inorganic ions (Na^+, K^+, Cl^-) and then by the production of organic osmolites such as proline and glycenebetaine (Tyerman, 1989). A cDNA clone corresponding to the gene ZHAI was isolated from *Zostera marina* epidermal cells, where it was expressed only after the leaves were exposed to seawater (Fukuhara et al., 1996). This gene is known to control the plasma membrane H^+-ATPase activity and may be important for excretion of salt.

ECOLOGY AND ECOLOGICAL ROLES

The complex ecology and multiple roles seagrass communities carry out are reasons for maintaining and improving seagrass communities. Like mangrove and salt marsh communities, seagrasses are important primary producers, stabilize the substrata, serve as habitats and nurseries, and are direct and indirect food for diverse fauna. Further, these submerged flowering plants can be used to monitor the "health" of coastal and estuarine communities.

Sediment Ecology

Seagrasses serve as sediment traps by acting as baffles (Ward et al., 1984; Komatsu, 1996) and stabilizing bottom sediments, which will improve water quality (den Hartog, 1967). Except for siphonaceous green algae, seagrasses are the only submerged marine macrophytes that are rooted and depend on the sediment for their nutrients (Duarte and Sand-Jensen, 1996). The below-ground portion (roots, rhizomes, short shoots) may reach 90% of their total biomass. Seagrasses physically and chemically change the sediment through release of oxygen, decomposition of subterranean parts, and by bioturbation through growth of their roots and rhizomes (Moriarty and Boon, 1989). Chemical changes wrought by the below-ground components are critical for survival because the high levels of sulfides in the anoxic sediments are toxic to the plant. For example, the die-off of *Thalassia testudinum* in Florida Bay may be due to the lowering of the plant's resistance to a parasitic slime mold by the stress of higher sediment sulfides (Durako, 1994). In addition to the oxidation of sediments, seagrasses release organic matter. For example, 6% to 28% of ^{14}C fixed by *Halodule wrightii* blades during photosynthesis was released in the sediments by the rhizome and roots within six hours (Moriarty et al., 1986). The excreted organic compounds support bacterial activity, causing elemental cycles like nitrification in the sediments.

Ammonium is the most abundant form of nitrogen in the sediment. It is dissolved in the interstitial water (5 to 100 μM to 15mM), bound with organic matter and clays (exchangeable), and fixed irreversibly into the clay lattice. The first two sources of ammonium are available to seagrasses, so that analyses for available ammonium should include extraction techniques to measure the dissolved and exchangeable portions. Ammonium regeneration can be rapid, with rates ranging from 56 to 490 mg N m^{-2} d^{-1} (Moriarty and Boon, 1989). Based on growth rates, seagrasses require large amounts of fixed nitrogen (10 to 450 mg N m^{-2} d^{-1}), with ammonium being the preferred source. By contrast, nitrate and nitrite concentrations are usually low (<5 μM) in the anaerobic sediments, presumably due to their rapid utilization by denitrifying and other anaerobic bacteria. Nitrogen fixation occurs in seagrass rhizospheres, with rates of 0.1 to 7.3 mg N m^{-1} d^{-1} for *Zostera noltii* (Welsh et al., 1996) and 20 to 40 mg N m^{-2} d^{-1} for *Thalassia testudinum* (Patriquin and Knowles, 1972). Further, nitrogen-fixing *Klebsiella* has been isolated from the root and rhizomes of *Halodule wrightii* (Schmidt and Hayaska, 1985).

Seagrasses show a variety of responses to limiting nutrients. Phosphorus (usually <20 μmol) may be the limiting nutrient in seagrass meadows (Moriarty and Boon, 1989), particularly where calcium carbonate binds the dissolved PO_4. In Florida Bay, phosphorus apparently controls the successional sequence between the early stage *Halodule wrightii* and the late-stage *Thalassia testudinum* (Fourqurean et al., 1995). *Halodule* was found to have a higher nutrient requirement than *Thalassia* and can replace turtle grass in areas of natural or anthropogenic eutrophication. However, in the Indian River Lagoon, ammonium was found to be the limiting nutrient during peak growth of *Syringodium filiforme* (Short et al., 1993). In another study, Pulich (1985) postulated that *Ruppia maritima* can grow on low-nutrient sediments, whereas *H. wrightii* preferred organic-rich substrates having substantial sulfate reduction. In Denmark, it was reported that *Zostera marina* can conserve nitrogenous compounds, take them up from the water column and sediment, and maintain high growth rates even when internal nitrogen is low (Pedersen and Borum, 1992).

Bacteria in seagrass communities have similar roles to those of terrestrial habitats. Bacterial activity is significantly higher in *Halodule wrightii* beds than in adjacent bare sand areas and is correlated with seagrass biomass rather than bacterial biomass (Moriarty et al., 1986). Such correlations demonstrate the role of bacteria in modifying seagrass detritus and in the utilization of organic exudates. Diel variations of dissolved ammonia and phosphate occur in estuarine sediment pore water; similar patterns for bacterial growth rates occur in many seagrass communities (Moriarty and Boon, 1989). Such fluctuations apparently follow the exudate cycle, as described above (see Moriarty et al., 1986), which is linked to photosynthesis.

Seagrass Dynamics

Disturbance can be defined as the source (abiotic or biotic) of a *perturbation* (Rykiel, 1985; Chap. 3). For example, the abiotic disturbance of ice scouring on *Zostera marina* (Robertson and Mann, 1984) and the biotic disturbance of grazing by dugongs on *Halodule uninveris* (Longh et al., 1995) both cause perturbations in seagrass meadows. Seagrasses are adapted to natural disturbances such as hurricanes in Hervery Bay, Queensland (Preen et al., 1996), and wave-induced blowouts at Barbados, West Indies (Patriquin, 1975). Seagrasses also tolerate varying levels of salinity, light fluctuations, and other abiotic factors, suggesting that they are *stress-adapted* (Grime, 1977; Chap. 3).

Seasonal fluctuations and patch dynamic studies indicate that seagrass meadows are not static even under "stable" environmental conditions. Monthly fluctuations in species abundance and biomass in North Queensland was linked to wet and dry season cycles (Lanyon and Marsh, 1995), to tidal exposure and water motion in South Sulawesi, Indonesia (Erftemeijer and Herman, 1994), and to periphyton shading and mudsnail grazing of *Zostera noltii* in the Dutch Wadden Sea (Philippart, 1995). Competition has also been noted, although stress appears to be more significant in controlling seagrass communities. Williams (1987) noted that shading by *Thalassia testudinum* resulted in reduced *Syringodium filiforme*, and Stewart (1989) found the intertidal seagrass *Phyllospadix torreyi* could outcompete a turf of *Corallina pinnatifolia*.

Den Hartog (1967) recognized six growth forms of seagrasses that show adaptations to specific environments. Species with simple morphologies such as *Halodule* are early (pioneer) plants that stabilize the substrata. More complex forms such as *Thalassia* or *Zostera* can follow, resulting in stratification of the meadow. All seagrasses show two types of reproductive strategies: vegetative (established) and reproductive (regenerative). Typically, the reproductive phase will dominate under optimal environmental conditions, whereas the vegetative phase is most critical after disturbances or at the extremes of its distribution. Both phases must be considered when a seagrass is evaluated in terms of stress, competition, and disturbance (Grime, 1977).

Succession in seagrass communities has been demonstrated by their recolonization after perturbations such as hurricanes (Preen et al., 1995) and blowouts (Patriquin, 1975). Long-term persistence of *Phyllospadix scouleri* continued over a three-year study in spite of removal and colonization by macroalgae on a rocky intertidal shoreline in Oregon; thus, it is a climax species (Turner, 1985). Williams (1990) proposed that succession of seagrasses in the Caribbean is dependent on sediment nutrient supply. She found that rhizophytic algae colonized bare sediment, followed by *Syringodium filiforme*, and then by *Thalassia testudinum*, which outcompeted the first seagrass for light. Throughout this succession, the level of sediment nutrient rose. Studies by Fourquean et al. (1995) in Florida Bay suggest that *Halodule wrightii* will replace *T. testudinum* as sediment nutrient levels rise.

The effects of *disease* have not been well documented for most marine plant communities except for seagrasses. The *wasting disease* of *Zostera marina*, caused by a pathogenic slime mold *Labyrinthula zosterae*, has resulted in significant losses of eelgrass meadows (Muehlstein, 1989). By the fall of 1932, two years after the discovery of the disease, more than 99% of eelgrass on the Atlantic coast had died. The symptoms of the disease were initial dark areas on the blades that increased in size at the rate of 3 to 5 dm d^{-1}. Loss of photosynthetic leaves then resulted in death of the rhizome. Recovery by eelgrass was slow, with almost complete recovery by 1944 when a second, less extensive recurrence of wasting disease occurred. A significant decline in eelgrass was reported for New England in 1984, with lesser effects on the European coast (Short and Wyllie-Echeverria, 1996). The wasting disease appears to be linked to stressful abiotic factors (light intensity, temperature, salinity, pollution), lowering phenolic compounds in *Z. marina* (Vergeer et al., 1995). It is thought that phenolic compounds are effective in prevention of the disease; thus, environmental stresses can cause a decline in these compounds, allowing the disease to develop.

A die-off of *Thalassia testudinum* in Florida Bay began in 1987, resulting in the loss of 4000 ha. A parasite, another species of *Labyrinthula*, was found on the blade surface of turtle grass and caused extensive lesions [Figs. 11-10(A) and 11-10(B)]. Various species of slime molds occur in seagrass blades throughout the world, indicating that the relationship is more common than initially thought (Vergeer and den Hartog, 1994). For example, *Halodule wrightii* is commonly infected by another slime mold, *Plasmodiophora diplantherae*, which causes enlarged internodes and white galls (Braselton and Short, 1985). The turtle grass die-off in Florida Bay has been linked to a species of *Labyrinthula*, and Durako and Kuss (1994) propose that the necrosis of turtle grass blades by *Labyrinthula* would reduce photosynthesis, with the lower oxy-

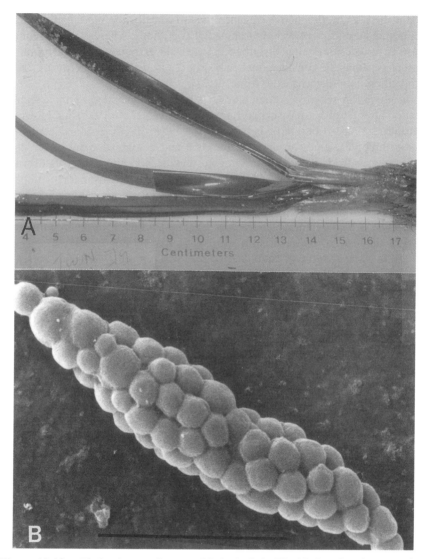

Figure 11-10. *Thalassia testudinum* and *Labyrinthula*. The blades of turtle grass taken from Florida Bay show lesions due to an infection by the slime mold *Labyrinthula* (A). The sporangia of the slime mold (B) have been found on the blade surface (courtesy of J. Landsberg and B. Blakesley, Florida Marine Research Institute). Unit mark equals 10 μm.

gen resulting in hypoxia and sulfide toxicity in the rhizosphere. Reduction of photosynthesis in another seagrass, *Zostera marina*, does cause root anoxia and limitation of carbon transfer from the blades (Zimmerman and Alberte, 1996). However, the question as to why *Labryinthula* has been able to infect turtle grass in Florida Bay is not easily understood. Infection has been explained on the basis of stress to the seagrass due to hypersalinity (Roblee et al., 1991), although eutrophic water from the Everglades agricultural area in South Florida (Belanger et al., 1989) may be a likely stressor (Lapointe et al., 1996; see also Chap. 5).

Biomass and Productivity

Characterizations of seagrass biomass includes determination of standing stock (g d wt m^{-2}) and short-shoot densities (#ss m^{-2}). Biomass data are usually limited to the total of above-ground material or only the photosynthetic (green blade) tissue. Such data are available for many of the larger seagrass taxa like *Thalassia* (McRoy and McMillan, 1977; Dawes, 1981; Hillman et al., 1989), but not for the smaller species like *Halophila* (Table 11-5). In general, the below-ground biomass usually constitutes 50 to 90% of the standing stock, with smaller seagrasses having lower ratios (*H. ovalis*: 1:1; Hillman et al., 1989) and larger seagrasses having higher ratios (*T. testudinum*: 1:3; Zieman and Zieman, 1989). Standing stock is based on the environmental conditions, particularly irradiance, so that populations found at the depth limit of a species will have lower biomass and short-shoot densities (Duarte, 1991).

The amount of leaf area available for photosynthesis can be used to determine the potential productivity in seagrasses. Leaf area indices (LAI; m^2 m^{-2}), or the leaf area of one side expressed relative to the ground area covered, correlate with increased PAR received (McRoy and McMillian, 1977). Terrestrial plants have LAIs ranging from 4 to 6 (broad-leaved trees), to 9 to 10 (grasses and cereal crops), to 20 (tropical rainforest broad-leaved trees; McRoy and McMillian, 1977; Hillman et al., 1989). Seagrasses either equal or exceed these values, as shown for Australian (*Posidonia australis*, 4.9 to 6.5; *Amphibolis antarctica*, 3.3 to 18.4), Mediterranean (*Posidonia oceanica*, 6.9 to 8.1; *Cymodocea podosa*, 8 to 11), Caribbean (*Thalassia testudinum*, 18.6), and Alaskan species (*Zostera marina*, 12 to 21). The large LAIs of seagrasses reflect the high density of blade cover in relation to ground area, their adaptation to submarine (lower) irradiance, and the need for water movement to enhance nutrient and gas diffusion, and to overcome the effect of shading by epiphytization.

Production studies in seagrasses have been primarily limited to above-ground productivity (leaf growth, photosynthesis; Table 11-5) with little data available for below-ground production. Leaf growth rates have been measured by punching a hole at the base of the leaves (Zieman, 1974) or by tagging larger species. Blade growth of small species can be followed by marking with an underwater pen or time-lapse photography (Hillman et al., 1989). The short shoots with marked blades are harvested one to two weeks later, and leaf growth is determined by separating the new and old portions of the marked blades, and drying and weighing them. Leaf growth can be expressed per short shoot of biomass (g dwt ss^{-1} d^{-1}) or in terms of new blade growth (g dwt leaf d^{-1}); data from the

TABLE 11-5. Standing Stock (Biomass) and Productivity of Different Seagrasses

Species	Locality	Biomass (g dwt m^{-2})	Productivity (gC m^{-2} d^{-1})
Amphibolis			
A. antarctica	W. Australia	340–680	
A. griffthii	W. Australia	220–1150	
Halodule			
H. uninervis	New Guinea	150	
H. wrightii	North Carolina	105–200	0.5–2.0
	Texas	220–250	
Halophila			
H. ovalis	India	48	
	W. Australia	40–60	
Posidonia			
P. australis	S. & W. Australia	140–453	
P. oceanica	Mediterranian	543–1072	1.5–5.0
P. sinuosa	W. Australia	500–600	
Syringodium			
S. filiforme	Florida	200	
S. isoetifolium	New Guinea	327	
Thalassia			
T. hemprichii	Queensland	70	
T. testudinum	Florida	20–8100	0.4–2.5
	Cuba	20–800	9.3–12.5
Zostera			
Z. marina	E. North America	15–2062	0.2–2.6
	W. North America	62–1840	0.7–3.8
	W. Europe	70–443	0.4–7.9
	Japan	70–235	

Sources: Modified from McRoy and McMillan (1977); Hillman et al. (1989).

latter technique can be used to determine turnover rates. New leaf production also can be expressed as g C d^{-1} by ashing the blade material. As seen in Table 11-5, blade productivity is in the range of 0.2 to 18.7 g C m^{-2} d^{-1}, with the upper ranges of productivity equal to that of phytoplankton (see Table 7-1).

Measurement of leaf photosynthesis also can be used to determine above-ground productivity. The idea that oxygen released during photosynthesis and stored in the lacunae (<10%) is internally recycled (Zieman, 1974) is no longer considered a limiting factor to using oxygen meters. Below-ground production and turnover rates in seagrasses are less well known (Hillman et al., 1989). Pioneer species (*Halophila ovalis*) show higher rates, with 50% of its below-ground biomass turned over within a year, which equals its total above-ground biomass. By contrast, slower-growing "climax" species have reduced rates of below-ground turnover; *Thalassia hemprichii* (11% to 27%, with 100% turnover

TABLE 11-6. Types of Dominant Faunal Groups of Mangals Worldwide

Faunal Group	Taxa Present
Infauna	
Unicellular	Protozoa, bacteria
Meiofauna	Nematodes, polychaetes, harpacticoid copepods
Macrofauna	Polychaetes, bivales, amphipods, holothurians
Motile epifauna	
Microfauna	Protozoans
Meiofauna	Ostracods, nematodes, rotifers
Macrofauna	Amphipods, isopods, gastropods, echinoderms
Sessile epifauna	Hydroids, bivalves, bryozoans, sponges
Epibenthic fauna	Fish, decapods, cephalopods

Source: After McRoy and Helfferich (1977).

in 2.6 y), *Zostera marina* (22% to 31%, with 100% turnover in 140 to 193 d), and *Cymodocea rotundata* (33%, with 100% turnover in 135 days).

Nursery and Habitat

The importance of seagrass communities as *nurseries* and *habitats* for diverse animals are described in several publications (Zieman, 1982; Heck and Thoman, 1984; Phillips, 1984; Thayer and Fonseca, 1984; Larkum et al., 1989a; Zieman and Zieman, 1989; Seeliger, 1992; Kuo et al., 1996); that is, seagrasses provide food and refuge (Virnstein, 1987) for a variety of fish, invertebrates, and turtles. They also serve as habitats for invertebrates (Virnstein, 1987), fish (Stoner, 1984), turtles (Williams, 1988), dudongs (Longh et al., 1995), and manatees (Lefebvre et al., 1989); all major faunal groups are also present within these communities (Table 11-6). The infauna and epipfauna of seagrass beds are known to serve as prey for larger invertebrates and fish (Virnstein, 1987). For example, in a *Halodule wrightii* meadow in Apalachicola Bay, Florida, 58 species of infaunal and epifaunal species showed seasonal fluctuations due to influxes of juvenile fishes and crabs that preyed on them (Sheridan and Livingston, 1983). Eelgrass beds in lower Chesapeake Bay serve as an important nursery for juvenile blue crabs with total seagrass-associated secondary production being around 200 g dwt m^{-2} y^{-1} (Fredette et al., 1990). Similarly, eelgrass beds cover over 10 km^2 of Great Bay, New Hampshire, where they also serve as nurseries and habitats for scallops, crabs, finfish, geese, and ducks (Short, 1992). The abundance and diversity of icthyofauna in seagrass meadows is well known (Stoner, 1984) and the data have been used to argue for protection of these communities (Ogden, 1980). The role of benthic algae in seagrass meadows is less understood, although drift species are known to serve as habitats and food sources for gammaridean amphipods (Virnstein and Howard, 1987).

Seagrasses also serve as basiphytes for microalgal and macroalgal *epiphytes* (McRoy and McMillan, 1977; Harlin, 1980; Borowitzka and Lethbridge, 1989). Epiphytism may solve the space crunch, as described in Chap. 3, but it can create problems of nutrient and light limitation for the basiphytes. Complex relationships between epiphytic algae and the seagrass host occur with regard to production and irradiance (Libes, 1986). Transfer of organic carbon, nitrogen, or phosphorus from the seagrass to the epiphytic community is suggested by the lower levels of alkaline phosphatase activity in epiphytes (Lobban and Harrison, 1994). Seagrass epiphytes can result in biomass reduction and lower short-shoot densities where eutrophication occurs. That is, epiphytes increase drag on the blades, resulting in their removal; they shade the leaf, reducing available irradiance (up to 63% on *Posidonia australis*; Borowitzka and Lethbridge, 1989); and they limit the diffusion of CO_2 or other nutrients to the blades. It is no wonder that most seagrasses have rapid blade turnover, ranging from 11 to 100+ days. Small grazers (e.g., copepods) can play a role in epiphyte control (Fry, 1984; Virnstein, 1987).

Seagrass epiphytes serve as primary producers, as a habitat for small invertebrates, and as direct food (Fry, 1984). Most important, epiphytes are highly productive. Based on productivity studies of seagrasses blades and their epiphytes, the epiphytic community was found to contribute significantly to primary production in meadows of *Thalassia testudinum* (25% to 33%), *Halodule wrightii* (48% to 56%), and *Posidonia australis* (> 60%; see Borowitzka and Lethbridge, 1989). Seagrass epiphytes range from bacteria and microalgae (diatoms, dinoflagellates; Fig. 7-20) to macroalgae [Fig. 11-11(A)] that attach directly on the cuticular surface [Figs. 11-7(A), 11-7(B), and 11-11(B)]. Harlin (1980) lists over 400 macroalgal and 150 microalgal epiphytes of seagrasses, and Borowitzka and Lethbridge (1989) describe their biology and ecology. Older, apical regions of blades tend to show the highest epiphytization and increasing diversity. Encrusting coralline algae (e.g., *Fosliella, Lithoporella, Pneophyllum*) tend to be early colonizers. Epiphyte-basiphyte specificity is lacking for most macroalgal epiphytes, although *Smithora naiadum* appears to be specific to *Zostera marina* and *Phyllosphadix scouleri* (Harlin, 1980).

Figure 11-11. Epiphytes of seagrass blades. The blades of *Thalassia testudinum* collected in Tampa Bay, Florida (A), are epiphytized by diverse macroalgae, including *Chondria tenuissima*; filamentous "fuzzes" of red (*Polysiphonia subtilissima*), green (*Enteromorpha chaetomorphoides*), and brown (*Hummia onusta*) algae; microalgae including diatoms and blue-green algae are also common epiphytes. The electron micrograph (B) shows an epiphytic filamentous brown alga (b), bacteria, and a red algal spore (r) growing on a leaf of *R. maritima*. The epidermal cell contains many chloroplasts and a nucleus (n). Unit mark equals 5 μm.

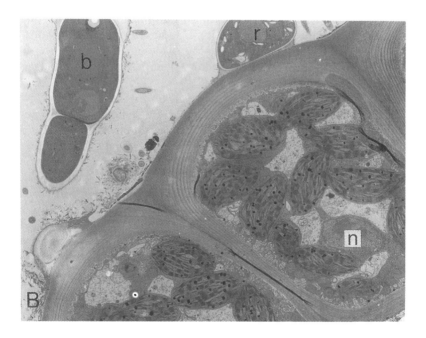

Secondary Production

Using stable carbon-isotope ratios, Fry and Parker (1979) found that seagrasses and benthic algae in Texas bays are major sources of nutrition for juvenile shrimp and fish. It appears that seagrass epiphytes, not detritus or living seagrass tissue, are important sources of food for invertebrates. In turn, the smaller invertebrates are prey for fish and decapod crustaceans (Virnstein, 1987). Seagrasses serve as direct food sources for only a few invertebrates, fish, turtles, water fowl, and manatees (Odgen, 1980; Thayer et al., 1984; Klumpp et al., 1989). However, they have a larger role as an indirect food source via detritus production and as a substrate for epiphytes (Klumpp et al., 1989). A variety of birds, including Brent Geese, Widgeon, and Teal, are known to eat *Zostera noltii* in southern England and Ireland (Tubbs and Tubbs, 1983). Thayer et al. (1984), in a review of grazing by larger herbivores, concluded that the paucity of direct seagrass grazers may be a function of the low levels of nitrogen in the plants, as well as their extensive cellulose cell walls, and the occurrence of antiherbivory compounds such as phenolics. Of the 154 grazers listed by McRoy and Helfferich (1977), the most important invertebrates were echinoderms, molluscs, and crustaceans. Seagrasses constituted up to 50% of the total diet for crustaceans (Klumpp et al., 1989). Crabs, isopods, and sea urchins are the primary direct grazers on seagrasses, whereas other invertebrates use seagrass detritus and their various algal epiphytes. Fry and Parker (1979) and Fry (1984) demonstrated through carbon isotopic studies ($\Delta\ ^{13}C:^{12}C$ ratios) that benthic microalgae of the sediment and seagrass blades are a major source of carbon for grazers. In some seagrass meadows, microalgae and macroalgae can dominate the food chains (Virnstein, 1987).

Long-term impacts of grazers have been shown for the sea urchin (*Lytechinus variegatus* on *Thalassia testudinum* (Heck and Valentine, 1995), the dugong on *Halodule uninvervis* (Longh et al., 1995), and the limpet *Tectura depicta* on *Zostera marina* (Zimmerman et al., 1996). In the first study, Heck and Valentine (1995) proposed that urchins and seagrass coexist in Gulf of Mexico beds because there is a balance among intensive grazing, resultant loss of habitat, and predation by fish on the urchins. After protection from predator fish via enclosures, there was intense grazing by the urchins that was most destructive in the winter when turtle grass could not recover as rapidly. Significant reductions in above- and below-ground biomass occurred and these losses were apparent 3.5 years after grazing had ceased. In the second study, Longh et al. (1995) reported a seasonal pattern with *H. uninervis* in the Moluccas, East Indonesia. Thus, regrowth of grazed areas took five months in the wet season (winter) and it did not occur during the dry season (summer) due to intense grazing by dugongs that removed 93% of the shoots and 75% of the below-ground biomass. Dugongs prefer to graze plants with high levels of organic carbohydrate (i.e., in the summer). Perhaps, the dense seagrass beds now present in the Gulf of Mexico reflect the loss of manatees, and historically these beds were less abundant due to intense grazing. However, the historical losses of seagrasses due to anthropogenic causes confuses the issue. Zimmerman et al. (1996) report that the northward extension of the eelgrass limpet has resulted in a catastrophic decline of *Z. marina* due to removal of the chlorophyll-rich epidermis and subsequent carbon limitation to the rhizome. Thus, the limpet is a top–down controller via grazing on a bottom–up mechanism, namely, photosynthesis of eelgrass.

Methods

The FAO publication, *Seagrass Research Methods* (Phillips and McRoy, 1990) gives detailed procedures for the study of seagrasses. The text includes field and laboratory techniques for study of above- and below-ground biomass; short-shoot density; leaf, root, and rhizome production; epiphyte load; and proximate constituents. Standard measurement techniques for marine plant communities are given in App. I.

SEAGRASSES AND HUMAN AFFAIRS

The need for conservation and management (Chap. 5) of coastal seagrass meadows is evident when their extensive ecological roles are considered. Thus, their declines, which have occurred worldwide, have been linked to natural and human-induced disturbances (Short and Wyllie-Echeverria, 1996). Reports of seagrass declines include studies in western Australia (Cambridge et al., 1986); Chesapeake Bay, United States (Orth and Moore, 1983); Great Bay, New Hampshire (Short, 1992); and Florida Bay (Durako, 1994). Natural disturbance to seagrass communities range from volcanic activity (Fortes, 1991) to blowouts by waves and currents (Patriquin, 1975), and tropical storms (Birch and Birch, 1984; Pulich and White, 1991; van Tussenbrock, 1994; Preen et al., 1996). Biological damage to seagrass beds include overgrazing by urchins (Camp et al., 1973), competition (Villele and Verlaque, 1995), bioturbidation (Short and Wyllie-Echeverria, 1996), and disease (see above). As on land, natural fluctuations in seagrass biomass and development occur (Philippart, 1995), which may reflect seasonal cycles (Baron et al., 1993) and must be monitored in ports and harbors (Long et al., 1996).

Uses and Transplantation

Direct uses of seagrasses in the past were limited to harvesting of wrack for fodder, production of paper pulp, and harvesting the rhizome (e.g., *Zostera marina*) as a direct food. More recently, seagrass communities have been transplanted to stabilize sediments after dredging events (McRoy and Helfferich, 1977) and to reestablish damaged beds (Fonseca, 1994). The Seri Indians on the Gulf of California use *Z. marina* to make dolls by using the dried grass blades; they also harvest the fruit and make flour from the seeds (Felger and Moser, 1973).

The need for a source of transplants has resulted in culture studies using flasks, aquaria, and outdoor tanks. For example, Caribbean seagrasses have been grown in 1.5 m² tanks for two or more years (Short, 1985) and a variety of species have been cultured in aquaria (McMillan et al., 1981). All of these culture programs were experimental and are too costly to be used as a nursery for transplantation. The one exception is *Ruppia maritima*, the growth of which has been carried out from seed (Koch and Dawes, 1991) using defined media (Koch and Durako, 1991; Bird et al., 1996). Thus, *R. maritima* is now considered to be a viable cultivar for restoration of seagrass communities (Koch and Durako, 1991).

A critical aspect in the restoration or creation of a marine plant community is the availability of cultivars for transplantation. Seedlings of salt marsh (*Juncus* spp., *Spartina* spp.) and mangrove plants (*Rhizophora* spp., *Avicennia* spp.) can be purchased from nurseries. With the exception of *Ruppia maritima*, there is no such source for seagrasses, so that transplants are obtained by damaging a "donor" meadow. By using donor material from other beds, transplantation has been successful with *Zostera marina* and *Halodule wrightii* (Kenworthy and Fonseca, 1992), *Thalassia testudinum* (Tomasko et al., 1991), *Posidonia oceanica* (Molenaar and Meinesz, 1995), and *P. australis* and *Z. capricorni* (West et al., 1990). Transplants may be via turf plugs, bare rhizome and root, or seeds. In a guide to transplanting Caribbean seagrasses, Fonseca (1994) describes methods for plugs and bare roots (with metal staples) and recommends *H. wrightii* as a colonizer species. Methods to improve the survival of seagrasses include comparisons of planting procedures, addition of fertilizers, and the use of enclosure cages (Fonseca et al., 1994). As noted, the removal of transplants causes damage to donor beds, and survivorship of planting units varies with species and technique: 47% for *H. wrightii* (Fonseca et al., 1996), 25% or 80% for single or multiple short shoots of *T. testudinum* (Tomasko et al., 1991), and up to 98% for *Z. marina* (Short, 1992). The results from seagrass transplantation indicate a need for development of nurseries rather than continued damage to natural beds.

Anthropogenic Effects

Most studies on seagrass damage have focused on human affairs (Thayer et al., 1975; Short and Wyllie-Echeverria, 1996) because they may be controlled and are important to management decisions regarding estuaries and coastal habitats (Chap. 5). Seagrass diebacks have resulted from oil spills (Zieman and Zieman, 1989), thermal pollution by power plants (Zieman, 1975), and eutrophication. The last may result from reduced illumination because of increased epiphyte and phytoplankton loads (Lapointe et al., 1994; Dawes et al., 1995b). Further, direct mechanical losses due to dredging also cause dieback due to sediment resuspension and increased turbidity; (Larkum and West, 1990; Onuf, 1991). Other forms of mechanical damage to seagrass beds are net trawling (Fonseca et al., 1984), propeller cuts (Sargent et al., 1994; Dawes et al., in press), and boat moorings (Walker et al., 1989). In addition to the abiotic impacts on seagrasses, there is at least one example of an exotic green alga (*Caulerpa taxifolia*) outcompeting the seagrass *Posidonia oceanica* in the Mediterranean (Villele and Verlaque, 1995; also see Chap. 5).

The loss of seagrass meadows due to anthropogenic activities has been most severe in estuaries and coastal communities. For example, Tampa Bay, with 1036 km^2 of surface water, may have supported 30,970 ha of seagrass meadows in 1870 (Johansson, 1991). By 1982, there were only 8763 ha of seagrass meadows. As a result of legal controls and management procedures, the seagrass meadows have increased to 10,386 ha (18.5%) by 1992. The expansion of seagrass beds was linked to a variety of controls, not the least of which was the building of a tertiary (nitrogen removal) sewage-treatment plant. In Tampa Bay, the cessation of secondarily treated sewage loads, with its high nutrient content, resulted in significant declines in dissolved nitrogen, which in turn resulted in lower phytoplankton-caused turbidity and increased submarine illumination (Johansson, 1991). Other important controls include the decline in urban runoff around Tampa Bay with use of catchment ponds and (in the future) the reduction of air nitrate loads through pollution-abatement equipment on power plant stacks. The 5344 ha of filled bay bottom and destroyed coastal habitats in Tampa Bay will never be recovered. However, by creation of new beds around spoil islands as well as protection of the remaining wetlands, a return to at least 50% of the original ecosystem should be possible (Lewis, 1982; Short 1992).

Management

An understanding of the successional stages of seagrass meadows will help identify species that are rapid colonizers (e.g., *Halodule wrightii* in the Caribbean). Seagrasses are sensitive to shifts in water quality, particularly factors that affect submarine illumination. Thus, changes in meadow size, depth distribution, biomass, growth rates, and epiphyte loads can be used as *biomonitors* of estuarine health (Short, 1992; Dennison et al., 1993; Tomasko et al., 1996). In a review, Short and Wyllie-Echeverria (1996) listed 28 publications that described human-induced pollution and water-quality disturbances of 24 seagrass species worldwide. Several studies, including those of Silberstein et al. (1986) with *Posidonia australis* and Short et al. (1995) with *Zostera marina*, have correlated seagrass dieback with eutrophication and increased epiphyte biomass. However, the eutrophication-epiphyte load relationship may not always be so closely connected. Lin et al. (1996) reported that the epiphyte biomass on *Zostera marina* was not a good indicator of nutrient loading or level of eutrophication. What does seem evident is that reduction in seagrass biomass is probably due to a combination of factors, including water transparency (turbidity, phytoplankton biomass, epiphytes) and anoxic conditions.

CHAPTER 12

Marine Plants of Coral Reefs

Coral reefs are complex and highly diverse ecosystems (Fig. 12-1) that exhibit a gross productivity 50 to 100 times greater than the surrounding tropical ocean waters (Sorokin, 1990b, 1993). It is sometimes overlooked that the microalgae and macroalgae contribute more to coral reef development than the larger marine angiosperms, namely the seagrasses and mangroves, and thus are instrumental in both reef formation and its primary production. The gross primary production of coral reefs ranges from 5 to 20 g C m^{-2} d^{-1} versus 0.5 to 0.3 g C^{-2} m^{-1} d found in tropical phytoplankton. Further, primary production also is high year-round, in contrast to other temperate and boreal marine ecosystems that show periodic (seasonal) fluctuations (Sorokin, 1993). The diversity of species on coral reefs is well known and has intrigued marine biologists for years. In part, the highly diverse benthic fauna and flora of coral reefs are due to the delicate partitioning of resources, which supports pelagic and deep-sea benthic populations. Further, the high rates of organic production and calcification on coral reefs seem to be independent of nutrient availability in nearby oligotrophic tropical waters.

The basis for high gross productivity (yet low net productivity), in spite of low nutrients, is biological regulation of a positive balance of nutrients for growth and calcification (Dubinsky, 1990). According to Sorokin (1990b), the primary biological processes responsible for nutrient uptake in coral reefs are fourfold:

1. Intense biofiltration of inorganic nutrients by a complex periphytic microflora, including bacteria, cyanobacteria (blue-green algae), diatoms, and filamentous macroalgae.

2. A wide variety of filter-feeding reef fauna, including sponges, corals, polychaetes, and bivalves.

3. A variety of mutualistic relationships that result in efficient exchange of inorganic nutrients for organic nutrients between host and symbiont.

4. The semienclosed to closed (tight) nutrient cycling that occurs on coral reefs.

Sorokin notes that filter feeders play a major role in reef productivity through

Figure 12-1. An aerial photo of Heron Island, which is a southern extension of the Great Barrier Reef in Australia. The reef rim (r), flat (f), and moat (m) are visible on the edge facing the prevailing wind (courtesy of R. A. Davis, University of South Florida, Tampa).

their use of reef phytoplankton, bacterioplankton (with a biomass of 100 to 500 mg m^{-3}), and particulate organic detritus (pseudoplankton) that is released from the coral community. In this chapter, the role of microalgae and macroalgae in the coral reef ecosystem will be examined; the reader is also referred to the texts of Dubinsky (1990) and Sorokin (1993).

PHYSICAL AND BIOGEOCHEMICAL ASPECTS

Coral reefs are marine, biogenic, and wave-resistant carbonate structures that are composed of shells and skeletons of *hermatypic* (reef-building) organisms (Dubinsky, 1990; Sorokin, 1993). Coral reefs cover about 15% of seabeds in shallow (0 to 30 m), tropical seas accounting for 0.2% of oceanic surface (Smith, 1978). The primary reef builders are cnidarian corals and calcified red algae. The cnidarians, particularly the scleractinian corals and hydrocorallians, are the most important animals contributing to coral-reef development. The high rates of calcification by cnidarians is attributed to their symbiotic relationship with dinoflagelate *zooxanthellae*, which support "light-enhanced calcification" (Goreau and Goreau, 1959). Because of their beauty and complex structure, coral reefs have been extensively studied, including their various symbiotic relationships. Sediment production is primarily due to calcified green algae

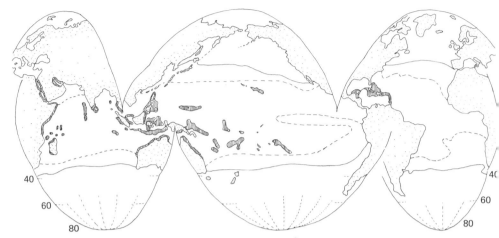

Figure 12-2. The world distribution of coral reefs. The hatched areas exaggerate the coral reefs that fall within the summer (solid) and winter (dashed) 21°C isotherms.

(e.g., *Halimeda*), foraminifera, and a variety of invertebrates, including molluscs, sponges, and polychaetes.

Distribution and Types

Coral reefs show a pantropical distribution, being most abundant in the Indo-Pacific Ocean. They cover 2×10^6 km^2 of tropical oceans and are found within the 18 to 20°C isotherms (Dubinsky and Stambler, 1966; Fig. 12-2). The Great Barrier Reef alone extends over 2000 km along the eastern coast of Australia, being the largest and most spectacular of barrier reefs. Indo-Pacific coral reefs extend into the Red Sea, exist along the East African coast, and are found as atolls in the Indian and Pacific oceans eastward to Hawaii. In the Atlantic Ocean, coral reefs are limited to the Caribbean Basin. They are absent from the western shores of Africa due to sedimentation and upwelling; similarly, they are largely absent from the eastern shores of South America due to extensive discharge of rivers such as the Amazon.

Darwin (1842) recognized three primary types: fringing, atoll, and barrier coral reefs (Fig. 12-3), and each of these can be subdivided into a number of subforms (Sorokin, 1993). Darwin proposed that atolls originate from *fringing reefs* that develop along the coast of a volcanic island. When volcanic activity ceases, the island begins to subside. If the reefs grow and maintain themselves near sea level, the lower parts of the sinking island fill with water, forming a lagoon between the reef and the central core of the island. The main reef is now a *barrier reef*. With further subsidence, the island becomes fully submerged, resulting in a large lagoon and the formation of an *atoll*. If the fringing reef develops along a larger land mass (e.g., continent) and submergence occurs, it

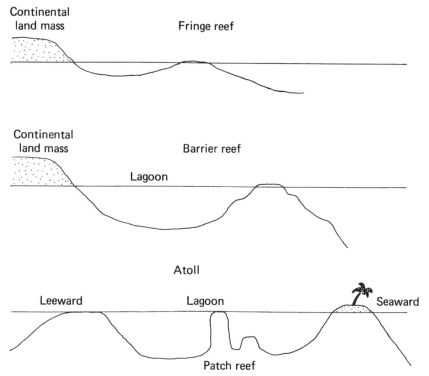

Figure 12-3. The three main types of coral reef. Fringe and barrier reefs develop next to or offshore a land mass, and atolls form on volcanos or fault blocks and lack any coastal connection.

can be transformed into a barrier reef that is separated from the mainland by deeper water. Although Darwin's explanations are valid today, because plate tectonics provides an explanation for subsidence, theories of reef formation also must consider historical changes in sea level and tectonics. Thus, barrier reefs can shift back to a fringing reef, and atolls can be drowned or increasingly exposed due to shifts in geological plates and rapid changes in sea level.

Fringing reefs, which are usually separated from the coast by a reef flat or shallow body of water from the coast, are common in the Caribbean, Red Sea, and East Africa. Barrier reefs are also associated with land masses, but they occur further offshore and have a deep lagoon between the reef and the land. For example, portions of the Great Barrier Reef occur 100 km off the Australian coast. Barrier reefs tend to be ribbon-shaped strips 300 to 1000 m wide. Darwin's ideas (1842) regarding the evolution of an atoll (Fig. 12-4) have been confirmed; thus, cores taken at Eniwetak Atoll showed a base of volcanic rock at 1400 m, with the earliest reef material dating back to the Eocene. Atolls also can develop by subsidence of continental shelf, as seen in Caribbean atolls off Belize (Stoddard, 1962). Atolls are most common in the Indo-Pacific, with only

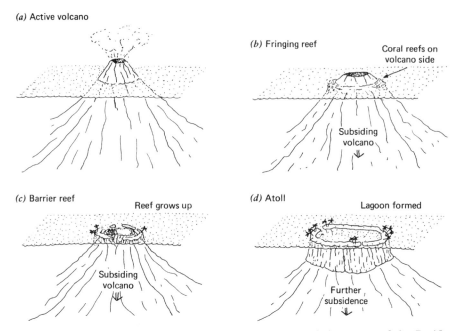

Figure 12-4.   The formation of an atoll. Darwin proposed that many of the Pacific atolls developed on subsiding volcanos.

9 of the world's 330 atolls occurring elsewhere, and range in diameter from a few to 70 km (2240 km^2); they also contain from 0 to 40 islands. The lagoons of atolls may contain patch reefs (knolls). For example the lagoon of the Caribbean atoll Glover's Reef is 108 km^2, is shallow (10 to 15 m deep), and has over 200 patch reefs (Fig. 12-5). Other reefs include *patch* or open-ocean reefs without an atoll (Mulaku) and *sinuous irregular reefs*, which are characterized by many small lagoons, as seen in Madagascar (Stoddard, 1973).

Development and Evolution

The basic coral-reef biota is similar throughout the Indo-Pacific, reflecting long-term dispersal of coral larvae over great distances (Richmond, 1982). The wide dispersal and similarity of Indo-Pacific biota is probably due to the spread from one island cluster to another wherever substrate was available (Dubinsky, 1990), a type of "stepping-stone" process as well as some "filtration" or separation of species (Veron, 1995). The long-dispersal concept is in contrast to the earlier proposal by Heck and McCoy (1978), who suggested that the deep oceanic trenches were natural barriers and acted as biological filters.

The latitudinal distribution of hermatypic corals was thought to be primarily due to water temperature, with no reef formation occurring where temperatures are below 18°C. However, coral reefs in the Persian Gulf are exposed to 8 to

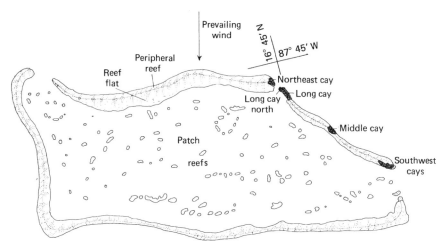

Figure 12-5. Glover's Reef, Belize, Central America. The diagram shows a reef (12 × 9 km) with a lagoon (10 to 20 m deep) and patch reefs. Four islands occur on the windward side of the atoll that is forming on a fault block.

9°C in the winter, and some shallow reefs in more tropical regions are exposed daily to 45°C. Temperature may be overrated as a single controlling factor, as in many areas with high stable water temperatures, corals do not develop in mass if they are exposed to annual pulses of nutrients (Margalef, 1968). It appears that there must exist a combination of oligotrophic water, high-water transparency, and high-water movement to support reef development. Several authors have considered why corals dominate tropical reefs but are rare or absent on temperate reefs (Miller and Hay, 1996). They studied temperate reefs off the North Carolina coast, including some dominated by seaweeds (well illuminated) and others by corals (low irradiances due to turbid or deep water). Using enclosures, they found that seaweed competition and animal grazing, along with irradiance and nutrients, can limit coral recruitment, growth, and abundance, and suppress reef development, which supports the earlier ideas of Margalef (1968).

The bathymetric distribution of hermatypic corals is determined by light and water transparency (Dubinsky, 1990). Thus, there is a positive correlation between the number of hermatypic species and the level of irradiance, with most reefs occurring at irradiances above 10% transmittance and the greatest biomass being at depths 20 m or above. Few hermatypic corals tolerate exposure to air, with most intertidal forms only growing where waves prevent desiccation. The coincidence of extreme low tides, high air temperatures, and calm water results in coral diebacks (Yonge, 1963). Their recovery may take 10 or more years. Other abiotic factors that can damage corals include low salinities and sedimentation. Thus, the southerly extension of Caribbean reefs is 5° N latitude, due to low salinity and sedimentation from the Orinoco and Amazon rivers.

Biogenic reefs have existed over much of Earth's history yet have varied considerably as to the dominant constructors (see Hallock, 1997). Further, biogenic reefs have been dominated by stromatolite structures constructed by cyanobacteria. The oldest stromatolites are approximately 3.5 billion years old. During the Proterozoic Era (Era of Bacteria), which lasted from about 2.5 billion years ago until about 570 million years ago, stromatolites were widespread in shallow seas. Constructed by cyanobacteria, stromatolites can still be found in shallow seas, namely Exuma Sound in the Bahamas (Dill et al., 1986) and Shark Bay in Western Australia (Logan et al., 1974). The abundance and diversity of stromatolitic reefs sharply declined with the evolution of multicellular animal life around 600 million years ago. Beginning in the Lower Paleozoic, a variety of calcareous plants and animals have contributed to the construction of biogenic reefs. Sponges, including important groups such as the archaeocyathids (ancient cups) and sclerosponges, were among the first animals to build biogenic reefs. The sponges were joined by bryozoans, tabulate and rugose corals, brachiopods, foraminiferans, and calcareous algae and cyanobacteria through the Paleozoic Era. Following perhaps the largest extinction in Earth history at the end of the Paleozoic, reef communities did not exist for several million years during the beginning of the Mesozoic. However, during the Triassic Period, many modern orders of cnidaria, calcified algae, and foraminifera evolved; by the Late Triassic, corals and some forminifera had developed symbiotic relationships with dinoflagellates (Stanley, 1992; Stanley and Swat, 1995) and other algae, enabling more rapid growth and calcification, and therefore a greater potential to construct biogenic reefs. During the Jurassic Period, coral reefs were similar to present-day ones, with hermatypic corals and coralline red algae being important constructors. With the rise of carbon dioxide levels in the atmosphere and the rise of sea level, an extreme Greenhouse condition prevailed during the Cretaceous Period and an unusual combination of pelecypods called "rudistids" dominated as principal constructors of biogenic reefs. Although a diverse community of corals continued to flourish, they did not construct reefs during this period, perhaps due to the high CO_2 levels and hot summer temperatures (due to the Greenhouse Effect; Chap. 5).

Another major environmental crisis occurred at the end of the Cretaceous Period (e.g., end of Mesozoic Era). The rudistids suffered complete extinction; corals and reef-dwelling larger foraminifera lost complete families of species. Greenhouse conditions persisted into the early Cenozoic (modern) Era, with corals and larger foraminifera again showing expanded diversity. Foraminifera, along with coralline red algae and bryozoans, became the major reef builders in the Paleocene and Eocene. In the late Eocene, as Australia moved away from Antarctica, the Earth began to change from Greenhouse conditions to the modern glacial-interglacial conditions. Declining concentrations of CO_2 in the atmosphere helped to trigger this cooling and made conditions more favorable for corals and calcareous green algae, which secrete the aragonite crystal form of calcium carbonate. Aragonite is stronger than calcite and energetically easier to produce in warm waters (> 25°C) when CO_2 is being removed via photosynthesis that results in high pH of seawater. By the Oligocene, despite the shrinking size of tropical oceans due to the cooling of temperate and polar climates, corals were building vast reefs for the first time since the Jurassic (Frost, 1977).

With the closure of the Central American Seaway in the Miocene, the smaller western Atlantic and Caribbean rapidly lost tropical species, especially scleractinian corals and larger foraminifera. Edinger (1991) observed that shelf-edge and slope-dwelling coral genera were nearly eliminated in the Atlantic in the early Miocene, and Frost (1977) estimates that the Caribbean lost at least 75% of the formerly cosmopolitan coral and larger foraminiferal species. By contrast, extinction levels of corals in the Indo-Pacific were much lower. With the present alternation of glacial and interglacial conditions over the past 2 million years, coral taxa have been reduced to about 2500 species of the original 7500 taxa (Dubinsky, 1990; Sorokin, 1993). Some, like the fast-growing species of *Acropora* and *Montipora*, have diversified and now account for about 25% of the Indo-Pacific coral species.

Formation and Erosion

Glacial advances result in sea level falls of 100 to 140 m; and, after melting of the Northern Hemisphere glaciers, the level rises to present levels. Thus, coral reefs must "migrate" up and down the continental shelves with each sea level fluctuation. For example, the sea level has risen about 130 m since the last glacial advance 18,000 years ago. Modern shallow water reefs are typically less than about 5000 years old and are only about 5 to 20 m in thickness (Sorokin, 1993), and were formed on older reef platforms built during Pleistocene interglacial times.

Although appearing to be stable, biogenic ecosystems such as coral reefs show dynamic growth and erosion. Their survival and growth depend on sedimentation and accretion of biogenic carbonates. An annual production of 4 to 5 kg $CaCO_3$ m^{-2} occurs even with stable sea levels, due to the need for replacement from erosion (Sorokin, 1993). Globally, the present annual production of $CaCO_3$ is about 2.5×10^9 mt reaching 20 kg m^{-2} y^{-1}. Biogenic carbonate production is primarily dependent upon hermatypic corals (see "Calcification"), as well as calcified green and red macroalgae, and foraminifera. According to Sorokin (1993), rates of calcification vary from 143, to 8.9, to 2.4 mg $CaCO_3$ g dwt^{-1} h^{-1} in macroalgae (*Galaxaura* spp., *Halimeda opuntia*, *Lithothamnion* sp., respectively) and 3.3 to 0.26 mg $CaCO_3$ g dwt^{-1} h^{-1} for hermatypic corals (*Acropora cervicornis*, *Millepora complanata*).

The annual production of $CaCO_3$ on coral reefs is very similar geographically, varying from 13.0, to 11.0, to 9.4 g m^{-2} d^{-1} for One Tree Island, Lizard Island, and Rib Reef, respectively, on the Great Barrier Reef flat (Sorokin, 1993). Higher rates are known for reef flats on Eniwetok Island in the Pacific, as well as Barbados in the Caribbean (31.0 and 42.0 g m^{-2} d^{-1}, respectively). Much of this biogenic production is eroded, with resultant sediment and rubble being washed into the back reef or down the outer slope. Annual rates of erosion of reef flat surfaces range from 1 (Bikini atoll) to 3 to 13 kg m^{-2} (Florida). Most of the erosion results from boring by animals, plants, and fungi, although the mechanical effects of waves and scouring are also critical. Of particular interest are species of boring green (*Ostreobium*) and blue-green algae (*Entophysalis*, *Hlyella*, *Mastigocoleus*).

Nutrient Cycles

Nitrogen and phosphorus, which are essential for plant growth, are usually in lower abundance in seawater than other elements. Silicon is an essential element for diatoms and other photosynthetic organisms (Chap. 2). Thus, biogeochemical control of these elements affects the productivity of coral reefs (D'Elia and Wiebe, 1990). As pointed out, coral reefs exhibit high productivity and biomass, yet they occur in oligotrophic seawater that supports low plankton productivity and biomass per unit area. Thus, the "action" is in the benthic community, not the overlying water column (D'Elia and Wiebe, 1990). Coral reef ecosystems also exhibit high gross but low net primary productivity, indicating that nutrient cycling and conservation are important. The sources of nutrients to coral reefs may be from terrestrial runoff or flowing seawater.

Availability of nutrients is a controlling factor in reef development and distribution. Further, eutrophication is considered to be a major factor in reef decline [see "Anthropogenic Stresses" and reviews by Hallock and Schlager, 1986; Hallock et al., 1993; Dubinsky and Stambler, 1996)]. The effect of sewage outfall in Kaneohe Bay, Hawaii, demonstrates how critical oligotrophic conditions are for coral reef survival. Discharge of domestic wastes between the 1960s and 1970s (up to 20,000 m^3 d^{-1}) caused radical changes in the benthic communities. The green macroalga *Dictyosphaeria cavernosa* overgrew and excluded reef corals; increased phytoplankton productivity also caused a decline in water transparency, reducing rates of calcification when turbidity was maximal (Maragos et al., 1985). After a reduction of sewage discharge (late 1970s), the process began to reverse, although the original reef structure was still not evident by the late 1980s.

Nitrogen is most likely the primary growth-limiting element, as dissolved inorganic nitrogen (DIN) compounds in reef waters are very low, greater than 0.4 to 1.0 μM L^{-1}; ammonium-N is maximal followed by nitrate-N and then nitrite-N. Levels of DIN are higher in sediment pore water and coral skeletons (D'Elia and Wiebe, 1990). The nitrogen cycle is mostly a biological process, with all stages occurring on coral reefs (Chap. 2, Fig. 2-16). Due to the high rates of nitrogen fixation on coral reefs by cyanobacteria and bacteria, a large amount of dissolved nitrogen is exported. Thus, coral reefs can be viewed as open rather than closed ecosystems (Chap. 3). High rates of nitrogen fixation occur in both Indo-Pacific and Caribbean reefs, with major sites being the windward forereef and spur and groove zones (see "Macroalgal Zonation"). In addition to benthic blue-green algae and bacteria, nitrogen is fixed via cyanobacteria epiphytic on macroalgae plus endophytes of sponges and ascidians (e.g., *Prochloron*).

Dissolved inorganic phosphorus (DIP) concentrations range from greater than 0.1 to 0.6 μM L^{-1} in overlying waters on coral reefs. Thus, they are extremely low when compared with coastal or deep-ocean water (D'Elia and Wiebe, 1990). The phosphorus cycle is affected by both chemical and biological processes, with major pools, including organic matter, DIP, and mineral phosphorus (Fig. 2-17). The tight cycling of phosphorus by coral symbionts is shown by the fact that hermatypic corals excrete less and take up more DIP than invertebrates of the same size. The physiochemical processes of precipitation, absorption, and chemiso-absorption in the phosphorus cycle partition silicon between dissolved and sediment particulates.

Because siliceous organisms (diatoms, silicoflagellates, siliceous sponges) are not abundant on coral reefs, silicon may not be a limiting factor (D'Eila and Wiebe, 1990). Although the data are limited, it appears that concentrations of dissolved silicic acid are about 2 μM L^{-1} in reef waters; these values can be higher with input from ground water and terrestrial runoff, as noted in Chap. 2. The silicon cycle has both chemical and biological phases; its major pools include amorphous, biogenic silica, dissolved silicic acid, and mineralized, crystalline silica (Fig. 2-18).

SYMBIOTIC ALGAE AND CORAL REEFS

The frequency of symbiotic associations between plants and animals in the tropics is probably a result of lower nutrients, higher rates of predation, and the long period for evolution of coral reefs (Lüning, 1990). Levels of calcification in corals ranges from about 90% in octocorals to 98% for scleractinians; values for coral polyps range from 0.3 to 3 mg wet weight. Polyps of hermatypic corals contain 1 to 5×1010^6 endosymbiont cells (*zooxanthellae*) cm^{-2} of surface area. The most important endosymbionts are dinoflagellates, and the most common species is *Gymnodinium microadriaticum* (*Symbiodinium microadriaticum*), although unicellular green algae and filamentous blue-green algae also occur. Corals may be viewed as living greenhouses; endosymbionts aid in calcification and supply photosynthates, while obtaining nutrients from the animal. In fact, the survival and development of coral reefs in oligotropic tropical seas are largely due to the microalgal zooxanthellae. The unique and important mutualistic relationship has been well described (Yonge, 1963; Taylor, 1983; Dubinsky, 1990; Sorokin, 1993).

Zooxanthellae are primarily known from their symbiotic stage in the coral polyp. The symbiont has a well-developed periplast (amphisma) or complex set of membranes that may contain cellulose (Chap. 7; "Division Phyrrophyta"). The periplast protects the cell from digestion by host enzymes, but it is permeable to metabolite exchange. The endosymbiont contains a large chloroplast with 10 to 12 bands of thylakoids in groups of 3. Pigment composition is similar to that in free-living dinoflagellates: chlorophylls *a* and *c*, peridinin, and dinoxanthin, as well as other cartenoids. In addition to the endosymbiotic stage, zooxanthellae can be induced into a motile form in the laboratory, with these having a typical dinoflagellate morphology and flagella. "Infection" of coral polyps by zooxanthellae occurs via coral larvae and with mature polyps. Motile zooxanthellae swim into the mesenterial (digestive) cavity and are ingested. The zooxanthellae, with their protective periplast, are not digested but are phagocytosed by mesenterial phagocytes and transferred to the polyp's tissues (Sorokin, 1993).

Foraminifera, which are shelled protozoans, are usually very abundant in coral reefs and include several large species that may reach a cm or more in diameter. Foraminifera can be so abundant that they have been called "living sands," and their shells can contribute in excess of 90% of the sand on some Pacific atolls (Maxwell, 1968). Larger foraminifera, like many scleractinian corals, host algal endosymbionts that aid in nutrition and calcification (Lee and Anderson, 1991). However, unlike corals, which appear to be limited to dinoflagellate symbionts, larger foraminifera host a diverse array of algal symbionts, including red and green algae, diatoms, and dinoflagellates. Like corals, the large, orbitoid Soritidae host *Symbodinum* spp. endosymbionts and recent genetic studies indicate that these foraminifera can host the same strains as the local corals. The porcellaneous Archaiasines host chlorophyte symbionts of the genus *Chlamydomonas*. The peneroplids, which were classified by Linnaeus as tiny nautiloids, are the only organisms known to have endosymbiotic associations with red algae. Finally, members of the families Alveolinidae, Nummulitidae, and Amphisteginidae host a variety of diatom symbionts. The types of symbionts, along with the light transmissivity of the shell structure of the foraminifera, appear to influence the reef and open shelf habitats that each taxon can occupy (Hallock, 1988).

Irradiance and Photosynthesis

The most important factor affecting carbon fixation by endosymbiotic zooxanthellae is light. Thus, zooxanthellate corals are restricted to the euphotic zone (Falkowski et al., 1990), and the shallow-water coral populations are phototrophic, whereas deep-water forms are obligate heterotrophs. The highest diversity of hermatypic corals is at 20 m, and their bathymetric distribution correlates with light-harvesting strategies and abundance of zooxanthellae. The carbon sources translocated from the zooxanthellae to the coral polyps lack nitrogenous compounds; the polyps must obtain nitrogen via filtration of seawater for zooplankton or dissolved organics (Muscatine, 1990). Changes in colony morphology, such as flattening of branches, is an adaptation by deep-water corals to increase light exposure for their zooxanthellae compared with the terete branches of shallow-water forms. All coral polyps expand during daylight, exposing the zooxanthellae to light; but deep-water corals also expand specialized globular tentacular bubbles, which contain numerous zooxanthellae (Fricke and Vareschi, 1981). Many shallow-water corals also tend to have small polyps as compared with the large heterotrophic polyps of deep-water individuals.

Zooxanthellae acclimate to lower irradiances in deep-water corals through increased concentrations of chlorophylls (*a* and *c*) and cartenoids (peridinin), that is, by increasing the size of their photosynthetic units (PSUs, Chap. 4; Barnes and Chalker, 1990). Thus, the light-harvesting PSUs of deep-water endosymbionts are larger and more efficient. Another photoacclimation process for zooxanthellae is increased concentrations of β-carotene, dinoxanthin, and diadinoxanthin (yellow xanthophylls) under high light (shallow corals), which prevents photoinhibition. Varying levels of UV-absorbing compounds (mycosporine) and shifts in *P-I* patterns also occur (Chap. 4). Zooxanthellae from deep-water corals have low levels of UV-absorbing pigments and low I_c and I_k values, whereas α shows a steep slope, indicating more efficient photosynthesis under low irradiances (Barnes and Chalker, 1990). Zooxanthellae of shallow (0.5 to 3 m) hermatypic corals show saturation irradiances (I_k) of 400 to 2100 μmol photons m^{-2} s^{-1} and P_{max} of 0.05 to 0.08 mg Q h^{-1} (Muscatine, 1990). The large range in I_k values measured in shallow-water zooxanthellae may reflect seasonal changes. However, the data are too limited to be certain.

Zooxanthellae obtain their required carbon for photosynthesis from CO_2 respired by coral polyps and from H_2CO_3 dissolved in seawater and translocated into the endosymbiont. Zooxanthellae apparently utilize the C_3 pathway for carbon fixation and undergo photorespiration when concentrations of O_2 are high and CO_2 low (Muscatine, 1990). Thus, as in other C_3 plants, the ribulose-bis-phosphate of zooxanthellae is an oxygenase as well as a carboxalyase (Chap. 5). However, zooxanthellae show a high level of C_4 enzyme activity, including malate dehydrogenase and phosphoenolpyruvate carboxylase (PEP carboxylase), which is similar to that in other dinoflagellates.

Although zooxanthellae use and respire fixed carbon, a large portion (78 to 95% in shallow-water corals) is translocated to and used by coral polyps. Translocated carbon is mostly composed of water-soluble compounds having low molecular weights, especially glycerol and, to a lesser extent, fatty acids or esterified lipid droplets (Muscatine, 1990). The coral polyps use the photosynthates for cellular energy, enhancement of calcification, growth, and mucus production, much of which is released as particulate or dissolved organic carbon (up to 45 to 50%).

Calcification

Hallock (1997) has listed three primary mechanisms of calcification in present-day oceans in her detailed review on reef limestone history. The first mechanism is geochemical precipitation of $CaCO_3$ in response to CO_2 uptake by photosynthesis, a process particularly effective under relatively high atmospheric concentrations of CO_2. The second process is biomineralization by protozoan and animal cells that occurred when CO_2 concentrations dropped so that shell construction and maintenance became energetically feasible. The third process uses calcification to provide CO_2 for photosynthesis, again probably linked to the drop in atmospheric CO_2 concentrations that limited aquatic photosynthesis in warm, shallow seas. It is the third process that probably arose independently in various algal groups (reds, greens, coccolithophorids) and also in algae symbiotic in calcified animals and protists (zooxanthellae). In present-day reefs, the bulk of coral reef $CaCO_3$, has been biologically precipitated. Scleractinian corals, calcified red and green algae, and larger foraminifera are the main producers of reef calcium carbonate, with aragonite production by the corals and green algae accounting for over 50% of all reef carbonate. As noted in Chap. 2 (Eqs. 2-12 and 2-13), the process of calcification is enhanced in reef water by a rise in pH due to photosynthetic removal of CO_2, the supersaturation of Ca^{+2} and CO^{-2} ions, and subsequent nucleation and crystallization (Barnes and Chalker, 1990). Both the "stony" scleractinian corals and calcified green algae (e.g., *Halimeda*, *Udotea*) precipitate aragonite crystals outside their organic structure. By contrast, coralline red algae deposit calcite crystals within their preexisting organic cell walls.

Zooxanthellae play a major role in coral calcification. The high gross (3 to 6 kg m^{-2} y^{-1}) and net (4 ± 1 kg $CaCO_3$ m^{-2} y^{-1}) rates of calcification, and high gross production (7 ± 1 g C m^{-2} d^{-1}), result in rapid coral reef growth. The high rate of calcification by corals is dependent on zooxanthellae photosynthesis, although, as shown by skeletal growth, the process will continue in the dark (Barnes and Chalker, 1990). Calcification in both corals and tropical algae increases in the light, probably due to production of photosynthate, which serves as an energy source for transporting calcium, building the skeleton, and taking up inorganic nutrients. Depletion of CO_2 via photosynthesis increases cellular pH. Further, calcification will also remove CO_2 and release protons (Barnes and Chalker, 1990), as shown in Eq. 12-1:

$$Ca^{+2} + H_2O + CO_2 \leftrightarrow CaCO_3 + 2H^+ \qquad (12\text{-}1)$$

The results of CO_2 depletion and production of protons will inhibit photosynthesis. However, hydrolysis of urea at sites of calcification in corals may provide CO_2 for skeletal carbonate and ammonia to neutralize the protons. The zooxanthellae use carbonic anhydrase to convert HCO_3^- to CO_2 within their cells, which avoids an increase in pH as well as a reduction of dissolved carbon dioxide. The lack of an increase in pH is because the uptake of HCO_3^- also requires simultaneous uptake of a proton and the removal or neutralization of the hydroxide (Borowitzka, 1982). In coccolithophorids, light-enhanced calcification results in one carbon atom incorporated into carbonate for each carbon atom fixed in photosynthesis (Borowitzka, 1982). In corals and calcified reef algae, the ratio favors fixation into organic compounds (via photosynthesis) by a factor of 4 to 8 over skeletal development; thus, the two processes do not appear to be linked. Even so, photosynthetically derived OH$^-$ may be used at sites of calcification in corals and calcified algae without this linkage (Barnes and Chalker, 1990). The role of organic matter in calcification of scleractinian corals is unclear, although transport of organic material to the area of calcification appears necessary for the precipitation of $CaCO_3$ (Yonge, 1968) and may control the morphology of its crystals.

The limited data from calcareous algae, particularly the green alga *Halimeda*, suggest that calcification is inhibited in the dark, perhaps due to release of respiratory CO_2 and subsequent acidification (Borowitzka, 1982). In the light, algae, like corals, show increased calcification. The light-enhanced increase is thought to be due to the photosynthetic removal of CO_2 that causes a shift in the carbonate equilibrium (Borowitzka, 1982). The intercellular compartments of *Halimeda* are probably isolated from the surrounding seawater, so that the carbonate equilibrium is shifted (i.e., there is no influx of CO_2) and carbonate is precipitated. Thus, on reaching a threshold level of photosynthesis where sufficient CO_2 has been removed, the carbonate equilibrium will shift in the macroalga, resulting in precipitation of $CaCO_3$. If this is correct, then use of carbonic anhydrase and organic compounds to control calcification by macroalgae parallels that reported for corals (Borowitzka, 1982; Barnes and Chalker, 1990).

CORAL REEF ALGAE

In addition to the endosymbiotic microalgae, which include the zooxanthellae, other coral reef algae include phytoplankton (Chap. 7), mat-forming and boring microfilamentous algae, and calcified, fleshy, and turf macroalgae (Berner, 1990; Lüning, 1990). The reef flat and crest sometimes appear almost devoid of macroalgae, yet the surface may consist of crustose corallines. In contrast, the unconsolidated sediment in reef lagoons contains psammophytic green (e.g., *Caulerpa*) and red algae (e.g., *Amphiroa*, *Galaxura*), plus seagrass beds that support an abundant epiphytic flora. Fleshy algae (e.g., *Codium*, *Padina*, *Halymenia*) are also common in lagoons, but they may be limited to crevices in coral rubble or under coral heads of patch reefs. The cover of macroalgae and seagrasses can range from 10 to 100% in lagoons, depending on the intensity of grazing, nutrient supply, as well as submarine illumination and water movement.

Microalgae

Endosymbiotic algae can account for about two-thirds of the primary production (Wanders, 1976) in a Caribbean fringing reef, exclusive of the bacterioplankton, phytoplankton, and boring microalgae. Several genera of microfilamentous blue-green (*Plectonema*, *Phormidium*), green (*Ostrebium*), and red algae (*Audouinella*) bore into the corals, forming up to a 2-mm thick layer 13 mm below the surface (Berner, 1990). Boring algae are adapted to low irradiances of 2 to 3 μmol photons m^{-2} s^{-1} (Lüning, 1990), which represents the lower range of the photic zone. Some blue-green species (e.g., *Calothrix*) that grow as epiphytes on the turf-forming macroalgae play a major role in nitrogen fixation, contributing 985 kg N ha^{-1} y^{-10} (Berner, 1990).

Macroalgae

Species lists of macroalgae in oceanic reef plant communities contain from 100 to 250 red, green, and brown taxa. This number may double or triple on fringing coral reefs, where neritic waters have higher levels of nutrients. By contrast, there may be 700 to 1000 macroalgal species in temperate coastal regions (Sorokin, 1993). Brown macroalgae are only represented by a few species (>30), some of the most common genera being *Sargassum*, *Padina*, *Dictyota*, and *Turbinaria*. In terms of biomass, green macroalgae are dominated by calcified and psammophytic species, whereas red algae are primarily crustose and articulated forms. Fleshy and turf-forming macroalgae are less abundant on coral reefs due to extensive grazing but are known to be critical primary producers (Adey and Goeremiller, 1987; Carpenter and Williams, 1996).

Primary production of coral reef macroalgae varies according to their morphology, as noted in the functional-form and function (FF) model outlined by Littler et al. (1983) and in Chap. 3. The productivity of crustose coralline algae varies from 40 to 150 μg C m^{-2} d^{-1}, whereas erect fleshy species exhibit rates of 8 g C m^{-2} d^{-1} (up to 40 g C m^{-2} d^{-1}), which is similar to that of terrestrial communities. The high productivity of turf algal communities in nutrient-poor waters of coral reefs has been demonstrated (Adey and Goertemiller, 1987). The large range of productivity just noted may reflect differences in measurement (Table 12-1). The photosynthesis-to-respiration (P/R) ratio for coral reef metabolism is close to 1 or higher, suggesting a balanced ecosystem or one where some export (loss) is occurring (Kinsey, 1983). With limited export and rapid recycling by benthic communities, any nutrient limitations will more likely affect the phytoplankton in the water column.

As noted earlier, calcareous macroalgae (Table 12-2) play significant roles in reef building (see also Littler, 1976). Calcified green macroalgae produce the aragonite form of calcium carbonate and include species of *Halimeda*, *Udotea*, *Penicillus*, *Dasycladus*, *Neomeris*, and *Acetabularia*. Over half of the unconsolidated sediments of Pacific atoll lagoons may be derived from *Halimeda* and *Penicillus*. It is no wonder that calcified green algae are some of the most important sources of reef sediment (Hillis-Colinvaux, 1980), as plants like *Halimeda* can double their size in 15 days and produce 7 g dwt m^{-2} d^{-1}. A similar statement can be made for foraminifera that produce calcite. Coralline red algae produce a calcite form of carbonate and are well represented by both articulated and crustose species. Crustose coralline macroalgae or "lithothamnia" (e.g., *Porolithon* and *Lithothamnion*) are the "cementers" of many coral reefs. They overgrow the softer limestone reef material and produce a harder calcite form of carbonate, consisting of a magnesium (18%) and calcium (82%) covering of 1 to 2 mm month^{-1} with an annual increase of 1 to 5 mm^{-1} y (Berner,

TABLE 12-1. Productivity (g C m^{-2} d^{-1}) of Coral Reefs, Reef Algae, and Other Communities

	Berner (1990)	Wanders (1976)	Dawes (1981)
Open sea	—	0.2–0.3	0.1–0.4
Coral reefs			
Total Reef	8	17	5–10
Zooxanthellae	—	11	2.7
Crustose algae	0.7–2.4	2.7	—
Turf algae	1–6	—	—
Fleshy algae	0.1–5.7	3.3	—
Thalassia bed	—	—	11
Terrestrial communities			
Sugarcane	—	—	19
Wheat	—	—	5

1990; Lüning, 1990). The importance of lithothamnia to coral reef development is the rapid production of calcites with high levels of magnesium that have similar solubilities as aragonite. With their rapid growth and production of calcite crusts, crustose corallines can form intertidal algal ridges on the exposed windward side of coral reefs where they cement the carbonate sediments. For

TABLE 12-2. Calcareous Algae: Examples of Various Types and Sites of Deposition for Calcium Carbonate in Microalgae and Macroalgae

Taxonomic Group (Chap.)	Generic Example	Type of CaCo$_3$	Site of Deposition
Cyanophyta (7)			
Nostocales	*Oscillatoria*	Calcite and aragonite	Sheath, surface of cell
Rhodophyta (7)			
Nemalionales	*Galaxura*	Aragonite	Surface of cell
Gigartinales	*Titanophora*	Aragonite	Surface of cell
Corallinales	*Corallina*	Calcite	In cell wall
Chlorophyta (6)			
Caulerpales	*Halimeda*	Aragonite	Surface of cell
Dasycladales	*Acetabularia*	Aragonite	Surface of cell
Phaeophyta (6)			
Dictyotales	*Padina*	Aragonite	Surface of cell
Chrysophyta (8)			
Coccosphaerales	*Hymenomonas*	Calcite and aragonite	Coccoliths form in the cell and move to wall

Source: After Littler (1976).

example, 17 to 40% of cores taken from Great Barrier Reef flats consisted of sediments from coralline red algae and *Halimeda* spp. (Sorokin, 1993).

ECOLOGY OF REEF MACROALGAE

The zonation and distribution of macroalgae from one reef cannot be easily extrapolated to another part of the world. The reason is that reef structure depends on many factors, including historical, abiotic, and biotic conditions (Berner, 1990). In reefs that are not dominated by hermatypic corals, nonarticulate coralline and filamentous algae substitute. For example, on One Tree Island in the Great Barrier Reef, Borowitzka (1981) found 180 species of turf algae, and 40 fleshy and 25 calcified taxa. By contrast, Benayau and Loya (1977) found that the turf communities in the Gulf of Eilat contained few species, with the filamentous brown alga *Sphacelaria tribuloides* dominating.

Abiotic Factors

As might be expected for marine plant communities in oligotrophic waters, nutrient (e.g., P^{+2}, N^+, Ca^{+2}) limitations influence the distribution of macroalgae, with nitrogen usually being the most critical. By contrast, excess nutrients (nutrification) can cause overgrowth of coral reefs (Chap. 5, Fig. 5-4) by benthic macroalgae (Berner, 1990). Intertidally, temperature and desiccation are significant factors that influence macroalgal reef habitats. High temperatures and exposure during low tide will limit algal development. By contrast, temperature and salinity are usually not limiting factors subtidally, as they usually show small seasonal variations. One exception is where deep-water species are exposed to cold-water upwelling. Irradiance (quantity and quality) controls the depth distribution of macroalgae, just as it does with zooxanthellae in hermatypic corals. Water motion is a critical factor affecting photosynthesis (Carpenter and Williams, 1996) and the distribution of reef-building algae. The zonation and morphology of shallow-water hermatypic corals and crustose coralline algae are also influenced by the intensity of water motion at the windward and leeward sides of coral reefs (Fig. 12-8).

Biotic Factors

Both competition and grazing play major roles in macroalgal development in coral reefs. In the case of *grazing*, 67 to 95% of total seaweed production can be removed by herbivores (Hay and Fenical, 1988; see also Chaps. 3 and 8). Thus, the effects of grazing on macroalgae in coral reefs have been studied in the Virgin Islands (van Steveninck and Bak, 1986), Panama (Hay, 1981a, b), Hawaii and Belize (Littler and Littler, 1988), Bahamas (Adey and Goertemiller, 1987), and the Great Barrier Reef (Borowitzka, 1981). The major reef grazers are fish, sea urchins, molluscs, and micrograzers (John et al., 1992); fish are the most important in the Pacific, while echinoderms dominate in the Caribbean. Damselfish (e.g., *Eupomacentrus*) are often cited as examples of grazers that control the algal community within their territory, in some cases weeding out unwanted algae (John and Lawson, 1990). The abundance of herbivores in the tropics (20% of fish; high diversity of sea urchins) can be contrasted to those found in temperate waters (10% of fish; low diversity of sea urchins). The differences suggest that grazing and not the detrital food chain removes the macroalgal biomass in coral reefs (Lüning, 1990). Of the antigrazing adaptations listed in Chap. 3 (temporal, spatial, structural, chemical), calcification and chemical defenses appear to play major roles in coral reef macroalgae. The majority of calcified green and red macroalgae are tropical and can be found in coral reefs. Further, over 500 secondary metabolites (terpenes, aromatic compounds, acetogenins, amino-acid-derived substances, polyphenolics) have been isolated from mostly tropical seaweeds (Hay and Fenical, 1988). However, only a few have been studied (John and Lawson, 1990), and the alleged chemical defenses of a number of tropical algae are open to question (Lewis, 1985).

The intensity of grazing is shown by a study of a Caribbean fringing reef, where only 10% was occupied by small (2-to-3-mm) turf-forming macroalgae (Wanders, 1976). The same author also found that inhibition of grazing via caging or removal of predators resulted in a decline in algal diversity but increased algal biomass (Wanders, 1977). Caging experiments on coral reefs have demonstrated the important role herbivores play in epiphyte removal, as well as maintenance of the algal turfs and the limiting of fleshy algae. Most filamentous, epiphytic species grow and recruit rapidly. Thus, if micrograzers like ampiphods are absent, the epiphytes will rapidly cover and shade out corals. The same is true for macroalgae on coral reefs, as shown by the 1983–1984 mass mortality that killed 95 to 99% of most Caribbean populations of the black sea urchin *Diadema antillarum*. Because overfishing had removed most herbivorous fish, the urchin dieback resembled a caging experiment. Three effects were evident within one year: algal turf biomass increased, biomass-specific community productivity decreased, and algal diversity shifted to macroalgae (Glynn, 1990). Grazing affects the growth rate, reproduction, and survival of macroalgae and there is a strong relationship among grazing, nutrients, and the type of algal flora (Littler and Littler, 1988). Namely, whereas grazing controls algal biomass, nutrients set the upper limits to that biomass, and both play a role in the type of species present (Fig. 12-6). Thus, an increase in nutrients in oligotrophic coral reefs will result in a shift from filamentous to frondose algae if grazing is suppressed (e.g., via overfishing).

Although interspecific *competition* between macroalgae and scleractinian corals exists on coral reefs, it has not been extensively studied (Berner, 1990; Tanner, 1995). The green alga *Dictyosphaeria cavernosa* will compete with corals for space on fringing and patch reefs in Hawaii (Stimson et al., 1996). It forms sacs or chambers as it grows, which contain higher levels of inorganic nitrogen and phosphorus than within the adjacent water column. The nutrients are obtained from the sediments beneath the thalli, giving the species

HERBIVORY

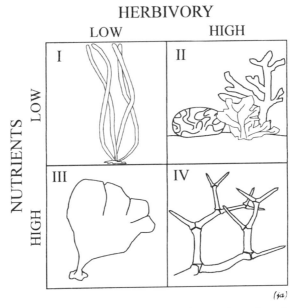

Figure 12-6. Interaction between nutrients, grazing, and coral-reef organisms. The existence of three morphological groups of algae (filaments, I; fleshy algae, II; calcarious algae, IV) and corals (II) is dependent in part on the intensity of grazing and nutrient levels (modified from Littler and Littler, 1988; with their permission).

an advantage in oligotrophic reef waters. After macroalgal removal from a Great Barrier Reef community, the cover of *Acropora* spp. increased significantly when compared to areas where the algae remained (Tanner, 1995). Thus, coral growth, fecundity, fission, survivorship, recruitment, and cover were negatively affected by macroalgae. Encrusting algae can be overgrown by erect species and thus do best in regions where grazing or wave activity removes the epiphytes.

Macroalgal Zonation

Three examples from atolls and one from a fringing reef are given in what follows. The intertidal zones on a coral reef could begin with beach rock of an island that extends into a moat formed by tidal currents, a reef flat with tidal pools and channels, an algal ridge that forms at the windward edge of the reef, and a subtidal seaward plain beyond the ridge (Fig. 12-1). Whereas microalgae, especially blue-green taxa, dominate the intertidal beach rock and coral rubble, frondose macroalgae grow on the reef flat, usually found under coral heads and in tidal channels. Coenocytic green algae (*Udotea, Halimeda, Caulerpa*) and seagrasses are more common closer to the island away from high-wave activity growing in the moat and tidal pools, whereas tougher brown algae (e.g.,

Turbinaria, Sargassum) grow nearer the ridge. The algal ridge is an extreme habitat if wave activity is high and consists of crustose coralline algae (e.g., *Porolithon* on Banika Island, *Lithophyllum* on Guadalcanal) or a mix of coral and algae (Glover's Reef). The subtidal algal plain occurs below the spur and groove communities (see what follows) if the slope is not extreme. The plain consists of coral rubble and coarse sand, and usually is dominated by psammophytic green algae (Dahl, 1973). The effect of grazing around coral reefs was demonstrated in Lameshur Bay, St. John, in the U.S. Virgin Islands (Mathieson et al., 1975) and Galeta Reef in Panama (Hay, 1981b). In the former study, diversity and biomass of macroalgae increased with increasing distance (10 spp. at 10 m; 38 spp. at 120 m) and depth (11 spp. at 1 m; 58 spp. at 20 m) from the algal ridge. In the latter study, Hay hypothesized that intensive herbivory excluded sand-plain species from the resource-rich shallower reef slope. The almost barren zone around Caribbean patch reefs, called the "Randall zone," is due to diurnal fish grazers that feed close to the reefs and the nocturnal forays of sea urchins that are restricted to about 10 m from their shelter. The increased diversity of macroalgae away from the reef reflects the removal of herbivores by offshore predators (Glynn, 1990).

1. *Pacific Atolls.* Womersley and Bailey (1969) described the zonation of macroalgae in the Solomon Islands on four types of coral reefs, which varied with respect to water activity, naming a reef with moderate energy at Mamara Island [Fig. 12-7(A), Guadacanal Atoll] and a high-energy reef at Banika Island [Fig. 12-7(B), Russell Islands]. The seaward rim consists of crustose coralline algae (predominantly *Lithophyllum*) and corals, and behind this the coralline alga *Porolithon onkodes* dominates. Several noncalcified algae occur on the adjacent reef flats. The reef flat gradually drops toward the island, forming a "moat" with a variety of fleshy and filamentous subtidal algae. On shallow intertidal rubble near the island shore, crustose and endolithic green algae and cyanobacteria dominate.

2. *Caribbean Fringing Reef.* Detailed community studies of fringing reefs in Curacao (van den Hoek et al., 1975; Wanders, 1976) showed a conspicuous zonation of scleractinian corals and macroalgae (Fig. 12-8). The reef extends to approximately 30 m, which is shallow compared with Caribbean atolls such as Glover's Reef off Belize, Central America (to 50 m; Tsuda and Dawes, 1974). Analysis showed seven zones dominated by scleractinian corals and seaweeds. A total of 142 macroalgae were recorded with species richness reaching 52 spp. 25 cm^{-2}. The first zone occurred on intertidal rubble and extended to about 0.6 m depth and was greatly influenced by exposure [Fig. 12-8(A)], and the second zone (0.6 to 1.2 m) was influenced by wave action. The third and fourth zones extended to 2.6 m and included the shallow-water corals *Millepora* and *Acropora* [Fig. 12-8(B)], and the fifth zone (at about 5.1 m) contained both soft and hard (*Montastrea*) corals [Fig. 12-8(C)]. The sixth zone was a steep sloping area with deep-water corals (*Agaricia, Siderastrea*); it ended at about

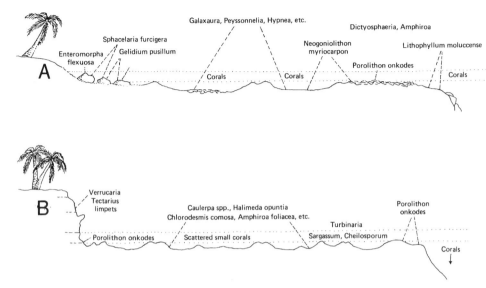

Figure 12-7. Transects across two Indo-Pacific fringing reefs. Mamara Island on Guadalcanal Atoll (A) experiences moderate wave activity, whereas Banika Island in the Russell Island group (B) has extreme wave action. Both reefs are 25 to 30 m wide, have a coralline ridge of *Porolithon* and *Lithophyllum*, and a variety of macroalgae on their reef flats. The upper and lower intertidal zones are about 0.7 m on both islands (after Womersley and Bailey, 1969).

20 m, where the final region of sediment and coral rubble dominated [Fig. 12-8(D)].

3. *Caribbean Atoll.* Glover's Reef in Belize has four northwest-facing islands on the side of the atoll facing the prevailing wind and a lagoon of about 100 km² (Fig. 12-4). A total of 100 taxa were reported for the reef, including 8 cyanobacteria and 39 green, 19 brown, and 34 red macroalgae, as well as 4 sea-grasses (Tsuda and Dawes, 1974). This number was estimated to be about 70% of the actual taxa present, as seen in the study of the Belize barrier island Carrie Bow Cay (165 algal species; Norris and Bucher, 1982). Five zones were distinguished on the seaward side off the main island, Long Cay. The intertidal zone (Fig. 12-9, zone 1) ranged from an upper region of coral rubble and beach rock (coquina) bordering the sand shore [Fig. 12-10(A)] to a narrow reef flat [Fig. 12-10(B)]. Blue-green algae dominated the upper region and were graded into a multispecies turf of *Cladophorophsis*, *Centeroceras*, and *Pterocladia* in pockets on the reef flat. Brown algae such as *Turbinaria*, *Sargassum*, and *Colpomenia* could be found in the lower intertidal reef flat (see also Fig. 8-5). The breaker zone (Fig. 12-9, zone 2) ranged from 0.1 to 1.3 m and was dominated by highly branching corals such as *Acropora palmata* and *Millipora alcicornis*. Most of the macroalgae were filamentous (*Giffordia*, *Polysiphonia*) or turf-forming species (*Gelidiella*, *Pterocladia*) that were intensely grazed by parrot

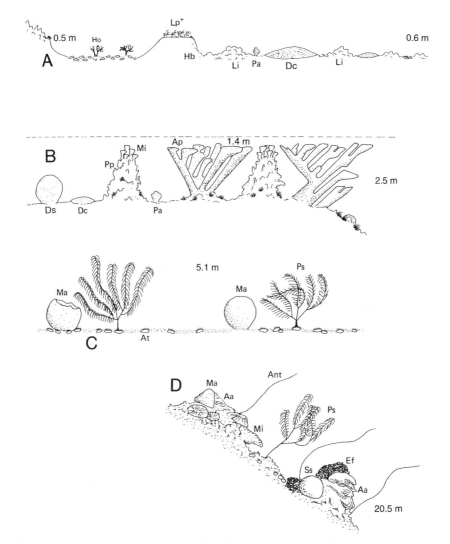

Figure 12-8. A transect through a Caribbean fringing reef. Four sections are taken from the study by van den Hoek et al. (1975) at Curacao. The intertidal and subtidal zones to 0.6 m (A) contain a variety of seaweeds, including *Halimeda opuntia* (Ho), *Laurencia papillosa* (Lp+), *Hydrolithon boergesenii* (Hb), *Lithophyllum intermedium* (Li), plus the corals *Porites astreoides* (Pa) and *Diploria clivosa* (Dc). The shallow subtidal zone (B: 1.2 to 2.5 m) includes the crustose coralline alga *Porolithon pachydermum* (Pp) and the corals *Diplora strigosa* (Ds), *Millepora* sp. (Mi), *Acropora palmata* (Ap) and Pa. At intermediate depths (C: 5 to 5.3 m), the transect includes the coral *Montastrea annularis* (Ma) and the soft corals *Pseudopterogorgia acerosa* and *P. americana* (Ps), and an algal turf (At). The end of the transect (D: 17 to 20.5 m) includes hermatypic corals *Agaricia agaricites* (Aa), *Antipatharian* sp. (Ant), *Siderastrea siderea* (Ss), *Eusmilia fastigiata* (Ef), Mi, and soft corals (Ps).

fish and the sea urchin *Diadema*. The third zone of 3 to 5 m (Fig. 12-9, zone 3) was characterized by *A. palmata* and buttress-type corals, including *Montastrea, Agaricia, Diploria, Porites,* and *Millepora*. The third zone also had filamentous and turf macroalgae, as well as *Valnoia* and articulated corallines (*Jania, Amphiroa*) growing on the underside of coral rubble. The fourth zone at 5 to 15 m (Fig. 12-9, zone 4) featured elkhorn coral *A. cervicornis*, which formed dense thickets on a steep slope (Fig. 12-11). Here *spurs* of coral heads were separated by *groves* of sand on which sediment and debris moved downward. Macroalgae included dwarfed forms of psammophytic green (*Udotea, Penicillus, Halimeda, Caulerpa*), red filamentous (*Ceramium, Heterosiphonia*), and large brown algae (*Sargassum, Zonaria, Spatoglossum*). The fifth zone, which was dominated by *Montastrea* (Fig. 12-9, zone 5), was on a 40° slope, where spurs of *M. annularis* and sand grooves were pronounced (Fig. 12-12); it diminished as the slope increased between 15 and 40 m. At 30 m depth, hermatypic corals were limited to small individuals with flattened morphologies, including species of *Agaricia, Diploria,* and *Siderastrea*. Foliose algal morphologies were also common and consisted of *Pocockiella, Zonaria*, flattened chainlike branches of *Halimeda* and *Amphiroa*, and crustose noncalcified (*Hildenbrandtia, Peyssonnelia*) and calcified (*Lithophyllum, Neogoniolithon*) red algae (Fig. 12-13).

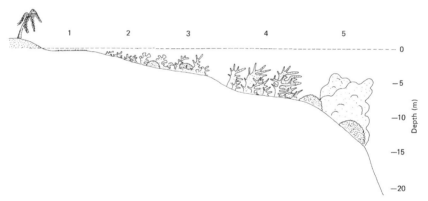

Figure 12-9. Transect through the seaward reef of Long Cay, Glover's Reef, Belize, Central America. The reef is about 25 m wide and can be divided into five zones: (1), the reef flat (0.0 to 0.3 m) has coralline algae (e.g., *Lithothamnion, Goniolithon*), *Thalassia testudinum* growing in small depressions, and small hermatypic corals (e.g., *Millepora*); (2) the breaker zone (0.3 to 1.0 m) includes the corals *Acropora palmata* and *M. alcicornis*; (3) a surge zone is divided into two areas, an upper *A. palmata* (1 to 3 m) and lower buttress (3 to 5 m) consisting of *Agaricia, Montastrea, Siderstrea, Porites,* and *Millepora*; (4) the elkhorn coral *Acropora cervicornis* grows in dense clusters below (5 to 12 m); and (5) at the lower edge (12 to 30 m) *Montastrea annularis* forms large (to 5 m tall) multilobed heads as well as other corals (*Diploria, Siderastrea, Porites, Agaricia*) and tubular sponges. Shingle corals (e.g., *Agaricia*), soft corals, and leafy alga occur below the zone 5.

Figure 12-10.   The intertidal zone at Glover's Reef, Belize. The coquina (beach rock) and broken limestone of the spray zone (A) is dark due to blue-green algae. A seaward view of the reef flat (B) shows shallow pools and channels that contain macroalgae (courtesy of R. A. Davis, University of South Florida, Tampa).

STRESSES, MANAGEMENT, AND REEF RESTORATION

The rich diversity of species and unique biological feedback of coral reefs belies their fragile balance with nature. Coral reefs have been extensively exploited in many parts of the world, although preservation and management have become an international concern. Although some coral reefs have not recovered after natural or anthropogenic stresses and overexploitation, typically, these systems are "well adapted to recovery from a variety of sources of natural stress" (Grigg

Figure 12-11. A bed of *Acropora cervicornis* at Glover's Reef, Belize. The bed at 6 m includes small heads of *Millepora* (m), soft corals (arrows), and a fine covering of filamentous algae growing on the dead branches.

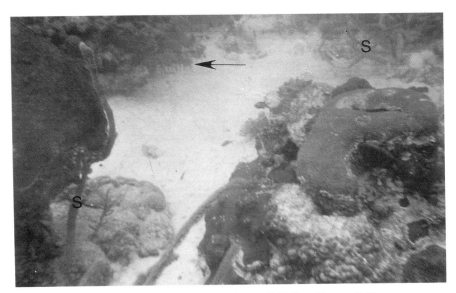

Figure 12-12. The butress zone at Glover's Reef, Belize. Spurs of *Montastrea* separated by sand grooves occur at 20 m. Tubular sponges (s) and psammophytic green algae (*Udotea* and *Penicillus*) are seen in the uppermost portion of the photo (arrow).

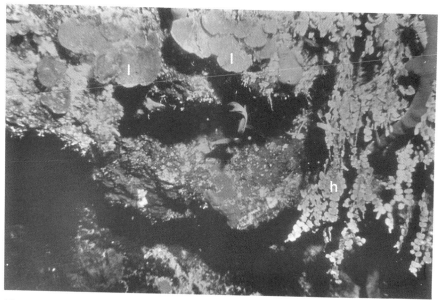

Figure 12-13. Deep-water algae at Glover' Reef, Belize. The brown alga *Lobophora variegata* (l) and two calcareious green algal species of *Halimeda* (h) show foliose morphologies at 34 meters in order to trap light. A variety of crustose algae (e.g., corallines, *Peyssonnelia, Hilldenbrandia*) and sponges cover the surface. Note the damsel fish (center) defending its territory.

and Dollar, 1990). Further, large-scale stresses, such as the mass mortality events of the black sea urchin *Diadema antillarum* in the Caribbean, can result in long-lasting effects (Lessios, 1988, 1995). As Hallock (1997) points out, the ability of a reef to recover is a function of water quality. For example, there has been little recovery of the Molassas Coral Reef in the Florida Keys after the grounding of the ship *Wellwood*, and this is attributed to the poor water quality due to anthropogenic effects.

Natural Stresses

Stress, defined in Chap. 3 as any restriction on biomass, can be divided into *limited* (short-term) and *disruptive* (long-term) effects (Grigg and Dollar, 1990). Natural stresses can range from trivial to massive events and include local storms, typhoons, El Niño events, massive sedimentation episodes, population explosions (e.g., crown-of-thorns starfish, *Acanthaster planci*), and population losses (*Diadema antillarum* dieoff). Volcanic eruptions (Galapagos Islands, Hawaii), heavy rains coupled with low tides (Hawaii), and water cooling (Persian Gulf) may also be significant. Reviews of disturbances (Brown and Howard, 1985; Grigg and Dollar, 1990) include coral-reef responses from insignificant changes to total destruction. The effects of natural stresses can be seen in modifications in species diversity. On the Kona coast of Hawaii, one coral, *Porites* spp., dominates the area due to the low frequency of disturbance In contrast, at Gardner Pinnacles, Hawaii, intense wave disturbances result in low cover and high diversity (Grigg and Dollar, 1990). However, species of *Porities* are usually considered to be pioneer (early) species, and thus the Hawaiian example may reflect the colonization of the islands by opportunistic corals rather than that of a climax community. As indicated in the previous para graph, the damage caused by natural events is greatly exacerbated by human influence, with the leading problem being nutrient flux to coastal waters. Thus a "natural event" such as a hurricane, which can cause massive coastal runoff may trigger nutrification events through the introduction of hydrocarbons (from roads, buildings) and nutrients (from fertilizers, sewage; Hallock and Schlager 1986; Hallock et al., 1993).

Imbalances or loss of keystone species demonstrate how important biological processes are in controlling coral-reef development. Overgrazing of Guam reef corals by predatory *Acanthaster planci* (crown-of-thorns starfish) allowed epi phytic algae to colonize the dead *Acropora* branches in less than 24 hours (Bell and Belk, 1975). The initial algal species were two blue-greens (*Microcoleus lyn byaceus, Hormothamnion solutum*) that were replaced by a brown algae (*Giffor dia indica*). Disturbances, like removal of a herbivore via a disease, will lowe grazing pressure and allow competition to occur between macroalgae and corals For example, with the massive die-off of the sea urchin *Diadema antillarum* i the Caribbean, macroalgae such as the brown seaweeds *Lobophora variegata* and *Dictyota* spp. increased their cover over crustose coralline algae and scleractinia corals like *Agaricia agaricites* (van Steveninck and Bak, 1986). Beginning i 1983, the black urchin die-off progressed throughout the Caribbean, the Gulf c Mexico, and the Bahamas, covering an area of 3.5 million km² by early 198 (Lessios, 1988). Mortality, which approached 93%, was thought to be a host-spe cific pathogen, and the black urchin had not recovered after 10 years (Lessios 1995). The immediate response to this die-off was an increase in algal biomass (e.g., a fourfold increase in algal biomass in Jamaica) in areas where fishing pres sure kept herviborous fish populations low. With the increase in algal biomass crustose corallines and corals were overgrown and declined. One might question whether the die-off of *Diadema antillarum* was a natural disturbance (e.g., dis ease) or whether it was anthropogenic due to the probable transport of a virus vi the man-made Panama Canal. Further, corals, by virtue of their symbiotic zoox anthellae are adapted to "nutrient deserts" and, like a desert, adding water (= nutri ents) does not produce larger desert plants (= corals) but rather grasses and wild flowers. Thus, with more nutrients, the corals do not grow larger; rather, algae an sponges and other bioeroders of limestone do, thus, outcompeting them.

Cyanobacterial like *Phormidium corallicum* and various bacterial species *Beggiatoa* and *Desulfovibrio* cause bandlike tissue necrosis and colony death Caribbean and Great Barrier Reef corals (Sorokin, 1993). The diseases appa ently infect damaged or stressed corals, such as those handled by divers, scrape by anchors, or growing in nutrified waters.

Anthropogenic Stresses

The four most common types of man-made stresses are sedimentation, nutrient and organic pollution, thermal pollution, and oil spills (Brown and Howard, 1985; Grigg and Dollar, 1990; Dubinsky and Stambler, 1996; Hallock et al., 1997). Other human-induced stresses include nuclear tests (Eniwetok), the use of dynamite for "fishing" (Philippines), anchoring and boat damage (Florida Keys), coral mining (India), and shell collecting (everywhere). Sedimentation is probably the single most serious anthropogenic stress on coral reefs (Grigg and Dollar, 1990), because dredging, blasting, and harbor construction occur throughout the world. For example, about 29% of the coral reefs in Kaneohe Bay, Hawaii, were removed or killed due to dredging in 1939. Further, sediment loading can result from coastal and interior activities on land, including clear-cut logging, mining, and agriculture, which result in terrestrial erosion (Philippines, Indonesia, Kenya).

Nutrification is an increase in the nutrient load in the water column, resulting in a shift in community structure (see Chap. 5) and is probably either first or second in causing the demise of coral reefs throughout the world (Hallock and Schlager, 1986; Hallock et al., 1997). While mechanical damage is usually more localized (although sedimentation can be widespread), nutrification can spread for tens to hundreds of kilometers from the point source. Further, nutrification (via terrestrial runoff, sewage outfalls, groundwater seepage) promotes vast changes in the ecosystem structure so that the coral communities may cease to exist. The increase in available nutrients will stimulate the growth of algal epiphytes, which outcompete corals for space and light. A secondary effect of nutrification is the trapping of sediment by the algae, which then smother the coral polyps. If the input of nutrients is sufficiently high, then the community experiences eutrophication (e.g., sewage outfalls, high levels of organic pollution), which results in oxygen depletion primarily by stimulating growth of microbes (Dubinsky and Stambler, 1996) and may result in increased levels of toxic contaminants (Pastorek and Bilyard, 1985).

Partial recovery of corals can be rapid. In Kaneohe Bay, Oahu, Hawaii, after 14 years, a point-source discharge (peak 1.9×10^4 m^3 d^{-1}) was ended, and 5 years later the total live coral coverage had doubled. However, they have not fully recovered, and it has been questioned whether coral reef communities can recover in less than human life spans after cessation of organic pollution. Even so, it is encouraging to witness the adaptations of these ecosystems to disturbances (Grigg and Dollar, 1990).

Overfishing removes the predators and competitors, as has occurred in the Indo-Pacific and the Caribbean (e.g., Jamaica). With the loss of grazers, filamentous algae overgrow the corals. The black sea urchin *Diadema antillarium* increased in numbers in the 1970s in the Caribbean, probably due to lack of predators that had been removed by overfishing. With its dieback in the late 1980s, there were no herbivores to remove the algae that overgrow the coral reefs in the Caribbean.

Catastrophic oil spills on coral reefs are not as common but do occur, as seen in the Panama oil spill (Ornitz, 1996), and single-event episodes appear to have limited effects (National Research Council, 1985). In contrast, chronic exposures to petroleum products (e.g., nearby harbors) result in high mortality of hermatypic corals, as also seen in mangals (Chap. 5), as well as reduction of macroalgal diversity and development of pioneer algal floras (blue-green, green mats). Most of this damage occurs on the reef or back reef, where the surface comes directly into contact with the oil. Thus, in the Panama Canal Zone, subtidal corals were unaffected from a diesel oil spill, whereas the intertidal organisms were coated and killed.

Thermal stress from heated water can cause significant diebacks on coral reefs, in part because lethal temperatures are close to ambient ones. Thermally enriched cooling water from a power plant in Biscayne Bay, Florida ($+ 4°C$), resulted in a dieback of 50 ha of seagrass and corals; in Guam, 10 ha of reef was lost, and in Oahu, Hawaii (+4 to 5°C), 0.71 ha of reef was killed (Grigg and Dollar, 1990). Reorientation of the Oahu thermal plume to deep water resulted in recovery of the shallow-water coral reef (Coles, 1984), again demonstrating the ability of coral reef communities to respond.

Management and Restoration

Preservation and management of coral reefs have become international concerns. In developing countries (e.g., French Polynesia, New Guinea, Philippines), reef resources such as fish, molluscs, and coral are intensively exploited, resulting in overfishing. Damage can be very extensive, due to the illegal use of dynamite or heavy nets to catch fish. Laws protecting the reefs exist in most countries, but they are not enforced because of the high cost of keeping officers in the field, limited income, and opposition by local populations accustomed to taking what they wish from the sea. Many developing and developed countries now allocate funds for law enforcement, education, and value of coral reefs, including their use as tourist attractions (Hallock et al., 1997). Coral reefs have been incorporated into national parks (Great Barrier Reef Marine Park) and sanctuaries (Florida Keys Reef Sanctuary). The Great Barrier Reef Marine Park is an outstanding example of how a plan was developed in conjunction with local fisherman, the public at large, and the government (Sorokin, 1993).

Artificial reefs can provide new and usable habitats in areas of damaged or recovering coral reefs. There is controversy as to whether artificial reefs result in increased fish production or they simply attract local fish. Regardless, their construction in tropical regions does result in new substrata for algal and coral development, as well as a variety of reef fish such as chaetodontids, apogonids, and pomocentrids (Sorokin, 1993). In the Gulf of Mexico, where there are over 3000 underwater structures used in oil pumping, these platforms function as artificial reefs and have doubled the level of fishing to about $85 million.

In summary, coral-reef ecosystems with their marine plant communities have evolved with natural stresses, and, if water quality does not deteriorate, they can recover after a reasonable disturbance, although the recovery period may be 20 to 100 years. Coral reefs are sensitive to frequent or chronic damage, and thus all types of impacts must be considered. Further, "indirect" effects ranging from overfishing to land use in adjacent coasts must be incorporated in a management plan for coral reefs. The ability of reefs to recover indicates that management schemes can work. However, having evolved in nutrient deserts, coral reefs cannot recover if subjected to nutrification. If these highly productive, diverse ecosystems are to survive, it is imperative to rationally manage, use, and protect the reefs and their resources.

Selected Methods for Study of Marine Plants

The methods described here are for field or laboratory studies of marine plants that are not readily available in the literature. A number of texts are available detailing field and laboratory procedures for the study of seaweeds (Hellebust and Craigie, 1978; Littler and Littler, 1985; Lobban et al., 1988), phytoplankton (Sournia, 1978), seagrasses (Phillips and McRoy, 1990), and mangroves (Snedaker and Snedaker, 1984). Collections of procedures for measuring chemical and physical characteristics of seawater are available in a variety of texts (Strickland and Parsons, 1968; Schlieper, 1972; Parsons et al., 1984).

FIELD METHODS

Collection

Whether making collections using snorkle and SCUBA equipment or by wading, similar water-resistant equipment will be required. If wading, collecting shoes, or boots, and gloves should be available in a waterproof backpack. Many of the following items are standard for field work in marine plant communities: collecting pail or mesh bag, scraper or prying bar, small knife or clippers, digging tool (for seagrasses), plastic bags with water-resistant labels or marker, and recording material (encased clipboard with wet-dry paper or SCUBA board and attached pencil). Standard measurements at field sites usually include sediment cores (modified plastic syringe), water temperature (thermometer), salinity (water bottle), Secchi disc (water transparency), and dissolved oxygen (field oxygen kit or BOD bottles and preservation chemicals). Water samples can be also taken for nutrients and pH (store on ice) or measured in the field using commercial kits (e.g., Hach Co., Loveland Co.).

Transect Sampling

Transects are used in plant zonation studies of intertidal communities (Fig. A-1) or where line quadrats are used (e.g., across seagrass beds). Easily seen

Figure A-1. A line transect. Dr. and Mrs. Mathieson are using a calibrated, 100-m tape and elevation rod to determine marsh size and drop in elevation across a stunted upper tidal salt marsh at Thompson Island, Maine.

stakes can be aligned from the highest to lowest zones and a metric tape or marked line stretched between them. Samples for identification can be taken along the transect in each zone or every unit of measurement (every centimeter to every few meters, depending on slope and detail required). Elevations can be measured using a sighting or line levels (Geological Methods; Fig. A-5). Percent species is determined by dividing the number of individuals within a zone by the total present along the entire transect. Percent species cover is calculated by dividing the length (cm, m) of the transect (or zone) the species covers by the total length of the transect (or zone). Typical abiotic data from a zonation study using transects would include sediment samples or description of substrata, water temperature, dissolved oxygen, salinity, and pH of tide pools and coastal water. Biological data would include species identification, position (elevation, distance), and morphological variations. In addition, frequency and abundance, as shown in Eqs. A-1 and A-2 can be calculated.

$$\text{Frequency} = \frac{\text{number of occupied quadrats/units}}{\text{total number of quadrats/units}} \quad \text{(A-1)}$$

$$\text{Abundance} = \frac{\text{total number of individual plants}}{\text{number of occupied quadrats/units}} \quad \text{(A-2)}$$

Quadrats

Unit-area measurement can be done using quadrats ranging in size from 1 m^2 [Fig. A-2(A)] or smaller [e.g., 25 cm^2; Fig. A-2(B)], rings (e.g., hoola hoops),

Figure A-2. Use of quadrats. A 1-m² quadrat divided into 10-cm subunits is used to determine abundance and frequency on a "uniform" limestone surface (A). Smaller, 25-cm² quadrats can be used along a transect running through a mangrove or salt marsh communities (B).

or other regular shapes. Unit-area samplers are useful in regions where there is little to no zonation or major shifts in abiotic factors such as a subtidal algal or seagrass communities. They also can be used in zonation studies (strip quadrats, band transects) to develop a more accurate determination of percent cover, fre-

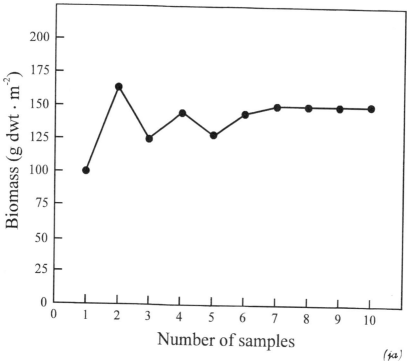

(fa)

Figure A-3.   A performance curve to determine the number of samples needed. The curve plots the cumulative unit to be measured (e.g., biomass) against the number of samples taken. The minimum number of samples needed to avoid possible variations in the data (e.g., 7) can then be determined, and further sampling is avoided.

quency, and abundance. Quadrats may be subdivided [Fig. A-2(A)] or small ones used [Fig. A-2(B)] if detailed sampling is required. To avoid bias in sampling, random or haphazard methods can be used for quadrat placement. To carry out a *random* study, the area is subdivided into appropriate-sized units and a random-numbers table used to select which subunits will be measured. Lines and stakes are used in the field to subdivide the region and the preselected random numbers are used to select the subunits. An alternative is a *haphazard* selection of the sampling sites in which one throws the quadrat or hoop in the field while traversing the community. The minimum number of unit areas that should be measured to ensure adequate sampling can be determined through the use of a performance curve (Fig. A-3).

Quadrant samplings usually count all individuals by species found in each subunit. Percent cover can be estimated using broad categories (e.g., 100%, 75%, 50%, 25%, <5%, 0%) for dominant species. In addition to frequency and abundance (Eqs. A-1 and A-2), density can be determined, as shown in Eq. A-3.

$$\text{Density} = \frac{\text{number of individuals species}^{-1} \text{ unit area}^{-1}}{\text{total number of unit areas}} \qquad (A\text{-}3)$$

Phytosurvey

Survey techniques include creation of landscape and vegetational maps through remote sensing (aerial, satellite photography; Belsher et al., 1985) and phytosociology (Boudouresque, 1971). Phytosurveys are particularly useful for the study of large areas (e.g., kilometers of coastline) and have been used in studies of algal and seagrass communities of the Mediterranean. The procedures are simple and yield repeatable results in studies of seaweed, seagrass, and tidal marsh communities. Sample plots (1 to 10 m^2) are selected along the coast based on visual observation and determination of associations. Phytosurveys allow rapid coverage of large areas and so the dominant plants must be easily identifiable. Transects should first be done to determine if zonation is present and unit-area measurements carried out to obtain statistical data. Each dominant species is ranked for abundance, cover, and growth form (sociability) using the following rankings:

1. *Abundance*: r = rare, usually only one specimen per plot; + = sparse, a few specimens per plot.
2. *Cover*: 1 = covering less than 25% of the plot; 2 = covering 25 to 50% of the plot; 3 = numerous, covering 50 to 75% of the plot; 4 = covering 75 to 100% of the plot.
3. *Sociability*: 1 = single plants; 2 = small tufts or clusters; 3 = small patches and distinct groups; 4 = a dense carpet or mat, but not homogeneous, closed vegetation; 5 = a homogeneous and dense carpet, little else present.

GEOLOGICAL METHODS

Location

Latitude and longitude of a site can be determined using Loran, navigational aids, or plotting on nautical charts. Loran relies on coastal transmitting stations that beam coded signals. By triangulation, the Loran instrument determines the exact location of a ship. Navigational aids range from hand-held to mounted instruments that determine the coordinates of a position through triangulation of transmissions from U.S. Global Positioning Satellites (GPS). Use of nautical charts such as those produced by NOAA (U.S. National Oceanic and Atmospheric Administration) is probably the simplest and less costly method. Such charts are produced by most countries for their coasts. Latitude and longitude can be determined by triangulation. A straight line is run from the site horizontally to the latitude scale marked on the edge of the chart. A second straight

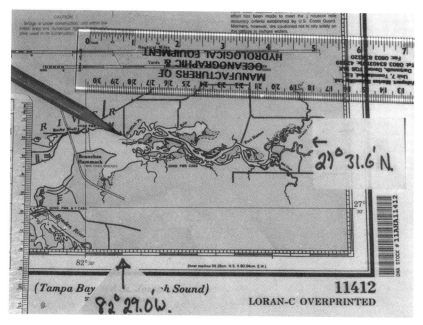

Figure A-4. Site location using a chart. Latitude and longitude can be determined for a coastal site using rulers and coast and geodetic charts. For example, Redfish Point (at pencil point) is located on NOAA Chart 11412 (27° 31.6′ N, 82° 29.0′ W).

line is run vertically to the longitude scale and the two sets of coordinates are determined (Fig. A-4).

Elevation

In studies of intertidal plant communities, elevation should be determined for each zone. Elevations can be measured using standard survey equipment (e.g., Philadelphia Level and Rod) that give exact elevations and are related to geodetic benchmarks. This was done in salt and mangrove communities to determine if small (±2 mm) berms existed that would retain saltwater during ebb flow, resulting in the high soil salinities of salterns (Hoffman and Dawes, 1996). Usually, studies in marine plant communities do not require such precision and line or sighting levels are sufficient (Fig. A-5). A sighting level is mounted at a known height against a rod and focused on a second rod some distance away [Fig. A-5(A)]. The bubble in the sighting level is brought to the center of the scale while focusing. The person holding the second rod marks and measures the height above ground at the leveled focal point. The difference in elevation from the upper and lower rod is the drop or rise in elevation. A (carpenter's) line level also can be used. It is suspended on a string tightly stretched between two metric rods and the bubble in the line level is centered [Fig. A-5(B)].

Figure A-5. Techniques for determining elevation. A sighting level (A) or line level (B, arrow) A-5 can be used to determine differences in elevation. In both cases, a point that is level between the two vertical rulers is marked and the increase or decrease in elevation is determined.

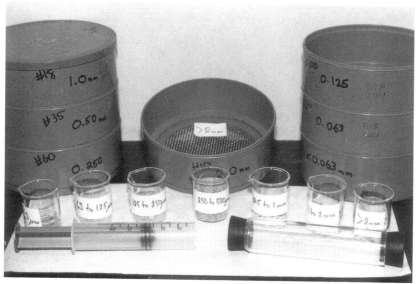

Figure A-6. Particle-size determination. A set of sieves with screen sizes of 2.0 to 0.063 mm (125 µm) can be used to determine particle size of sediments. Two types of homemade corers are also shown: a plastic syringe (25 cc) with the tip cut off and a plexiglass tube with rubber corks. Both the syringe and tube have a beveled end for cutting into the sediment.

Particle Size

Edaphic features that help characterize tidal and seagrass communities include nutrients of pore water, redox level of the sediment, and particle-size distribution. The last abiotic feature can be used to identify levels of water movement and type of sediment. Sediment samples can be obtained using a 50-cc plastic syringe that has had its tip cut off. The syringe tube is pushed into the sediment (ca. 10 cm) and the rubber-tipped plunger inserted to create a vacuum. The syringe is pulled from the sediment and the plunger used to eject the core into a plastic bag. The sediment core can be subdivided if desired. After the sediment core is dried, it is sieved using a set of standard screens (2.0-mm to 0.063-mm pore size (Fig. I-6). Particle-size distribution is obtained by dividing the dry weights of each size class by the total dry weight of the sample. Drying will destroy any biologically produced sediment clumps (e.g., feces), so the effect of bioturbidation or deposition will not be known.

SEAWATER ANALYSIS

Methods for determining levels of nutrients (N, P, Si), salinity, and dissolved oxygen are detailed in a number of seawater handbooks (Strickland and

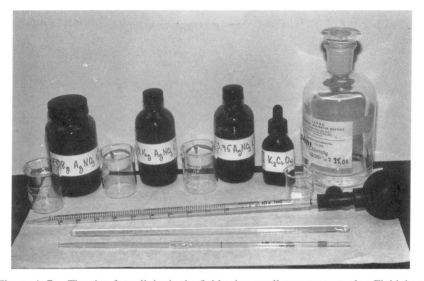

Figure A-7.   Titration for salinity in the field using small seawater samples. Field determination of salinity can be done with solutions of $AgNO_3$ (23.95, 19.16, 9.58 g L^{-1}) an end-point solution of 5% K_2CrO_2, a standard seawater source for calibration, calibrated 10-mL and volumetric 1-mL pipettes, a stirring rod, assorted beakers, and a pipette bulb. The $AgNO_3$ and K_2CrO_2 solutions are in dark bottles.

Parsons, 1968; Parsons et al., 1989). The techniques given here are modifications of published methods that have proven useful for direct field measurements.

Salinity

Standard field measurements of salinity include using a refractometer, which determines specific gravity by measuring light refraction, and a conductivity meter, which measures the electrical resistance of saltwater and calculates ion concentration. Other methods are usually done in the laboratory, namely, titration of the chlorine ion with $AgNO_3$ and use of a hydrometer (see Chap. 2). Refractometers, conductivity meters, and hydrometers may not be available. Thus a field titration method for small samples is given that can be used on a shore or in a laboratory. Three concentrations of silver nitrate in dark 100-mL bottles, a 5% potassium chromate (K_2CrO_4) solution in a dropper bottle, and distilled water in a plastic squeeze bottle are needed. Glassware includes a 5-mL pipette graduated to 0.1 mL, a volumetric 1-mL pipette, a stirring rod, and a 20-mL beaker (Fig. A-7).

1. The three aqueous concentrations of $AgNO_3$ are given.

AgNO$_3$ (g L^{-1})	Molarity	Chloride Equivalent (mg mL^{-1})
23.95	0.14	5 (salinities above 25 ppt)
19.16	0.11	4 (salinities 10 to 25 ppt)
9.58	0.05	2 (salinities below 10 ppt)

2. To 1 mL of seawater sample, add 5 mL of distilled water (to increase volume) and 1 drop of 5% K$_2$CrO$_4$ in the beaker.

3. Titration (drop by drop from a calibrated pipette) is carried out using the appropriate silver nitrate solution (based on estimate of salinity). Any white AgCl precipitate is broken up with constant stirring.

4. Titration ceases when the orange AgCrO$_4$ color remains; and the titrations are done in duplicate or triplicate and averaged.

5. Salinity is determined by calculating molarity of chlorine (Eq. A-6), chlorosity (Eq. A-5), and then chlorinity (Eq. A-6). Density at 20°C for Eq. A-6 is interpolated based on the chlorosity calculated in Eq. A-5. Salinity also can be interpolated from chlorosity, but is more accurately determined using Eq. A-7, which expresses its relationship with chlorinity.

$$\text{Molarity of Cl} = \frac{\text{vol. AgNO}_3 \times \text{molarity AgNO}_3}{\text{volume of sample (1 mL)}} \qquad (A\text{-}4)$$

$$\text{Chlorosity} = \text{molarity of Cl} \times 35.5 \qquad (A\text{-}5)$$

where 35.5 is the atomic weight of chlorine.

Density @ 20°C	Chlorosity	Salinity
1.000	0.00	0.0
1.006	5.53	9.9
1.014	11.21	19.9
1.024	17.50	30.8
1.029	22.78	39.9
1.034	28.06	49.0

$$\text{Chlorinity} = \frac{\text{chlorosity}}{\text{density of sample at 20°C}} \qquad (A\text{-}6)$$

$$\text{Salinity} = 0.03 + (1.805 \times \text{chlorinity}) \qquad (A\text{-}7)$$

Dissolved Oxygen

Oxygen production or consumption is usually measured via manometric, titration, or current-flow techniques (Chap. 2). The first is limited to the laboratory and the others can be done in the field. The manometric procedure measures pressure changes that result from uptake or release of oxygen. The Gilson Differential Respirometer (Gilson Medical Electronics; Fig. A-8) measures changes in

Figure A-8.   The Gilson Respirometer (model GRP-20). The manometric device can be used to measure respiration and photosynthesis (Dawes, 1985). The reference flask (218 mL, left) is used to balance the gases in the 20 small (e.g., 20 mL) reaction flasks. Each reaction flask has an individual manometer and digital scaler (d). All flasks are held in the water bath, where temperature and lights (below) are controlled. Other gases (e.g., oxygen) can be connected to the system.

gas volume of small manometers with digital micrometers. This method uses reaction flasks in a constant-pressure *respirometer* (volumeter) and is very useful in the studies of seagrass and seaweed productivity (Dawes, 1985). Respirometers may have from 12 to 20 independent reaction flasks, allowing multiple replicates and are highly accurate in photosynthetic and respiratory studies.

Oxygen meters (Chap. 2; Thomas, 1988) use standard biological oxygen demand (BOD) bottles (Fig. A-9), as does the third procedure, titration (Winkler titration; Dawes, 1988; Chap. 2) of iodine (I_2) with sodium thiosulfate ($Na_2S_2O_3$). The titration technique has been modified for use in field kits (e.g., Hach Co.). The accuracy of field kits (usually to 1 mg O_2 L^{-1}) can be improved (to ± 0.1 mg O_2 L^{-1}) by using a 10-mL syringe that has been calibrated against the standard Winkler procedure.

Conversions

Conversion of moles to mg O_2 (ppm O_2) per standard (300-mL) BOD bottle can be carried out as shown in the example, where 40 m mol O_2 g dwt h^{-1} is changed to 4.26 mg O_2 g dwt h^{-1} 300 mL^{-1} (Eq. I-8).

Figure A-9. Use of oxygen meters. Orbisphere (a) and YSI (b) oxygen meters are shown along with a magnetic mixer, and 60- and 300-mL BOD bottles. The tweezers are used to change algal samples and remove the magnetic rod from the bottles, and the aluminum pan is used to dry the algal sample after measurement.

$$40 \text{ n mol } O_2 \text{ d dwt h}^{-1} = 0.4 \times 3.2 = 1.28 \text{ mg L}^{-1} \text{ g dwt h}^{-1} \times 3.33$$
$$= 4.26 \text{ mg } O_2 \text{ g dwt}^{-1} \text{ h}^{-1} 300 \text{ mL}^{-1} \qquad \text{(A-8)}$$

where 3.2 mg = 1 m mol O_2 L^{-1}, and 3.33 is a volumetric conversion from 1 L to 300 mL or the volume of a standard BOD bottle that is normally used in studies of oxygen production or consumption.

Conversion from μL to mg O_2 uses the gas volume, temperature, and pressure relationship, where 1 M of any gas at standard temperature and pressure (STP) (273 K; 760 mm Hg) occupies 22.4 L. Therefore 1 mole of O_2, which weighs 32 g at STP, occupies 22,400 mL (Eq. A-9):

$$1 \mu\text{L } O_2 = \frac{32,000,000 \ \mu\text{L}}{22,400,000 \ \mu\text{g@STP}} = 1.4285 \mu\text{L}/\mu\text{g} \qquad \text{(A-9)}$$

For other temperatures (T_x), the volume (μL) is multiplied by the conversion factor obtained in Eq. A-10.

$$\text{Conversion for } T_x = 1.4285 \times \frac{273}{°\text{C} + 273} \qquad \text{(A-10)}$$

PHYSIOLOGICAL METHODS

The simplified procedures that follow are useful in physiological ecology studies of marine plants. More varied procedures are found in the texts listed at the beginning of this appendix.

Dry Weight

Percent dry weight is a simple yet useful method for comparing biomass, dominance of a species, or population, as well as allocation of resources to different plant parts. Plants are weighed after blotting (wet weight), dried, and reweighed (dry weight). Drying may be done in the sun, suspending over incandescent light bulbs (wire screen over an open box with side holes for air movement), or in a drying oven. The last drying method should not exceed 60°C because combustion may occur. The plants are cleaned of epiphytes and debris and rinsed in clean seawater from the site (ambient salinity). Rinsing in freshwater is not recommended because it can result in different dry weights due to variations in loss of surface salt. The dried material can be ground (e.g., Wiley Mill; 40-mesh screening or mortar and pestle) and stored in capped vials for determination of proximate constituents.

Pigment Extraction

Comparison of photosynthetic pigment levels are useful in studies of shade and sun plants, effects of nutrients in eutrophication studies, and comparisons of photosynthetic rates. Levels of all pigments (chlorophylls, cartenoids, phycoblins) can be determined but require chromatographic procedures (Stewart, 1974; Lobban et al., 1988). In this section, simplified extraction techniques are given for measurement of chlorophylls *a* and *b* and phycoerythrin:

1. Freshly collected samples ($n = 5$ to 10) are weighed (wet; use 0.1 to 0.5 g sample^{-1}) and the pigment extracted.

2. The fresh samples can be ground using a mortar and pestle or a ground-glass homogenizer (Tissue Grinders). All grinding and extraction should be done under low light and at 10 to 12°C (cold room, ice bath), because the pigments are easily degraded by light and heat. Grinding is greatly enhanced if liquid nitrogen is used; gloves and safety glasses should be worn.

3. Chlorophylls *a* and *b* are extracted by grinding in 80% spectroanalyzed acetone, and phycoerythrin is extracted in a 0.1-M phosphate buffer (pH 6.5). Small samples of most seaweeds and seagrasses can be ground within 1 min in 2 to 3 mL of the extracting fluid.

4. The slurry of powder and solvent is poured into centrifuge tubes marked to 6 mL and the grinding apparatus is flushed with the solvent until the tube is

filled to the mark. The extract is centrifuged (45,000 rpm, 2 to 5 min) in the cold using a tabletop centrifuge.

5. The supernatant is decanted into a cuvette and the absorption is measured at appropriate wavelengths on a calibrated spectrophotometer.

Measuring Phycobilins

Two sets of calculations are given for calculation of phycobilins. Beer and Eshel (1985) used the following calculations for determination of phycoerthrin (PE: Eq. A-11) and phycocyanin (PC: Eq. A-12) using a 0.1-M PO_4 buffer (pH 6.8). The absorption peaks measured are 455, 564, 592, 618, and 654 nm.

$$PE \ (mg \ mL^{-1}) = [(A_{564} - A_{592}) - (A_{455} - A_{592}) \times 0.20] \times 0.12 \quad (A-11)$$

$$PC \ (mg \ mL^{-1}) = [(A_{618} - A_{645}) - (A_{592} - A_{645}) \times 0.51] \times 0.15 \quad (A-12)$$

The mg/g dwt^{-1} can be obtained by dividing the volume of buffer used by the g dwt of the sample and using that factor to multiply the mg/mL. Percent dry weight (% dwt) is calculated by dividing the wet weight by the dry weight of at least 10 extra samples. Then the dry weight (g/dwt) of the sample used in pigment extraction can be obtained by multiplying the wet weight by the known percent dry weight.

Phycoerythrin can also be determined using O'Carra's (1955) absorption coefficient obtained from *Ceramium*. A visible scan of the phycerythrin extract shows three absorption peaks at 495, 540, and 565 nm. Absorption is measured at 565 nm (A_{565}) using Eq. A-13 to determine the level of phycoerythrin (PE g dwt^{-1}).

$$PE \ (mg/g \ dwt^{-1}) = 12.4 \times A_{565} \times \frac{vol. \ (mL)}{g \ wwt \times \% \ dwt \times 1000} \quad (A-13)$$

where 12.4 is the absorption coefficient, and vol. is the total volume of 0.1-M phosphate extracting fluid used.

Measuring Chlorophylls

The absorption peaks of chlorophyll *a* at 663 nm (A_{663}) and chlorophyll *b* at 645 nm (A_{645}) are measured, and chlorophyll content is calculated using Eqs. A-14 and A-15 (Smith and Benitez, 1955). A correction for turbidity is made by subtracting the absorption obtained at 725 nm.

$$mg \ chl \ a \ g \ dwt^{-1} = 11.9A_{663} - A_{725} \times \frac{vol. \ (mL)}{g \ dwt \times 1000} \quad (A-14)$$

$$mg \ chl \ b \ g \ dwt^{-1} = 22.9A_{645} - A_{725} \times \frac{vol. \ (mL)}{g \ dwt \times 1000} \quad (A-15)$$

Other absorption coefficients of chlorophylls can be found in Meeks (1974). Sternman (1994) extracted in 90% acetone, measured the absorption peaks of 647 and 664 nm, and used the following calculations for chlorophyll a (Eq. A-16) and chlorophyll b (Eq. A-17).

$$\text{mg chl } a \text{ g dwt}^{-1} = 11.93A_{664} - 1.93A_{647} \times \frac{\text{vol. (mL)}}{\text{dwt} \times 1000} \qquad \text{(A-16)}$$

$$\text{mg chl } b \text{ g dwt}^{-1} = 20.36A_{647} - 5.50A_{664} \times \frac{\text{vol. (mL)}}{\text{dwt} \times 1000} \qquad \text{(A-17)}$$

Use of Light and Dark BOD Bottles

Biological oxygen demand (BOD) bottles are typically used in oxygen measurements employing an oxygen meter or the Winkler titration (Thomas, 1988), and they are reliable for measuring photosynthesis or respiration (Patten et al., 1964). Further, the bottles can be used in the field so that in situ studies are possible. The bottles have a tapered ground glass stopper that prevents air being trapped when stoppered (Fig. I-9) and are available in different volumes. The 60-mL size is useful for small samples or short exposures, whereas the standard 300-mL size is effective with large plants. Transparent bottles are used and kept in the dark for respiration studies, although opaque ones are available. The bottles should be acid-cleaned, rinsed, and dried before use.

A minimum of five replicate BOD bottles should contain plant material along with two blanks to correct for background variations in oxygen level. Sufficient macroalgae material (e.g., 0.5 to 1 g wet wt) or seagrass blades (ten to twenty 1-mm segments) are placed in each BOD bottle (300 mL). A test run using increasing levels of biomass will detect low or excessive levels of oxygen production and shading. The dissolved oxygen (DO) background level is measured, samples placed in the bottles, and the bottles placed in the dark with the blanks to measure respiration. Duration of the experiment can range from 30 min to 2 h, depending on the rate of respiration. Temperature should be stable (room temp, water bath) throughout the experiment.

At the end of the dark run, the oxygen level is measured in the bottles. Because the Winkler procedure fixes the oxygen, the plant material must be transferred to a new set of bottles before addition of the chemicals. Further, the initial oxygen level in the new bottles must be determined for each run. With the oxygen meter, the same bottle and plant material can be run through a sequential set of dark and light exposures. Use of an oxygen meter requires constant mixing (i.e., magnetic stirrer) of the BOD bottle to ensure even distribution of oxygen. Photosynthesis can be measured at a single irradiance or a series of increasing irradiances series to form a P-I curve (Chap. 4). Either natural or artificial light (e.g., 1000-watt lamp) can be used. Irradiance levels can controlled by changing the number of neutral density screens (e.g., black fiberglass window screening) and monitored with a light meter (e.g., Li-Cor

Quantum Meter). The bottles are placed on their side in a tray with running water in the field or lab and exposed for 30 min to 1 h. An hour or more is usually needed in the dark and under low irradiances (0 to 200 μmol photons $m^{-2} s^{-1}$).

The difference between the initial and final oxygen levels for each run indicates respiration (–) or photosynthesis (+), which are then corrected using the levels of the blanks. Respiration is corrected by adding (– O_2) or subtracting (+ O_2) the changes in the blanks, and the reverse is carried out for photosynthesis. At the end of the experiment, the samples are dried, weighed, and production or consumption is expressed as mg O_2 g dwt h^{-1} by dividing the change by the dry weight. *Net photosynthesis* reflects the use of oxygen in respiration, whereas *gross photosynthesis* is the sum of corrected respiration and net photosynthesis.

PROXIMATE CONSTITUENTS

The determination of proximate constituents (protein, carbohydrate, lipid) and energy levels in plant parts (caloric) has proven very useful, among other things, showing allocation of resources, seasonal shifts in growth patterns, and nutritional value. Detailed procedures are available for seagrasses (Phillips and McRoy, 1990) and seaweeds (Hellebust and Craigie, 1978; Littler and Littler, 1985).

Ash

The inorganic fraction is determined by differences in dry weight before and after combustion in a muffle furnace at 500°C for 4 h. Longer periods or higher temperatures should be avoided to avoid combustion of carbonate. The samples should be dry and ashing crucibles clean and weighed to ±0.01 mg if small plant samples (25 to 100 mg) are used. A minimum of five replicates should be measured.

Protein

Spectrophotometric assays for plant protein include the Bradford or Coomassie stains (Bradford, 1976) and copper reactions with proteins, including the Lowry procedure (Lowry et al., 1951) and Biuret reaction (Layne, 1957). The Lowry analysis is in two steps using small (e.g., 10-mg) samples. There is an initial reaction of the alkaline copper with the protein, which is followed by a reduction of the phosphomolybdic (Folin-Ciocalteau) reagent by the copper-treated protein, which reacts with protein molecules to form a dark-blue color. The test is most sensitive to proteins containing tryptophane and tyrosine.

A related technique uses bicinchoninic acid (Smith et al., 1985). In a comparison of the Lowry, Bradford, and bicinichoninic acid assays, the Lowry procedure was found to agree with the last method, which gave levels 20% higher than the Bradford technique (Berges et al., 1993). The Lowry procedure is sensitive to the presence of tannins (Marks et al., 1985), but it has been the most commonly used technique and so data can be compared with earlier studies. A minimum of two replicates of each sample should be analyzed.

1. About 10 mg (±0.01 mg) of dried plant powder are placed in a 15-mL test tube.

2. Exactly 5 mL of 1 N NaOH is added to the tube, capped with a marble, and allowed to extract the protein at room temperature for 24 h.

3. Exactly 0.5 mL of the extract are pipetted into a clean 10-mL test tube to which 5 mL of alkaline copper tartrate reagent is added. The solution is mixed (vortex mixer) and allowed to stand for 10 min. The reagent is freshly made each day in the sequence that follows or a precipitate will form: 2 mL of 2% $CuSO_4 \cdot 5H_2O$ to 2 mL of 4% sodium tartrate to 96 mL of 3% $NaCO_3$ in 0.1 N NaOH.

4. Exactly 0.5 mL of the 1 N Folin-Ciocalteu phenol reagent is added, mixed, and the solution allowed to stand for 20 min. The Folin reagent is available commercially in a 2-N concentration.

5. The mixture is decanted into a cuvette and absorption at 660 nm is measured in a spectrophotometer.

6. The percent protein is calculated as shown in Eq. A-18. The mg protein for 0.5 mL of solution is determined by comparing the absorption of the sample with a standard curve using 0.00- to 0.25-mg concentrations of bovine serum albumin (reagent grade) in 1 N NaOH.

$$\text{Percent protein} = \frac{\text{mg protein} \times 10}{\text{mg tissue}} \times 100 \qquad \text{(A-18)}$$

where the dilution of the original sample is the mg protein in 0.5 mL × 10.

Soluble Carbohydrate

The phenol-sulfuric acid method of Dubois et al. (1956) is a colorimetric procedure for dissolved carbohydrates having free or potentially free reducing groups. An orange to yellow color develops when the sugars are treated with phenol and concentrated sulfuric acid. Insoluble carbohydrates can be estimated by subtraction after determining protein, lipid, ash, and soluble carbohydrate.

1. About 5 mg (±0.01) of dried powder is placed in a 15-mL centrifuge tube and 10 mL of 5% trichloroacetic acid (TCA) is added. The level is marked on the tube for replenishment with distilled water after heating.

2. After standing overnight, the tubes are heated in a hot-water bath for 3 h at 80 to 90°C and shaken gently at least three times to ensure adequate extraction.

3. After removal, cooling, and addition of distilled water to return levels to the initial mark, the tubes are centrifuged for 5 min at about 45,000 rev/min (tabletop centrifuge).

4. Exactly 0.2 mL of the solution is pipetted into a 10 mL test tube from below the surface (to avoid scum) to which 1 mL of 5% phenol is added and the tubes mixed (vortex mixer).

5. Then, 5 mL of concentrated reagent grade H_2SO_4 is rapidly added and the solution is carefully mixed (e.g., Vortex mixer). The reaction is exothermic, so care must be taken. The tubes are allowed to cool for 30 min before measuring the absorption at 490 nm.

6. The percent carbohydrate is calculated according to Eq. A-19 based on absorptions taken from a standard curve of 0.00 to 0.25 mg glycogen (reagent grade) dissolved in 5% TCA.

$$\text{Percent carbohydrate} = \frac{\text{mg carbohydrate} \times 50}{\text{mg plant tissue}} \times 100 \qquad \text{(A-19)}$$

where the dilution of the original sample is the mg carbohydrate in 0.2 mL × 50.

Lipid

The gravimetric procedure for total lipid in dried plant material is modified from Freeman et al. (1957) and Sperry and Brand (1955). The plant material is homogenized in a 2:1 chloroform-methanol (C:M) mixture, washed with water, and the lower lipid phase is separated, dried, and weighed.

1. About 100 mg (±0.1) of dried powder is placed in a 30-mL screw-cap vial that is marked at the 25-mL level to which 25 mL 2:1 (volume:volume) of C:M mixture is added.

2. The vial is loosely capped and heated at 60°C for 15 min, cooled, and the fluid returned to the 25-mL mark with the C:M mixture.

3. The cap is tightened, the vial is shaken, and the solution is filtered through a Whatman No. 541 filter paper with sufficient C:M to return to the 20-mL mark of a new 30-mL screw-cap vial.

4. Exactly 4 mL of distilled water is added, the vial is capped, shaken for 5 min., and centrifuged to separate the aqueous and solvent phases (ca. 1500 rev/min). The upper aqueous phase is decanted and the solvent evaporated at 60°C under a filtered (cotton-filled vacuum flask) air stream.

5. The lipid extract is transferred with a minimum of C:M rinses of the vial

into a preweighed shell-glass vial, the solvent evaporated at 60°C under a stream of air, and the shell-glass vial reweighed.

6. The initial dry weight of the shell-glass vial is subtracted from the final weight to obtain lipid weight. The percent lipid is calculated using Eq. A-20.

$$\text{Percent lipid} = \frac{\text{lipid weight} \times 3/2}{\text{mg dried plant}} \times 100 \qquad \text{(A-20)}$$

where the numerator is two-thirds of the original sample.

Cellulose and Lignin

A method that uses detergents to separate soluble compounds from cell-wall constituents is also effective for measurement of cellulose and lignin levels in seagrasses and tidal plants (Goering and van Soest, 1970; Dawes and Kenworthy, 1990). By using neutral detergents, plant cell walls can be analyzed for neutral detergent fiber (NDF), which contains hemicelluloses, cellulose, and lignin. With acid detergents, the wall can be separated into cellulose (acid detergent fiber; ADF) and lignin; the latter then can be isolated from the former. The ADF and lignin procedures are given here.

1. About 1 g of powder is placed in a 600-mL tall beaker (Berzelius) and 100 mL acid detergent (AD) is added. The AD is made up with 40.04-g 1-N reagent grade H_2SO_4 to which 20 g of technical grade cetyl trimethyl ammonium bromide is added. The AD mixture is brought to 1 L with distilled water.

2. The mixture is brought to a boil, the heat is reduced to avoid foaming, and then refluxed for 60 min using either a reflux condensor or a 600-mL Kjeldahl flask filled with ice and mounted on top of a beaker. The boiling is adjusted to a slow, even roll.

3. The heated solution is filtered through a preweighed (± 0.01 mg) Gooch crucible that is placed in a filter manifold consisting of a rubber crucible adapter on a glass funnel set in a 500-mL vacuum flask. There is sufficient vacuum from a two-stage mechanical pump to run five of these Gooch setups along with at least two in-line water traps. A high-form, 30-mL Pyrex Gooch crucible with a frittered glass filter having coarse porosity should be used for this step.

4. The filtrate trapped on the glass filter is broken up with a glass rod and rinsed twice with boiling water while under vacuum. The mat is washed with acetone until no color remains, and any fiber lumps are broken with the rod. The crucible is vacuumed dry and then dried at 80°C over night and reweighed.

5. The ADF weight is determined using Eq. A-21, where W_o is the weight of the oven-dried crucible including ADF fiber, W_t is the initial weight of the empty, oven-dry crucible, and S is the weight of the initial sample.

$$\text{ADF} = \frac{(W_o - W_t) \times 100}{S} \qquad \text{(A-21)}$$

Lignin can be determined using the dried ADF sample after weighing.

1. Sufficient 72% H_2SO_4 is added to cover the ADF in the Gooch crucible and is stirred to form a smooth paste. The crucible should be mounted in the filter manifold using the rubber adapter. Sulfuric acid is added as it drains away and a vacuum is applied if draining is slow. The process is continued for 2 h at room temperature.

2. The paste is rinsed with hot water under vacuum until all of the acid is removed (pH = 6), and the crucible is dried at 80°C overnight and reweighed to obtain the level of lignin and inorganic material in the ADF sample.

3. The sample is then ashed by heating the crucible in a furnace at 500°C for 3 h and *slowly* cooled and reweighed. Rapid cooling or overheating will crack the Gooch crucible at the position of the frittered glass filter.

4. The quantity of lignin is determined using Eq. A-22, where L is the loss after ashing and treatment with 72% H_2SO_4, and S is the reweighed oven-dried sample.

$$\text{Lignin} = \frac{L \times 100}{S} \qquad \text{(A-22)}$$

Caloric Levels

Calories can be calculated using the conversion factors of Brody (1964): protein, grams $\times$ 5.65; starch and soluble carbohydrate, grams $\times$ 4.1; and lipid grams $\times$ 9.45. Conversion to joules is done by multiplying calories by 4.184.

GROWTH

The ultimate biological measurement of marine plant responses is growth because it is the sum of all metabolic and physiological processes. Growth is usually calculated using wet or dry weights or, in some cases, length or surface area. Specific growth rates of prokaryotic and eukaryotic microbes are standard approaches in determining the uptake and utilization kinetics of nutrients (Admiraal, 1977) or the effect of abiotic factors.

Growth Kinetics and Rates

The kinetics of specific growth rates should reflect the effect of any limiting factor if all other factors are in excess. Further, growth should be compatible with Michaelis-Menten kinetics (Eq. A-23), where S is the substrate concentration, and K_S is the substrate concentration at which the growth rate (m) is half of the maximum (m_{max}).

$$m = m_{max} \frac{S}{S + K_S} \tag{A-23}$$

The same equation, recalculated and plotted as S/m over S, produces a straight line that allows K_S and m_{max} to be evaluated. Growth has been plotted for many seaweeds, including *Codium fragile* ssp. *tomentosoides* (Hanisak, 1979) and *Gracilaria foliifera*, and *Neoagardhiella baileyi* (DeBoer et al., 1978).

Measurement of Growth

Daily growth rates (DGR) of seaweeds (m = DGR = percent growth d^{-1}) can be calculated in terms of linear (Eq. A-24) or exponential (Eq. A-25) growth depending on the observed increase in biomass. Use of linear growth is more common in laboratory culture, whereas exponential growth is used more frequently in field studies (Dawes et al., 1994). The initial weight (W_i) is subtracted from (Eq. A-24) or divided into (Eq. A-25) the final weight (W_f) and that is divided by the number of days (d) of the growth experiment.

$$DGR = \frac{\frac{W_f - W_i}{W_i} \times 100}{d} \tag{A-24}$$

$$DGR = \frac{\frac{\log_n(W_f)}{(W_i)} \times 100}{d} \tag{A-25}$$

Uses of Algae

Uses of salt marsh, mangrove, and seagrass communities are dealt with in Chaps. 9 through 11; thus, the uses of algae, particularly seaweeds and their maricul-ture, are briefly presented here. Macroalgae and microalgae are used extensively in domestic, industrial, and farming activities. The primary use of *microalgae* is as food for a variety of shell fish and larvae, which, in turn, are used as food in fish farms. However, there have been a number of new products from microalgae. Docosahexaenoic acid (DHA) is obtained from heterotrophically grown micro-algae (Kyle et al., 1992). The compound is present in human milk, but not eas-ily available in infant formulas, and is vital for formation of gray matter of the brain and retina. Another microalga grown in mass culture is the blue-green *Spir-ulina* (Radmer, 1996), which was consumed by the Aztec's since ancient times. It thrives under alkaline conditions and may form natural unialgal cultures in some lakes. The alga is dried and eaten as a high-protein source with a present value of U.S. $80 million. The salt-tolerant green flagellate, *Dunaliella* is now being farmed as a source of beta carotene, which is used as a nutritional supplement. The coccoid green alga *Chlorella* is also cultured under high CO_2 conditions in order to increase biomass production and is used in the Far East as a food with a value of U.S. $100 million. *Chlorella pyrenoidosa* is being considered as a major source of vitamin C or L-ascorbic acid (Radmer, 1996).

In 1987, the world production of *macroalgae* was estimated at 3,243,400 metric tons wet weight of which brown seaweeds accounted for 67%, reds 30%, and others 3% (Food and Agriculture Organization statistics). The level of sea-weed production has increased gradually since 1983, showing a steady increase that is probably due to improved mariculture efforts. Thus, "seaweed ranching" (Ohno and Critchley, 1993) has resulted in a stable source of macroalgae for products versus earlier fluctuations of wild stock harvests. Detailed reviews of seaweed uses and mariculture are given by several authors (Levring et al., 1969; Chapman, 1970; Bird and Benson, 1987; Akatsuka, 1990; Guiry and Blunden, 1991; Ohno and Critchley, 1993).

Uses of Seaweeds

Of the 500 species of seaweeds used as food, fodder, and chemicals, less than 20 species are farmed (Lobban and Harrison, 1994). Five genera dominate mariculture efforts, including three red seaweeds (*Eucheuma, Gracilaria, Porphyra*) and two brown (*Laminaria, Undaria*). The principal uses of seaweeds are as direct food and as a source for the extraction of soluble carbohydrate products (phycocolloids) with a production of 454,730 and 50,000 metric tonnes dry weight (mt dwt), respectively (Ohno and Critchley, 1993). In addition, seaweeds are used as sources of medicine, animal fodder, fertilizer, and fuel.

Sea Vegetables

Consumption of seaweeds is recorded in a 600-B.C. Chinese book of poetry, in which the importance of these marine plants is proclaimed. The preparation of dishes containing sea vegetables and usable species are detailed in a cookbook (Madlener, 1977). Presently, human consumption of green (0.4%), brown (66.5%), and red (33%) seaweeds is high in Asia, particularly Japan, China, and Korea. Almost 94% of edible seaweeds are produced by mariculture rather than harvesting of wild stock, and prices vary from U.S. $7500 to U.S. $10,000 per dry metric tonne (Ohno and Critchley, 1993). In Korea, 75% all seaweed consumption involved the kelp *Undaria pinnatifida* (wakame), whereas in Japan the red alga *Porphyra* (nori) constitutes 45% of all seaweeds eaten.

The nutritional value of sea vegetables is similar to that of terrestrial crops (Madlener, 1977). The fresh weight of seaweeds consists of 75% to 90% water; the dry weight is about 75% organic matter and 25% ash, consisting mostly of potassium, magnesium, and calcium ions (Lüning, 1990; Indergaard and Minsaas, 1991). The dry organic portion consists of 5 to 15% protein (except *Porphyra yezoensis*: 33 to 47% protein), 1 to 2% lipid, 32 to 60% soluble carbohydrates, and 4% to 12% fiber (Chapman, 1970). Iodine content is high (18 to 1600 ppm), whereas vitamins are similar to or less than those of terrestrial plants (Indergaard and Minsaas, 1991). The vitamins in *Ascophyllum nodosum* include 35 to 80 ppm carotene (vitamin A), 550 to 1650 ppm ascorbic acid (C), 1 to 30 ppm niacin (B), 5 to 10 ppm riboflavin (B_2), and 1 to 5 ppm thiamine (B_1).

Green algal sea vegetables account for the smallest portion of seaweeds grown as food. *Monostroma latissimum* (aonoriko) is widely farmed within estuarine habitats of south central Japan, whereas species of *Enteromorpha* (tai tyau) are harvested from natural populations and cultured on nets at mouths of rivers and bays (Ohno, 1993b). The heteromorphic life history of *M. latissimum* is exploited in mariculture. Its haploid blades are collected in April, desiccated overnight to induce gamete discharge, and floated over boards in tanks to allow settlement of the new zygotes after gametic fusion. The unicellular zygotes (*Codiolum* phase) attach and grow on the boards during the summer under conditions of low light and enriched seawater. In September, a dark treatment triggers zygotic meiosis, resulting in haploid zoospores that settle on suspended nets in tanks. After settlement, the nets are transferred to the field and suspended on bamboo poles or from rafts where the gametophytic blades of *M. latissimum* develop.

Figure B-1. *Caulerpa lentillifera* as sold in a Philippine fresh market.

The isomorphic life history of *Enteromorpha* spp. allows the use of either haploid or diploid plants grown from zooids that had settled on lines that are suspended in mid-September over "seed beds." The beds are regions of sandy bottom and broken rock that are rich in natural populations of the species and a source of zoospores. By early November, the plants are 2 to 3 cm in length and are moved to the cultivation areas where they reach maximum length (82 to 136 cm). The green alga *Caulerpa lentillifera* is farmed in Philippine coastal ponds (Fig. B-1) and in cages suspended in bays on Okinawa. The plant is sold fresh in the markets and used in salads; it is prized for its "peppery" flavor.

Dried aonoriko is boiled with sugar in soy sauce and bottled as "nori-jam" or used fresh in "miso-soup." Dried tai tyau feen are toasted, crushed, or powdered and used as a topping for foods, soups, and as a green coloring. Annual yield of these two genera is about 2,700 mt dwt, with a 1992 value of about U.S. $3,600,000 (Patwary and van der Meer, 1992; Ohno and Critchley, 1993).

A variety of *brown* seaweeds are eaten in Japan and Asia, including the kelps *Eisenia bicyclis* (Arame), *Kjellmaniella gyrata* (Tororo kombu), and species of kombu or *Laminaria* (Madlener, 1977). However, two genera of brown seaweeds are the most common sea vegetables, namely, ma-kombu (true *Laminaria*) in China and Japan (*Laminaria japonica*; 294,600 mt dwt, US$3.0 $\times$ 10^8) and wakame in South Korea (*Undaria pinnatifida*, *U. undarioides*, and

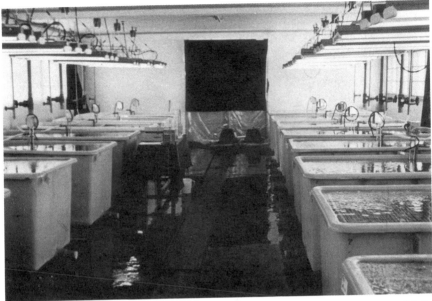

Figure B-2.   A culture facility for kombu (*Laminaria*) sporelings (courtesy of John West, University of Melbourne).

U. peteroseniana; 81,400 mt dwt, US$2.0 × 10^8). Two other species of brown algae, mozuku (*Cladosiphon okamuranus*) and hiziki (*Hizikia fusiforme*; 9,600 mt dwt), are farmed and used as sea vegetables in Okinawa and South Korea, respectively, but do not approach the level of use of kombu and wakame (Patwary and van der Meer, 1992; Ohno and Critchley, 1993). Two brown seaweeds are known to cause poisonings, *Cladosiphon okamurauus* and *Nemacystus decipens* (Yasumoto, 1993). The life history of kombu and wakame is typical of kelps (Chap. 7), with haploid zoospores germinating into microscopic gametophyte filaments that produce eggs and sperm. After sexual fusion, the zygote develops into a large sporophyte. Kombu is a biennial; its intercalary meristem produces fronds two years in a row. Reproductive sporophytic blades are suspended over ropes in the laboratory to allow release and attachment of haploid zoospores and subsequent development of the gametophytes. After sexual reproduction, the diploid sporelings on the ropes are grown in the laboratory (Fig. B-2) in the winter and the ropes transferred to farms in the early spring. Dried kombu can be made into cakes, resoaked in weak vinegar or soy sauce, boiled with meat, fish, soups, used as a vegetable, or with rice.

Wakame is an annual. The control of zoospore release, development of gametophytes, and production of sporophytes after attachment to "seeded ropes" in the fall and winter are similar to kombu culture. The sporophytes are then transferred to field farms in early spring after they reach 2 to 3 cm in length. Wakame is considered to be a luxury food among Japanese and Koreans; it is sold either boiled or dried and used in soups (vegetable, bean paste) and salads (cucumber, shrimp).

Of the *red* sea vegetables, nori (*Porphyra tenera, P. yezoensis*) is the most important, with 69,130 mt dwt (U.S.$2.5 × 10^9) being farmed in Japan, Korea, and China. In 1993, nori of high quality sold wholesale for more than U.S.$24 kg^{-1}. Other red macroalgae that are eaten directly include Irish moss (*Chondrus crispus*) in Ireland and British Isles, agar-agar (*Eucheuma* spp.) in the Philippines, and dulse (*Rhodymenia palmata*) in the eastern maritime provinces of Canada. Two species of *Gracilaria* and one of *Chondria* have caused human poisonings (Yasumoto, 1993). The farming of nori, as kombu and wakame, must take into account its heteromorphic life history (Fig. B-3), water temperature, and photoperiod in order to induce reproduction (Oohusa, 1993). Diploid carpospores are induced to discharge from mature nori blades using long days and higher temperature; they then germinate on oyster shells (in tanks) during the summer. After two weeks, carpospores grow into *Conchocelis* filaments that penetrate the shells; these are then suspended from horizontal poles in indoor tanks (Fig. B-4). In October, when water temperature is cooler and

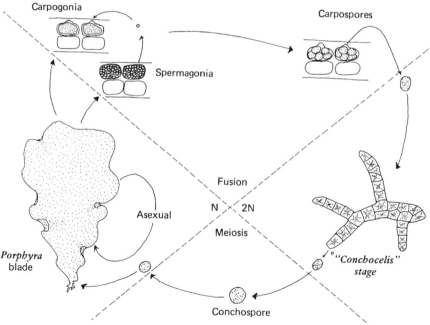

Figure B-3. The heteromorphic life history of *Porphyra*. The diploid conchocelis filaments grow on shells and produce haploid conchospores that attach to ropes in tanks (Fig. B-4). The ropes are placed in the field, where blades develop, and are harvested as nori (Fig. B-5). Blades undergo sexual reproduction in tanks, producing carpogonia, which are fertilized by spermagonia. The zygote produces diploid carpospores that produce conchocelis filaments.

Figure B-4. Shells covered by pink conchocelis filaments produced in tank culture.

there are short photoperiods, the *Conchocelis* filaments produce haploid con-chocelis spores that attach to nets suspended below the oyster shells. Subsequently, the nets can be stored under refrigeration and moved to field (Fig. B-5) next spring when day length and water temperature support the growth of nori blades.

Phycocolloids

The cell-wall polysaccharides of seaweeds represent about 30 to 40% of the plant's dry matter (Craigie, 1990). As noted in Chap. 6, the cell walls of red and brown algae consist of a matrix of cellulose microfibrils, which contains amorphous polysaccharides that include the phycocolloids. The latter substances, which are also called hydrocolloids, form colloidal suspensions and gels at low concentrations (0.1 to 5%); further, some form gels that are stable at room temperature. About 900,000 mt wet weight (50,000 mt dwt) of seaweed are harvested for the extraction of agar, alginate, and carrageenan (Lüning, 1990). Detailed discussions of the phycocolloids and their uses are available (Gacesa, 1988; Lewis et al., 1988; Cosson et al., 1995) and Table B-1 lists some of the more common uses.

Alginic acid is a polyuronic (polysugar) polymer consisting of varying proportions of manuronic and glucuronic acid residues with a carboxyl group on C-6 (Fig. B-6). The former is a beta 1,4 linked D-manuronic acid, and the latter is alpha 1,4 linked L-glucuronic acid. Alginic acid polymers consists of blocks

TABLE B-1. Examples of Uses of Seaweed Phycocolloids

	Agars	Aginates	Carrageenans
Food Applications in Water Processes			
Frozen foods	×	×	×
Pastry fillings		×	×
Icings	×	×	×
Candies	×	×	
Sauces and gravies		×	×
Salad dressings		×	×
Food Applications in Milk Processes			
Whipped toppings		×	×
Chocolate milk			×
Cheeses	×	×	×
Flans and custards		×	×
Yogurt	×		×
Ice cream		×	×
Industrial Uses			
Paper sizings and coatings	×	×	
Adhesives	×	×	
Textile printing and dyeing	×	×	
Air-freshener gels		×	×
Polishes		×	
Castings and impressions	×	×	
Welding rods		×	
Medical and Pharmaceutical Uses			
Laxatives	×		×
Capsules and tablets	×	×	×
Anticoagulants	×		
Lotions and creams		×	×
Shampoos			×
Toothpastes			×

Sources: After Dawes (1981); Lewis et al. (1988).

of each acid. The viscosity and gel strength of the alginate salts depend on the level of Ca^{+2} binding (Gacesa, 1988; Craigie, 1990).

Alginates have the largest use of all phycocolloids (22,000 to 25,000 dwt mt per year), with demands continuing to rise at a rate of 10 to 30% per year (Ohno and Critchely, 1993). The brown algae most commonly harvested for alginate are species of *Ecklonia, Eisenia, Laminaria, Macrocystis, Ascophyllum,* and *Sargassum.* The harvesting of different seaweeds varies, with *Macrocystis pyrifera* being cut by harvesting ships (Fig. B-7) and other species are hand har-

Figure B-5. A nori farm in Japan. The hibi sticks allow the plants to be exposed at low tide.

vested. The annual value of alginate production is over U.S. $100 million. The freshly cut brown seaweeds are milled and washed at the factory. The insoluble alginate salts are then converted to soluble sodium alginate via heating in an alkali bath. Sedimentation, screening, filtration, and centrifugation separate the solid from the liquid phase, which is then clarified (acidified or electrolysis) and dried to a white powder. Alginates are primarily used in the paper and textile industries (50%), with the phycocolloid improving ink holdout and sheet smoothness in paper. The textile industry uses the compound with fiber-reactive

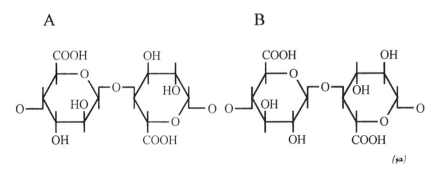

Figure B-6. The structure of alginic acid. The compound consists of (A) 1,4 beta-linked D-manuronic acid and (B) 1,4 alpha-linked L-glucuronic acid.

Figure B-7. A kelp harvester cutting the upper 2 to 3 meters of *Macrocystis pyrifera* off the coast of California (courtesy of K. Bird, University of North Carolina, Wilmington).

dye pastes to print sharp lines and conserve dyes. The food industry (30%) uses alginates in production of frozen desserts, salad dressings, and dairy and bakery products. Medicinal uses for alginates include protective coatings, suspension agents, and dental impression materials.

Agar is a mixture of polymers (Craigie, 1990), which includes (1) agarose (Fig. B-8) with a repeating beta D-galactopyranose and 3,6 linked anhydro (alpha-linked) L-galactopyranose (3,6 AG); (2) ionized agarose that is a highly sulfated beta D-galactopyranose; (3) and galactans with most of the 3-6 AG being replaced by 6-sulfated, L-galactose (Cosson et al., 1995). The first type of agarose lacks sulfation, it has a low level of ionization, and produces strong gels. By contrast, the second type of agarose is highly ionized, sulfated, and charged with its gel strength being inversely correlated with the level of OSO$_3$. The galactans are nongelling.

Species of *Gelidium* and *Gracilaria* are the primary sources of agar in a world market that is valued at over U.S.$200 million. Asian and South American species of *Gracilaria* now account for more than 70% of all agar produced because of the expansion of mariculture techniques during the past 20 years (Dawes, 1987; Critchley, 1993). Agar is extracted by dissolving dried plant powder in hot water; the water-soluble agar is separated by centrifugation, filtration, and chemically treated to remove color. The extract is then gelled by cooling and then squeezed to exude the water, impurities, and salts. The solution

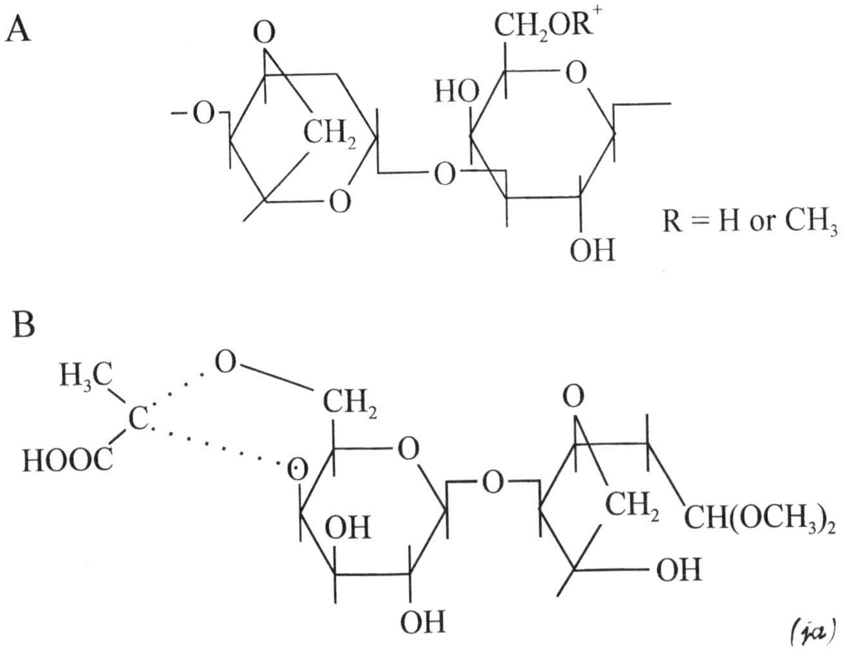

A

R = H or CH₃

B

(ja)

Figure B-8. The structure of agarose (A) and ionized agarose of agar (B).

is dried as flakes or sheets and ground into a white powder. Aqueous concentrations of 1 to 2% agar produce strong to brittle gels at room temperature. Agar is used in food processing (baked goods, candies, juices, wines, vinegar), industrial applications (adhesives, cosmetics, toolmaking, electrophoretic gels), diverse medicinal uses (dental casts, bulk laxative, formative structure in pills and capsules), and as a substratum for microbiological research.

Of the 14 types of *carrageenan*, three types are most important commercially, *lambda*, *kappa*, and *iota* (Fig. B-9). Carrageenans are linear polysaccharides with alternating 1,3 beta-linked D-galactopyranosyl and 1,4 alpha-linked D-galactophranosyl units, having varying levels of sulfation (Craigie, 1990; Cosson et al., 1995). Kappa-type carrageenans (kappa and iota) are identified by having the glactopyranosyl bonded to C-3 and they are always sulfated at C-4. Lambda-type carrageenans are always sulfated on C-2 of the glactopyranose unit.

Carrageenan has annual sales of over $200 million and accounts for about 15% of the world use of food hydrocolloids (Bixler, 1996). The primary sources of carrageenan are from *Chondrus crispus* and *Mastocarpus stellatus* in the North Atlantic, plus tropical species of *Eucheuma* and *Kappaphycus* in the Philippines and *Hypnea* in North Africa. Originally, each of these plants was used as a sea vegetable, making jellies, puddings (blancmange), and thickened boiled dishes. Although still true, the species are primarily harvested for carrageenan. Because of successful mariculture, Philippine species

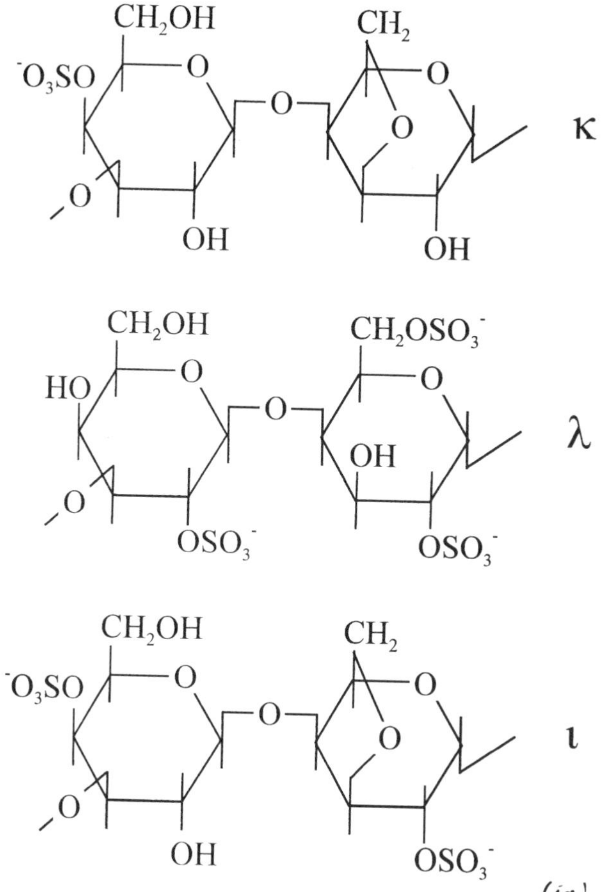

Figure B-9. The basic units of kappa, lambda, and iota carrageenan. Residues of 3,6 anhydrogalactose occur in kappa and iota (kappa type) but not lambda carrageenan.

are presently the source of about 50% of carrageenan used worldwide (40,000 tons). *Eucheuma* and *Kappaphycus* farming in the Philippines are simple and highly productive (Bixler, 1996). As described by Trono (1993), they are farmed using monolines tied between stakes or suspended on floating logs. Dried plants are washed to remove salts and debris and digested with hot water with alkali (KOH, CaOH). Alkaline extraction promotes swelling and removes sulfate groups, increasing 3,6-anhydro-D-galactose content. The latter substance increases gel strength and protein reactivity, an important use of carrageenan. Refined carrageenan is usually recovered by precipitation, washed with alcohol, dried, and powdered. Semirefined carrageenan is obtained by cooking the plants in aqueous KOH and then soaking in freshwater to extract most of the residual alkali, drying, and grinding to produce a flour. Because of their different chemical and physical features, carrageenans can be specifically tailored for a wide variety of uses, especially for dairy products, as well as deserts and medical and industrial applications (Table B-1).

Fodder, Fertilize, and Fuel

The term "kelp" in Europe was originally used to describe the burnt ash of seaweeds and has become the common name for members of the order Laminariales. Seaweed ash was used to obtain *soda* (sodium compounds: $NaCO_3$, NaOH), *potash* (potassium compounds: KCO_3, KOH), and *iodine* (Chapman, 1970). Levels of soda in brown seaweeds (kelps and rockweeds) range from 15 to 22%, and iodine content will range from 0.6 to 0.7%. The production of ash began around 1720 in Great Britain, and by the end of the eighteenth century, Scotland alone was producing 20,000 mt annually from about 400,000 mt of wet seaweeds. It is hard to imagine what ecological modifications might have occurred in the intertidal and shallow subtidal regions due to this massive removal of the large seaweeds. The soda and potash obtained from seaweed ash was initially used in the production of glass and glazes in Europe until the importation of Barilla soda in 1810. By the 1850s, the primary product from seaweed ash was iodine; by the 1940s, this was replaced by imported mineral deposits. Between 1910 and 1930, kelps were intensively harvested and processed in California to produce gunpowder for World War I (Neushul, 1987). Potash and acetone were also produced; the former as a basis for black gunpowder and the latter as a source of acetone used to make cordite (Chapman, 1970). Potash was also used in fertilizers because of the difficulty of obtaining overseas materials during World War I.

Seaweeds have been used as additives and direct foods in *animal feeds* well back into medieval times (Round, 1981). Improvement in animal growth was noted by adding up to 10% seaweed meal to cattle, horse, and poultry feeds, which was apparently due to the mineral content, especially trace elements (Chapman, 1970). Commercial mixes or seaweed additives ("Algavit L," Philippines) are presently available for animal feeds in some countries. In general, the level of protein is lower, and vitamin content is similar in dried seaweeds when compared to hay (Chapman, 1970). There is some question how digestible seaweed carbohydrates are for animals although use as fodder supplements continues today in Ireland, Iceland, and coastal areas of Norway, France, and Scotland.

Seaweed *manure* has also been used since early (fourth century) times in coastal areas of Europe. Much of the seaweed is raked after storms and then spread over the fields or plowed under during planting. The advantage of using seaweed manure is that it is cheap (if next to a source), contains trace mineral elements, and acts as a soil modifier by improving texture. Commercial seaweed fertilizers are mostly liquid supplements including the British "Maxicrop" and the Philippine "Algafer." In France, loose coralline algae called *maerl* (marl), which contain up to 80% lime ($CaCO_3$), are collected and used to "sweeten" or neutralize the acidic humic and peaty soils. The maerl, composed of rhodoliths and dead corallines of the species *Lithothamnion coralloides* and *Phymatolithon calcareum*, is raked from depths of 1 to 8 m as a gravelly mud and spread on the soil once every two to three years.

The use of seaweeds as a source of methane for *fuel* has been proposed and cost analyses have been carried out (Bird, 1987). Biomass-to-energy systems using nearshore kelp or tidal flat farms could produce methane or alcohol through anaerobic fermentation. Biogas production is well understood and has been tested with farm waste and excess crop plant biomass. Thus, there are possibilities for small coastal communities to obtain their fuel for electricity production. The problem lies in keeping the cost of gas production within a useful range (Bird, 1987).

References

Abel, K. M., and E. A. Drew. 1989. Carbon metabolism. In *Biology of Seagrasses*, A. W. D. Larkum, A. J. McComb, and S. A. Shepherd (eds.), 760–796. Elsevier, New York.

Aberg, P. 1990. Population ecology of *Ascophyllum nodosum:* Demography and reproductive effort in stochastic environments. Ph.D. dissertation, University of Götborg, Götborg, Sweden.

Adam, P. 1990. *Saltmarsh Ecology*. Cambridge University Press, Cambridge.

Adams, D. A. 1963. Factors influencing vascular plant zonation in North Carolina salt marshes. *Ecology* 44: 445–456.

Adams, J. B., and G. C. Bate. 1994. The ecological implications of tolerance to salinity by *Ruppia cirrhosa* (Petagna) Grande and *Zostera capensis* Setchell. *Bot. Mar.* 37: 449–456.

Adey, W. H., and T. Goertemiller. 1987. Coral reef algal turfs: Master producers in nutrient poor seas. *Phycologia* 26: 374–386.

Admiraal, W. 1977. Influence of various concentrations of orthophosphate on the division rate of an estuarine benthic diatom, *Navicula arenaria* in culture. *Mar. Biol.* 42: 1–8.

Aikanathan, S., and A. Sasekumar. 1994. The community structure of macroalgae in a low shore mangrove forest in Selangor, Malaysia. *Hydrobiologia* 285: 131–137.

Akatsuka, I. (ed.). 1990. *Introduction to Applied Phycology*. SPB Academic, The Hague.

Allen, J. R. L., and K. Pye (eds.). 1992. *Saltmarshes: Morphodynamics, Conservation, and Engineering Significance*. Cambridge University Press, Cambridge.

Alongi, D. M. 1994. Zonation and seasonality of benthic primary production and community respiration in tropical mangrove forests. *Oecologia* 98: 320–327.

Alongi, D. M., K. G. Boto, and A. I. Robertson. 1992. Nitrogen and phosphorus cycles. In *Coastal and Estuarine Studies 41. Tropical Mangrove Ecosystems*, A. I. Robertson and D. M. Alongi (eds.), 251–292. American Geophysical Union, Washington, D.C.

Alongi, D. M., and A. Sasekumar. 1992. Benthic communities. In *Coastal and Estuarine Studies 41. Tropical Mangrove Ecosystems*, A. I. Robertson and D. M. Alongi (eds.), 137–171. American Geophysical Union, Washington, D.C.

Almodovar, L. R., and F. A. Pagan. 1971. Notes on a mangrove lagoon and mangrove channels at La Parguera, Puerto Rico. *Nova Hedwigia* 21: 2451–2459.

Al-Thukair, A. A., and K. Al-Hinai. 1993. Preliminary damage assessment of algal mats, sites located in the Western Fulg following the 1991 oil spill. *Mar. Pollut. Bull.* 27: 229–238.

Ancibar, E. 1979. Systematic anatomy of vegetative organs of the Hydrocharitaceae. *Bot. J. Linn. Soc.* 78: 237–266.

Anderson, C. E. 1974. A review of structure in several North Carolina salt marsh plants. In *Ecology of Halophytes*, R. J. Reimold and W. H. Queen (eds.), 307–344. Academic Press, New York.

Andrews, T. J., and G. J. Muller. 1985. Photosynthetic gas exchange of the mangrove, *Rhizophora stylosa* Griff., in its natural environment. *Oecologia* 65: 449–455.

A.P.H.A. 1989. *Standard Methods for the Examination of Water and Wastewater*, 17th ed. American Public Health Association, New York.

Arbor, A. 1920. *Water Plants, a Study of Aquatic Angiosperms.* Cambridge University Press, London.

Arrontes, J. 1990. Composition, distribution on host and seasonality of epiphytes on three intertidal algae. *Bot. Mar.* 33: 205–211.

Asia-Pacific Symposium on Mangrove Ecosystems. 1995. *Hydrobiologia* 295: 1–355.

Bach, S. D., and M. N. Josselyn. 1978. Mass blooms of the alga *Cladophora* in Bermuda. *Mar. Pollut. Bull.* 9(3): 34–37.

Bachmann, R. W., and E. P. Odum. 1960. Uptake of Zn^{65} and primary productivity of marine benthic algae. *Limnol. Oceanogr.* 5: 349–355.

Bagnis, R., J.-M. Hurtel, S. Chanteau, E. Chungue, A. Inoue, and Y. Yasumoto. 1979. Le dinoflagelle *Gambierdiscus toxicus* Adachi *et* Fukuyyo; agent causal probable de la ciguatera. *C. R. Acad. Sci. Paris* 289: 671–674.

Bakker, J. P. 1985. The impact of grazing on plant communities, plant populations and soil conditions in salt marshes. *Vegetatio* 38: 77–87.

Baldwin, A. H., K. L. McKee, and I. A. Mendelssohn. 1996. The influence of vegetation, salinity, and inundation on seed banks of oligohaline coastal marshes. *Amer. J. Bot.* 83: 470–479.

Ball, M. C. 1980. Patterns of secondary succession in a mangrove forest of southern Florida. *Oecologia* 11: 226–235.

———. 1988. Salinity tolerance in the mangroves *Aegiceras corniculatum* and *Avicennia marina*. I. Water use in relation to growth, carbon partitioning and salt balance. *Austral. J. Plant Physiol.* 15: 263–276.

———. 1996. Comparative ecophysiology of mangrove forest and tropical lowland moist rainforest. In *Tropical Forest Plant Ecophysiology*, S. S. Mulkey, R. L. Chazdon, and A. P. Smith (eds.), 461–496. Chapman and Hall, New York.

Ballantine, D. L. 1979. The distribution of algal epiphytes on macrophyte hosts offshore from La Parguera, Puerto Rico. *Bot. Mar.* 22: 107–111.

Balsmo, R. A., M. E. Adams, and W. W. Thomson. 1995. Electrophysiology of the salt glands of *Avicennia germinans*. *Intern. J. Plant Sci.* 156: 658–667.

Barber, H. G., and E. Y. Hayworth. 1981. *A Guide to the Morphology of the Diatom Frustule*, Science Publication 44. Freshwater Biological Association, Cumbria, England.

Barnes, D. J., and B. E. Chalker. 1990. Calcification and photosynthesis in reef-building corals and algae. In *Ecosystems of the World 25. Coral Reefs*, Z. Dubinsky (ed.), 109–131. Elsevier, Amsterdam.

Barnett, M. R., and D. W. Crewz (eds.). 1990. *An Introduction to Planting and Maintaining Selected Common Coastal Plants in Florida*, Report 97. Florida Sea Grant, Gainesville, Florida.

Baron, J., J. Clavier, and B. A. Thomassin. 1993. Structure and temporal fluctuations of two intertidal seagrass-bed communities in New Caledonia (SW Pacific Ocean). *Mar. Biol.* 117: 139–144.

Barry, J. P., C. H. Baxter, R. D. Sagarin, and S. E. Gilman. 1995. Climate-related, long-term faunal changes in a California rocky intertidal community. *Science* 267: 672–675.

Baydoun, E. A.-H., and C. T. Brett. 1985. Comparison of cell wall compositions of the rhizomes of three seagrasses. *Aquat. Bot.* 23: 191–196.

Beanland, W. R., and W. J. Woelkerling. 1982. *Avicennia* canopy effects on mangrove algal communities in Spencer Gulf, South Australia. *Aquat. Bot.* 17: 309–313.

Beer, S. 1996. Inorganic carbon transport in seagrasses. In *Seagrass Biology: Scientific Discussion from an International Workshop*, J. Kuo, D. I. Walker, and H. Kirkman (eds.), 43–47. University of Western Australia, Nedlands.

Beer, S., and A. Eshel. 1985. Determining phycoerythrin and phycocyanin concentrations in aqueous crude extracts of red algae. *Austral. J. Mar. Freshwat. Res.* 36: 785–792.

Beer, S., and E. Koch. 1996. Photosynthesis of marine macroalgae and seagrasses in globally changing CO_2 environments. *Mar. Ecol. Prog.*, Ser. 141: 199–204.

Beer, S., A. Shomer-Ilan, and Y. Waisel. 1980. Carbon metabolism in seagrasses. 2. Patterns of photosynthetic CO_2 incorporation. *J. Exp. Bot.* 31: 1019–1026.

Beer, S., and Y. L. Waisel, 1979. Some photosynthetic carbon fixation properties of seagrasses. *Aquat. Bot.* 7129–7138.

Belanger, T. V., D. J. Scheidt, and J. R. Platko, II. 1989. Effects of nutrient enrichment on the Florida Everglades. *Lake Reserv. Mgt.* 5: 101–111.

Belk, M. S., and D. Belk. 1975. An observation of algal colonization on *Acropora aspera* killed by *Acanthaster planci*. *Hydrobiologia* 46: 29–32.

Bell, E. C. 1993. Photosynthetic response to temperature and desiccation of the intertidal alga *Mastocarpus papillatus*. *Mar. Biol.* 117: 337–346.

———. 1995. Environmental and morphological influences on thallus temperature and desiccation of the intertidal alga *Mastocarpus papillatus* Kützing. *J. Exp. Mar. Biol. Ecol.* 191: 29–55.

Bellamy, D. J., M. D. John, and A. Whittick. 1968. The "kelp forest ecosystem" as a "phytometer" in the study of pollution of the inshore environment. In *Underwater Association Report. 1968*, 79–82. Underwater Association, London.

Belsher, T., L. Loubersac, and G. Belbeoch. 1985. Remote sensing and mapping. In *Handbook of Phycological Methods. Ecological Field Methods: Macroalgae*, M. M. Littler and D. S. Littler (eds.), 177–198. Cambridge University Press, Cambridge.

Benayahu, Y., and Y. Loya. 1977. Seasonal occurrence of benthic-algae communities and grazing regulation by sea urchins at the coral reefs of Eliat, Red Sea. *Proceedings of the Third International Coral Reef Symposium, Miami*: 383–389.

Benz, M. C., N. J. Eiseman, and E. E. Gallaher. 1979. Seasonal occurrence and variation in standing crop of a drift algal community in the Indian River, Florida. *Bot. Mar.* 22: 413–420.

Berges, J. A., A. E. Fisher, and P. J. Harrison. 1993. A comparison of Lowry, Bradford and Smith protein assays using different protein standards and protein isolated from the marine diatom *Thalassiosira pseudonoana*. *Mar. Biol.* 115: 187–193.

Berner, T. 1990. Coral-reef algae. In *Ecosystems of the World 25. Coral Reefs*, Z. Dubinsky (ed.), 253–264. Elsevier, Amsterdam.

Bertness, M. D. 1991. Interspecific interactions among high marsh perennials in a New England salt marsh. *Ecology* 72: 125–137.

Bertness, M. D., and A. M. Ellison. 1987. Determinants of pattern in a New England salt marsh plant community. *Ecol. Monograf.* 57: 129–147.

Bertness, M. D., and S. W. Shumway. 1993. Competition and facilitation in marsh plants. *Amer. Nat.* 142: 718–724.

Bertness, M. D. K. Wikler, and T. Chatkupt. 1992. Flood tolerance and the distribution of *Iva frutescens* across New England salt marshes. *Oecologia* 91: 171–178.

Besada, E. G., L. A. Loeblich, and A. R. Loeblich III. 1982. Observations on tropical, benthic dinoflagellates from ciguatera-endemic areas: *Coolia, Gambierdiscus*, and *Ostreopsis*. *Bull. Mar. Sci.* 32: 725–735.

Biebl, R. 1970. Vergleichende Untersuchungen zur Temperaturresistenz von Meeresalgen entlang der pazifischen Kuste Nordamericas. *Protoplasma* 69: 61–83.

Birch, W. R., and M. Birch. 1984. Succession and pattern of tropical intertidal seagrasses in Cockle Bay, Queensland, Australia: A decade of observations. *Aquat. Bot.* 19: 343–367.

Bird, K. T. 1987. Cost analysis of energy from marine biomass. In *Seaweed Cultivation for Renewable Resources*, K. T. Bird and P. H. Benson (eds.), 327–350. Elsevier, Amsterdam.

Bird, K. T., and P. H. Benson (eds.). 1987. *Seaweed Cultivation for Renewable Resources*. Elsevier, Amsterdam.

Bird, K. T., M. S. Brown, T. T. Henderson, C. E. O'Hara, and J. M. Robbie. 1996. Culture studies of *Ruppia maritima* L. in bicarbonate- and sucrose-based media. *J. Exp. Mar. Biol. Ecol.* 199: 153–164.

Bischoff, B., and C. Wiencke. 1995. Temperature adaptation in strains of the amphi-equatorial green alga *Urospora penicilliformis* (Acrosiphonales): Biogeographical implications. *Mar. Biol.* 122: 681–688.

Bixler, H. J. 1996. Recent developments in manufacturing and marketing carrageenan. *Hydrobiologia* 326/327: 35–57.

Blackman, F. F., and A. G. Tansley. 1903. A revision of the classification of the green algae. *New Phytol.* 1: 17–224.

Blanchard, J., and G. Prado. 1995. Natural regeneration of *Rhizophora mangle* in strip clearcuts in northwest Ecuador. *Biotropica* 27: 160–167.

Blanchette, C. A. 1996. Seasonal patterns of disturbance influence recruitment of the sea palm, *Postelsia palmaeformis*. *J. Exp. Mar. Biol. Ecol.* 197: 1–14.

Blum, J. 1968. Salt marsh Spartinas and associated algae. *Ecol. Monograf.* 38: 213–221.

Boer, B. 1993. Anomalous pneumatophores and adventitous roots of *Avicennia marina*

(Forssk.) Vierh. mangroves two years after the 1991 Gulf War oil spill in Saudi Arabia. *Mar. Pollut. Bull.* 27: 207–211.

Boesch, D. R., and R. E. Turner. 1984. Dependence of fishery species on salt marshes: The role of food and refuge. *Estuaries* 7(4A): 460–468.

Bogorov, B. G. 1958. Perspectives in the study of seasonal changes of phytoplankton and the number of generations at different latitudes. In *Perspectives in Marine Biology*, A. A. Buzzati-Traverso (ed.), 145–158. University of California Press, Berkeley.

Bolach, G. T. 1993. The introduction of non-indigenous marine species in Europe: Planktonic species. In *Introduced Species in European Coastal Waters*, C. F. Boudouresque, F. Briand, and C. Nolan (eds.), 28–31. European Commission for Science, Research, and Development Ecosystems Research Report 8, Paris.

Bold, H. C., and M. J. Wynne. 1985. *Introduction to the Algae*, 2nd ed. Prentice-Hall, Englewood Cliffs, N.J.

Bolton, J. J. 1983. Ecoclinal variation in *Ectocarpus siliculosus* (Phaeophyceae) with respect to temperature growth optima and survival limits. *Mar. Biol.* 73: 131–138.

Borowitzka, M. A. 1981. Algae and grazing in coral reef ecosystems. *Endeavour* 5(3): 99–106.

―――. 1982. Mechanisms in algal calcification. *Prog. Phycol. Res.* 1: 137–177.

Borowitzka, M. A., and R. C. Lethbridge. 1989. Seagrass epiphytes. In *Biology of Seagrasses*, A. W. D. Larkum, A. J. McComb, and S. A. Shepherd (eds.), 458–499. Elsevier, Amsterdam.

Boserice, D. W. J. 1991. Coralline algae: Mineralization, taxonomy, and paleoecology. In *Calcareous Algae and Stromatolites*, R. Riding (ed.), 98–113. Springer-Verlag, Berlin.

Boudouresque, C. F. 1971. Contribution a l'etude phytosociologique des peuplements algaux des cotes varoises. *Vegetatio* 22: 83–184.

Boudouresque, C. F., A. Meinesz, and V. Gravez. 1996. *Scientific Documents Dealing with the Alga Caulerpa taxifolia Introduced to the Mediterranean*, 5th ed. GIS Posidone, Marseilles.

Boudouresque, C. F., A. Meinesz, M. A. Ribera, and E. Ballesteros,. 1995. Spread of the green alga *Caulerpa taxifolia* (Caulerpales, Chlorophyta) in the Mediterranean: Possible consequences of a major ecological event. *Sci. Mar.* 59(Suppl. 1): 21–29.

Bougis, P. 1976. *Marine Plankton Ecology*. North-Holland, Amsterdam.

Bourrelly, P. 1966. *Les Algues d'Eau Douce*. Vol. I. *Les Algues Vertes*. Boubee, Paris.

―――. 1968. *Les Algues d'Eau Douce*. Vol. II. *Les Algues Jaunes et Brunes*. Boubee, Paris.

―――. 1970. *Les Algues d'Eau Douce*. Vol. III. *Les Algues Bleues et Rouges*. Boubee, Paris.

Bradford, M. M. 1976. A rapid and sensitive method for the quantitation of microgram quantities of protein using the principle of protein-dye binding. *Anal. Biochem.* 72: 248–254.

Braselton, J. P., and F. T. Short. 1985. Karyotpic analysis of *Plasmodiophora diplantherae*. *Mycologia* 77: 940–945.

Bràten, T. 1975. Observations on mechanisms of attachment in the green alga *Ulva mutabilis* Foyn. *Protoplasma* 84: 161–173.

Brawley, S. H. 1992. Mesoherbivores. In *Plant-Animal Interactions in the Marine Benthos*, D. M. John, S. J. Hawkins, and J. H. Price (eds.), 235–263. Oxford Science, Oxford.

Bremmner, J. W. 1965. Inorganic forms of nitrogen. In *Methods of Soil Analysis 2*. C. A. Black (ed.), 1179–1232. American Society of Agronomy, Madison, Wis.

Briens, M., and F. Larher. 1982. Osmoregulation in halophytic higher plants: A comparative study of soluble carbohydrates, polyols, betaines, and proline. *Plant, Cell Environ.* 5: 287–292.

Briggs, J. C. 1974. *Marine Zoogeography*. McGraw-Hill, New York.

Brinkhuis, B. H. 1976. The ecology of temperate salt-marsh fucoids. I. Occurrence and distribution of *Ascophyllum nodosum* ecads. *Mar. Biol.* 34: 325–338.

Brinkhuis, B. H., N. R. Tempel, and R. F. Jones. 1976. Photosynthesis and respiration of exposed salt-marsh fucoids. *Mar. Biol.* 34: 349–359.

Brock, M. A. 1981. Accumulation of proline in a submerged aquatic halophyte, *Ruppia maritima* L. *Oecologia* 51: 217–219.

Brody, S. 1964. *Bioenergetics and Growth*. Hafner, New York.

Broome, S. W., E. D. Seneca, and W. W. Woodhouse Jr. 1988. Tidal salt marsh restoration. *Aquat. Bot.* 32: 1–22.

Brown, B. E., and L. S. Howard. 1985. Assessing the effects of "stress" on reef corals. *Adv. Mar. Biol.* 22: 1–63.

Buggeln, R. G. 1981. Morphogenesis and growth regulators. In *The Biology of Seaweeds*, C. S. Lobban and M. J. Wynne (eds.), 627–660. Blackwell, Oxford.

Bulthuis, D. A. 1983. Effects of temperature on the photosynthesis irradiance curve of the Australian seagrass, *Heterozostera tasmanica*. *Mar. Biol. Lett.* 4: 47–57.

Bunt, J. S. 1995. Continental scale patterns in mangrove litter fall. *Hydrobiologia* 295: 135–140.

Burdick, D. M. 1989. Root aerenchyma development in *Spartina patens* in response to flooding. *Amer. J. Bot.* 76: 777–780.

Burdick, D. M., and I. A. Mendelssohn. 1990. Relationship between anatomical and metabolic responses to soil waterlogging in the coastal grass *Spartina patens*. *J. Exp. Bot.* 41: 223–228.

Burdick, D. M., I. A. Mendelssohn, and K. L. McKee. 1989. Live standing crop and metabolism of the marsh grass *Spartina patens* as related to edaphic factors in a brackish, mixed marsh community in Louisiana. *Estuaries* 12: 195–205.

Burger-Wiersma, T., M. Veenhuis, H. J. Korthals, C. C. M. van der Weil, and L. R. Mur. 1986. A new prokaryote containing chlorophylls a and b. *Nature* 320: 262–264.

Burkholder, P. R., and L. R. Almodovar. 1973. Studies on mangrove algal communities in Puerto Rico. *Fla. Scient.* 36: 66–74.

Burkholder, P. R., A. Repak, and J. S. Sibert. 1965. Studies on some Long Island Sound littoral communities of micro-organisms and their primary productivity. *Bull. Torrey Bot. Club* 92: 378–402.

Burr, F. A., and J. A. West. 1971. Protein bodies in *Bryopsis hypnoides*, their relationship to the wound healing and branch septum development. *J. Ultrast. Res.* 35: 476–498.

Butcher, R. W. 1959. *An Introductory Account of the Smaller Algae of the British*

Coastal Waters. I. Introduction and Chlorophyceae. Ministry of Agriculture, Fisheries and Food, HMS Stationery Office, London.

————. 1967. *An Introductory Account of the Smaller Algae of British Coastal Waters.* Part VIII. *Cryptophyceae.* Ministry of Agriculture, Fisheries and Food, HMS Stationery Office, London.

Buth, G. J., and C. de Wolf. 1985. Decomposition of *Spartina anglica, Elytrigia pungens* and *Hamlimione portulacoides* in a Dutch salt marsh in association with faunal and habitat influences. *Vegetatio* 62: 337–355.

Cain, M. L. 1994. Consequences of foraging in clonal plant species. *Ecology* 75: 933–944.

Callaway, J. C., and M. N. Josselyn. 1992. The introduction and spread of smooth cordgrass (*Spartina alterniflora*) in south San Francisco Bay. *Estuaries* 15: 218–226.

Cambridge, M. L., A. W. Chiffings, C. Brittan, L. Moore, and A. J. McComb. 1986. The loss of seagrass in Cockburn Sound, Western Australia: II. Possible causes of seagrass decline. *Aquat. Bot.* 24: 269–286.

Camp, D. K., S. P. Cobb, and J. F. van Breedveld. 1973. Overgrazing of seagrasses by a regular urchin *Lytechinus variegatus. Bioscience* 23: 37–38.

Carlson, P. R., L. A. Yarbro, C. F. Zimmermann, and J. R. Montgomery. 1983. Pore water chemistry of an overwash mangrove island. *Fla. Scient.* 46: 239–249.

Carlton, J. M., and M. D. Moffler. 1978. Propagation of mangrove by air-layering. *Envir. Conserv.* 5: 147–150.

Carlton, J. T., and J. A. Scanlon. 1985. Progression and dispersal of an introduced alga: *Codium fragile* ssp. *tomentosoides* (Chlorophyta) on the Atlantic coast of North America. *Bot. Mar.* 28: 155–165.

Carpenter, E. J. 1986. Partitioning herbivory and its effect on coral reef algal communities. *Ecol. Monograf.* 56: 345–363.

Carpenter, R. C., and S. L. Williams. 1996. Effects of oscillatory flow on rates of photosynthesis of coral reef algal turf communities. *Abstract of the 24th Benthic Ecology Meeting* Univ. South Carolina, 25 Columbia.

Carr, N. G., and B. A. Whitton (eds.). 1973. *The Biology of Blue-Green Algae.* Blackwell, Oxford.

Cattolico, R. A., J. C. Botthroyd, and S. P. Gibbs. 1976. Synchronous growth and plastid replication in the naturally wall-less *Olisthodiscus luteus. Plant Physiol.* 57: 497–503.

Chapman, A. R. O. 1974. The ecology of macroscopic marine algae. *Ann. Rev. Ecol. Syst.* 5: 65–80.

————. 1979. *Biology of Seaweeds. Levels of Organization.* University Park Press, London.

Chapman, V. J. 1970. *Seaweeds and Their Uses.* Methuen, London.

————. 1974. *Salt Marshes and Salt Deserts of the World*, 2nd ed. J. Cramer, Berlin.

————. 1976. *Mangrove Vegetation.* J. Cramer, Berlin.

————. (ed.). 1977. *Ecosystems of the World. 1. Wet Coastal Ecosystems.* Elsevier, Amsterdam.

Chisholm, J. R. M., J. M. Jaubert, and G. Giaccone. 1995. *Caulerpa taxifolia* in the northwest Mediterranean: Introduced species or migrant from the Red Sea? *C. R. Acad. Sci. Paris*, Ser. III, 318: 1219–1226.

Chock, J. S., and A. C. Mathieson. 1976. Ecological studies of the salt marsh ecad *scorpiodes* (Hornemann) Hauck of *Ascophyllum nodosum* (L.) Le Jolis. *J. Exp. Mar. Biol. Ecol.* 23: 171–190.

———. 1978. Physiological ecology of *Ascophyllum nodosum* (L.). Le Jolis and its detached ecad *scorpioides* (Hornemann) Hauck (Fucales, Phaeophyta). *Bot. Mar.* 22: 21–26.

Chopin, T., A. Hournam, J.-Y. Floc'hs, and M. Penot. 1990. Seasonal variations in the red alga *Chondrus crispus* on the Atlantic French coast. II. Relations with phosphorus concentration in seawater and intertidal phosphorylated fractions. *Can. J. Bot.* 68: 512–517.

Cintron, G. A., A. E. Lugo, and R. Martinez. 1985. Structural and functional properties of mangrove forests. In *Monographs in Systematic Botany 10. The Botany and Natural History of Panama*, W. G. D'Arcy and M. D. Correa (eds.), 53–66. Missouri Botanical Garden, St. Louis.

Cintron, G. A., and Y. S. Novelli. 1984. Methods for studying mangrove structure. In *The Mangrove Ecosystem: Research Methods*, S. C. Snedaker and J. G. Snedaker (eds.), 91–113. UNESCO, Paris.

Clirsen, J. 1992. Estudio multitemporal de mangalares, camaroneras y areas salinas de la costa ecuatoriana, mediante el empleo de la informatcion de senores remotos actualizado a 1991. Memoria Tecnica, Guayaquil, Ecuador.

Clough, B. F. (ed.). 1982. *Mangrove Systems in Australia, Structure, Function and Management*. Australian National Press, Canberra.

Cole, K. M., and R. G. Sheath (eds.). 1990. *Biology of the Red Algae*. Cambridge University Press, Cambridge.

Coles, S. L. 1984. Colonization of Hawaiian reef corals on new and denuded substrata in the vicinity of a Hawaiian power station. *Coral Reefs* 3: 123–130.

Colinveaux, L. H. 1974. Productivity of the coral reef alga *Halimeda* (Order Siphonales). *Proceedings of the 2nd International Coral Reef Symposium*, Brisbane: 35–42.

Committee on Biological Diversity. 1995. *Understanding Marine Biodiversity*. National Academy Science Press, Washington, D.C.

Connell, J. H., and R. O. Slayter. 1977. Mechanisms of succession in natural communities and their role in community stability and organization. *Amer. Nat.* 111: 1119–1144.

Conover, J. T. 1958. Seasonal growth of benthic marine plants as related to environmental factors in an estuary. *Univ. Texas Inst. Mar. Sci.* 5: 97–147.

Corlett, R. T. 1986. The mangrove understory: Some additional observations. *J. Trop. Ecol.* 2: 93–94.

Cosson, J., E. Deslandes, M. Zinoun, and A. Mouradi-Givernaud. 1995. Carrageenans and agars, red algal polysaccharides. *Prog. Phycol. Res.* 11: 270–324.

Craigie, J. S. 1990. Cell walls. In *Biology of the Red Algae*, K. M. Cole and R. G. Sheath (eds.), 221–257. Cambridge University Press, Cambridge.

Critchley, A. T. 1993. *Gracilaria* (Rhodophyta, Gracilariales): An economically important agarophyte. In *Seaweed Cultivation and Marine Ranching*, M. Ohno and A. T. Critchley (eds.), 89–112. Japan International Cooperative Agency, Nagai.

Critchley, A. T., W. F. Farnham, T. Yoshida, and T. A. Norton. 1990. A bibliography of the invasive alga *Sargassum muticum* (Yendo) Fensholt (Fucales: Sargassaceae). *Bot. Mar.* 33: 551–562.

Croley, F. C., and C. J. Dawes. 1970. Ecology of the algae of a Florida Key. I. A preliminary list of the marine algae including zonation and seasonal data. *Bull. Mar. Sci.* 20: 165–185.

Culkin, F., and J. Smed. 1979. The history of standard seawater. *Oecologica* 2: 355–364.

Curtiss, A. H. 1888. How the mangrove forms islands. *Garden and Forest* 1: 100.

Cushing, D. H. 1959. The seasonal variation in oceanic production as a problem in population dynamics. *J. Cons. Intern. Explor. Mer.* 24: 455–464.

Czerny, A. B., and K. H. Dunton. 1995. The effects of *in situ* light reduction on the growth of two subtropical seagrasses, *Thalassia testudinum* and *Halodule wrightii*. *Estuaries* 18: 418–427.

Dahl, A. L. 1973. Benthic algal ecology in a deep reef and sand habitat off Puerto Rico. *Bot. Mar.* 16: 171–175.

Daiber, F. C. 1977. Salt-marsh animals: Distributions related to tidal flooding, salinity and vegetation. In *Ecosystems of the World. 1. Wet Coastal Ecosystems*, V. J. Chapman (ed.), 79–108. Elsevier, Amsterdam.

———. 1982. *Animals of the Tidal Marsh*. Van Nostrand Reinhold, New York.

———. 1986. *Conservation of Tidal Marshes*. Van Nostrand Reinhold, New York.

Daly, M. A., and A. C. Mathieson. 1977. The effects of sand movement on intertidal seaweeds and selected invertebrates at Bound Rock, New Hampshire, USA. *Mar. Biol.* 43: 45–55.

Darley, W. M. 1982. *Algal Biology: A Physiological Approach*. Blackwell, Oxford.

Darley, W. M., C. L. Montague, F. G. Plumley, W. W. Sage, and A. T. Psalidas. 1981. Factors limiting edaphic algal biomass and productivity in a Georgia salt marsh. *J. Phycol.* 17: 122–128.

Darwin, C. R. 1842. *The Structure and Distribution of Coral Reefs*. Smith Elder, London.

Davey, A., and W. J. Woelkerling. 1985. Studies on Australian mangrove algae. III. Victorian communities: Structure and recolonization in western Port Bay. *J. Exp. Mar. Biol. Ecol.* 85: 177–190.

Davis, J. H. 1940. *The Ecology and Geologic Role of Mangroves in Florida*, Publication No. 517. Carnegie Institute, Washington, D.C.

Davis, M. A., and C. J. Dawes. 1981. Seasonal photosynthetic and respiratory responses of the intertidal red alga, *Bostrychia binderi* Harvey (Rhodophyta, Ceramiales) from a mangrove swamp and a salt marsh. *Phycologia* 20: 165–173.

Davis Jr., R. A. 1986. *Oceanography. An Introduction to the Marine Environment*. W. C. Brown, Dubuque, Iowa.

Davison, I. R., and J. O. Davison. 1987. The effect of growth temperature on enzyme activities in the brown alga *Laminaria saccharina*. *British Phycol. J.* 22: 77–87.

Davison, I. R., and G. A. Pearson. 1996. Stress tolerance in intertidal seaweeds. *J. Phycol.* 32: 197–211.

Dawes, C. J. 1966. A light and electron microscope study of algal cell walls. II. Chlorophyta. *Ohio J. Sci.* 66: 317–326.

———. 1969. A study of the ultrastructure of the green alga *Apjohnia laetevirens* Harvey with emphasis on cell wall structure. *Phycologia* 8: 77–84.

———. 1974. *Marine Algae of the West Coast of Florida.* University of Miami Press, Coral Gables, Fla.

———. 1979. Physiological and biochemical comparisons of *Eucheuma* spp. (Florideophyceae) yielding *iota*-carrgeenan. In *Ninth International Seaweed Symposium,* A. Jensen and J. R. Stein (eds.), 188–207. Science Press, Princeton.

———. 1981. *Marine Botany.* John Wiley, New York.

———. 1985. Respirometry and manometry. In *Handbook of Phycological Methods. Ecological Field Methods: Macroalgae,* M. M. Littler and D. S. Littler (eds.), 329–348. Cambridge University Press, Cambridge.

———. 1986. Seasonal proximate constituents and caloric values in seagrasses and algae on the west coast of Florida. *J. Coast. Res.* 2: 25–32.

———. 1987. The biology of commercially important tropical marine algae. In *Seaweed Cultivation for Renewable Resources,* K. T. Bird and P. H. Benson (eds.), 155–190. Elsevier, Amsterdam.

———. 1988. The Winkler procedure for measurement of dissolved oxygen. In *Experimental Phycology: A Laboratory Manual,* C. S. Lobban, D. J. Chapman, and B. P. Kremer (eds.), 78–82. Cambridge University Press, Cambridge.

———. 1996. Macroalga diversity, standing stock and productivity in a northern mangal on the west coast of Florida. *Nova Hedwigia* 112: 525–535.

Dawes, C. J., J. Andorfer, C. Rose, C. Uranowski, and E. Ehringer. 1997. Regrowth of the seagrass *Thalassia testudinum* into propeller scars. *Aquat. Bot.*

Dawes, C. J., and D. C. Barilotti. 1969. Cytoplasmic organization and rhythmic streaming in growing blades of *Caulerpa prolifera. J. Phycol.* 3: 117–127.

Dawes, C. J., S. S. Bell, R. A. Davis Jr., E. D. McCoy, H. R. Mushinsky, and J. L. Simon. 1995a. Initial effects of Hurricane Andrew on the shoreline habitats of southwestern Florida. *J. Coast. Res.* (Special Issue) 21: 103–110.

Dawes, C. J., K. Bird, M. Durako, R. Goddard, W. Hoffman, and R. McIntosh. 1979. Chemical fluctuations due to seasonal and cropping effects on an algal-seagrass community. *Aquat. Bot.* 6: 79–86.

Dawes, C. J., M. Chan, R. Chinn, E. W. Koch, A. Lazar, and D. Tomasko. 1986. Proximate composition, photosynthetic and respiratory responses of the seagrass *Halophila englemannii* from Florida. *Aquat. Bot.* 27: 195–201.

Dawes, C. J., and R. H. Goddard. 1978. Chemical composition of the wound plug and entire plants for species of the coenocytic green alga, *Caulerpa. J. Exp. Mar. Biol. Ecol.* 35: 259–263.

Dawes, C. J. and M. D. Guiry. 1992. Proximate constituents in the seagrasses *Zostera marina* and *Z. noltii* in Ireland: Seasonal changes and the effect of blade removal. *P.S.Z.N.I. Mar. Ecol.* 13: 307–315.

Dawes, C. J., M. O. Hall, and R. K. Riechert. 1985. Seasonal biomass and energy content in seagrass communities on the west coast of Florida. *J. Coast. Res.* 3: 255–262.

Dawes, C. J., D. Hanisak, and W. J. Kenworthy. 1995b. Seagrass biodiversity in the Indian River Lagoon. *Bull. Mar. Sci.* 57: 59–66.

Dawes, C. J., and W. J. Kenworthy. 1990. Organic constituents. In *Seagrass Research Methods*, R. C. Phillips and C. P. McRoy (eds.), 87–96. UNESCO, Paris.

Dawes, C. J., and E. W. Koch. 1991. Branch, miropropagule and tissue culture of the red algae *Eucheuma denticulatum* and *Kappaphycus alverzii* farmed in the Philippines. *J. Appl. Phycol.* 3: 247–257.

Dawes, C. J., and Lawrence, J. M. 1979. Effects of blade removal on the proximate composition of the rhizome of the seagrass *Thalassia testudinum* Banks ex Konig. *Aquat. Bot.* 7: 255–266.

———. 1980. Seasonal changes in the proximate constituents of the seagrass *Thalassia testudinum*, *Halodule wrightii*, and *Syringodium filiforme*. *Aquat. Bot.* 8: 371–380.

———. 1990. Seasonal changes in limestone and sand plant communities off the Florida west coast. *P.S.Z.N.I. Mar. Ecol.* 11: 97–104.

Dawes, C. J., C. S. Lobban, and D. A. Tomasko. 1989. A comparison of the physiological ecology of the seagrasses *Halophila decipiens* Ostenfeld and *H. johnsonii* Eiseman from Florida. *Aquat. Bot.* 33: 149–154.

Dawes, C. J., A. O. Lluisma, and G. C. Trono. 1994. Laboratory and field growth studies of commercial strains of *Eucheuma denticulatum* and *Kappaphycus alvarezii* in the Philippines. *J. Appl. Phycol.* 6: 21–24.

Dawes, C. J., E. D. McCoy, and K. L. Heck Jr. 1991. The tropical western Atlantic including the Caribbean sea. In *Ecosystems of the World 24 Intertidal and Littoral Ecosystems*, A. C. Mathieson and P. H. Nienhuis (eds.), 215–233. Elsevier, Amsterdam.

Dawes, C. J., and R. P. McIntosh. 1981. The effect of organic material and inorganic ions on the photosynthetic rate of the red alga *Bostrychia binderi* from a Florida estuary. *Mar. Biol.* 64: 213–218.

Dawes, C. J., A. C. Mathieson, and D. P. Cheney. 1974. Ecological studies of Floridian *Eucheuma* (Rhodophyta, Gigartinales). I. Seasonal growth and reproduction. *Bull. Mar. Sci.* 24: 235–273.

Dawes, C. J., R. E. Moon, and M. A. Davis. 1978. The photosynthetic and respiratory rates and tolerances of benthic algae from a mangrove and salt marsh estuary: A comparative study. *Estuar. Coast. Mar. Sci.* 6: 175–185.

Dawes, C. J., R. E. Moon, and J. LaClaire III. 1976. Photosynthetic responses of the red alga *Hypnea musciformis* (Wulfen) Lamouroux (Gigartinales). *Bull. Mar. Sci.* 26: 467–473.

Dawes, C. J., and E. L. Rhamstine. 1967. An ultrastructural study of the giant green algal coenocyte *Caulerpa prolifera*. *J. Phycol.* 3: 117–127.

Dawes, C. J., F. M. Scott, and E. Bowler. 1961. A light- and electron-microscopic survey of algal cell walls. I. Phaeophyta and Rhodophyta. *Amer. J. Bot.* 48: 925–934.

Dawes, C. J., and D. A. Tomasko. 1988. Depth distribution of *Thalassia testudinum* in two meadows on the west coast of Florida: A difference in effect of light availability. *P.S.Z.N.I. Mar. Ecol.* 9: 123–130.

Dawsn, N. S., and P. L. Walne. 1994. Evolutionary trends in Euglenoids. *Arch. Protistenkd.* 144: 221–225.

Dawson, E. Y. 1960. Marine red algae of Pacific Mexico. 3. Cryptonemiales, Corallinaceae subf. Melobesioideae. *Pac. Natur.* 2: 3–124.

———. 1966. *Marine Botany. An Introduction.* Holt, Rinehart and Winston, New York.

Day Jr., J. W., C. A. S. Hall, W. M. Kemp, and Y. Yanez-Arancibia. 1989. *Estuarine Ecology.* John Wiley, New York.

Dayton, P. K. 1973. Dispersion, dispersal, and persistence of the annual intertidal alga, *Postelsia palmaeformis* Ruprecht. *Ecology* 54: 433–438.

———. 1975a. Experimental studies of algal canopy interactions in a sea-otter dominated kelp community at Amchitka Island, Alaska. *U.S. Dept. Comm. Fish. Bull.* 73: 230–237.

———. 1975b. Experimental evaluation of ecological dominance in a rocky intertidal community. *Ecol. Monograf.* 45: 137–159.

Dayton, P. K., V. Currie, T. Gerrodette, B. D. Keller, R. Rosenthal, and D. V. Tresca. 1984. Patch dynamics and stability of some California kelp communities. *Ecol. Monograf.* 54: 253–289.

DeBoer, J. A. 1981. Nutrients. In *The Biology of Seaweeds*, C. S. Lobban and M. J. Wynne (eds.), 356–391. Blackwell, Oxford.

DeBoer, J. A., H. J. Guigli, T. L. Israel, and C. F. D'Elia, 1978. Nutritional studies of two red algae. I. Growth rate as a function of nitrogen source and concentration. *J. Phycol.* 14: 261–266.

D'Elia, C. F., and W. J. Wiebe. 1990. Biogeochemical nutrient cycles in coral-reef ecosystems. In *Ecosystems of the World 25. Coral Reefs*, Z. Dubinsky (ed.), 49–74. Elsevier, Amsterdam.

DeFraine, E. 1912. The anatomy of the genus *Salicornia*. *J. Linn. Soc. Bot.* 41: 317–348.

de Lacerda, L., D. V. José, C. E. de Rezende, M. C. F. Francisco, J. C. Wasserman, and J. C. Martins. 1986. Leaf chemical characteristics affecting herbivory in a new world mangrove forest. *Biotropica* 18: 350–355.

Delgado, O., C. Rodriguez-Prieto, E. Gacia, and E. Ballesteros. 1996. Lack of severe nutrient limitation in *Caulerpa taxifolia* (Vahl) C. Agardh, an introduced seaweed spreading over the oligotrophic northwestern Mediterranean. *Bot. Mar.* 39: 61–67.

de Molenaar, J. G. 1974. Vegetation of the Angmagssalik district in southeast Greenland. I. Littoral vegetation. *Meddelelser om Gronland* 198: 1–79.

den Hartog, C. 1967. The structural aspect in the ecology of sea-grass communities. *Helgol. Wiss. Meeresunter.* 15: 648–659.

———. 1970. *The Seagrasses of the World.* North Holland, Amsterdam.

Denley, E. J., and P. K. Dayton. 1985. Competition among macroalgae. In *Handbook of Phycological Methods: Ecological Field Methods: Macroalgae.* M. M. Littler and D. S. Littler (eds.), 511–530. Cambridge University Press, Cambridge.

Dennison, W. C., R. J. Orth, K. A. Moore, J. C. Stevenson, V. Carter, S. Koller, P. W. Bertgstrom, and R. A. Batiuk. 1993. Assessing water quality with submersed aquatic vegetation. *Bioscience* 43: 86–94.

Denny, M. W. 1988. *Biology and Mechanics of the Wave-Swept Environment.* Princeton University Press, Princeton.

Desikachary, T. V. 1959. *Cyanophyta.* Indian Council Agricultural Research, New Delhi.

de Villele, X., and M. Verlaque. 1995. Changes and deregulation in a *Posidonia ocean-*

ica bed invaded by the introduced tropical alga *Caulerpa taxifolia* in the northwestern Mediterranean. *Bot. Mar.* 38: 79–87.

Devinny, J. S. 1980. Effects of thermal effluents on communities of benthic marine macroalgae. *J. Env. Manag.* 11: 225–242.

Dill, R. F., E. A. Shinn, and A. T. Jones. 1986. Giant subtidal stromatolites forming in normal salinity waters. *Nature* 324: 55–58.

Dijkema, K. S. 1984. Western-European salt marshes. In *Salt Marshes in Europe*, K. S. Dikema (ed.), 82–103. European Commission for Conservation of Nature and Natural Resources, Strasbourg.

Dodge, J. D. 1983. Dinoflagellates: Investigation and phylogenetic speculation. *Brit. Phycol. J.* 18: 335–356.

Doty, M. S. 1946. Critical tide factors that are correlated with the vertical distribution of marine algae and other organisms along the Pacific Coast. *Ecology* 27: 315–328.

———. 1957. Rocky intertidal surfaces. *Geol. Soc. Amer. Memoir* 67, 1: 535–585.

Drew, K. M. 1955. Life histories in the algae with special reference to the Chlorophyta, Phaeophyta, and Rhodophyta. *Biol. Rev.* 30: 343–390.

Dring, M. J. 1982. *The Biology of Marine Plants*. Edward Arnold, London.

———. 1988. Photocontrol of development in algae. *Ann. Rev. Plant Physiol. Plant Mol. Biol.* 39: 157–174.

Drouet, F. 1981. Revision of the Stigonemataceae with a summary of the classification of the blue-green algae. *Nova Hedwigia* 66: 1–221.

Duarte, C. M., 1990. Seagrass nutrient content. *Mar. Ecol. Prog.*, Ser. 67: 201–207.

———. 1991. Seagrass depth limits. *Aquat. Bot.* 40: 363–377.

Duarte, C. M. and K. Sand-Jensen. 1996. Nutrient constraints on establishment from seed and on vegetative expansion of the Mediterranean seagrass *Cymodocea nodosa*. *Aquat. Bot.* 54: 279–286.

Dubinsky, Z., (ed.). 1990. *Ecosystems of the World 25. Coral Reefs*. Elsevier, Amsterdam.

Dubinsky, Z. and N. Stambler. 1996. Marine pollution of coral reefs. *Global Change Biol.* 2: 511–526.

Dubois, M., K. A. Gilies, J. K. Hamilton, P. A. Rebers, and F. Smith. 1956. Colorimetric methods for determination of sugars and related substances. *Anal. Chem.* 28: 350–356.

Ducker, S. C., and R. B. Knox. 1984. Epiphytism at the cellular level with special references to algal epiphytes. In *Encyclopedia of Plant Physiology 17*, H.-F. Linskens and J. Heslop-Harrison (eds.), 113–133. Springer, Berlin.

Dudgeon, S. R., I. R. Davison, and R. L. Vadas. 1989. Effect of freezing on photosynthesis of intertidal macroalgae: Tolerance of *Chondrus crispus* and *Mastocarpus stellatus* (Rhodophyta). *Mar. Biol.* 101: 107–114.

Duffy, J. E., and M. E. Hay. 1990. Seaweed adaptations to herbivory. *Bioscience* 40: 368–375.

Dugdale, R. C., and J. J. Goering. 1967. Uptake of new and regenerated forms of nitrogen in primary productivity. *Limnol. Oceanogr.* 12: 196–206.

Duggins, D. O. 1980. Kelp beds and sea otters: An experimental approach. *Ecology* 61:447–453.

Duke, N. C. 1992. Mangrove floristics and biogeography. In *Coastal and Estuarine Studies 41. Tropical Mangrove Ecosystems*, A. I. Robertson and D. M. Alongi (eds.), 63–100. American Geophysical Union, Washington, D.C.

Duke, N. C. and Z. S. Pinzon. 1992. Aging *Rhizophora* seedlings from leaf scar nodes: A technique for studying recruitment. *Biotropica* 24:173–186.

Duncan, W. H. 1974. Vascular halophytes of the Atlantic and Gulf coasts of North America north of Mexico. In *Ecology of Halophytes*, R. J. Reimold and W. H. Queen (eds.), 23–50. Academic Press, New York.

Dunton, K. H. and D. A. Tomasko. 1994. *In situ* photosynthesis in the seagrass *Halodule wrightii* in a hypersaline subtopical lagoon. *Mar. Ecol. Prog.*, Ser. 107:281–293.

Durako, M. J. 1994. Seagrass die-off in Florida Bay (USA): Changes in shoot demographic characteristics and population dynamics in *Thalassia testudinum. Mar. Ecol. Prog.* Ser. 110: 59–66.

Durako, M. J., and K. M. Kuss. 1994. Effects of *Labyrinthula* infection on the photosynthetic capacity of *Thalassia testudinum. Bull. Mar. Sci.* 54: 727–732.

Durako, M. J., R. C. Phillips, and R. R. Lewis III (eds.). 1987. *Proceedings of the Symposium on Tropical and Subtropical Seagrasses of the Southeastern U.S.*, Florida Marine Research Publication 42. Florida Department of Natural Resources, St. Petersburg.

Edwards, P. 1972. Cultured red alga to measure pollution. *Mar. Pollut. Bull.* 3: 184–188.

Edwards, P. 1975. An assessment of possible pollution effects over a century on the benthic marine algae of Co. Durham, England. *Bot. J. Linn. Soc.* 70: 269–305.

Edinger, E. N. 1991. Mass Extinction of Caribbean Coars at the Oligocene-Miocene Boundary: Paleoecology, Paleooceanography, Paleobiogeography. M.Sc. thesis, McMaster University, Hamilton, Ontario.

Eilers, II. P. 1979. Production ecology in an Oregon coastal salt marsh. *Estuar. Coast. Mar. Sci.* 8: 399–401.

Einay, R., S. Breckle, and S. Beer. 1995. Ecophysiological adaptative strategies of some intertidal marine macroalgae of the Israeli Mediterranean coast. *Mar. Ecol. Prog. Ser.* 125: 219–228.

Eleutherius, L. N. 1980. *An Illustrated Guide to Tidal Marsh Plants of Mississippi and Adjacent States*. Mississippi-Alabama Sea Grant Consortium, Gulf Coast Research Laboratory, Ocean Springs, Mississippi.

Ellison, A. M., and E. J. Farnsworth. 1992. The ecology of Belizean mangrove-root fouling communities: Patterns of epibiont distribution and abundance, and effects on root growth. *Hydrobiologia* 247: 87–98.

———. 1993. Seedling survivorship, growth and response to disturbance in Belizean mangal. *Amer. J. Bot.* 80: 1137–1145.

———. 1996a. Anthropogenic disturbance of Caribbean mangrove ecosystems: Past impacts, present trends, and future predictions. 1996. *Biotropica* 28: 549–565.

———. 1996b. Spatial and temporal variability in growth of *Rhizophora mangle* saplings on coral cays: Links with variation in insolation, herbivory, and local sedimentation rate. *J. Ecol.* 84:717–731.

Ellison, J. C., and D. R. Stoddart. 1991. Mangrove ecosystem collapse during predicted sea-level rise: Holocene analogues and implications. *J. Coast. Res.* 7: 151–165.

Elmqvist, T., and P. A. Cox. 1996. The evolution of vivipary in flowering plants. *Oikos* 77:3–9.

Elner, R. W. and R. L. Vadas, Sr. 1990. Inference in ecology: The sea urchin phenomenon in the northwest Atlantic. *Amer. Nat.* 136:108–125.

Englemann, T. W. 1884. Untersuchungen Über dei Quantitativen Beziehungen Zwischen Absorption des Lichtes und Assimilation in Pflanzenzellen. *Bot. Zert.* 42: 81–93.

Eppley, R. W., and B. J. Peterson. 1979. Particulate organic matter flux and planktonic new production in the deep ocean. *Nature* 282: 677–680.

Erftemeijer, P. L. A., and P. M. J. Herman. 1994. Seasonal changes in environmental variables, biomass production and nutrient contents in two contrasting tropical intertidal seagrass beds in South Sulawesi, Indonesia. *Oecologia* 99:45–49.

Eshel, A. 1985. Response of *Suaeda aegyptiaca* to KCl, NaCl, and Na_2SO_4 treatments. *Physiologia Plantarum* 64: 308–315.

Evans, L. V., and A. O. Christie. 1970. Studies on the ship-fouling alga *Enteromorpha*. I. Aspects of the fine-structure and biochemistry of swimming and newly settled zoospores. *Ann. Bot.* 34: 451–466.

Evans, L. V., and K. D. Hoagland. 1986. *Algal Biofouling*. Elsevier, Amsterdam.

Fagerberg, W. R., and C. J. Dawes. 1976. Studies on *Sargassum*. I. A light microscopic examination of the wound regeneration process in mature stipes of *S. filipendula*. *Amer. J. Bot.* 63: 110–119.

Fahn, A. 1988. Secretory tissue in vascular plants. *New Phytol.* 108: 229–257.

Fahn, A., and D. F. Cutler. 1992. *Handuch der Pflanzenanatomie 13 (3). Xerophytes.* Gebruder Borntraeger, Berlin.

Falkner, D. J. 1977. Interesting aspects of marine natural products chemistry. *Tetiar. Lett.* 33: 1421–1443.

Falkowski, P. B., and J. LaRoche. 1991. Acclimation to spectral irradiance in algae. *J. Phycol.* 27: 8–14.

Falkowski, P. B., P. L. Jokiel, and R. A. Kinzie III. 1990. In *Ecosystems of the World 25. Coral Reefs*, Z. Dubinsky (ed.), 89–107. Elsevier, Amsterdam.

Farnsworth, E. J., and A. M. Ellison. 1993. Dynamics of herbivory in Belizean mangal. *J. Trop. Ecol.* 9:435–453.

Farnsworth, E. J., and A. M. Ellison. 1996a. Sun-shade adaptability of the red mangrove, *Rhizophora mangle* (Rhizophoraceae): Changes through ontogeny at several levels of biological organization. *Amer. J. Bot.* 83: 1131–1143.

Farnsworth, E. J., and A. M. Ellison. 1996b. Scale-dependent spatial and temporal variability in biogeography of mangrove root epibiont communities. *Ecol. Monogr.* 66: 45–66.

Farrell, T. M. 1991. Models and mechanisms of succession: An example from a rocky intertidal community. *Ecol. Monograf.* 61: 95–113.

Felger, R., and M. B. Moser. 1973. Eelgrass (*Zostera marina* L.) in the Gulf of California: Discovery of its nutritional value by the Seri Indians. *Science* 181: 355–356.

Feller, I. C. 1995. Effects of nutrient enrichment on growth and herbivory of dwarf red mangrove (*Rhizophora mangle*). *Ecol. Monograf.* 65: 477–505.

————. 1996. Effects of nutrient enrichment on leaf anatomy of dwarf *Rhizophora mangle* L. (Red Mangrove). *Biotropica* 28: 13–22.

Femino, R. J., and A. C. Mathieson. 1980. Investigations of New England algae. IV. The ecology and seasonal succession of tide pool algae at Bald Head Cliff, York Maine, USA. *Bot. Mar.* 23: 319–332.

Fisher, A. E., and P. J. Harrison. 1996. Does carbohydrate content affect the sinking rates of marine diatoms? *J. Phycol.* 32: 360–365.

Fishman, J. R., and R. J. Orth. 1996. Effects of predation on *Zostera marina* L. seed abundance. *J. Exp. Mar. Biol. Ecol.* 198: 11–26.

Fitzpatrick, J., and H. Kirkman, 1995. Effects of prolonged shading stress on growth and survival of seagrass *Posidonia australis* in Jervis Bay, New South Wales, Australia. *Mar. Ecol. Prog.*, Ser. 127: 279–289.

Fletcher, R. L. 1995. Epiphytism and fouling in *Gracilaria* cultivation: An overview. *J. Appl. Phycol.* 7: 225–333.

Floc'h, J.-Y., R. Pajot, and V. Mouret. 1996. *Undaria pinnatifida* (Laminariales, Phaeophyta) 12 years after its introduction into the Atlantic Ocean. *Hydrobiologia* 326/327: 217–222.

Flowers, T. J., M. A. Hajibagheri, and M. J. W. Clipson. 1986. Halophytes. *Quat. Rev. Biol.* 61: 313–337.

Flowers, T. J., and A. R. Yeo. 1986. Ion relation of plants under drought and salinity. *Austral. J. Plant Physiol.* 13: 75–91.

Fogg, G. E. and B. Thake. 1987. *Algal Cultures and Phytoplankton Ecology.* Althone Press, London.

Fonseca, M. S. 1989. Regional analysis of the creation and restoration of seagrass systems. In *Wetland Creation and Restoration: The Status of the Science. I. Regional Reviews*, EPA/600/3-89/038a, J. A. Kusler and M. E. Kentula (eds.), 175–198. U.S. Environmental Protection Agency, Corvallis, Oregon.

————. 1994. *A Guide to Planting Seagrasses in the Gulf of Mexico*, TAMU-SG-94-601. Texas A&M University Sea Grant, College Station.

Fonseca, M. S., W. J. Kenworthy, and F. X. Courtney. 1996. Development of planted seagrass beds in Tampa Bay, Florida, USA. I. Plant components. *Mar. Ecol. Prog.*, Ser. 132: 127–139.

Fonseca, M. S., W. J. Kenworthy, F. X. Courtney, and M. O. Hall. 1994. Seagrass planting in the southeastern United States: Methods for accelerating habitat development. *Restor. Ecol.* 2: 198–212.

Fonseca, M. S., G. W. Thayer, A. J. Chester, and C. Foltz. 1984. Impact of scallop harvesting on eelgrass (*Zostera marina*) meadows: Implications for management. *North Amer. J. Fish. Manag.* 4: 286–293.

Fortes, M. D. 1987. Structure of algal communities epiphytic on the mangroves of Puerto Galera, Oriental Mindoro. *Philip. J. Sci.* 1987: 43–53.

————. 1988. Mangrove and seagrass beds of East Asia: Habitats under stress. *Ambio* 17: 207–213.

————. 1991. Seagrass-mangrove ecosystems management: A key to marine coastal conservation in the ASEAN region. *Mar. Pollut. Bull.* 22: 113–116.

Foster, M., M. Neushul, and R. Zingmark. 1971. The Santa Barbara oil spill.

2. Initial effects on intertidal and kelp bed organisms. *Envir. Pollut.* 2: 115–134.

Fourqurean, J. W., G. V. N. Powell, W. J. Kenworthy, and J. C. Zieman. 1995. The effects of long-term manipulation of nutrient supply on competition between the seagrasses *Thalassia testudinum* and *Halodule wrightii* in Florida Bay. *Oikos* 72: 349–358.

Fourqurean, J. W. and J. C. Zieman. 1991. Photosynthesis, respiration and whole plant carbon budgets of *Thalassia testudinum, Halodule wrightii* and *Syringodium filiforme*. In *The Light Requirements of Seagrasses: Proceedings of a Workshop*, NOAA Technical Memorandum NMFS-SEFC-287, W. J. Kenworthy and D. E. Haunert (eds.), 59–70. U.S. National Marine Fisheries Service, Beaufort, North Carolina.

Fredette, T. J., R. J. Diaz, J. van Montvrans, and R. J. Orth. 1990. Secondary production within a seagrass bed (*Zostera marina* and *Ruppia maritima*) in lower Chesapeake Bay. *Estuaries* 13: 431–440.

Freeman, N. K., F. T. Lindgren, Y. C. Ng, and A. V. Nichols. 1957. Infrared spectra of some lipoproteins and related lipids. *J. Biol. Chem.* 293: 293–304.

Freshwater, D. W., S. Fredericq, B. S. Butler, M. H. Hommersand, and M. W. Chase. 1994. A gene phylogeny of the red algae (Rhodophyta) based on plastid rbcL. *Proc. Natl. Acad. Sci. U.S.A.* 91: 7281–7285.

Frey, R. W., and P. B. Basan. 1985. Coastal salt marshes. In *Coastal Sedimentary Environments*, 2nd ed., R. A. Davis Jr. (ed.), 225–302. Springer-Verlag, New York.

Fricke, H., and E. Vareschi. 1981. A scleractinian coral (*Plerogyra sinuosa*) with "photosynthetic organs." *Mar. Ecol.* 7: 273–278.

Friedlander, M. 1992. *Gracilaria conferta* and its epiphytes: The effect of culture conditions on growth. *Bot. Mar.* 35: 423–428.

Friedlander, M., and A. Ben-Amotz. 1991. The effect of outdoor culture conditions on growth and epiphytes of *Gracilaria conferta. Aquat. Bot.* 39: 315–333.

Friedlander, M., and C. J. Dawes. 1985. *In situ* uptake kinetics of ammonium and phosphate and chemical composition of the red seaweed *Gracilaria tikvahiae. J. Phycol.* 21: 448–453.

Fritsch, F. E. 1935. *Structure and Reproduction of the Algae.* Vol. I. Cambridge University Press, Cambridge.

———. 1945. *Structure and Reproduction of the Algae.* Vol. II. Cambridge University Press, Cambridge.

Frost, B. W. 1980. Grazing. In *The Physiological Ecology of Phytoplankton*, I. Morris (ed.), 465–491. University of California Press, Berkeley.

Frost, S. H. 1977. Cenozoic reef systems of the Caribbean—Prospects for paleoecologic synthesis. In *Reefs and Related Carbonates-Ecology and Sedimentology*, S. H. Frost and J. B. Saunders (eds.), 93–110. Amer. Assoc. Petrol. Geol., Tulsa, Oklahoma.

Fry, B. 1984. $^{13}C/^{12}C$ ratios and the trophic importance of algae in Florida *Syringodium filiforme* seagrass beds. *Mar. Biol.* 79: 11–19.

Fry, B., and P. L. Parker. 1979. Animal diets in Texas seagrass meadows: $\Delta^{13}C$ evidence for the importance of benthic plants. *Estuar. Coastal Mar. Sci.* 8: 499–509.

Fuge, R., and K. H. James. 1974. Trace metal concentrations in *Fucus* from the Bristol Channel. *Mar. Pollut. Bull.* 5: 9–12.

Fukuhara, T., J.-Y. Pack, Y. Ohwaki, H. Tsujimura, and T. Nitta. 1996. Tissue-specific expression of the gene for a putative plasma membrane H^+-ATPase in a seagrass. *Plant Physiol.* 110: 35–42.

Gacesa, P. 1988. Alginates. *Carbo. Polym.* 8: 161–182.

Galgani, F., T. Burgeot, G. Bocquene, F. Vincent, J. P. Leaute, J. Labastie, A. Forest, and R. Guichet. 1995. Distribution and abundance of debris on the continental shelf of the Bay of Biascy and in Seine Bay. *Mar. Pollut. Bull.* 30: 58–62.

Gallagher, J. L., W. J. Pfeiffer, and L. R. Pomeroy. 1976. Leaching and microbial utilization of dissolved organic carbon from leaves of *Spartina alterniflora*. *Estuar. Coast. Mar. Sci.* 4: 467–471.

Gallagher, J. L., G. F. Somers, D. M. Grant, and D. M. Seliskar. 1988. Persistent differences in two forms of *Spartina alterniflora*: A common garden experiment. *Ecology* 69: 1005–1008.

Gantt, E. 1980. Photosynthetic cryptophytes. In *Phytoflagellates*, E. R. Cox (ed.), 381–406. Elsevier/North Holland, New York.

Garbary, D. J., J. Burke, and T. Lining. 1991. The *Asophyllum-Polysiphonia-Mycosphaerella* symbiosis. II. Aspects of the ecology and distribution of *Polysiphonia lanosa* in Nova Scotia. *Bot. Mar.* 34: 391–401.

Geitler, L. 1932. Cyanophyceae. In *Kryptogamenflora von Deutschland, Osterrich under der Schweiz*, L. Rabenhorst (ed.), Vol. 14, 673–1056. Akademische Verlags Gesellschaft, Leipzig.

Gentry, F. 1982. Phytogeographic patterns as evidence for a Choco refuge. In *Biological Diversification in the Tropics*, Gh. T. Prince (ed.), 112–136. Columbia University Press, New York.

Gessner, F. 1970. Temperature: Plants. In *Marine Ecology*, O. Kinne (ed.), Vol. I, Pt. 1, John Wiley, New York.

———. 1971. The water economy of the sea grass *Thalassia testudinum*. *Mar. Biol.* 10: 258–260.

Gibbs, S. P. 1978. The chloroplasts of *Euglena* may have evolved from symbiotic green algae. *Canad. J. Bot.* 56: 2883–2889.

Giblin, A. E., G. W. Luther III, and I. Valiela. 1986. Trace metal solubility in salt marsh sediments contaminated with sewage sludge. *Estuar. Coast. Shelf Sci.* 23: 477–498.

Giesen, W. B. J. T., M. M. van Katwijk, and C. den Hartog. 1990. Eelgrass condition and turbidity in the Dutch Wadden Sea. *Aquat. Bot.* 37: 71–85.

Gillott, M. 1990. Phylum Cryptophyta. In *Handbook of Protoctista*, L. Margulis, J. O. Corliss, M. Melkonian, and D. J. Chapman (eds.), 139–151. Jones and Bartlett, Boston.

Gilmartin, M. 1966. Ecology and morphology of *Tydemania expeditionis*, a tropical deepwater siphonous green alga. *J. Phycol.* 2: 100–105.

Glynn, P. W. 1990. Feeding ecology of selected coral-reef macroconsumers: Patterns and effects on coral community structure. In *Ecosystems of the World 25. Coral Reefs*, Z. Dubinsky (ed.), 365–400. Elsevier, Amsterdam.

Goering, H. K., and P. J. van Soest. 1970. *Forage Fiber Analyses, Apparatus, Reagents, Procedures and Some Applications*, Handbook 379. U.S. Department of Agriculture. Washington, D.C.

Goff, L. J. (ed.). 1983. *Algal Symbiosis*. Cambridge University Press, Cambridge.

Gomez, I., and C. Wiencke. 1996. Photosynthesis, dark respiration and pigment contents of gametophytes and sporophytes of the Antarctic brown alga *Desmarestia menziesii*. *Bot. Mar.* 39: 149–158.

Goodman, J. L., K. A. Moore, and W. C. Dennison. 1995. Photosynthetic responses of eelgrass (*Zostera marina* L.) to light and sediment sulfide in a shallow barrier island lagoon. *Aquat. Bot.* 50: 37–48.

Gordon, D. M., K. A. Grey, S. C. Chase, and S. J. Simpson. 1994. Changes to the structure and productivity of a *Posidonia sinuosa* meadow during and after imposed shading. *Aquat. Bot.* 47: 265–273.

Goreau, T. F., and N. I. Goreau. 1959. The physiology of skeleton formation in corals. II. Calcium deposition by hermatypic corals under various conditions in the reef. *Biol. Bull.* 117: 239–250.

Gornitz, V., S. Lebendeff, and J. Hansen. 1982. Global sea level trend in the past century. *Science*, 215: 1611–1614.

Gosselink, J. G. 1984. *The Ecology of Delta Marshes of Coastal Louisiana: A Community Profile*, FWS/Obs-84/09. U.S. Fish and Wildlife Service, Washington, D.C.

Graham, A. 1995. Diversification of Gulf/Caribbean mangrove communities through cenozoic time. *Biotropica* 20: 20–27.

——. 1996. Green algae to land plants: An evolutionary transition. *J. Pl. Res.* 109: 241–251.

Gray, A. J. 1986. Do invading species have definable genetic characteristics? *Phil. Trans. Roy. Soc. London Series B* 314: 665–674.

——. 1992. Saltmarsh plant ecology: Zonation and succession revisited. In *Saltmarshes. Morphodynamics, Conservation and Engineering Significance*, J. R. L. Allen and K. Pye (eds.), 63–79. Cambridge University Press, Cambridge.

Gray, A. J., D. F. Marshall, and A. F. Raybould. 1991. A century of evolution in *Spartina anglica*. *Adv. Ecol. Res.* 21: 1–54.

Gray, A. J., and R. Scott. 1977. The ecology of Morecambe Bay. VII. The distribution of *Puccinellia maritima*, *Festuca rubra* and *Agrostis stolonifera* in the saltmarshes. *J. Appl. Ecol.* 14: 229–241.

Green, J. C., K. Perch-Nielsen, and P. Westbroek. 1990. Phylum Pyrmnesiophyta. In *Handbook of Protoctista*, L. Margulis, J. D. Corliss, M. Melkonian, and D. J. Chapman (eds.), 293–317. Jones and Bartlett, Boston.

Grigg, R. W., and S. J. Dollar. 1990. Natural and anthropogenic disturbance on coral reefs. In *Ecosystems of the World 25. Coral Reefs*, Z. Dubinsky (ed.), 439–452. Elsevier, Amsterdam.

Grime, J. P. 1977. Evidence for the existence of three primary strategies in plants and its relevance to ecological and evolutionary theory. *Amer. Nat.* 111: 1169–1194.

Grinnell, J. 1917. The niche-relationship of the California Thrasher. *Auk* 34: 427–433.

Grobe, C. W., and T. M. Murphy. 1994. Inhibition of growth of *Ulva expansa* (Chlorophyta) by ultraviolet-B radiation. *J. Phycol.* 30: 783–790.

Gronitz, V., S. Lebedeff, and J. Hansen. 1982. Global sea level trend in the past century. *Science* 190: 1611–1614.

Gross, M. F., M. A. Hardisky, P. L. Wolf, and V. Klemas. 1991. Relationship between aboveground and belowground biomass of *Spartina alterniflora* (Smooth Cordgrass). *Estuaries* 14: 180–191.

Gross-Custard, S., J. Jones, J. A. Kitching, and T. A. Norton. 1979. Tide pools at Carrigthorna and Barlodge creed. *Phil. Trans. R. Soc. Ser. B.* 287: 1–44.

Guenegou, M. C., J. Citharel, and J. E. Levasseur. 1988. The hybrid status of *Spartina anglica* (Poaceae). Enzymatic analysis of the species and of the presumed parents. *Canad. J. Bot.* 66: 1830–1833.

Guiry, M. D., and G. Blunden (eds.). 1991. *Seaweed Resources in Europe: Uses and Potential* John Wiley, Chickseley, England.

Guiry, M. D., and C. J. Dawes. 1992. Daylength, temperature and nutrient control of tetrasporogenesis in *Asparagopsis armata* (Rhodophyta). *J. Exp. Mar. Biol. Ecol.* 158: 197–218.

Guiry, M. D., and L. M. Irvine. 1989. Sporangial form and function in the Nemaliophycidae (Rhodophyta). In *Phycotalk I*, H. D. Kumar (ed.), 155–184. Rastogi and Co., Subhash Bazar, Meerut, India.

Haberlandt, G. 1914. *Physiological Plant Anatomy*. Macmillan, London.

Hackney, J. M., R. C. Carpenter, and W. H. Adey. 1989. Characteristic adaptations to grazing among algal turfs on a Caribbean coral reef. *Phycologia* 28: 109–119.

Hader, D.-P., and E. Hoiczyk. 1992. Gliding motility. In *Algal Cell Motility*, M. Melkonian (ed.), 1–38. Chapman and Hall, New York.

Haglund, K., M. Bjorklund, S. Gunnare, A. Sandberg, U. Olander, and M. Pedersen. 1996. New method for toxicity assessment in marine and brackish environments using the macroalga *Gracilaria tenuistipitata* (Gracilariales, Rhodophyta). *Hydrobiologia* 326/327: 317–325.

Halle, F., R. A. A. Oldeman, and P. B. Tomlinson. 1978. *Tropical Trees and Forests—An Architectural Analysis*. Springer-Verlag, Berlin.

Hallock, P. 1988. Diversification in algal symbiont-bearing foraminifera: A response to oligotrophy? *Revue Palèobiol.* Sp. Vol. 2. 789–797.

Hallock, P., F. E. Müller-Karger, and J. C. Halas. 1997. Coral reef decline. *Natl. Geogr. Res. & Explor.* 9: 358–378.

Hallock, P., and W. Schlager. 1986. Nutrient excess and the demise of coral reefs and carbonate platforms. *Palaios* 1: 389–398.

Hamilton, L. S., and D. H. Murphy. 1988. Use and management of Nipa Palm (*Nypa fruticans*, Arecaceae): A review. *Econ. Bot.* 42: 206–213.

Han, T., and J. M. Kain. 1992. Bluelight sensitivity of UV-irradiated young sporophytes of *Laminaria hyperborea*. *J. Exp. Mar. Biol. Ecol.* 158: 219–230.

Hanisak, M. D. 1979. Growth patterns of *Codium fragile* ssp. *tomentosoides* in response to temperature, irradiance, salinity, and nitrogen sources. *Mar. Biol.* 50: 319–332.

———. 1987. Cultivation of *Gracilaria* and other macroalgae in Florida for energy production. In *Seaweed Cultivation for Renewable Resources*, K. T. Bird and P. H. Benson (eds.), 191–218. Elsevier, Amsterdam.

Hanisak, M. D., and S. M. Blair. 1988. The deep-water macroalgal community of the east Florida continental shelf (USA). *Helgol. Wiss. Meeresunter.* 42: 133–163.

Hardwick-Witman, M. N. 1984. Ice-rafting in a northern New England salt marsh community. M.S. thesis, University of New Hampshire, Durham.

Harlin, M. M. 1980. Seagrass epiphytes. In *Handbook of Seagrass Biology, An Ecosystem Perspective*, C. P. McRoy (ed.), 117–152. Garland, New York.

Harris, G. P. 1986. *Phytoplankton Ecology, Structure, Function and Fluctuation*. Chapman and Hall, London.

Harris, L. G. 1996. A new community state in the southwestern Gulf of Maine? In *Gulf of Maine Ecosystem Dynamics*, Presentation to the Regional Association for Research in the Gulf of Maine, University of Maine, Orono.

Harrison, W. G., T. Platt, and M. R. Lewis. 1987. f-ratio and its relationship to ambient nitrate concentration in coastal waters. *J. Plank. Res.* 9: 235–248.

Harvey, W. H. 1841. *A Manual of the British Algae*. Reeve Brothers, London.

Harvey, H. W., L. H. N. Cooper, M. V. Lebour, and P. S. Russell. 1935. Plankton production and its control. *J. Mar. Biol. Assoc. U.K.* 20: 407–441.

Hawkins, S. J., and R. G. Hartnoll. 1985. Factors determining the upper limits of intertidal canopy-forming algae. *Mar. Ecol. Prog.*, Ser. 20: 265–271.

Hay, M. E. 1981a. The functional morphology of turf-forming seaweeds: Persistence in stressful marine habitats. *Ecology* 62: 739–750.

———. 1981b. Herbivory, algal distribution, and the maintenance of between-habitat diversity on a troical fringing reef. *Amer. Nat.* 118: 520–540.

———. 1986. Functional geometry of seaweeds: ecological consequences of thallus layering and shape in contrasting light environments. In *On the Economy of Plant Form and Function*, T. J. Givenish (ed.), 635–666. Cambridge University Press, Cambridge.

Hay, M. E., and W. Fenical. 1988. Marine plant-herbivore interactions: The ecology of chemical defense. *Ann. Rev. Ecol. Syst.* 19: 111–145.

Heck, K. L., and E. D. McCoy. 1978. Long distance dispersal and the reef building corals of the eastern Pacific. *Mar. Biol.* 48: 349–356.

Heck Jr., K. L., and T. A. Thoman. 1984. The nursery role of seagrass meadows in the upper and lower reaches of the Chesapeake Bay. *Estuaries* 7: 70–92.

Heck Jr., K. L., and J. F. Valentine. 1995. Sea urchin herbivory: Evidence for long-lasting effects in subtropical seagrass meadows. *J. Exp. Mar. Biol. Ecol.* 189: 205–217.

Heijs, F. M. L. 1985. The macroalgal component in monspecific seagrass beds from Papua, New Guinea. *Aquat. Bot.* 22: 291–324.

Heimdal, B. R., J. K. Egge, M. J. W. Veldhuis, and P. Wesbroek. 1994. The 1992 Norwegian *Emiliania huxleyi* experiment: An overview. *Sarsia* 79: 285–290.

Hellebust, J. A. 1970. Light: Plants. In *Marine Ecology*, O. Kinne (ed.), Vol. 1, 125–158. John Wiley, New York.

Hellebust, J. A., and J. S. Craigie (eds.). 1978. *Handbook of Phycological Methods. Physiological and Biochemical Methods*. Cambridge University Press, Cambridge.

Henriquez, P., A. Candia, R. Norambuena, M. Silva, and R. Zemelman. 1979. Antibiotic properties of marine alge. II. Screening of Chilean marine algae for antimicrobial activity. *Bot. Mar.* 22: 451–453.

Herke, W. H., E. E. Knudsen, P. A. Knudsen, and B. D. Rogers. 1992. Effects of semi-impoundment of Louisiana marsh on fish and crustacean nursery use and export. *North Amer. J. Fish. Manag.* 12: 151–160.

Hester, M. W., K. L. McKee, D. M. Burdick, M. S. Koch, K. M. Flynn, S. Patterson, and I. A. Mendelssohn. 1994. Clonal integration in *Spartina patens* across a nitrogen and salinity gradient. *Canad. J. Bot.* 72: 767–770.

Heywood, P. 1990. Phylum Raphidophyta. In *Handbook of Protoctista*, L. Margulis, J. O. Corliss, M. Melkonian, and D. J. Chapman (eds.), 318–325. Jones and Bartlett, Boston.

Hibberd, D. J. 1980a. Prymnesiophytes. In *Phytoflagellates*, E. R. Cox (ed.), 273–318. Elsevier/North Holland, New York.

———. 1980b. Eustigmatophytes. In *Phytoflagellates*, E. R. Cox (ed.), 319–334. Elsevier/North Holland, New York.

———. 1990a. Phylum Xanthophyta. In *Handbook of Protoctista*, L. Margulis, J. O. Corliss, M. Melkonian, and D. J. Chapman (eds.), 686–697. Jones and Bartlett, Boston.

———. 1990b. Phylum Eustigmatophyta. In *Handbook of Protoctista*, L. Margulis, J. O. Corliss, M. Melkonian, and D. J. Chapman (eds.), 326–333. Jones and Bartlett, Boston.

Hillis-Colinveaux, L. 1980. Ecology and taxonomy of *Halimeda*: Primary producer of coral reefs. *Adv. Mar. Biol.* 17: 1–327.

Hillman, K., D. I. Walker, A. W. D. Larkum, and A. J. McComb. 1989. Productivity and nutrient limitation. In *Biology of Seagrasses*, A. W. D. Larkum, A. J. McComb, and S. A. Shepherd (eds.), 635–685. Elsevier, Amsterdam.

Hixon, M. A., and W. N. Brostoff. 1996. Succession and herbivory: Effects of differential fish grazing on Hawaiian coral-reef algae. *Ecol. Monograf.* 66: 67–90.

Hodgson, L. M. 1980. Control of the intertidal distribution of *Gastroclonium coulteri* in Monterey Bay, California, USA. *Mar. Biol.* 57: 121–126.

Hoffman, B. A., and C. J. Dawes. 1997. Vegetational and abiotic analysis of the salterns of mangals and salt marshes of the west coast of Florida. *J. Coast. Res.* 13: 147–154.

Hoffman, W. E., and C. J. Dawes. 1980. Photosynthetic rates and primary production by two Florida benthic red algal species from a salt marsh and a mangrove community. *Bull. Mar. Sci.* 30: 358–364.

Hoffman, W. E. and J. A. Rodgers Jr. 1981. Cost-benefit aspects of coastal vegetation establishment in Tampa Bay, Florida. *Envir. Conserv.* 8: 39–43.

Holmes, M. J., and R. J. Lewis. 1993. The origin of ciguatera. *Mem. Queensland Mus.* 34: 497–504.

Hopkins, W. G. 1993. *Introduction to Plant Physiology.* John Wiley, New York.

Horn, M. H. 1989. Biology of marine herviborous fishes. *Oceanogr. Mar. Biol. Ann. Rev.* 27: 167–272.

Howes, B. L., J. W. H. Dacey, and D. D. Goehringer. 1986. Factors controlling the growth form of *Spartina alterniflora*: Feedbacks between above ground production. *J. Ecol.* 74: 881–898.

Huggett, J., and C. L. Griffiths. 1986. Some relationships between elevation, physicochemical variables and biota of intertidal rock pools. *Mar. Ecol. Prog.*, Ser. 29: 189–197.

Humm, H. J., and S. R. Wicks. 1980. *Introduction and Guide to the Marine Bluegreen Algae.* John Wiley, New York.

Huntsman, S. A., and W. G. Sunda. 1980. The role of trace metals in regulating phytoplankton growth. In *The Physiological Ecology of Phytoplankton*, I. Morris (ed.), 285–328. University of California Press, Berkeley.

Hussain, M. I., and T. M. Kohja. 1993. Intertidal and subtidal blue-green algal mats of open and mangrove areas in the Farasan archipelago (Saudi Arabia), Red Sea. *Bot. Mar.* 36: 377–388.

Hutchings, P., and P. Saenger. 1987. *Ecology of Mangroves.* University of Queensland Press, St. Lucia.

Hutchinson, G. E. 1958. Concluding remarks. *Cold Spring Harbor Symposium on Quantative Biology.* 22: 415–427.

Indergaard, M., and J. Minsaas. 1991. Animal and human nutrition. In *Seaweed Resources in Europe: Uses and Potential*, M. D. Guiry and G. Blunden (eds.), 21–64. John Wiley, Chickseley, England.

Ish-Shalom-Gordon, N., and Z. Dubinsky. 1990. Possible modes of salt secretion in *Avicennia marina* in the Sinai. *Plant Cell Physiol.* 31: 27–32.

Ish-Shalom-Gordon, N., and Z. Dubinsky. 1992. Ultrastructure of the pneumatophores of the mangrove *Avicennia marina. S. Afr. J. Bot.* 58: 358–362.

Iyer, V., and A. D. Barnabas. 1993. Effects of varying salinity on leaves of *Zostera capensis* Setchell. I. Ultrastructural changes. *Aquat. Bot.* 46: 141–153.

Jackson, E. F. 1968. *Algae, Man, and the Environment.* Syracuse University Press, Syracuse.

Jackson, G. A. 1980. Phytoplankton growth and zooplankton grazing in oligotrophic oceans. *Nature* 284: 439–440.

Jagels, R. 1983. Further evidence for osmoregulation in epidermal leaf cells of seagrasses. *Amer. J. Bot.* 70: 327–333.

Jagels, R., and A. Barnabas. 1989. Variation in leaf ultrastructure of *Ruppia maritima* L. along a salinity gradient. *Aquat. Bot.* 33: 297–221.

Janzen, D. H. 1985. Mangroves: Where's the understory? *J. Trop. Ecol.* 1: 89–92.

Jerlov, N. G. 1976. *Marine Optics.* Elsevier, Amsterdam.

Johansson, J. O. R. 1991. Long-term trends of nitrogen loading, water quality and biological indicators in Hillsborough Bay, Florida. In *Second Tampa Bay Area Scientific Information Symposium*, S. F. Treat and P. A. Clark (eds.), 157–176. Lewis Environmental Systems, Tampa.

John, D. M., S. J. Hawkins, and J. H. Price. 1992. *Plant-Animal Interactions in the Marine Benthos.* Clarendon Press, Oxford.

John, D. M., and G. W. Lawson. 1990. The effects of grazing animals on algal vegetation. In *Introduction to Applied Phycology*, I. Akatuska (ed.), 307–345. SPB Academic, The Hague.

Johnson, D. S. and A. F. Skutch. 1928. Littoral vegetation on a headland of Mt. Desert Island. Maine. I. Submersible or strictly littoral vegetation. *Ecology* 9: 188–215.

Jones, R. C. 1980. Productivity of algal epiphytes in a Georgia salt marsh: Effect of inundation frequency and implications for total marsh productivity. *Estuaries* 3: 315–317.

Jones, R. F. 1960. The accumulation of nitrosyl ruthenium by fine particles and marine organisms. *Limnol. Oceanograf.* 5: 312–325.

Jones, W. E., and A. Demetropolus. 1968. Exposure to wave action: Measurements of an important ecological parameter on rocky shores on Anglesey. *J. Exp. Mar. Biol. Ecol.* 2: 46–63.

Joshi, G. V., S. Sontakke, I. Bhosale, and A. P. Waghmode. 1984. Photosynthesis and photorespiration in mangroves. In *Physiology and Management of Mangroves*, H. J. Teas (ed.), 210–238. Dr. W. Junk, The Hague.

Josselyn, M. 1983. *The Ecology of San Francisco Bay Tidal Marshes: A Community Profile*, FWS/OBS-83/23. U.S. Fish and Wildlife Service, Washington, D.C.

Kanwisher, J. W. 1957. Freezing and drying in intertidal algae. *Biol. Bull.* 113: 275–285.

————. 1966. Photosynthesis and respiration in some seaweeds. In *Some Contemporary Studies in Marine Sciences*, H. Barnes (ed.), 407–420. Allen and Unwin, London.

Karp-Boss, L., E. Boss, and P. A. Jumars. 1996. Nutrient fluxes to planktonic osmotrophs in the presence of fluid motion. *Oceanogr. Mar. Biol. Ann. Rev.* 34: 71–107.

Karsten, U., C. Wiencke, and G. O. Kirst, 1989. The β-dimethylsulphonio propionate (DMSP) content of macroalgae from Antarctica and southern Chile. *Bot. Mar.* 32: 143–146.

Keats, D. W., G. R. South, and D. H. Steele. 1985. Algal biomass and diversity in the upper subtidal at a pack-ice disturbed site in eastern Newfoundland. *Mar. Ecol. Prog.*, Ser. 25: 151–158.

Kemp, W. M., R. R. Twilley, J. C. Stevenson, W. R. Boynton, and J. C. Means. 1983. The decline of submerged vascular plants in upper Chesapeake Bay: Summary of results concerning possible causes. *Mar. Tech. Soc. J.* 17: 78–89.

Kempers, A. J., and A. Zweers. 1986. Ammonium determination in soil extracts by the salicylate method. *Comm. Soil Sci. Plant Anal.* 17: 715–723.

Kennelly, S. J. 1987. Inhibition of kelp recruitment by turfing algae and consequences for an Australian kelp community. *J. Exp. Mar. Biol. Ecol.* 112: 49–60.

Kenworthy, W. J., M. J. Durako, S. M. R. Fatemy, H. Valavi, and G. W. Thayer. 1993. Ecology of seagrasses in Northeastern Saudi Arabia one year after the Gulf War oil spill. *Mar. Pollut. Bull.* 27: 213–222.

Kenworthy, W. J., and M. S. Fonseca. 1992. The use of fertilizer to enhance growth of transplanted seagrasses. *Zostera marina* L. and *Halodule wrightii* Aschers. *J. Exp. Mar. Biol. Ecol.* 163: 141–161.

Kerby, N. W., and J. A. Raven. 1985. Transport and fixation of inorganic carbon by marine algae. *Adv. Bot. Res.* 11: 71–123.

Ketchum, B. H. (ed.). 1983. *Ecosystems of the World 26. Estuaries and Enclosed Seas.* Elsevier, Amsterdam.

Kidman, B. 1995. Cobscook Bay ecosystem study. *Gulf Maine News* 3(1): 1–5.

Kiirikki, M. 1996. Mechanisms affecting macroalgal zonation in the northern Baltic Sea. *European J. Phycol.* 31: 225–232.

Kiirikki, M., and A. Ruuskanen. 1996. How does *Fucus vesiculosus* survive ice scraping? *Bot. Mar.* 39: 133–140.

Kim, H. G., R. S. Kang, and C. H. Sohn. 1992. Effects of thermal effluents on the marine algal community at the coast of Kori Nuclear Power Plant. *Korean J. Phycol.* 7: 269–279.

Kim, J. H., and R. E. DeWreede. 1996. Effects of size and season of disturbance on alga patch recovery in a rocky intertidal community. *Mar. Ecol. Prog.*, Ser. 133: 217–228.

King, R. J. 1990. Macroalgae associated with the mangrove vegetation of Papua, New Guinea. *Bot. Mar.* 33: 55–62.

King, R. J., and M. D. Wheeler. 1984. Composition and geographic distribution of mangrove macroalgal communities in New South Wales. *Proc. Linn. Soc. N.S.W.* 108: 97–117.

Kinne, O. (ed.). 1970. *Marine Ecology*, Vol. 1, Pt. 1. John Wiley, New York.

———. 1971. *Marine Ecology*, Vol. 2, Pt. 2, 3. John Wiley, New York.

———. 1976. *Marine Ecology*, Vol. 3, Pt. 1. John Wiley, New York.

Kinsey, D. W. 1983. Standards of performance in coral reef primary production and carbon turnover. In *Perspectives on Coral Reefs*, D. J. Barnes (ed.), 209–220. Brian Clouston, Brisbane.

Kirk, J. J. O. 1994. *Light & Photosynthesis in Aquatic Ecosystems*, 2nd ed. Cambridge University Press, Cambridge.

Kirst, G. O. 1979. Osmotische Adaptation der marinen Planktonalgae *Platymonas subcordiformis* (Hazen). *Deutsch. Bot. Ges* (Berlin) 92: 31–42.

Kirst, G. O., and M. A. Bisson. 1979. Regulation of turgor pressure in marine algae: Ions and low molecular weight organic compounds. *Austral. J. Plant Physiol.* 6: 539–556.

Kiyosawa, H., and J. Ishizaka. 1995. Distribution of the prochlorophyte *Prochlorococcus* in the central Pacific Ocean as measured by HPLC. *Limnol. Oceanograf.* 40: 983–989.

Klekowski Jr., E. J., J. E. Corredor, J. M. Morell, and C. A. del Castillo. 1994. Petroleum pollution and mutation in mangroves. *Mar. Pollut. Bull.* 28: 166–169.

Klinger, T. 1985. Allocation of blade surface area to meiospore production in annual and perennial representatives of the genus Laminaria. M.Sc. thesis, University of British Columbia, Vancouver.

Kloareg, B., and R. S. Quatrano. 1988. Structure of the cell walls of marine algae and ecophysiological functions of the matric polysaccharides. *Oceanograf. Mar. Biol. Ann. Rev.* 26: 259–315.

Klugh, A. B. 1924. Factors controlling the biota of tide-pools. *Ecology* 5: 180–196.

Klumpp, D. W., R. K. Howard, and D. A. Pollard. 1989. Trophodynamics and nutritional ecology of seagrass communities. In *Biology of Seagrasses*, A. W. D. Larkum, A. J. McComb, and S. A. Shepherd (eds.), 394–457. Elsevier, Amsterdam.

Knight, M., and M. W. Parke. 1931. *Manx Algae. Mem. Liverpool Mar. Biol. Comm.* 301: 1–147.

Knox, G. A. 1986. *Estuarine Ecosystems: A Systems Approach*, Vols. 1 and 2. CRC Press, Boca Raton, Fla.

Koch, E. W. 1993. Hydrodynamics of flow through seagrass canopies: Biological, physical, and geochemical interactions. Ph.D. dissertation, University of South Florida, St. Petersburg.

———. 1994. Hydrodynamics, diffusion-boundary layers and photosynthesis of the seagrass *Thalassia testudinum* and *Cymodocea nodosa. Mar. Biol.* 118: 767–776.

Koch, E. W., and C. J. Dawes. 1991. Ecotypic differentiation in populations of *Ruppia maritima* L. germinated from seeds and cultured under algae-free laboratory conditions. *J. Exp. Mar. Biol. Ecol.* 152: 145–159.

Koch, E. W., and M. J. Durako. 1991. *In vitro* studies of the submerged angiosperm *Ruppia maritima*: Auxin and cytokinin effects on plant growth and development. *Mar. Biol.* 110: 1–6.

Kockelman, W. J., T. J. Conomos, and A. E. Leviton (eds.). 1982. *San Francisco Bay Use and Protection.* California Academy of Science, San Francisco.

Komarek, J., and K. Anagnostidis. 1986. Modern approach to the classification system of the cyanophytes. 4. Nostocales. *Algol. Stud.* 56: 247–345.

Komatsu, T. 1996. Influence of a *Zostera* bed on the spatial distribution of water flow over a broad geographic area. In J. Kuo, D. I. Walker, and H. Kirkman (eds.), 123–125. *Seagrass Biology: Scientific Discussion from an International Workshop.* University of Western Australia, Nedlands.

Kooistra, W. H. C. F., A. M. T. Joosten, and C. van den Hoek. 1989. Zonation patterns in intertidal pools and their possible causes: A multivariate approach. *Bot. Mar.* 32: 9–26.

Kristensen, E., F. O. Andersen, and L. H. Kofoed. 1988. Preliminary assessment of benthic community metabolism in a south-east Asian mangrove swamp. *Mar. Ecol. Prog.*, Ser. 48: 137–145.

Kristiansen, J. 1982. Chrysophyceae. In *Synopsis and Classification of Living Organisms*, S. P. Parker (ed.), 81–91. McGraw-Hill, New York.

Kruczynski, A. 1994. Anatomical and Cytological Adaptations to Salinity in the Leaves of Ruppia maritima. M.S. thesis, University of South Florida, Tampa.

Krugens, P., and R. E. Lee. 1988. Ultrastructure of fertilization in a cryptomonad. *J. Phycol.* 24: 385–393.

Kumar, H. D. 1990. *Introductory Phycology.* Affiliated East-West Press, New Delhi.

Kuo, J., and A. J. McComb. 1989. Seagrass taxonomy, structure and development. In *Biology of Seagrasses*, A. W. D. Larkum, A. J. McComb, and S. A. Shepherd (eds.), 6–73. Elsevier, Amsterdam.

Kuo, J., R. C. Phillips, D. I. Walker, and H. Kirkman (eds.). 1996. *Seagrass Biology, Proceedings of an International Workshop.* University of Western Australia, Nedlands.

Kuramoto, R. T., and D. E. Brest. 1979. Physiological response to salinity by four salt marsh plants. *Bot. Gaz.* 140: 295–298.

Kurtz, H., and K. Wagner. 1957. Tidal marshes of the Gulf and Atlantic coasts of northern Florida and Charleston, South Carolina. *Fla. St. Univ. Stud.* 24: 1–168.

Kutkuhn, J. H. 1966. The role of estuaries in the development and perpetuation of commercial shrimp resources. *Amer. Fish. Soc. Sp. Publ.* 3: 16–36.

Kyle, D. J., K. D. B. Boswell, R. M. Glaude, and S. E. Reeb. 1992. Designer oils from microalgae as nutritional supplements. In *Biotechnology and Nutrition*, D. D. Bills and S.-D. Kung (eds.), 451–468. Butterworth-Heinemann, Boston.

Lacerda, L. D., V. Ittekkot, and S. R. Patchineelam. 1995. Biogeochemistry of mangrove soil organic matter: A comparison between *Rhizophora* and *Avicennia* soils in southeastern Brazil. *Estuar. Coast. Shelf Sci.* 40: 713–720.

Lanyon, J. M., and H. Marsh. 1995. Temporal changes in the abundance of some tropical intertidal seagrasses in North Queensland. *Aquat. Bot.* 49: 217–237.

Lapointe, B. E. In Press. Nutrient thresholds for bottom-up control of macroalgal blooms on coral reefs in Jamaica and southeast Florida. *Limnol. Oceanograf.*

Lapointe, B. E., W. R. Matzie, D. A. Tomasko, and M. A. Voorhees. 1996. Deterioration of water quality in the south Florida coastal ecosystem: Hypersalinity or eutrophication? In *Proceedings of the 8th International Coral Reef Symposium*, (Unpublished Presentation).

Lapointe, B. E., D. A. Tomasko, and W. R. Matzie. 1994. Eutrophication and trophic state classification of seagrass communities in the Florida Keys. *Bull. Mar. Sci.* 54: 696–717.

Lapointe, B. E., and J. D. O'Connell. 1989. Nutrient-enhanced growth of *Cladophora prolifera* in Harrington Sound, Bermuda: Eutrophication of a confined, phosphorus-limited ecosystem. *Estuar. Coast. Shelf. Sci.* 28: 347–360.

Larkum, A. W. D., and C. den Hartog. 1989. Evolution and biogeography of seagrasses. In *Biology of Seagrasses*, A. W. D. Larkum, A. J. McComb, and S. A. Shepherd (eds.), 112–156. Elsevier, Amsterdam.

Larkum, A. W. D., A. J. McComb, and S. A. Shepherd (eds.). 1989. *Biology of Seagrasses.* Elsevier, Amsterdam.

Larkum, A. W. D., G. Roberts, J. Kuo, and S. Strother. 1989. Gaseous movement in seagrasses. In *Biology of Seagrasses*, A. W. D. Larkum, A. J. McComb, and S. A. Sherperd (eds.), 686–722. Elsevier, Amsterdam.

Larkum, A. W. D. and R. J. West. 1990. Long-term changes of seagrass meadows in Botany Bay, Australia. *Aquat. Bot.* 37: 55–70.

Larkum, A. W. D., and W. F. Wood. 1993. The effect of UV-B radiation on photosynthesis and respiration of phytoplankton, benthic macroalgae and seagrasses. *Photosyn. Res.* 36: 17–23.

Lawrence, J. M. 1975. On the relationships between marine plants and sea urchins. *Oceanograf. Mar. Biol. Ann. Rev.* 13: 213–286.

Lawrence, J. M., Ch.-F. Boudouresque, and F. Maggiore. 1989. Proximate energy in *Posidonia oceanica* (Potomogetonaceae). *P.S.Z.N.I. Mar. Ecol.* 10: 263–270.

Layne, E. 1957. Spectrographic and turbidimetric methods for measuring proteins. In *Methods in Enzymology*, Vol. 3, S. P. Kolowich and N. O. Kaplan (eds.), 447–454. Academic Press, New York.

Lazar, A. C., and C. J. Dawes. 1991. A seasonal study of the seagrass *Ruppia maritima L.* in Tampa Bay, Florida. Organic constituents and tolerances to salinity and temperature. *Bot. Mar.* 34: 265–269.

Lee, J. J., and O. R. Anderson. 1991. Symbiosis in foraminifera. In *Biology of Foraminifera*, J. J. Lee and O. R. Anderson (eds.), 157–220. Academic Press, New York.

Lee, R. W. 1989. *Phycology*, 2nd ed. Cambridge University Press, Cambridge.

Leedale, G. R. 1967. Euglenida/Euglenophyta. *Ann. Rev. Microbiol.* 21: 31–48.

Lefebvre, L. W., T. J. O'Shea, G. B. Rathbun, and R. C. Best. 1989. Distribution, status, and biogeography of the West Indian Manatee. *Biogeo. West Indies* 1989: 567–610.

Lembi, C. A., and J. R. Waaland (eds.). 1989. *Algae and Human Affairs.* Cambridge University Press, Cambridge.

Lesser, M. P. 1996. Acclimation of phytoplankton to UV-B radiation: Oxidative stress and photoinhibition of photosynthesis are not prevented by UV-absorbing compounds in the dinoflagellae *Prorocentrum micans. Mar. Ecol. Prog.,* Ser. 132: 287–297.

Lessios, H. A. 1988. Mass mortality of *Diadema antillarum* in the Caribbean: What have we learned? *Ann. Rev. Ecol. Syst.* 19: 371–393.

―――. 1995. *Diadema antillarum* 10 years after mass mortality: Still rare despite help from a competitor. *Proc. Roy. Soc. Lon. Ser. B* 259: 331–337.

Levin, P. S. and A. C. Mathieson. 1991. Variation in a host-epiphyte relationship along a wave exposure gradient. *Mar. Ecol. Prog.,* Ser. 77: 271–278.

Levine, H. G. 1984. The use of seaweeds for monitoring coastal waters. In *Algae as Ecological Indicators,* L. E. Shubert (ed.), 189–210. Academic Press, New York.

Levine, J. M., and M. D. Bertness. 1996. Nutrients, competition and plant segregation in a New England salt marsh. In *Proceedings of the 24th Benthic Ecology Meeting,* 53. Univ. South Carolina, Columbia.

Levring, T., H. A. Hoppe, and O. J. Schmid. 1969. *Marine Algae.* Cram, De Gruyter, Hamburg.

Lewin, R. A. 1976. Prochlorophyta as a proposed new division of algae. *Nature* 261: 697–698.

Lewin, R. A., and L. Cheng (eds.). 1989. *Prochloron A Microbial Enigma.* Chapman and Hall, New York.

Lewis, J. G., N. F. Stanley, and G. G. Guist. 1988. Commercial production and applications of algal hydrocolloids. In *Algae in Human Affairs,* C. A. Lembi and J. R. Waaland (eds.), 205–236. Cambridge University Press, Cambridge.

Lewis, J. R. 1964. *The Ecology of Rocky Shores.* English University Press, London.

Lewis III, R. R. (ed.). 1982. *Creation and Restoration of Coastal Plant Communities.* CRC Press, Boca Raton, Fla.

Lewis III, R. R., and E. D. Estevez. 1988. *The Ecology of Tampa Bay, Florida: An Estuarine Profile,* Biological Report 85. U.S. Fish and Wildlife Service, Washington, D.C.

Lewis, S. M. 1985. Herbivory on coral reefs: Algal susceptibility to herbivorous fishes. *Oecologia* 65: 370–375.

Levitt, G. J., and J. J. Bolton. 1991. Seasonal patterns of photosynthesis and physiological parameters and the effects of emersion in littoral seaweeds. *Bot. Mar.* 34: 403–410.

Libes, M. 1986. Productivity-irradiance relationship of *Posidonia oceanica* and its epiphytes. *Aquat. Bot.* 26: 285–306.

Li-Cor Inc. 1985. *LI-1000 Datalogger Instruction Manual.* Li-Cor Inc., Lincoln, Nebraska.

Lin, G., and L. de S. L. Sternberg. 1994. Utilization of surface water by red mangrove (*Rhizophora mangle* L.): An isotopic study. *Bull. Mar. Sci.* 54: 94–102.

Lin, H.-J., S. W. Nixon, E. I. Taylor, S. L. Granger, and B. A. Buckley. 1996. Responses of epiphytes on eelgrass, *Zostera marina* L., to separate and combined nitrogen and phosphorus enrichment. *Aquat. Bot.* 52: 243–258.

Linden, O., and A. Jernelov. 1980. The mangrove swamp-An ecosystem in danger. *Ambio* 9: 81–88.

Linskens, H. F. 1963. Beitrag zur Frage der Beziehungen zwischen Epiphyt und Basiphyt bei marinen Algen. *Publ. Staz. Zool. Napoli* 33: 274–293.

Lipkin, Y., S. Beer, and A. Eshel. 1993. The ability of *Porphyra linearis* (Rhodophyta) to tolerate prolonged periods of desiccation. *Bot. Mar.* 36: 517–523.

Little, C., and J. A. Kitching. 1996. *The Biology of Rocky Shores.* Oxford University Press, Oxford.

Littler, M. M. 1976. Calcification and its role among macroalgae. *Micronesica* 12: 27–41.

Littler, M. M., and D. S. Littler. 1980. The evolution of thallus form and survival strategies in benthic marine macroalgae: Field and laboratory tests of a functional form model. *Amer. Nat.* 116: 25–44.

————. 1984. Deepest known plant life discovered on an uncharted seamount. *Science* 227: 57–59.

————. 1988. Structure and role of algae in tropical reef communities. In *Algae and Human Affairs*, C. A. Lembi and J. R. Waaland (eds.), 29–56. Cambridge University Press, Cambridge.

———— (eds.). 1985. *Handbook of Phycological Methods. Ecological Field Methods: Macroalgae.* Cambridge University Press, Cambridge.

Littler, M. M., D. S. Littler, and P. R. Taylor. 1983. Evolutionary strategies in a tropical barrier reef system: Functional-form groups of marine macroalgae. *J. Phycol.* 19: 229–237.

Littler, M. M., and S. N. Murray. 1974. The primary productivity of marine macrophytes from a rocky intertidal community. *Mar. Biol.* 27: 131–135.

Lobban, C. S., D. J. Chapman, and B. P. Kremer (eds.). 1988. *Experimental Phycology, A Laboratory Manual.* Cambridge University Press, Cambridge.

Lobban, C. S., and P. J. Harrison. 1994. *Seaweed Ecology and Physiology.* Cambridge University Press, Cambridge.

Loeblich, A. R. III. 1977. Marine chloromonads: More widely distributed in neritic environments than previously thought. *Proc. Biol. Soc. Wash.* 90: 388–399.

————. 1982. Dinophyceae. In *Synopsis and Classification of Living Organisms 1*, S. P. Parker (ed.), 101–115. McGraw-Hill, New York.

Logan, B. W., J. F. Reed, and G. M. Hagan. 1974. Evolution and diagenesis of quaternary carbonate sequences, Shark Bay, Western Australia. *Amer. Assoc. Petrol. Geol. Mem.* 22: 1–358.

Long, S. P., and C. F. Mason. 1983. *Saltmarsh Ecology.* Chapman and Hall, London.

Long, W. J. L., L. J. McKenzie, M. A. Rasheed, and R. G. Coles. 1996. Monitoring seagrasses in tropical ports and harbors. In *Seagrass Biology: Proceedings of the International Workshop*, J. Kuo, R. C. Phillips, D. I. Walker, and H. Kirkman (eds.), 345–350. University of Western Australia, Nedlands.

Longh, H. H. de, B. J. Wenno, and E. Meelis. 1995. Seagrass distribution and seasonal biomass changes in relation to dugong grazing in Moluccas, East Indonesia. *Aquat. Bot.* 50: 1–19.

Lovett Doust, J., and L. Lovett Doust. 1988. *Plant Reproductive Ecology: Patterns and Strategies.* Oxford University Press, Oxford.

Lowry, O. H., N. J. Rosebrough, A. L. Farr, and R. J. Randall. 1951. Protein measurement with the folin phenol reagent. *J. Biol. Chem.* 193: 265–275.

Lubchenco, J. 1978. Plant species diversity in a marine intertidal community: Importance of herbivore food preference and algal competitive abilities. *Amer. Nat.* 112: 23–39.

———. 1982. Effect of grazers and algal competitors on fucoid colonization in tide pools. *J. Phycol.* 18: 544–550.

Lubchenco, J., and S. D. Gaines. 1981. A unified approach to marine plant-herbivore interactions. I. Populations and communities. *Ann. Rev. Ecol. Syst.* 12: 405–437.

Lugo, A. E. 1986. Mangrove understory: An expensive luxury? *J. Trop. Ecol.* 2: 287–288.

Lugo, A. E., and S. C. Snedaker. 1974. The ecology of mangroves. *Ann. Rev. Ecol. Sys.* 5: 39–64.

Lumbang, W. A., and V. J. Paul. 1996. Chemical defenses of the tropical green seaweed *Neomeris annulata* Dickie: Effects of multiple compounds on feeding by hervivores. *J. Exp. Mar. Biol. Ecol.* 201: 185–196.

Lundberg, P., R. G. Weich, P. Jensen, and H. J. Vogel. 1989. Phosphorus-31 and nitrogen-14 studies of the uptake of phosphorus and nitrogen compounds in the marine macroalga *Ulva lactuca. Plant Physiol.* 89: 1380–1387.

Lüning, K. 1990. *Seaweeds. Their Environment, Biogeography, and Ecophysiology,* C. Yarish and H. Kirkman (trans.). John Wiley, New York.

Lüning, K., and W. Freshwater. 1988. Temperature tolerance of northeast Pacific marine algae. *J. Phycol.* 24: 310–315.

Lüning, K., M. D. Guiry, and M. Masuda. 1987. Upper temperature tolerance of North Atlantic and North Pacific geographic isolates of the red alga *Chondrus. Helgol. Wiss. Meeresunter.* 41: 297–306.

Lüning, K., and M. Neushul. 1978. Light and temperature demands for growth and reproduction of laminarian gametophytes in southern and central California. *Mar. Biol.* 45: 297–309.

Mackey, A. P., and G. Smail. 1995. Spatial and temporal variation in litter fall of *Avicennia marina* (Forssk.) Vierh. in the Brisbane River, Queensland, Australia. *Aquat. Bot.* 52: 133–142.

MacNae, W. 1968. A general account of the fauna and flora of mangrove swamps and forests in the Indo-West-Pacific region. *Adv. Mar. Biol.* 6: 73–270.

Madlener, J. C. 1977. *The Seavegetable Book.* Clarkson Potter, New York.

Maegawa, M., M. Kunieda, and W. Kida. 1993. The influence of ultraviolet radiation on the photosynthetic activity of several red algae from different depths. *Japan J. Phycol.* 41: 207–214.

Mahall, B. E., and R. B. Park. 1976. The ecotone between *Spartina foliosa* Trin. and *Salicornia virginica* L. in salt marshes of northern San Francisco Bay. III. Soil aeration and tidal immersion. *J. Ecol.* 48: 811–819.

Malecki, R. A., B. Blossey, S. D. Hight, D. Schroeder, L. T. Kok, and J. R. Coulson. 1993. Biological control of purple loosestrife. *Bioscience* 43: 680–686.

Mann, K. H., and A. R. O. Chapman. 1975. Primary production of marine macrophytes. In *Photosynthesis and Productivity in Different Environments*, J. P. Cooper (ed.), IBP 3, 207–248. Cambridge University Press, Cambridge.

Mann, K. H., and J. R. N. Laziev. 1991. *Dynamics of Marine Ecosystems: Biological-Physical Interactions in the Ocean*. Blackwell Scientific Publications, Oxford, England.

Maragos, J. E., C. Evans, and P. Holthus. 1985. Reef corals in Kaneohe Bay six years before and after termination of sewage discharges (Oahu, Hawaiian Archipelago). *Proceedings of the 5th International Coral Reef Symposium*, 4: 189–194.

Marchant, H. J., A. T. Davidson, and G. J. Kelly. 1991. UV-B protecting compounds in the marine alga *Phaeocystis pouchetii* from Antarctica. *Mar. Biol.* 109: 391–395.

Margalef, R. 1963. Succession in marine populations. In *Advancing Frontiers of Plant Science*, 2nd ed., R. Vira (ed.), 137–188. Institute for the Advancement of Science and Culture, New Delhi.

Margalef, R. 1968. The pelagic ecosystem of the Caribbean Sea. In *Symposium on Investigations and Resources of the Caribbean Sea and Adjacent Regions*, 484–498. UNESCO, Paris.

Margulis, L., and K. V. Schwartz. 1988. *Five Kingdoms: An Illustrated Guide to the Phyla of Life on Earth*, 2nd ed. W. H. Freeman, New York.

Marks, D. L., R. Bushsbaum, and T. Swain. 1985. Measurement of total protein in plant samples in the presence of tannins. *Anal. Biochem.* 147: 136–143.

Martínez, E. A. 1996. Micropopulation differentiation in phenol content and susceptibility to herbivory in the chilean kelp *Lessonia nigresceus* (Phaeophyta, Laminariales). *Hydrobiologia* 326/327: 205–211.

Matheke, G. E. M., and R. Horner. 1974. Primary productivity of the benthic microalgae in the Chukchi Sea near Barrow, Alaska. *J. Fish. Res. Bd. Can.* 31: 1779–1786.

Mathieson, A. C. 1979. Vertical distribution and longevity of subtidal seaweeds in northern New England U.S.A. *Bot. Mar.* 30: 511–520.

———. 1989. Phenological patterns of northern New England seaweeds. *Bot. Mar.* 32: 419–438.

Mathieson, A. C., R. A. Fralick, R. Burns, and W. Flashive. 1975. Phycological studies during Tektite II, at St. John, U.S. V.I. In *Results of the Tektite Program: Coral Reef Invertebrates and Plants*, Science Bulletin No. 20, S. A. Earle and R. L. Lavenberg (eds.), 77–103. Natural History Museum, Los Angeles.

Mathieson, A. C., and Z. Guo. 1992. Patterns of fucoid reproductive biomass allocation. *British Phycol. J.* 271–292.

Mathieson, A. C., and E. J. Hehre. 1986. A synopsis of New Hampshire seaweeds. *Rhodora* 88: 1–139.

———. 1994. A comparison of marine algae from the Goleta Slough and adjacent open coast of Goleta/Santa Barbara, California with those in the southern Gulf of Maine. *Rhodora* 96: 207–258.

Mathieson, A. C., and P. H. Nienhuis (eds.). 1991. *Ecosystems of the World 24. Intertidal and Littoral Ecosystems*. Elsevier, Amsterdam.

Mathieson, A. C., C. A. Penniman, P. K. Busse, and E. Tveter-Gallagher. 1982. Effects of ice on *Ascophyllum nodosum* within the Great Bay Estuary System of New Hampshire–Maine. *J. Phycol.* 18: 331–336.

Mathieson, A. C., J. W. Shipman, J. R. O'Shea, and R. C. Hasevlat. 1976. Seasonal growth and reproduction of estuarine fucoid algae in New England. *J. Exp. Mar. Biol. Ecol.* 25: 273–284.

Mathieson, A. C., E. Tveter, M. Daly, and J. Howard. 1977. Marine algal ecology in a New Hampshire tidal rapid. *Bot. Mar.* 20: 277–290.

Mattox, K. R., and K. D. Stewart. 1984. Classification of the green algae: A concept based on comparative cytology. In *Systematics of the Green Algae*, D. E. G. Irvine and D. M. John (eds.), 29–72. Academic Press, New York.

Maxwell, W. G. H. 1968. *Atlas of the Great Barrier Reef.* Elsevier, Amsterdam.

Maze, J., P. Morand, and P. Potoky. 1993. Stabilization of 'Green tides' *Ulva* by a method of composting with a view to pollution limitation. *J. Appl. Phycol.* 5: 183–190.

McCarthy, J. J. 1980. Nitrogen. In *The Physiological Ecology of Phytoplankton*, I. Morris (ed.), 191–233. University of California Press, Berkeley.

McConchie, C. A., and R. B. Knox. 1989. Pollination and reproductive biology of seagrasses. In *Biology of Seagrasses*, A. W. D. Larkum, A. J. McComb, and S. A. Shepherd (eds.), 74–111. Elsevier, Amsterdam.

McCoy, E. D., and K. L. Heck Jr. 1976. Biogeography of corals, seagrasses, and mangroves. An alternative to the center of orgin concept. *System. Zool.* 25: 201–210.

McKee, K. L., I. A. Mendelssohn, and M. W. Hester. 1988. Reexamination of pore water sulfide concentrations and redox potentials near the aerial roots of *Rhizophora mangle* and *Avicennia germinans. Amer. J. Bot.* 75: 1352–1359.

McMillan, C. 1980. Reproductive physiology in the seagrass *Syringodium filiforme* from the Gulf of Mexico and the Caribbean. *Amer. J. Bot.* 67: 104–110.

McMillan, C., S. C. Williams, L. Escobar, and O. Zapata. 1981. Isozymes, secondary compounds and experimental cultures of Australian seagrasses *Halophila, Halodule, Zostera, Amphibolis* and *Posidonia. Austral. J. Bot.* 29: 247–260.

McRoy, C. P., and C. Helfferich (eds.). 1977, *Seagrass Ecosystems*. Marcel Dekker, New York.

McRoy, C. P. and C. McMillan. 1977. Production ecology and physiology of seagrasses. In *Seagrass Ecosystems. A Scientific Perspective*, C. P. McRoy and C. Helfferich (eds.), 53–81. Marcel Dekker, New York.

Meeks, J. C. 1974. Chlorophylls. In *Algal Physiology and Biochemistry*, W. D. P. Stewart (ed.), 161–175. University of California Press, Berkeley.

Meinesz, A., L. Benichou, T. Komatsu, R. Lemee, H. Molenaar, and X. Mari. 1995. Variations in the structure, morphology and biomass of *Caulerpa taxifolia* in the Mediterranean Sea. *Bot. Mar.* 38: 499–508.

Meinesz, A., and C.-F. Boudouresque. 1996. Sur l'origine de *Caulerpa taxifolia* en Mediterranee. *C.R. Acad. Sci. Paris*, Ser. III, 319: 603–613.

Meinesz, A., J. de Vaugelas, B. Hesse, and X. Mari. 1993. Spread of the introduced tropical green alga *Caulerpa taxifolia* in northern Mediterranean waters. *J. Appl. Phycol.* 5: 141–147.

Mendelssohn, I. A., and K. L. McKee. 1988. *Spartina alterniflora* die-back in Louisiana: Time-course investigation of soil waterlogging effects. *J. Ecol.* 76: 509–521.

Mendelssohn, I. A., and E. D. Seneca. 1980. The influence of soil drainage on the growth of salt marsh cordgrass *Spartina alterniflora* in North Carolina. *Estuar. Coastal Mar. Sci.* 11: 27–40.

Menge, B. A., T. M. Farrell, A. M. Olson, P. van Tamelen, and T. Turner. 1993. Algal recruitment and the maintenance of a plant mosaic in the low intertidal region on the Oregon coast. *J. Exp. Mar. Biol. Ecol.* 170: 91–116.

Menzel, D. 1980. Plug formation and peroxydase accumulation in two orders of siphonous green algae (Caulerpales, Dasycladales) in relation to fertilization and injury. *Phycologia* 19: 37–48.

Merrill, J., and R. Fletcher. 1991. Green tides cause major economic burden in Venice Lagoon, Italy. *Appl. Phycol. Forum* 8(3): 1–3.

Michanek, G. 1979. Phytogeographic provinces and seaweed distribution. *Bot. Mar.* 22: 375–391.

Miller, M. W., and M. E. Hay. 1996. Coral-seaweed-grazer-nutrient interactions on temperate reefs. *Ecol. Monograf.* 66: 323–344.

Miller, W. R., and E. E. Egler. 1950. Vegetation of the Wequetequock-Pawactcck tidal marshes, Connecticut. *Ecol. Monograf.* 20: 141–172.

Milne, D. H. 1995. *Marine Life and the Sea.* Wadsworth, Belmont, Ca.

Mitsch, W. J., and J. G. Gosselink. 1993. *Wetlands*, 2nd ed., Van Nostrand Reinhold, New York.

Mobberley, D. G. 1956. Taxonomy and distribution of the genus *Spartina. Iowa State Coll. J. Sci.* 30: 471–574.

Molenaar, H., and A. Meinesz. 1995. Vegetative reproduction in *Posidonia oceanica*: Survival and development of transplanted cuttings according to different spacings, arrangements and substrates. *Bot. Mar.* 38: 313–322.

Molina, X., and V. Montecino. 1996. Acclimation to UV irradiance in *Gracilaria chilensis* Bird, McLachlan and Oliverira (Gigartinales, Rhodophyta). *Hydrobiologia* 326/327: 415–420.

Mooney, H. A. 1976. Some contributions of physiological ecology to plant population biology. *System. Bot.* 1: 269–283.

Moriarty, D. J. W., and P. I. Boon. 1989. Interactions of seagrasses with sediment and water. In *Biology of Seagrasses*, A. W. D. Larkum, A. J. McComb, and S. A. Shepherd (eds.), 500–535. Elsevier, Amsterdam.

Moriarty, D. J. W., R. Iverson, and P. C. Pollard. 1986. Exudation of organic carbon by the seagrass *Halodule wrightii* and its effect on bacterial growth in the sediment. *J. Exp. Mar. Biol. Ecol.* 96: 115–126.

Morris, I. (ed.). 1980. *The Physiological Ecology of Phytoplankton.* University of California Press, Berkeley.

Morton, J. 1991. *Shore Life Between Fundy Tides.* Canadian Scholars Press, Toronto.

Mostaert, A. S., U. Karsten, and R. J. King. 1995. Inorganic ions and mannitol in the red alga *Caloglossa leprieurii* (Ceramiales, Rhodophyta): Response to salinity change. *Phycologia* 34: 501–507.

Muehlstein, L. K. 1989. Prospectives on the wasting disease of eelgrass *Zostera marina*. *Diseases Aquat. Org.* 7: 211–221.

Muller, D. G. 1972. Studies on reproduction in *Ectocarpus siliculosus*. *Soc. Bot. Fr. Mem.* 1972: 87–98.

————. 1977. Sexual reproduction in British *Ectocarpus siliculosus* (Phaeophyta). *British Phycol. J.* 12: 131–136.

Murphy, D. H. 1990. The natural history of insect herbivory on mangrove trees in and near Singapore. *Raffles Bull. Zool.* 38: 119–203.

Murren, C. J., and A. M. Ellison. 1996. Effects of habitat, plant size, and floral display on male and female reproductive success of the neotropical orchid *Brassavola nodosa*. *Biotropica* 28: 30–41.

Muscatine, L. 1990. The role of symbiotic algae in carbon and energy flux in reef corals. In *Ecosystems of the World 25. Coral Reefs*, Z. Dubinsky (ed.), 75–87. Elsevier, Amsterdam.

Naidoo, G. 1990. Effects of nitrate, ammonium and salinity on growth of the mangrove *Bruguiera gymnorrhiza* (L.) Lam. *Aquat. Bot.* 38: 209–219.

Nalewajko, C., and D. R. S. Lean. 1980. Phosphorus. In *The Physiological Ecology of Phytoplankton*, I. Morris, (ed.), 235–258. University of California Press, Berkeley.

National Research Council. 1985. *Oil in the Sea, Inputs, Fates and Effects*. National Academy Science Press, Washington, D.C.

Neushul, M. 1963. Studies on the giant kelp, *Macrocystis*. II. Reproduction. *Amer. J. Bot.* 50: 354–359.

————. 1972. Functional interpretation of benthic marine algal morphologies. In *Contributions to the Systematics of Benthic Marine Algae of the North Pacific*, I. A. Abbott and M. Kruogi (eds.), 47–73. Japanese Society of Phycology, Kobe.

Neushul, P. 1987. Energy from marine biomass: The historical record. In *Seaweed Cultivation for Renewable Resources*, K. T. Bird and P. H. Benson, (eds.), 1–37. Elsevier, Amsterdam.

Newell, R. I. E., N. Marshall, A. Sasekumar, and V. C. Chong. 1995. Relative importance of benthic microalgae, phytoplankton, and mangroves as sources of nutrition for penaeid prawns and other coastal invertebrates from Malaysia. *Mar. Biol.* 123: 595–606.

Nielsen, T. G., T. Kioboe, and P. K. Bjornsen. 1990. Effects of *Chrysochromulina polyplepis* subsurface bloom on the planktonic community. *Mar. Ecol. Prog.*, Ser. 62: 21–35.

Niering, W. A. 1990. Vegetation dynamics in relation to wetland creation. In *Wetland Creation and Restoration. The Status of the Science*, J. A. Kusler and M. E. Kentula (eds.), 479–486. Island Press, Washington, D.C.

Niering, W. A., and R. S. Warren. 1980. Vegetation patterns and processes in New England salt marshes. *Bioscience* 30: 301–307.

Niesembaum, R. A. 1988. The ecology of sporulation by the macroalga *Ulva lactuca* L. (Chlorophyceae). *Aquat. Bot.* 32: 221–227.

Nixon, S. W. 1982. *The Ecology of New England High Salt Marshes: A Community Profile*, FWS/OBS-81/55. U.S. Fish and Wildlife Service, Washington, D.C.

Nixon, S. W., and C. A. Oviatt. 1973. Ecology of a New England salt marsh. *Ecol. Monograf.* 43: 463–498.

Norris, J. N., and K. E. Bucher. 1982. Marine algae and seagrasses from Carrie Bow Cay, Belize. In *The Atlantic Barrier Reef Ecosystem of Carrie Bow Cay, Belize. I. Structure and Communities*, K. Rutzler and I. G. MacIntyre (eds.), 167–238. Smithsonian Institution Press, Washington, D.C.

Norton, T. A. and A. C. Mathieson. 1983. The biology of unattached seaweeds. *Prog. Phycol. Res.* 2: 333–386.

Norton, T. A., M. Melkonian, and R. A. Anderson. 1996. Algal biodiversity. *Phycologia* 35: 308–326.

Novaczek, I. 1984. Development and phenology of *Ecklonia radiata* at two depths in Goat Island Bay, New Zealand. *Mar. Biol.* 81: 189–197.

Novak, R. 1984. A study in ultra-ecology: Microorganisms on the seagrass *Posidonia oceanica* (L.) Delile. *Mar. Ecol. Prog.*, Ser. 5: 143–190.

Nultsch, W., J. Pfau, and U. Ruffer. 1981. Do correlations exist between chromatophore arrangement and photosynthetic activity in seaweeds? *Mar. Biol.* 62: 111–117.

Oates, B. R. 1986. Components of photosynthesis in the intertidal saccate alga *Halosaccion americanum* (Rhodophyta, Palmariales). *J. Phycol.* 22: 217–223.

———. 1988. Water relations of the intertidal saccate alga *Colpomenia peregrina* (Phaeophyta, Scytosiphonales). *Bot. Mar.* 31: 57–63.

Oates, B. R., and S. N. Murray. 1983. Photosynthesis, dark respiration, and desiccation resistance of the intertidal seaweeds *Hesperophycus harveyanus* and *Pelvetia fastigiata* f. *gracilis. J. Phycol.* 19: 371–380.

O'Carra, P. 1955. Purification and N-terminal analysis of algal biliproteins. *Biochem. J.* 94: 171–174.

Odgen, J. C. 1980. Faunal relationships in Caribbean seagrass beds. In *Handbook of Seagrass Biology: An ecosystem Perspective*, R. C. Phillips and C. P. McRoy (eds.), 173–198. Garland, New York.

Odum, E. P. 1974. Halophytes, energetics and ecosystems. In *Ecology of Halophytes*, R. J. Reimold and W. H. Queen (eds.), 599–602. Academic Press, New York.

Odum, W. E., C. C. McIvor, and T. J. Smith III. 1982. *The Ecology of the Mangroves of South Florida: A Community Profile*, FWS/OBS-81/24. U.S. Fish and Wildlife Service, Washington, D.C.

Ohno, M. 1984. Observation on the floating seaweeds of near-shore waters of southern Japan. *Hydrobiologia* 116/117: 408–412.

———. 1993a. Succession of seaweed communities on artificial reefs in Ashizuri, Tosa Bay, Japan. *Korean J. Phycol.* 8: 191–198.

———. 1993b. Cultivation of the green algae, *Monostroma* and *Enteromorpha* "Aonori." In *Seaweed Cultivation and Marine Ranching*, M. Ohno and A. T. Critchley (eds.), 7–15. Japan International Cooperative Agency, Nagai, Yokosuka.

Ohno, M., and A. T. Critchley (eds.), 1993. *Seaweed Cultivation and Marine Ranching.* Japan International Cooperative Agency, Nagai, Yokosuka.

Olsen, S. R., and L. E. Sommers. 1982. Phosphorus. In *Methods of Soil Analysis. 2. Chemical and Microbiological Properties*, A. L. Page, R. H. Miller, and D. R. Keenely (eds.), 403–427. American Society of Agronomy, Madison, Wisconsin.

Oltmanns, F. 1892. Ueber die Cultur- und Lebensbedingungen der Meeresalgen. *Jahr. Wiss. Bot.* 23: 349–440.

————. 1922a. *Morphologie und Biologie der Algen*, 2nd ed. Vol. I. Fischer, Jena, Germany.

————. 1922b. *Morphologie und Biologie der Algen*, 2nd ed. Vol. II. Fischer, Jena, Germany.

————. 1923. *Morphologie und Biologie der Algen*, 2nd ed. Vol. III. Fischer, Jena, Germany.

Onuf, C. P. 1991. Light requirement of *Halodule wrightii*, *Syringodium filiforme* and *Halophila englemanni* in a heterogeneous and variable environment inferred from long-term monitoring. In *The Light Requirements of Seagrasses: Proceeding of a Workshop*, NOAA Technical Memorandum NMFS-SEFC-287, W. J. Kenworthy and D. E. Hanuert (eds.), 95–105. U.S. National Marine Fisheries Service, Beaufort, North Carolina.

————. 1996. Seagrass responses to long-term light reduction by brown tide in upper Laguna Madre, Texas: Distribution and biomass patterns. *Mar. Ecol. Prog.*, Ser. 138: 219–231.

Oohusa, T. 1993. The cultivation of *Porphyra* "Nori." In *Seaweed Cultivation and Marine Ranching*, M. Ohno and A. T. Critchley, (eds.), 57–73. Japan International Cooperative Agency, Nagai, Yokosuka.

Oppenhiemer, C. H. 1970. Temperature: Bacteria, fungi and blue-green algae. In *Marine Ecology*, Vol. 1(1), O. Kinne (ed.), 347–362. John Wiley, London.

Ornitz, B. E. 1996. *Oil crisis in Our Oceans.* Tageh Press, Glenwood Springs, Colorado.

Orth, R. J., and K. A. Moore. 1983. Seed germination and seedling growth of *Zostera marina* (eelgrass) in Chesapeake Bay (USA). *Aquat. Bot.* 15:117–132.

Orth, R. J., and J. van Montfrans. 1984. Epiphyte-seagrass relationships with an emphasis on the role of micrograzing: A review. *Aquat. Bot.* 18: 43–69.

Paasche, E. 1980. Silicon. In *The Physiological Ecology of Phytoplankton*, I. Morris (ed.), 259–284. University of California Press, Berkeley.

Paine, R. T. 1979. Disaster, catastrophe, and local persistence of the sea palm, *Postelsia palmaeformis. Science* 205: 685–686.

Pakker, H., and A. M. Breeman. 1996. Temperature responses of tropical to warm-temperate Atlantic seaweeds. II. Evidence for ecotypic differentiation in amphi-Atlantic tropical-Mediterranean species. *European J. Phycol.* 31: 133–141.

Pakker, H., A. M. Breeman, W. F. P.-H. van Reine, M. J. H. van Oppen, and C. van den Hoek. 1996. Temperature responses of tropical to warm-temperate Atlantic seaweeds. I. Absence of ecotypic differentiation in amphi-Atlantic tropical-Canary Island species. *European J. Phycol.* 31: 123–132.

Parks, P. J., and M. Bonifaz. 1994. Nonsustainable use of renewable resources: Mangrove deforestation and mariculture in Ecuador. *Mar. Res. Econ.* 9: 1–18.

Parsons, T. R. 1982. The new physical definition of salinity: Biologists beware. *Limnol. Oceanograf.* 27: 384–385.

Parsons, T. R., Y. Maita, and C. M. Lalli. 1989. *A Manual of Chemical and Biological Methods for Seawater Analysis.* Pergamon Press, Oxford.

Pastorek, R. A., and G. R. Bilyard. 1985. Effects of sewage pollution on coral-reef communities. *Mar. Ecol. Prog.*, Ser. 21: 175–189.

Paterson, D. M. 1990. The influence of epipelic diatoms on the erodibility of an artificial sediment. In *Proceedings of the 10th International Symposium on Living and fossil Diatoms*, H. Simola (ed.), 345–355. Jenosuu, Königstein, Germany.

———. 1995. Biogenic structure of early sediment fabric visualized by low-temperature scanning electron microscopy. *J. Geol. Soc. London* 152: 131–140.

Patriquin, D. G. 1975. "Migration" of blowouts in seagrass beds at Barbados and Carriacou, West Indies, and its ecological and geological implications. *Aquat. Bot.* 1: 163–189.

Patriquin, D. G., and R. Knowles. 1972. Nitrogen fixation in the rhizosphere of marine angiosperms. *Mar. Biol.* 16: 49–58.

Patten, B. C., J. J. Norcross, D. K. Young, and C. L. Rutherford. 1964. Some experimental characteristics of dark and light bottles. *J. Cons. Intern. Pour L' Explor. Mer.* 28: 335–353.

Patwary, M. U., and J. P. van der Meer. 1992. Genetics and breeding of cultivated seaweeds. *Korean J. Phycol.* 7: 281–318.

Pedersen, M. F., and J. Borum. 1992. Nitrogen dynamics of eelgrass *Zostera marina* during a late summer period of high growth and low nutrient availability. *Mar. Ecol. Prog.*, Ser. 80: 65–73.

Penhale, P. A., and R. G. Wetzel. 1983. Structure and functional adaptations of eelgrass (*Zostera marina*) to the anaerobic sediment environment. *Canad. J. Bot.* 61: 1421–1428.

Pfeiffer, W. J., and R. G. Weigert. 1981. Grazers on *Spartina* and their predators. In *The Ecology of a Salt Marsh*, L. R. Pomeroy and R. G. Weigert (eds.), 88–112. Springer-Verlag, New York.

Philippart, C. J. M. 1995. Seasonal variation in growth and biomass of an intertidal *Zostera noltii* stand in the Dutch Wadden Sea. *Nethl. J. Sea Res.* 33: 205–218.

Phillips, A., G. Lambert, J. E. Granger, and T. D. Steinke. 1996. Vertical zonation of epiphytic algae associated with *Avicennia marina* (Forssk.) Vierh. pneumatophores at Beachwood Mangroves Nature Reserve, Durban, South Africa. *Bot. Mar.* 39: 167–176.

Phillips, R. C. 1984. *The Ecology of Eelgrass Meadows in the Pacific Northwest: A Community Profile*, FWS/OBS-84/24. U.S. Fish and Wildlife Service, Washington, D.C.

Phillips, R. C., and C. P. McRoy (eds.). 1980. *Handbook of Seagrass Biology: An Ecosystem Perspective*. Garland, New York.

———. 1990. *Seagrass Research Methods*. UNESCO, Paris.

Pianka, E. R. 1970. On r- and K-selection. *Amer. Nat.* 104: 592–597.

Pirc, H. 1985. Growth dynamics in *Posidonia oceanica* (L.) Delile. 1. Seasonal changes of soluble carbohydrates, starch, free amino acids, nitrogen and organic anions in different parts of the plant. *P.S.Z.N.I. Mar. Ecol.* 6: 141–165.

———. 1989. Seasonal changes in soluble carbohydrates, starch, and energy content in Mediterranean seagrasses. *P.S.Z.N.I. Mar. Ecol.* 10: 97–105.

Platt, T. 1981. *Physiological Bases of Phytoplankton Ecology.* Canadian Fisheries and Aquatic Sciences No. 210, Ottawa, Ontario.

Platt, T., and D. V. S. Rao. 1975. Primary production of marine microphytes. In *Photosynthesis and Productivity in Different Environments*, J. P. Cooper (ed.), 249–275. Cambridge University Press, Cambridge.

Pomeroy, L. R. 1959. Algal productivity in salt marshes of Georgia. *Limnol. Oceanograf.* 4: 386–397.

Pomeroy, L. R., W. M. Darley, E. L. Dunn, J. L. Gallager, E. B. Haines, and D. M. Whitney. 1981. Primary production. In *Ecology of a Salt Marsh*, L. R. Pomeroy and R. G. Wiegert (eds.), 39–67. Springer-Verlag, New York.

Pomeroy, L. R., and R. C. Wiegert (eds.). 1981. *The Ecology of a Salt Marsh.* Springer-Verlag, New York.

Post, A., and A. W. D. Larkum. 1993. UV-absorbing pigments, photosynthesis and UV exposure in Antarctica: Comparison of terrestrial and marine algae. *Aquat. Bot.* 45: 231–243.

Prager, E. J., and R. N. Ginsburg. 1989. Carbonate nodule growth on Florida's outer shelf and its implications for fossil interpretations. *Palaios* 4: 310–317.

Preen, A. R., W. J. L. Long, and R. G. Coles. 1995. Flood and cyclone related loss, and partial recovery, of more than 1000 km^2 of seagrass in Hervey Bay, Queensland, Australia. *Aquat. Bot.* 52: 3–17.

Preston, M. R. 1988. Marine pollution. In *Chemical Oceanography*, J. P. Riley (ed.), 53–196. Academic Press, Orlando.

Prezelin, B. B., and B. A. Boczar. 1986. Molecular basis of cell absorption and fluorescence in phytoplankton: Potential applications to studies in optical oceanography. *Prog. Phycol. Res.* 4: 351–464.

Prince, J. S., and J. M. Kingsbury. 1973. The ecology of *Chondrus crispus* at Plymouth, Massachusetts. III. Effect of elevated temperature on growth and survival. *Biol. Bull.* 145: 580–588.

Pringle, J. D., and A. C. Mathieson. 1987. Case study: *Chondrus crispus* Stackhouse. In *Studies of Seven Commercial Seaweed Resources*, M. S. Doty, J. Daddy, and B. Santelices (eds.), 49–122. FAO Fish Tech. Paper 281, Rome.

Proffit, C. E., D. J. Devlin, and M. Lindsey. 1995. Effects of oil on mangrove seedlings grown under different environmental conditions. *Mar. Pollut. Bull.* 30: 788–793.

Provost, M. W. 1973. Mean high water mark and use of tidelands in Florida. *Fla. Natural.* 37: 163–170.

———. 1976. Tidal datum planes circumscribing salt marshes. *Bull. Mar. Sci.* 26: 558–563.

Pueschel, C. M. 1989. An expanded survey of the ultrastructure of red algal pit plugs. *J. Phycol.* 25: 625–636.

Pulich Jr., W. M. 1985. Seasonal growth dynamics of *Ruppia maritima* L. and *Halodule wrightii* Aschers. in southern Texas and evaluation of sediment fertility status. *Aquat. Bot.* 23: 53–66.

———. 1986. Variations in leaf soluble amino acids and ammonium content in subtropical seagrasses related to salinity stress. *Plant Physiol.* 80: 283–286.

Pulich Jr., W. M., and W. A. White. 1991. Decline of submerged vegetation in the Galveston Bay system (Texas, USA): Chronology and relationships to physical processes. *J. Coast. Res.* 7: 1125–1138.

Quadir, A., P. J. Harrison, and R. E. DeWreede, 1979. The effects of emergence and submergence on the photosynthesis and respiration of marine macrophytes. *Phycologia* 18: 83–88.

Rabinowitz, D. 1978. Early growth of mangrove seedlings in Panama, and an hypothesis concerning the relationship of dispersal and zonation. *J. Biogeogr.* 5: 113–133.

Radmer, R. J. 1996. Algal diversity and commercial algal products. *Bioscience* 46: 263–270.

Ragan, M. A. 1976. Physodes and the phenolic compounds of brown algae. Composition and significance of physiods "in vivo." *Bot. Mar.* 14: 145–154.

Ragan, M. A., and R. R. Butell. 1995. Are red algae plants? *Bot. J. Linn. Soc.* 118: 81–105.

Ragan, M. A., and K. W. Glombitza. 1986. Pholorotannins, brown algal polyphenols. *Prog. Phycol. Res.* 4: 129–241.

Rai, L. C., J. P. Gaur, and H. D. Kumar. 1981. Phycology and heavy-metal pollution. *Biol. Rev.* 56: 99–151.

Ralph, P. J., and M. D. Burchett. 1995. Photosynthetic responses of the seagrass *Halophila ovalis* (R.Br.) Hook. f. to high irradiance stress, using chlorophyll *a* fluorescence. *Aquat. Bot.* 51: 55–66.

Ralph, P. J., M. D. Burchett, and A. Pulkownik. 1992. Distribution of extractable carbohydrate reserves within the rhizome of the seagrass *Posidonia australis* Hook. f. *Aquat. Bot.* 42: 385–392.

Ramus, J. 1978. Seaweed anatomy and photosynthetic performance: The ecological significance of light guides, heterogenous absorption and multiple scatter. *J. Phycol.* 14: 352–362.

———. 1981. The capture and transduction of light energy. In *The Biology of Seaweeds*, C. S. Lobban and M. J. Wynne (eds.), 29–46. Blackwell, Oxford.

Raven, J. A., and J. I. Sprent. 1989. Phototrophy, diazotrophy, and paleoatmospheres: Biological catalysis and the H, C, N. and O cycles. *J. Geol. Soc. London* 146: 161–170.

Raven, P. H., R. F. Evert, and S. E. Eichhorn. 1992. *Biology of Plants.* Worth Publishers, New York.

Raymont, J. E. G. 1980. *Plankton and Productivity in the Oceans*, 2nd ed. Pergamon, New York.

Redfield, A. C. 1965. Ontogeny of a salt marsh estuary. *Science* 147: 50–55.

Redeke, J. 1933. Uber den jetzigen Stand unserer Kenntnisse der Flora und Fauna des Brackwasseres. *Verh. Intern. Verein. Limnol.* 6: 46–61.

Reed, D. 1990. An experimental evaluation of density dependence in a subtidal algal population. *Ecology* 71: 2286–2296.

Reichert, R., and C. J. Dawes. 1986. Acclimation of the green alga *Caulerpa racemosa* var. *unifera* to light. *Bot. Mar.* 24: 533–537.

Reimold, R. J. 1972. The movement of phosphorus through the salt marsh cord grass *Spartina alterniflora* Loisel. *Limnol. Oceanograf.* 17: 606–611.

Remane, A., and C. Schlieper. 1971. *Biology of Brackish Water.* John Wiley, New York.

Ribera, M. A., and C.-F. Boudouresque. 1995. Introduced marine plants with special reference to macroalgae: mechanisms and impact. *Prog. Phycol. Res.* 11: 187–268.

Richardson, K., J. Beardall, and J. A. Raven. 1983. Adaptation of unicellular algae to irradiance: An analysis of strategies. *New Phytol.* 93: 157–191.

Richmond, R. H. 1982. Energetic considerations in the dispersal of *Pocillopora damicornis* (Linnaeus) planulae. *Proceedings of the 4th International Coral Reef Symposium*, 153–156.

Ridd, P., M. W. Sandstrom, and E. Wolanski. 1988. Outwelling from tropical tidal salt flats. *Estuar. Coast. Shelf Sci.* 26: 243–253.

Rivandeneyra, R. I. 1989. Ecology of the epibiosis on the submerged roots of *Rhizophora mangle* in Bahia de la Ascension, Quintana Roo, Mexico. *Cien. Mar.* 15: 1–20.

Rivera-Monroy, V. H., J. W. Day, R. R. Twilley, F. Ver-Herrera, and C. Coronado-Molina. 1995. Flux of nitrogen and sediment in a fringe mangrove forest in Terminos Lagoon, Mexico. *Estuar. Coast. Shelf Sci.* 40: 139–160.

Rivera-Monroy, V. H., and R. R. Twilley. 1996. The relative role of denitrification and immobilization in the fate of inorganic nitrogen in mangrove sediments. *Limnol. Oceanograf.* 41: 284–296.

Robertson, A. I., and D. M. Alongi (eds.). 1992. *Coastal and Estuarine Studies 41. Tropical Mangrove Ecosystems.* American Geophysical Union, Washington, D.C.

Robertson, A. I., D. M. Alongi, and K. G. Boto. 1992. Food chains and carbon fluxes. In *Coastal and Estuarine Studies 41. Tropical Mangrove Ecosystems*, A. I. Robertson and D. M. Alongi (eds.), 293–326. American Geophysical Union, Washington, D.C.

Robertson, A. I., and K. H. Mann. 1984. Disturbance by ice and life-history adaptations of the seagrass *Zostera marina*. *Mar. Biol.* 80: 131–141.

Robertson, D. E. 1972. Influence of the physico-chemical forms of radionucleides and stable trace elements in seawater in relation to uptake by the marine biosphere. In *OECD Marine Radioecology*, 21–76. Second ENEA Seminar, Washington, D.C.

Roblee, M. B., T. M. Barber, P. R. Carlson, M. J. Durako, J. W. Fourqurean, L. K. Muelstein, D. Porter, L. A. Yarbro, R. T. Zieman, and J. C. Zieman. 1991. Mass mortality of the tropical seagrass *Thalassia testudinum* in Florida Bay (USA). *Mar. Ecol. Prog.*, Ser. 71: 297–299.

Rodriguez, C., and A. W. Stoner. 1990. The epiphyte community of mangrove roots in a topical estuary: Distribution and biomass. *Aquat. Bot.* 36: 117–126.

Rollet, B. 1981. *Bibliography of Mangrove Research 1600–1975.* UNESCO, Rome.

Rosenthal, R. J., W. D. Clarke, and P. K. Dayton. 1974. Ecology and natural history of a stand of giant kelp, *Macrocystis pyrifera*, off Del Mar, California. *U.S. Dept. Comm. Fish. Bull.* 72: 670–684.

Roth, I. 1992. *Leaf Structure: Coastal Vegetation and Mangroves of Venezuela. Encyclopedia of Plant Anatomy*, Vol. 14, Part 2. Gebruder Borntraeger, Berlin.

Round, F. E. 1981. *The Ecology of Algae.* Cambridge University Press, Cambridge.

Round, F. E., R. M. Crawford, and D. G. Mann. 1990. *The Diatoms, Biology and Morphology of the Genera.* Cambridge University Press, Cambridge.

Rueness, J. 1989. *Sargassum muticum* and other introduced Japanese macroalgae: Biological pollution of European Coasts. *Mar. Pollut. Bull.* 20: 173–176.

Rugg, D. A., and T. A. Norton. 1987. In *Plant Life in Aquatic and Amphibious Habitats*, R. M. M. Crawford (ed.), 347–358. Blackwell, Oxford.

Russell, G. 1986. Variation and natural selection in marine macroalgae. *Oceanograf. Mar. Biol. Ann. Rev.* 24: 309–377.

———. 1988a. The seaweed flora of a young semi-enclosed sea: The Baltic. Salinity as a possible agent of a flora divergence. *Helgol. Wiss. Meeresunter.* 42: 243–250.

———. 1988b. Distribution and development of some Manx epiphyte populations. *Helgol. Wiss. Meeresunter.* 42: 477–492.

Russell, G., and J. J. Bolton. 1975. Euryhaline ecotypes of *Ectocarpus siliculosus* (Dillw.) Lyngb. *Estuar. Coastal Mar. Sci.* 3: 91–94.

Rutzler, K., and I. C. Feller. 1996. Caribbean mangrove swamps. *Sci. Amer.* (March): 94–99.

Rykiel Jr., E. J. 1985. Towards a definition of ecological disturbance. *Austral. J. Ecol.* 10: 361–365.

Ryther, J. H. 1969. Photosynthesis and fish production in the sea. *Science* 166: 72–76.

Saffo, M. B. 1987. New light on seaweeds. *Bioscience* 37: 654–664.

Salisbury, F. B., and C. W. Ross. 1978. *Plant Physiology.* Wadsworth Publishing Company, Belmont, CA.

Santelices, B. 1990. Patterns of reproduction, dispersal and recruitment in seaweeds. *Oceanograf. Mar. Biol. Ann. Rev.* 28: 177–276.

Sargent, F. J., T. J. Leary, D. W. Crewz, and C. R. Kruer. 1994. *Scarring of Florida's Seagrasses: Assessment and Management*, Technical Report 1H/94. Florida Marine Research Institute, St. Petersburg.

Schat, H. 1984. A comparative ecophysiological study on the effects of waterlogging and submergence on dune slack plants: Growth, survival, and mineral nutrition in sand culture experiments. *Oecologia* 62: 279–286.

Schiel, D. R., and M. S. Foster. 1986. The structure of subtidal algal stands in temperate waters. *Oceanograf. Mar. Biol. Ann. Rev.* 24: 265–307.

Schlieper, C. 1972. *Research Methods in Marine Biology.* University of Washington Press, Seattle.

Schmid, R. 1984. Blue light effects on morphogenesis and metabolism in *Acetabularia*. In *Blue Light Effects in Biological Systems*, H. Senger (ed.), 419–432. Springer-Verlag, Berlin.

Schmidt, M. A., and S. S. Hayaska. 1985. Localization of a dinitrogen-fixing *Klebsiella* sp. isolated from root-rhizomes of the seagrass *Halodule wrightii* Aschers. *Bot. Mar.* 28: 437–442.

Schmitz, K., and L. M. Srivastava. 1980. Long distance transport in *Macrocystis integrifoli*. III. Movement of THO. *Plant Physiol.* 66: 66–69.

Schneider, C. W. 1981. The effect of elevated temperature and reactor shutdown on the benthic marine flora of the Millstone Thermal Quarry, Connecticut. *J. Therm. Biol.* 6: 1–6.

Scholander, P. F. 1968. How mangroves desalinate seawater. *Physiol. Plant.* 21: 251–261.

Scholander, P. F., L. van Dam, and S. I. Scholander. 1955. Gas exchange in the roots of mangroves. *Amer. J. Bot.* 42: 92–98.

Scholten, M., and J. Rozema. 1990. The competitive ability of *Spartina anglica* on Dutch saltmarshes. In *Spartina anglica—A Research Review*, Research Publication No. 2, A. J. Gray and P. E. M. Benham (eds.), 39–47. HMS Stationery Office, Institute Terrestrial Ecology, London.

Scoffin, T. P. 1970. The trapping and binding of subtidal carbonate sediments by marine vegetation in Bimini Lagoon, Bahamas. *J. Sedim. Petrol.* 40: 249–273.

Seapy, R. R., and M. M. Littler. 1979. The distribution, abundance, community structure, and primary productivity of macroorganisms from two central California rocky intertidal habitats. *Pac. Sci.* 32: 293–314.

Searles, R. B. 1980. The strategy of the red algal life history. *Amer. Nat.* 115: 113–120.

Searles, R. B., and C. W. Schneider. 1987. Observations on the deep-water flora of Bermuda. *Hydrobiolgia* 151/152: 261–266.

Seeliger, U. (ed.). 1992. *Coastal Plant Communities of Latin America.* Academic Press, New York.

Seibert, R. W. 1969. Flowering patterns, germination, seed and seedling development of Juncus roemerianus. Ph.D. dissertation, North Carolina State University, Raleigh.

Sell, M. G. 1977. Modeling the response of mangrove ecosystems to herbicide spraying, hurricanes, nutrient enrichment, and economic development. Ph.D. dissertation, University of Florida, Gainesville.

Setchell, W. A. 1920a. The temperature interval in the geographical distribution of marine algae. *Science* 53: 187–190.

———. 1920b. Geographical distribution of the marine spermatophytes. *Bull. Torrey Bot. Club* 47: 563–579.

Shat, H. 1984. A comparative ecophysiological study on the effects of waterlogging and submergence on dune slack plants: Growth, survival and mineral nutrition in sand culture experiments. *Oecologia* 62: 279–286.

Sheath, R. G. 1984. The biology of freshwater red algae. *Prog. Phycol. Res.* 3: 89–157.

Shepherd, S. A., and H. B. S. Womersley. 1981. The algal and seagrass ecology of Waterloo Bay, South Australia. *Aquat. Bot.* 11: 305–371.

Sheridan, P. F., and R. J. Livingston. 1983. Abundance and seasonality of infauna and epipfauna inhabiting a *Halodule wrightii* meadow in Apalachicola Bay, Florida. *Estuaries* 6: 407–419.

Short, F. T. (ed.). 1992. *The Ecology of the Great Bay Estuary, New Hampshire and Maine: An Estuarine Profile and Bibliography*, NOAA, Ocean Coastal Program Publication. New Hampshire Fish and Game, Durham.

———. 1985. A method for the culture of tropical seagrasses. *Aquat. Bot.* 22: 187–193.

Short, F. T., D. M. Burdick, and J. E. Kaldy III. 1995. Mesocosm experiments quantify the effects of eutrophication on eelgrass, *Zostera marina. Limnol. Oceanograf.* 40: 740–749.

Short, F. T., J. Montgomery, C. F. Zimmermann, and C. A. Short. 1993. Production and nutrient dynamics of a *Syringodium filiforme* Kuetz. seagrass bed in Indian River Lagoon, Florida. *Estuaries* 16: 323–334.

Short, F. T., and S. Wyllie-Echeverria. 1996. Natural and human-induced disturbance of seagrasses. *Envir. Conserv.* 23: 17–27.

Sieburth, J. M., and J. T. Conover. 1965. *Sargassum* tannin, an antibiotic which retards fouling. *Nature* 208: 52–53.

Sieburth, J. M., V. Smetacek, and J. Lenz. 1978. Pelagic ecosystem structure: Heterotrophic compartments of the plankton and their relationship to plankton size fractions. *Limnol. Oceanogr.* 23: 1256–1263.

Silberstein, K., A. W. Chiffings, and A. J. McComb. 1986. The loss of seagrass in Cocburn Sound, Western Australia. III. The effect of epiphytes on productivity of *Posidonia australis* Hook. f. *Aquat. Bot.* 24: 355–371.

Skelton, N. J., and W. G. Allaway. 1996. Oxygen and pressure changes measured *in situ* during flooding in roots of the grey mangrove *Avicennia marina* (Forssk.) Vierh. *Aquat. Bot.* 54: 165–175.

Slankis, T., and S. P. Gibbs. 1972. The fine structure of mitosis and cell division in the Chrysophycean alga *Ochramonas danica. J. Phycol.* 8: 243–256.

Slocum, C. J. 1980. Differential susceptibility to grazers in two phases of an intertidal alga: Advantages of heteromorphic generations. *J. Exp. Mar. Biol. Ecol.* 46: 99–110.

Smayda, T. J. 1970. The suspension and sinking of phytoplankton in the sea. *Oceanogr. Mar. Biol. Ann. Rev.* 8: 353–414.

Smayda, T. J. 1980. Phytoplankton species succession. In *Physiological Ecology of Phytoplankton*, I. Morris (ed.), 493–570. University of California Press, Berkeley.

Smit, C. J., J. den Hollander, W. K. R. E. van Wingerden, and W. J. Wolf (eds.). 1981. *Terrestrial and Freshwater Fauna of the Wadden Sea Area*, Report 10. Wadden Sea Working Group, Balkema, Rotterdam.

Smith, J. H. C., and A. Benitez. 1955. Chlorophylls: Analysis in plant material. In *Modern Methods of Plant Analysis*, K. Paech and M. V. Tracey (eds.), 142–196. Springer-Verlag, Berlin.

Smith, K. K., R. E. Good, and N. F. Good. 1979. Production dynamics for above and below ground components of a New Jersey *Spartina alterniflora* tidal marsh. *Estuar. Coast. Mar. Sci.* 9: 189–201.

Smith, P. K., R. I. Krohn, T. I. Hermanson, A. K. Mallia, F. H. Gartner, M. D. Provenzano, E. K. Fujimoto, N. M. Goeke, B. J. Olson, and D. C. Klenck. 1985. Measurement of protein using bicinchoninic acid. *Anal. Biochem.* 150: 76–85.

Smith, S. V. 1978. Coral reef area and contributions to processes and resources of the world's oceans. *Nature* 273: 225–226.

Smith III, T. J. 1988. The influence of seed predators on structure and succession in tropical tidal forests. *Proc. Ecol. Soc. Aust.* 15: 203–211.

Smith III, T. J. 1992. Forest structure. In *Coastal and Estuarine Studies 41. Tropical Mangrove Ecosystems*, A. I. Robertson and D. M. Alongi (eds.), 101–136. American Geophysical Union, Washington, D.C.

Smith III, T. J., K. G. Boto, S. D. Frusher, and R. L. Giddins. 1991. Keystone species and mangrove forest dynamics: The influence of burrowing by crabs on soil nutrient status and forest productivity. *Estuar. Coast. Shelf Sci.* 33: 419–432.

Snedaker, S. C. 1995. Mangroves and climate change in the Florida and Caribbean region: Scenarios and hypotheses. *Hydrobiologia* 295: 43–49.

Snedaker, S. C., and J. G. Snedaker(eds.). 1984. *The Mangrove Ecosystem: Research Methods.* UNESCO, Paris.

Sokal, R. R., and F. J. Rohlf. 1981. *Biometry. The Principles and Practice of Statistics in Biological Research.* W. H. Freeman, New York.

Sommer, U. (ed.) 1989. *Plankton Ecology: Succession in Phytoplankton Communities.* Springer-Verlag, New York.

Sorokin, Y. I. 1990a. Plankton in the reef ecosystems. In *Ecosystems of the World 25. Coral Reefs*, Z. Dubinsky (ed.), 291–327. Elsevier, Amsterdam.

————. 1990b. Aspects of tropic relations, productivity and energy balance in coral-reef ecosystems. In *Ecosystems of the World 25. Coral Reefs.* Z. Dubinsky (ed.), 401–410. Elsevier, Amsterdam.

————. 1993. *Coral Reef Ecology.* Springer-Verlag, Berlin.

———— (ed.). 1978. *Phytoplankton Manual.* UNESCO, Paris.

Sournia, A. 1982. Form and function in marine phytoplankton. *Biol. Rev. Cambridge Philosoph. Soc.* 57: 347–394.

————. 1986. *Atlas du Phytoplankton Marin I.* Centre National Recherche Scientifique, Paris.

Sousa, W. P. 1979. Experimental investigation of disturbance and ecological succession on a rocky intertidal algal community. *Ecol. Monograf.* 49: 227–254.

South, G. R., and A. Whittick. 1987. *Introduction to Phycology.* Blackwell, Oxford.

Sperry, W. N., and F. C. Brand. 1955. The determination of total lipids in the blood serum. *J. Biol. Chem.* 213: 69–76.

Squires, E. R., and R. E. Good. 1974. Seasonal changes in the productivity, caloric content, and chemical composition of a population of salt-marsh cord-grass (*Spartina alterniflora*). *Chesapeake Sci.* 15: 63–71.

Stanier, R. Y. 1977. The position of the Cyanobacteria in the world of photrophs. *Carlsberg Res. Comm.* 42: 77–98.

Stanley Jr., G. D. 1992. Tropical reef ecosystems and their evolution. *Encyclop. Earth Syst. Sci.* 4: 375–388.

Stanley Jr., G. D., and P. K. Swart. 1995. Evolution of the coral-zooxanthellae symbiosis during the Triassic: A geochemical approach. *Paleobiol.* 21: 179–199.

State of Florida. 1986. Manasota 88 Inc. vs Wilbur Boyd Corp. Depositions of J. Carrol and G. Gilmore. Case 85-2904. Division of Administrative Hearings, Tallahassee.

Steele, J. C. 1993. Guam seaweed poisoning: Common marine toxins. *Micronesica* 26: 11–18.

Steemann-Nielsen, E. 1959. Primary production in tropical marine areas. *J. Mar. Biol. Assoc. India* 1: 7–12.

————. 1974. Light and primary production. In *Optical Aspects of Oceanography*, N. G. Jerlov (ed.), 331–388. Academic Press, New York.

Steidinger, K. A., and E. R. Cox. 1980. Free-living dinoflagellates. In *Phytoflagellates*, E. R. Cox (ed.), 407–432. Elsevier/North Holland, New York.

Steidinger, K. A., and G. A. Vargo. 1988. Marine dinoflagellate blooms: Dynamics and impacts. In *Algae and Human Affairs*, C. A. Lembi and J. R. Waaland (eds.), 373–402. Cambridge University Press, Cambridge.

Stein, J. R. 1973 (ed.). *Handbook of Phycological Methods. Culture Methods and Growth Measurements.* Cambridge University Press, Cambridge.

Steller, D. L., and M. S. Foster. 1995. Environmental factors influencing distribution and morphology of rhodoliths in Bahia Concepcion, B.C.S., Mexico. *J. Exp. Mar. Biol. Ecol.* 194: 201–212.

Steneck, R. S., and L. Watling. 1982. Feeding capabilities and limitation of herivorous molluscs: A functional approach. *Mar. Biol.* 68: 299–312.

Stephenson, T. A., and A. Stephenson. 1972. *Life Between Tidemarks of Rocky Shores.* W. H. Freeman, San Francisco.

Sternberg, L. da S. L., and P. K. Swart. 1987. Utilization of freshwater and ocean water by coastal plants of southern Florida. *Ecology* 68: 1898–1905.

Sternman, N. T. 1994. Spectrophotometric and fluorometric chlorophyll analysis. In *Experimental Phycology: A Laboratory Manual*, C. S. Lobban, D. J. Chapman, and B. P. Kremer (eds.), 35–46. Cambridge University Press, Cambridge.

Stevenson, J. C., L. G. Ward, and M. S. Kearney. 1986. Vertical accretion in marshes with varing rates of sea level rise. In *Estuarine Variability*, D. A. Wolfe (ed.), 241–259. Academic Press, New York.

Stevenson, J. C., L. G. Ward, and M. S. Kearney. 1988. Sediment transport and trapping in marsh systems: Implications of tidal flux studies. *Mar. Geol.* 80: 37–59.

Stewart, G. R., and T. O. Orebamjo. 1984. Studies of nitrate utilization by the dominant species of regrowth vegetation of tropical West Africa: A Nigerian example. In *Nitrogen as an Ecological Factor*, J. A. Lee, S. McNeil, and I. H. Rorison (eds.), 167–188. Blackwell, Oxford.

Stewart, G. R., and M. Popp. 1987. The ecophysiology of mangroves. In *Plant Life in Aquatic and Amphibious Habitats*, R. M. M. Crawford (ed.), 333–346. Blackwell, Oxford.

Stewart, J. G. 1989. Maintenance of a balanced, shifting boundary between the seagrass *Phyllospadix* and algal turf. *Aquat. Bot.* 33: 223–241.

Stewart, W. D. P. (ed.). 1974. *Algal Physiology and Biochemistry.* University of California Press, Berkeley.

Stimson, J., S. Larned, and K. McDermid. 1996. Seasonal growth of the coral reef macroalga *Dictyosphaeria cavernosa* (Forskal) Borgesen and the effects of nutrient availability, temperature and herbivory on growth rate. *J. Exp. Mar. Biol. Ecol.* 196: 53–77.

Stoddard, D. R. 1962. *Three Caribbean Atolls: Turneffe Islands, Lighthouse Reef, and Glover's Reef, British Honduras*, Atoll Research Bulletin No. 87. Smithsonian Institution, Washington, D.C.

———. 1973. Coral reefs of the Indian Ocean. In *Biology and Geology of Coral Reefs*, O. A. Jones and R. Endean (eds.), 51–92. Academic Press, New York.

Stoner, A. W. 1984. Distribution of fishes in seagrass meadows: Role of macrophyte biomass and species composition. *U.S. Dept. Comm. Fish. Bull.* 81: 837–846.

Storey, R., and R. G. Wyn Jones. 1979. Responses of *Atriplex spongiosa* and *Suaeda monoica* to salinity. *Plant Physiol.* 63: 156–162.

Stout, J. P. 1984. *The Ecology of Irregularly Flooded Salt Marshes of the Northeastern Gulf of Mexico: A Community Profile*, Biological Report 85 (7.1). U.S. Fish and Wildlife Service, Washington, D.C.

Strickland, J. D. H., and T. R. Parsons. 1968. *A Practical Handbook of Seawater Analysis*, Bulletin 167. Fisheries Research Board of Canada, Ottawa.

Sullivan, M. J. 1977. Edaphic diatom communities associated with *Spartina alterniflora* and *S. patens* in New Jersey. *Hydrobiolgia* 52: 207–211.

————. 1978. Diatom community structure, taxonomic and statistical analysis of a Mississippi salt marsh. *J. Phycol.* 14: 468–475.

Sullivan, M. J., and F. C. Daiber. 1975. Light, nitrogen, and phosphorus limitation of edaphic algae in a Delaware salt marsh. *J. Exp. Mar. Biol. Ecol.* 18: 79–88.

Surif, M. B., and J. A. Raven. 1989. Exogenous inorganic carbon sources for photosynthesis in seawater by members of the Fucales and the Laminariales (Phaeophyta): Ecological and taxonomic implications. *Oecologia* 78: 97–105.

Sverdrup, H. U., M. W. Johnson, and R. H. Fleming. 1964. *The Oceans, Their Physics, Chemistry, and General Biology.* Prentice-Hall, Englewood Cliffs, N.J.

Sweeney, B. M. 1977. Chronobiology (circadian rhythms). In *The Science of Photobiology*, K. C. Smith (ed.), 209–226. Plenum Press, New York.

Swift, D. G. 1980. Vitamins and phytoplankton growth. In *Physiological Ecology of Phytoplankton*, I. Morris (ed.), 329–368. University of California Press, Berkeley.

Sze, P. 1986. *A Biology of the Algae.* W. C. Brown, Dubuque, Iowa.

Tanner, J. E. 1995. Competition between scleractinian corals and macroalgae: An experimental investigation of coral growth, survival and reproduction. *J. Exp. Mar. Biol. Ecol.* 190: 151–168.

Taylor, D. L. 1983. The coral-algal symbiosis. In *Algal Symbiosis*, L. J. Goff (ed.), 19–36. Cambridge University Press, Cambridge.

Taylor, D. R., L. W. Aarssen, and C. Loehle. 1990. On the relationship between r/K selection and environmental carrying capacity: A new habitat templet for plant life history strategies. *Oikos* 58: 239–250.

Taylor, F. J. R. (ed.). 1987. *The Biology of Dinoflagellates.* Blackwell, Oxford.

Teal, J. M. 1969. *Life and Death of the Salt Marsh.* Little, Brown, Boston.

————. 1986. *The Ecology of Regularly Flooded Salt Marshes of New England: A Community Profile*, Biological Report 85 (7.4). U.S. Fish and Wildlife Service, Washington, D.C.

Teal, J. M. and I. Valiela. 1973. The salt marsh as a living filter. *Mar. Tech. Soc. J.* 7: 19–21.

Tefler, A., J. D. L. Rivas, and J. Barber. 1991. β-carotene within the isolated photosystem II reaction center: Photooxidation and irreversible bleaching of this chromophore by oxidized P680. *Biochem. Biophys. Acta* 1060: 106–114.

Tegner, M. 1980. Multispecies considerations of resource management in southern California kelp beds. In *Proceedings of the Workshop on the Relationship between Sea Urchin Grazing and Commercial Plant/Animal Harvesting*, Canada Technical Report 954, J. D. Pringle, G. J. Sharp, and J. F. Caddy (eds.), 125–143. Fisheries Aquatic Science, Ottawa.

Tenor, K. R., and R. B. Hanson. 1980. Availability of detritus of different types and ages to a polychaete macroconsumer, *Capitella capitata*. *Limnol. Oceanograf.* 25: 553–558.

Tett, P., and E. D. Barton. 1995. Why are there about 5,000 species of phytoplankton the sea? *J. Plankton Res.* 17: 1693–1704.

Thayer, G. W., K. A. Bjorndal, J. C. Ogden, S. L. Williams, and J. C. Zieman. 1984. Role of larger hervibores in seagrass communities. *Estuaries* 7: 351–376.

Thayer, G. W., and M. S. Fonseca. 1984. *The Ecology of Eelgrass Meadows of the Atlantic Coast: A Community Profile*, FWS/OBS-84/02. U.S. Fish and Wildlife Service, Washington, D.C.

Thayer, G. W., D. A. Wolfe, and R. B. Williams. 1975. The impact of man on seagrass systems. *Amer. Scient.* 63: 288–296.

Thibodeau, F. R., and N. H. Nickerson. 1986. Differential oxidation of mangrove substrate by *Avicennia germinans* and *Rhizophora mangle. Amer. J. Bot.* 73: 512–516.

Thom, B. G. 1967. Mangrove ecology and deltaic gemorphology: Tabasco, Mexico. *J. Ecol.* 55: 301–343.

Thomas, M. L. H. 1988. Photosynthesis and respiration of aquatic macro-flora using the light and dark bottle oxygen method and dissolved oxygen analyzer. In *Experimental Phycology: A Laboratory Manual*, C. S. Lobban, D. J. Chapman, and B. P. Kremer (eds.), 64–82. Cambridge University Press, Cambridge.

Thompson, E. Q., R. I. Stuckey, and E. B. Thompson. 1987. *Spread, Impact, and Control of Purple Loosestrife (Lythrum salicaria) in North American Wetlands*, Research Report 2. U.S. Fish and Wildlife Service, Washington, D.C.

Thompson, J. D., T. McNeilly, and A. J. Gray. 1991. Population variation in *Spartina anglica* C. E. Hubbard. *New Phytol.* 117: 115–128.

Tilden, J. 1910. *Minnesota Algae. I. The Myxophyceae of North American and Adjacent Regions Including Central America, Greenland, Bermuda, the West Indies and Hawaii*, Botanical Series 8. University of Minnesota Press, Minneapolis.

Tiner Jr., R. W. 1987. *A Field Guide to Coastal Wetland Plants of the Northeastern United States*. University of Massachusetts Press, Amherst.

Titlyanov, E. A., P. V. Kolmakov, V. A. Leletkin, and G. M. Voskoboinikov. 1987. A new type of light adaptation in aquatic plants. *Biol. Morya Vladivostov* 1987: 48–57 (in Russian).

Tokida, J. 1960. Marine algal epiphytes on Laminariales plants. *Bull. Fac. Fish. Hokkaido Univ.* 11: 73–105.

Tomas, C. R. 1978. *Olisthodiscus luteus* (Chrysophyceae) I. Effects of salinity and temperature on growth, motility and survival. *J. Phycol.* 14: 309–313.

Tomas, R. N., and E. R. Cox. 1973. Observations on the symbiosis of *Peridinum balticum* and its intracellular alga. I. Ultrastructure. *J. Phycol.* 9: 304–323.

Tomasko, D. A., and C. J. Dawes. 1989. Evidence for physiological integration between shaded and unshaded short shoots of *Thalassia testudinum. Mar. Ecol. Prog.*, Ser. 54: 299–305.

Tomasko, D. A., C. J. Dawes, and M. O. Hall. 1991. Effects of the number of short shoots and presence of the rhizome apical meristem on the survival and growth of transplanted seagrass *Thalassia testudinum. Contrib. Mar. Sci.* 32: 41–48.

————. 1996. The effects of anthropogenic nutrient enrichment on turtle grass (*Thalassia testudinum*) in Sarasota Bay, Florida (USA). *Estuaries* 19: 448–456.

Tomlinson, P. B. 1974. Vegetative morphology and meristem dependence. The foundation of productivity in seagrasses. *Aquacult.* 4: 107–130.

————. 1982. *Helobiae (Alismatidae)*. In *Anatomy of the Monocotyledons VII*, C. R. Metcalfe (ed.), Clarendon Press, Oxford.

————. 1986. *The Botany of Mangroves*. Cambridge University Press, Cambridge.

Trainor, F. R. 1978. *Introductory Phycology*. John Wiley, New York.

Trenhaile, A. S. 1987. *The Geomorphology of Rocky Coasts*. Clarendon Press, Oxford.

Trono Jr., G. C. 1993. *Eucheuma* and *Kappaphycus*: Taxonomy and cultivation. In *Seaweed Cultivation and Marine Ranching*, M. Ohno and A. T. Critchley (eds.), 75–88. Japan International Cooperative Agency, Nagai.

Trowbridge, C. D. 1995. Establishment of the green alga *Codium fragile* ssp. *tomentosoides* on New Zealand rocky shores: Current distribution and invertebrate grazers. *J. Ecol.* 83: 949–965.

Tsuda, R. R. T., and C. J. Dawes. 1974. *Preliminary Checklist of the Marine Benthic Plants from Glover's Reef. British Honduras*, Atoll Research Bulletin 173a. Smithsonian Institution, Washington, D.C.

Tubbs, C. R., and J. M. Tubbs. 1983. The distribution of *Zostera* and its exploitation by wildfowl in the Solent, Southern England. *Aquat. Bot.* 15: 223–239.

Turner, R. E. 1976. Geographic variations in salt marsh macrophytic production: A review. *Contrib. Mar. Sci.* 20: 47–68.

————. 1982. *Wetland Losses and Coastal Fisheries: An Enigmatic and Economically Significant Dependency*, FWS-OBS-82/59. U.S. Fish and Wildlife Service, Washington, D.C.

Turner, T. 1985. Stability of rocky intertidal surfgrass beds: Persistence, preemption, and recovery. *Ecology* 66: 83–92.

Twilley, R. R. 1995. Properties of mangrove ecosystems and their relation to the energy signature of coastal environments. In *Maximum Power*, C. A. S. Hall (ed.), 43–62. University of Colorado Press, Boulder, Colorado.

Tyerman, S. D. 1989. Solute and water relations of seagrasses. In *Biology of Seagrasses*, A. W. D. Larkum, A. J. McComb, and S. A. Shepherd (eds.), 723–760. Elsevier, Amsterdam.

Urbach, E., D. Robertson, and S. W. Chisholm. 1992. Multiple evolutionary origins of prochlorophytes within the cyanobacterial radiation. *Nature* 355: 267–269.

U.S. Army Corps of Engineers. 1977. *Chesapeake Bay Future Conditions Report*. 12 Vols. U.S. Army Corps of Engineers, Baltimore.

Vadas, R. L., M. Keser, P. C. Rusanowski, and B. R. Larson. 1976. The effects of thermal loading on the growth and ecology of a northern population of *Spartina alterniflora*. In *Thermal Ecology II*, G. W. Esch and R. W. MacFarlane (eds.), 54–63. ERDA Symposium Series, Augusta.

Vadas, R. L., and R. S. Steneck. 1988. Zonation of deep water benthic algae in the Gulf of Maine. *J. Phycol.* 24: 338–346.

Valiela, I. 1984. *Marine Ecological Processes*. Springer-Verlag, New York.

Valiella, W., K. Foreman, M. LaMontagne, D. Hersh, J. Costa, P. Peckol, B. DeMeo-Anderson, C. D'Avanzo, M. Babione, C.-H. Sham, J. Brawley, and K. Lajhtha. 1992. Couplings of watersheds and coastal waters: Sources and consequences of nutrient enrichment in Waquoit Bay, Massachusetts. *Estuaries* 15: 443–457.

Valiela, I., and C. S. Rietsma. 1995. Disturbance of salt marsh vegetation by wrack mats in Great Sippewissett Marsh. *Oecologia* 102: 106–112.

van den Hoek, C. 1984. World-wide latitudinal and longitudinal seaweed distribution patterns and their possible causes as illustrated by the distribution of Rhodophytan genera. *Helgol. Wiss. Meersunter.* 38: 227–257.

van den Hoek, C., F. Colijn, A. M. Cortel-Breeman, and J. B. Wanders. 1972. Algal vegetation type along the shores and inner bays and lagoons of Curacao and the lagoon Lac (Bonaire), Netherlands Antilles. *Ver. K. Ned. Akad. Wet.* 61: 1–72.

van den Hoek, C., A. M. Cortel-Breeman, and J. B. Wanders. 1975. Algal zonation in the fringing coral reef of Curacao, Netherlands Antilles, in relation to zonation of corals and gorgonians. *Aquat. Bot.* 1: 327–335.

van den Hoek, C., D. G. Mann, and H. M. Jans. 1995. *Algae. An Introduction to Phycology.* Cambridge University Press, Cambridge.

van den Hoek, C., W. T. Stam, and J. L. Olsen. 1988. The emergence of a new chlorophytan system and Dr. Kornmann's contribution thereto. *Helgol. Wiss. Meeresunter.* 42: 339–383.

Vandermeer, J. H. 1972. Niche theory. *Ann. Rev. Ecol. Syst.* 3: 107–132.

van Diggelen, J., J. Rozema, and R. Broekman. 1987. Mineral composition of and proline accumulation by *Zostera marina* L. in response to environmental salinity. *Aquat. Bot.* 27: 169–176.

van Eerdt, M. M. J. 1985. The influence of vegetation on erosion and accretion in salt marshes of the Oosterschelde, the Netherlands. *Vegetatio* 62: 367–373.

van Steveninck, E. D. de R., and R. P. M. Bak. 1986. Changes in abundance of coral-reef bottom components related to mass mortalithy of the sea urchin *Diadema antillarum. Mar. Ecol. Prog.*, Ser. 34: 87–94.

van Tamelen, P. G. 1987. Early successional mechanisms in the rocky intertidal: The role of direct and indirect interactions. *J. Exp. Mar. Biol. Ecol.* 112: 39–48.

———. 1996. Algal zonation in tidepools: Experimental evaluation of the roles of physical disturbance, herbivory and competition. *J. Exp. Mar. Biol. Ecol.* 201: 197–231.

van Tussenbroek, B. I. 1994. The impact of Hurricane Gilbert on the vegetative development of *Thalassia testudinum* in Puerto Morelos coral reef lagoon, Mexico: A restrospective study. *Bot. Mar.* 37: 421–428.

———. 1996. Techniques of rapid assessment of seagrass production yield, applied to *Thalassia testudinum* in a Mexican tropical reef lagoon. In *Seagrass Biology: Proceedings of an International Workshop*, J. Kuo, R. C. Philips, D. I. Walker, and H. Kirkman (eds.), 131–138. University of Western Australia, Nedlands.

van Valkenburg, S. D. 1980. Silicoflagellates. In *Phytoflagellates*, E. R. Cox (ed.), 335–350. Elsevier/North Holland, New York.

Vasquez, J. A., and N. Guerra. 1996. The use of seaweeds as bioindicators of natural and anthropogenic contaminants in northern Chile. *Hydrobiologia* 326/327: 327–333.

Veldhuis, M. J. W., R. Colijn, and L. A. H. Venekamp. 1986. The spring bloom of *Phaeocystis pouchetti* (Haptophyceae) in Dutch coastal waters. *Netherl. J. Sea Res.* 20: 37–48.

Vergeer, L. H. T., T. L. Aarts, and J. D. deGroot. 1995. The "wasting disease" and the effect of abiotic factors (light intensity, temperature, salinity) and infection with *Labyrinthula zosterae* on the phelolic content of *Zostera marina* shoots. *Aquat. Bot.* 52: 35–44.

Vergeer, L. H. T., and C. den Hartog. 1994. Omnipresence of Labyrinthulaceae in seagrasses. *Aquat. Bot.* 48: 1–20.

Verlaque, M., and P. Fritayre. 1995. Mediterranean algal communities are changing in face of the invasive alga *Caulerpa taxifolia* (Vahl) C. Agardh. *Oceanol. Acta* 17: 659–672.

Vernet, P., and J. L. Harper. 1980. The cost of sex in seaweeds. *Biol. J. Linnean Soc.* 13: 129–138.

Veron, J. E. N. 1995. *Corals in Space and Time.* Comstock/Cornell, Ithaca, New York.

Vincent, W. F., and S. Roy. 1993. Solar ultraviolet-B radiation and aquatic primary production: Damage, protection, and recovery. *Envir. Rev.* 1: 1–12.

Virnstein, R. W. 1987. Seagrass-associated invertebrate communities of the southeastern U.S.A.: A review. In *Proceedings of the Symposium in Subtropical-tropical Seagrasses of the Southeastern U.S.*, Publication 42, M. J. Durako, R. C. Phillips, and R. R. Lewis (eds.), 89–116. Bureau Marine Research, Florida Department of Natural Resources, St. Petersburg.

Virnstein, R. W., and P. A. Carbonara. 1985. Seasonal abundance and distribution of drift algae and seagrasses in the mid-Indian River Lagoon, Florida. *Aquat. Bot.* 23: 67–82.

Virnstein, R. W., and P. K. Howard. 1987. Motile epifauna of marine macrophytes in the Indian River Lagoon, Florida. II. Comparisons between drift algae and three species of seagrasses. *Bull. Mar. Sci.* 41: 13–26.

Wafar, S., A. G. Untawale, and M. Wafar. 1997. Litter fall and energy flux in a mangrove ecosystem. *Estuar. Coast. Shelf Sci.* 44: 111–124.

Waisel, Y. 1972. *Biology of Halophytes.* Academic Press, New York.

Walker, D. I., R. J. Lukatelich, G. Bastyan, and A. J. McComb. 1989. Effect of boat moorings on seagrass beds near Perth, Western Australia. *Aquat. Bot.* 36: 69–77.

Wallentinus, I. 1991. The Baltic Sea gradient. In *Ecosystems of the World 24. Intertidal and Littoral Ecosystems*, A. C. Mathieson and P. H. Nienhuis (eds.), 83–108. Elsevier, Amsterdam.

Walne, P. L. 1980. Euglenoid flagellates. In *Phytoflagellates*, E. R. Cox (ed.), 5–60. Elsevier/North Holland, New York.

Walne, P. L., and P. A. Kivic. 1990. Phylum Euglenida. In *Handbook of Protoctista*, L. Margulis, J. O. Corliss, M. Melkonian, and D. J. Chapman (eds.), 270–287. Jones and Bartlett, Boston.

Walsh, G. E. 1967. An ecological study of a Hawaiian mangrove swamp. In *Estuaries*, Publication 83, G. H. Lauff (ed.), 420–431. American Association for the Advancement of Science, Washington, D.C.

———. 1974. Mangroves, a review. In *Ecology of Halophytes*, R. J. Reimold and W. H. Queens (eds.), 51–174. Academic Press, New York.

———. 1977. Exploitation of mangal. In *Ecosystems of the World. 1. Wet Coastal Ecosystems*, V. J. Chapman (ed.), 347–362. Elsevier, Amsterdam.

Walsh, J. J. 1988. *On the Nature of Continental Shelves.* Academic Press, New York.

Wanders, J. B. W. 1976. The role of benthic algae in the shallow reef of Curacao, Netherlands Antilles. I. Primary productivity in the coral reefs. *Aquat. Bot.* 2: 235–270.

———. 1977. The role of benthic algae in the shallow reef of Curacao, Netherlands Antilles. II. The significance of grazing. *Aquat. Bot.* 3: 357–390.

Ward, L. G., W. M. Kemp, and W. R. Boynton. 1984. The influence of waves and seagrass communities on suspended particulates in an estuarine embayment. *Mar. Geol.* 59: 85–103.

Warming, E. 1883. Tropische Fragmente. II *Rhizophora mangle* L. *Bot. Jarb.* 4: 519–548.

———. 1909. *Ecology of Plants*, P. Groom and J. B. Balfour (trans.). Clarendon Press, Oxford.

———. 1925. *Oecology of Plants.* Oxford University Press, Oxford.

Warren, R. S., and W. A. Niering. 1993. Vegetation change on a northeastern tidal marsh: Interaction of sea-level rise and marsh accretion. *Ecology* 74: 96–103.

Webb, E. C., I. A. Mendelssohn, and B. J. Wilsey. 1995. Causes for vegetation dieback in a Louisiana salt marsh: A bioassay approach. *Aquat. Bot.* 51: 281–289.

Webster, J., and B. A. Stone. 1994. Isolation, structure and monosaccharide composition of the walls of vegetative parts of *Heterozostera tasmanica* (Martens ex Aschers.) den Hartog. *Aquat. Bot.* 47: 39–52.

Wells, B. W. 1928. Plant communities of the coastal plain of North Carolina and their successional relations. *Ecology* 9: 230–243.

Welsh, D. T., S. Bourgues, R. de Wit, and R. A. Herbert. 1996. Seasonal variations in nitrogen-fixation (acetylene reduction) and sulphate-reduction rates in the rhizosphere of *Zostera noltii*: Nitrogen fixation by sulphate-reducing bacteria. *Mar. Biol.* 125: 619–628.

Werner, D. (ed.). 1977. *The Biology of Diatoms.* University of California Press, Berkeley.

West, R. J., N. E. Jacobs, and D. E. Roberts. 1990. Experimental transplanting of seagrasses in Botany Bay, Australia. *Mar. Pollut. Bull.* 21: 197–203.

Wever, R. 1988. Ozone destruction by algae in the Arctic atmosphere. *Nature* 335: 501–502.

Wheeler, P. A. 1985. Nutrients. In *Handbook of Phycological Methods: Ecological Field Methods, Macroalgae*, M. M. Littler and D. S. Littler (eds.), 493–508. Cambridge University Press, Cambridge.

Whelan, P. M., and J. P. Cullinane. 1985. The algal flora of a subtidal *Zostera* bed in Ventry Bay, southwest Ireland. *Aquat. Bot.* 23: 41–51.

Whittaker, R. H. 1969. New concepts of kingdoms of organisms. *Science* 163: 150–162.

Wiegert, R. G., A. G. Chalmers, and P. R. Randerson. 1983. Productivity gradients in salt marshes: The response of *Spartina alterniflora* to experimentally manipulated soil water movement. *Oikos* 41: 1–6.

Wiegert, R. G., and B. J. Freeman. 1990. *Tidal Salt Marshes of the Southeastern Atlantic Coast: A Community Profile*, Biological Report 85 (7.29). U.S. Fish and Wildlife Service, Washington, D.C.

Wiegert, R. G., and L. R. Pomeroy. 1981. The salt-marsh ecosystem: A synthesis. In

The Ecology of a Salt Marsh, L. R. Pomeroy and R. G. Wiegert (eds.), 219–230. Springer-Verlag, New York.

Wijesinghe, D. K., and S. N. Handel. 1994. Advantages of clonal growth in heterogeneous habitats: An experiment with *Potentilla simplex. J. Ecol.* 82: 495–502.

Wilce, R. T., C. W. Schneider, A. V. Quinlan, and K. van den Bosch. 1982. The life history and morphology of a free living *Pyliella littorais* (L.) Kjellm. (Ectocarpaceae, Ectocarpales) in Nahant Bay, Massachusetts. *Phycologia* 21: 336–354.

Williams, S. L. 1987. Competition between the seagrass *Thalassia testudinum* and *Syringodium filiforme* in a Caribbean lagoon. *Mar. Ecol. Prog.*, Ser. 35: 91–98.

———. 1988. *Thalassia testudinum* productivity and grazing by green turtles in a highly disturbed seagrass bed. *Mar. Biol.* 98: 447–455.

———. 1990. Experimental studies of Caribbean seagrass bed development. *Ecol. Monograf.* 60: 449–469.

Wilson, K. E., P. L. Hoaker, and D. A. Hanan. 1977. Kelp restoration in southern California. In *The Marine Plant Biomass of the Pacific Northwest Coast*, R. Krauss (ed.), 183–202. Oregon State University Press, Corvalis.

Witman, J. D. 1987. Subtidal coexistence: Storms, grazing, mutualism, and the zonation of kelps and mussels. *Ecol. Monograf.* 57: 167–187.

Witz, M. J. A., and C. J. Dawes. 1995. Flowering and short shoot age in three *Thalassia testudinum* meadows of west-central Florida. *Bot. Mar.* 38: 431–436.

Woelkerling, W. M. 1990. An introduction. In *Biology of the Red Algae*, K. M. Cole and R. G. Sheath (eds.), 1–6. Cambridge University Press, Cambridge.

Wolfe, D. A., and B. Kjerfve. 1986. Estuarine variability: An overview. In *Estuarine Variability*, D. W. Wolfe (ed.), 3–17. Academic Press, New York.

Wolfe, J. M., and M. M. Harlin. 1988. Tidepools in southern Rhode Island, U.S.A. *Bot. Mar.* 31: 537–546.

Wood, W. F. 1987. Effect of solar ultra-violet radiation on the kelp *Ecklonia radiata. Mar. Biol.* 96: 143–150.

Womersley, H. B. S., and A. Bailey. 1969. The marine algae of the Solomon Islands and their place in biotic reefs. *Phil. Trans. Roy. Soc. Lond. Series B*, 255: 433–442.

Woodroffe, C. D. 1987. Pacific island mangroves: Distribution and environmental settings. *Pacific Sci.* 41: 166–185.

Woodroffe, C. D. 1992. Mangrove sediments and geomorphology. In *Coastal and Estuarine Studies 41. Tropical Mangrove Ecosystems*, A. I. Robertson and D. M. Alongi (eds.), 7–41. American Geophysical Union, Washington, D.C.

Wootton, J. T. 1995. Effects of birds on sea urchins and algae: A lower-intertidal trophic cascade. *Ecoscience* 2: 321–328.

Wyllie-Echeverria, S., and R. M. Thom. 1994. *Managing Seagrass Systems in Western North America*, Alaska Sea Grant Report 94-01. University of Alaska Press, Fairbanks.

Wyn Jones, R. G., and J. Gorham. 1983. Osmoregulation. In *Encyclopedia of Plant Physiology, n.s.*, O. L. Lange, P. S. Nobel, C. B. Osmond, and H. Ziegler. (eds.), Vol. 12C, 35–58. Springer-Verlag, Berlin.

Yapp, R. H., D. Johns, and O. T. Jones. 1917. The salt marshes of the Dovbery Estuary. *J. Ecol.* 5: 65–103.

Yasumoto, T. 1993. Guam seaweed poisoning: Biochemistry of the *Gracilaria* toxins. *Micronesica* 26: 53–57.

Yentsch, C. S. 1980. Light attenuation and phytoplankton photosynthesis. In *The Physiological Ecology of Phytoplankton*, I. Morris (ed.), 95–127. University of California Press, Berkeley.

Yokohoma, Y., A. Kageyama, T. Ikawa, and S. Shimura. 1977. A carotenoid characteristic of chlorophycean seaweeds living in deep coastal waters. *Bot. Mar.* 20: 433–436.

Yonge, C. M. 1963. The biology of coral reefs. *Adv. Mar. Biol.* 1: 209–260.

———. 1968. Living corals. *Proc. R. Soc. Lond. Series B*, 169: 329–344.

Young, B. M., and L. E. Harvey. 1996. A spatial analysis of the relationship between mangrove (*Avicennia marina* var. *australasica*) physiogomy and sediment accretion in the Hauraki Plains, New Zealand. *Estuar. Coastal Shelf Sci.* 42: 231–246.

Youssef, T., and P. Saenger. 1996. Anatomical adaptive strategies to flooding and rhizosphere oxidation in mangrove seedlings. *Austral. J. Bot.* 44: 297–313.

Zedler, J. B. 1977. Salt marsh structure in the Tijuana estuary, California. *Estuar. Coast. Mar. Sci.* 5: 39–53.

Zedler, J. B. 1982. *The Ecology of New England High Salt Marshes: A Community Profile*, FWS/OBS-81/55. U.S. Fish and Wildlife Service, Washington, D.C.

Zedler, J. B. 1980. Algal mat productivity: Comparison in a salt marsh. *Estuaries* 3: 122–131.

Zieman, J. C. 1974. Methods for the study of the growth and production of the turtle grass, *Thalassia testudinum. Aquaculture* 4: 139–143.

Zieman, J. C. 1975. Tropical seagrass ecosystems and pollution. In *Tropical Marine Pollution*, F. J. F. Wood and R. E. Johannes (eds.), 63–74. Elsevier, Amsterdam.

Zieman, J. C. 1982. *The Ecology of Seagrasses of South Florida: A Community Profile*, FWS/OBS-82/25. U.S. Fish and Wildlife Service, Washington, D.C.

Zieman, J. C., and R. T. Zieman. 1989. *The Ecology of the Seagrass Meadows of the West Coast of Florida: A Community Profile*, Biological Report 85 (7.25). U.S. Fish and Wildlife Service, Washington, D.C.

Zimmerman, R. C., and R. S. Alberte. 1996. Effect of light/dark transition on carbon translocation in eelgrass *Zostera marina* seedlings. *Mar. Ecol. Prog.*, Ser. 136: 305–309.

Zimmerman, R. C., D. G. Kohrs, and R. S. Alberte. 1996. Top-down impact through a bottom-up mechanism: The effect of limpet grazing on growth, productivity and carbon allocation of *Zostera marina* L. (eelgrass). *Oecologia* 107: 560–567.

Zimmerman, R. C., J. L. Reguzzoni, S. Wyllie-Echeverria, M. Josselyn, and R. S. Alberte. 1991. Assessment of environmental suitability for growth of *Zostera marina* L (eelgrass) in San Francisco Bay. *Aquat. Bot.* 39: 353–366.

SUBJECT INDEX

■■■■■■ TAXONOMIC INDEX

473